St. Olaf College

APR 1 8 1991

Science Library

B

# Einstein Studies

*Editors:* Don Howard   John Stachel

Published under the sponsorship
of the Center for Einstein Studies,
Boston University

*Volume 1:* Einstein and the History of General Relativity

Don Howard   John Stachel
Editors

# Einstein
## and the History of General Relativity

Based on the
Proceedings of the 1986 Osgood Hill Conference,
North Andover, Massachusetts
8–11 May 1986

Birkhäuser
Boston · Basel · Berlin

Don Howard
Center for Einstein Studies
Boston University
and
Department of Philosophy
University of Kentucky
Lexington, Kentucky 40506
U.S.A.

John Stachel
Center for Einstein Studies
and
Department of Physics
Boston University
Boston, Massachusetts 02215
U.S.A.

Library of Congress Cataloging-in-Publication Data
Einstein and the history of general relativity : based on the proceedings of the 1986 Osgood Hill Conference, North Andover, Massachusetts, 8–11 May 1986 / Don Howard, John Stachel, editors.
    p. cm. — (Einstein studies ; v. 1)
  "Organized under the auspices of the Boston University Center for Einstein Studies"—Pref.
  Includes index.
  ISBN 0-8176-3392-8 (alk. paper)
  1. General relativity (Physics)—History—Congresses.
2. Astrophysics—History—Congresses.  I. Howard, Don, 1949–
II. Stachel, John J., 1928–    III. Boston University. Center for Einstein Studies.  IV. Series.
QC173.6.E38 1989
530.1'1—dc19                                                           89-797

Printed on acid-free paper.

© The Center for Einstein Studies, 1989
The Einstein Studies series is published under the sponsorship of the Center for Einstein Studies, Boston University.

All rights reserved. No part of this publication may be reproduced, stored in a retrieval system, or transmitted in any form or by any means, electronic, mechanical, photocopying, recording or otherwise, without prior permission of the Center for Einstein Studies (745 Commonwealth Avenue, Boston University, Boston, Massachusetts 02215, U.S.A.).

ISBN 0-8176-3392-8
ISBN 3-7643-3392-8

Typeset by ASCO Trade Typesetting, Ltd., Hong Kong.
Printed and bound by Edwards Brothers, Inc., Ann Arbor, Michigan
Printed in the United States of America.

9 8 7 6 5 4 3 2 1

# Series Preface

Albert Einstein is the twentieth century's most influential and respected scientist. After 1919, when the results of the English eclipse expeditions began to make his name a household word, the steady stream of books and articles aiming to interpret, explain, and popularize his work soon became a flood that has hardly abated since. The interest in his life and work survived his death in 1955; but especially after the 1979 Einstein centennial and the opening of the Einstein Archive to the public in 1980, scholarly interest in Einstein has grown to the point where the various facets of his career have become the focus of more publishing activity than those of any scientist since Newton. Now that *The Collected Papers of Albert Einstein* have begun to appear (Volume One: *The Early Years, 1879–1902*, John Stachel et al., eds., was published by Princeton University Press in 1987; and Volume Two: *The Swiss Years: Writings, 1900–1909*, will appear in 1989), there is every reason to expect scholarly interest in Einstein to continue to grow.

*Einstein Studies* has been established with two principal aims: to stimulate further scholarly interest in research on Einstein's life, work, and times; and to establish and maintain high standards for such scholarship.

The series will publish the proceedings of the Osgood Hill conferences, organized under the auspices of the Boston University Center for Einstein Studies, the sponsor of the series, as well as the proceedings of other conferences focusing on Einstein. Books, monographs, and collections of papers based upon research carried out in the Einstein Archive will be published in the series, as will English translations of important publications in other languages.

*Einstein Studies* will be a multidisciplinary series, reflecting the wide variety of Einstein's own contributions to our century. The topics covered will include, of course, the history of twentieth-century science, as exemplified by the subject of this inaugural volume. But the series will also welcome work related to Einstein's activities in the philosophy of science and the history of philosophy, as well as social, cultural, and political history. Original research in physics, mathematics, and philosophy will also be published, in cases where such research explores the implications of Einstein's work for contemporary issues, or where developments and extensions of Einstein's work are concerned (Volume Two, for example, will comprise the proceedings of the Osgood Hill Conference on Conceptual Problems of Quantum Gravity, held in May 1988).

While the series will focus on Einstein's own life and work, and their place in the science and culture of the twentieth century, this focus is not narrowly conceived. Einstein's life can serve as a window through which to examine such diverse topics as: the history of German scientific institutions, the interactions between science and culture and between science and politics, the Jewish experience in Germany and the United States, and the development of the pacifist movement. The Einstein Archive, the use of which is encouraged by *Einstein Studies*, provides a rich source of documentation for these and for many other topics.

Don Howard
John Stachel

# Preface to Volume One

The conference that occasioned this volume is the third in a series. The two earlier ones were sponsored by the Boston University Institute for Relativity Studies (IRS), a previous avatar of the Center for Einstein Studies. The first conference, on Problems of Quantum Gravity, was held in 1972 at Boston University's Osgood Hill Conference Center in North Andover, Massachusetts. About thirty people lived, ate, discussed, debated, and recreated for almost a week in the former J.P. Stevens mansion and on the grounds of the attached estate. The organizer (J.S.), like the participants, was charmed by the lovely setting, convenient facilities, and good food and cheer provided by the staff, and resolved to hold many similar conferences there. Unfortunately, though the proceedings were taped, they were not published (but for an account of the meeting, see *Nature* 240 (1972): 382–383).

The second conference, held in 1973 to mark publication of the second edition of Adolf Grünbaum's *Philosophical Problems of Space and Time* (Dordrecht and Boston: D. Reidel, 1973), was devoted to Absolute and Relational Theories of Space and Space-Time. Like the first conference and the succeeding ones, this conference aimed to bring together a comparatively small group of people to listen to a minimum of formal reports on a sharply focused group of topics, and to provide the participants with a maximum of opportunity for formal and informal discussion of these topics. The proceedings are incorporated in Volume 8 of the Minnesota Studies in the Philosophy of Science, *Foundations of Space-Time Theories*, John S. Earman, Clark N. Glymour, and John J. Stachel, eds. (Minneapolis: University of Minnesota Press, 1977).

Planning for the continuation of the series resumed only after one of the editors (J.S.) returned from an eight-year stint in Princeton editing the Einstein Papers; it was at this time (1984) that the Center for Einstein Studies was established. The conference on The History of General Relativity was held at Osgood Hill from May 8–11, 1986. A list of the participants is appended. Several talks given at the meeting are not included in this volume, in particular one by Abhay Ashtekar, carrying forward the history of attempts at canonical quantization of general relativity from the point where Peter Bergmann's talk ends, and one by Josh Goldberg, telling the story of U.S. Air Force support for research in general relativity. We hope to include written versions of these talks in a future volume on the history of general relativity.

Conversely, not all of the papers printed here were delivered at Osgood Hill. John Stachel's paper, "Einstein's Search for General Covariance," was given at the 1980 Jena International Conference on General Relativity and Gravitation. This paper initiated a widespread and still current discussion on the significance of Einstein's "hole argument" against general covariance. It has circulated for many years in typescript and has been cited frequently in other publications. It is printed just before John Norton's paper on the hole argument, which also makes reference to it. George Ellis's important survey of relativistic cosmology, and Vladimir Vizgin's valuable review of the early attempts at a unified field theory are included, even though their authors were unable to attend the meeting.

As far as we have been able to ascertain, this conference was the first devoted exclusively to the history of general relativity. It met such a widely felt need among the participants that it was unanimously resolved to hold a second meeting two years later. Professor Jean Eisenstaedt, of the Institut Henri Poincaré, kindly agreed to undertake the organization of this meeting, which was held in Luminy in September 1988. The proceedings of that conference will appear as volume three of *Einstein Studies*.

The next volume of *Einstein Studies* will comprise the proceedings of the 1988 Osgood Hill conference on Conceptual Problems of Quantum Gravity, edited by Professor Abhay Ashtekar and John Stachel. A fifth conference, to be held in 1990, will attempt to assess the new information about the young Einstein that has been brought to light by the publication of the first two volumes of the Einstein Papers.

*Participants*

| | |
|---|---|
| Abhay Ashtekar | Joshua N. Goldberg |
| Peter G. Bergmann | Stanley Goldberg |
| Michel Biezunski | Peter Havas |
| Kenneth Brecher | Pierre Kerszberg |
| David Cassidy | A.J. Kox |
| Carlo Cattani | João Leão |
| Joy Christian | Alan Lightman |
| Michelangelo De Maria | John Norton |
| John Earman | Jürgen Renn |
| Jean Eisenstaedt | John Stachel |

# Acknowledgments

The editors extend heartfelt thanks to the following people, who have provided encouragement and assistance in establishing the *Einstein Studies* series and in seeing this first volume through to publication. Our good friend Robert S. Cohen, Director of the Boston University Center for the Philosophy and History of Science, an early supporter of the idea of establishing such a series, generously shared the fruits of his long editorial experience. George Adelman, Editorial Director at Birkhäuser Boston, was another early and enthusiastic supporter of the series, whose advice and gentle prodding are very much appreciated. Lauren Cowles, Physics Editor at Birkhäuser Boston, has provided valuable assistance in the completion of volume one and in the planning of future volumes. We also thank our production editor, James Kingsepp, for his many contributions and suggestions. At various stages in the production of this volume, essential help has been provided by Clelia Anderson, Margo Shearman, Deborah Stachel, and Kimberly D. Tolle.

To Boston University we extend out special thanks for its financial support of the *Einstein Studies* series through the Center for Einstein Studies. Special thanks are due also to the Dean S. Edmonds Foundation for its timely support of the Osgood Hill Conference on the History of General Relativity, out of which the present volume grew.

Finally, we thank The Hebrew University of Jerusalem, Israel, for permission to quote extensively from the writings and correspondence of Albert Einstein in this volume.

A NOTE ON SOURCES

In view of the frequent citations of unpublished correspondence or other items in the Einstein Archive, the editors have adopted a standard format for such citations. For example, a designation such as "EA 26-107" refers to item number 26-107 in the Control Index to the Einstein Archive. Copies of the Control Index can be consulted at the Jewish National and University Library (the Hebrew University), Jerusalem, where the Archive is housed; and at Mudd Manuscript Library, Princeton University, and Mugar Memorial Library, Boston University, where copies of the Archive are available for consultation by scholars.

# Contents

Series Preface ............................................................. v
Preface to Volume One ................................................. vii
Acknowledgments ....................................................... ix
A Note on Sources ....................................................... x

Introduction .............................................................. 1

*Einstein's Discovery of the General Theory*

What Was Einstein's Principle of Equivalence? ....................... 5
JOHN NORTON

The Rigidly Rotating Disk as the "Missing Link" in the History of
General Relativity ...................................................... 48
JOHN STACHEL

Einstein's Search for General Covariance, 1912–1915 ................ 63
JOHN STACHEL

How Einstein Found His Field Equations, 1912–1915 ................ 101
JOHN NORTON

*The Reception and Development of General Relativity*

Max Abraham and the Reception of Relativity in Italy: His 1912 and
1914 Controversies with Einstein ...................................... 160
CARLO CATTANI AND MICHELANGELO DE MARIA

The 1915 Epistolary Controversy between Einstein and
Tullio Levi-Civita ...................................................... 175
CARLO CATTANI AND MICHELANGELO DE MARIA

xii   Table of Contents

Hendrik Antoon Lorentz, the Ether, and the General Theory
of Relativity ............................................................. 201
A.J. Kox

The Early Interpretation of the Schwarzschild Solution ................. 213
Jean Eisenstaedt

The Early History of the "Problem of Motion" in General Relativity .. 234
Peter Havas

The Low Water Mark of General Relativity, 1925–1955 ............... 277
Jean Eisenstaedt

The Canonical Formulation of General-Relativistic Theories: The Early
Years, 1930–1959 ..................................................... 293
Peter G. Bergmann

*Unified Field Theories*

Einstein, Hilbert, and Weyl: The Genesis of the Geometrical Unified
Field Theory Program ................................................. 300
Vladimir P. Vizgin

Inside the Coconut: The Einstein-Cartan Discussion
on Distant Parallelism ................................................. 315
Michel Biezunski

*Cosmology*

The Einstein–de Sitter Controversy of 1916–1917 and the Rise of
Relativistic Cosmology ................................................ 325
Pierre Kerszberg

The Expanding Universe: A History of Cosmology from 1917
to 1960 ................................................................. 367
George F.R. Ellis

Contributors ........................................................... 433
Index .................................................................. 435

# Introduction

JOHN STACHEL

The volume is based on the proceedings of a conference on the history of general relativity held May 8–11, 1986, at the Boston University Conference Center at Osgood Hill (see the preface). As far as we known, this was the first conference exclusively devoted to the subject.

The aim in organizing the conference was to take a first step in remedying a deficiency in current research on the history of twentieth-century physics. For, in spite of the intense and continuing interest in the life and work of Albert Einstein, it is remarkable how little detailed work had been done until very recently on the origins and development of the general theory of relativity. In spite of the fact that Einstein considered this theory the major achievement of his life's work, most research on the history of relativity has concentrated on the special theory. The paucity of research on the general theory is especially striking in comparison to the large and still rapidly growing body of work on the development of the quantum theory. It is interesting to speculate on some possible reasons for the comparative neglect of general relativity.

1. *Technical Difficulty*: Compared to the special theory, work on "internal" aspects of the history of the general theory requires much greater preliminary technical training in several areas of mathematics, most notably differential geometry and tensor calculus. This preliminary hurdle may have served to deter historians of science, and even historically minded physicists who had not previously studied general relativity. It is no accident that the majority of papers in this volume are by physicists who have previously worked on the subject.

2. *Unfashionableness*: From the late 1920s until the late 1950s, general relativity was considered by most physicists a detour well off the main highway of physics, which ran through the quantum theory. (See Jean Eisenstaedt's paper in this volume, "The Low Water Mark of General Relativity, 1925–1955.") Special relativity had to be treated more respectfully, since at least the rudiments of that subject are required to understand relativistic quantum mechanics and quantum field theory or, at a more practical level, to design and operate such devices as klystrons and synchrotrons. By contrast, the

general theory seemed quite eccentric to the main body of physics, of little importance even to a theoretical physicist. The low estimate of general relativity was not unconnected with the prevalence of a pragmatic attitude toward physics among its practitioners: Only the calculation of a testable number counted as valid theoretical physics. This attitude often was associated with an uncritical acceptance of positivistic and operationalistic outlooks on science.

3. *Unavailability of Source Materials*: While the project "Sources for the History of Quantum Physics" made readily available a large body of manuscript and oral interview material, no comparable effort was made to gather material on the history of relativity, either special or general. This failure both reflects the neglect of the subject for other reasons, and has become a further cause for that neglect. The difficulties experienced by scholars until fairly recently in gaining ready access to the Einstein Archive, the main source of manuscript documents on the early history of general relativity, undoubtedly contributed to the problem.

In recent years, the situation has changed with respect to each of these reasons for the neglect of the history of general relativity. Since the late 1950s, the general theory has moved closer to the mainstream of modern physics. Relativistic astrophysics emerged as a new subdiscipline in the 1960s as more and more cosmic objects were discovered, the properties of which could not be fully explored without the use of general-relativistic methods. The post-World War II rise of relativistic quantum field theory convinced a number of physicists that it was time to reexamine the relationship between the general-relativistic and quantum field-theoretical approaches, embodying as they do the two most profound conceptual structures so far proposed for probing the foundations of physics. Many quantum field theorists looked upon the general theory as no more than a new world to conquer. As the best available theory of gravitation, it naturally demanded attention, and attempts to quantize the Einstein field equations, hitherto rather unfashionable, increased in number and popularity—if not in notable accomplishments. Other physicists, initially those with a background in "classical" general relativity, began to explore the possibility that the relationship between the two theories might not be that simple (see Peter Bergmann's paper in this volume, "The Canonical Formulation of General-Relativistic Theories: The Early Years, 1930–1959"). Difficulties encountered in the quantum field theory program made field theorists more sympathetic to such explorations. Suggestions that the foundations of quantum mechanics might be subject to critical scrutiny and alteration were no longer automatically taken as signs of mental incompetence. The recent dramatic successes of gauge theories of elementary particles have helped the search for a unified field theory to emerge from the underworld of physics, where it had barely survived for the last fifty years (see Vladimir Vizgin's paper in this volume, "Einstein, Hilbert, Weyl: The Genesis of the Geometrical Unified Field Theory Program"), onto the center stage.

Renewed discussion among physicists probably played some part in directing the attention of a number of philosophers of science in recent decades to problems connected with the general theory. Since the latter first developed, interest in the philosophy of space and time has been closely associated with interest in the general theory. The first such efflorescence of interest took place in the 1920s, and there has recently been a resurgence of work in this area. Fortunately for the history of general relativity, this renewed interest has been contemporaneous with a growing rapprochement between the philosophy and the history of science.

Moreover, a new generation of philosophers and historians of science has emerged, one that takes for granted the need to develop technical expertise before starting work on a topic. Having taken Hilbert spaces in stride in order to work on quantum mechanics, the current generation is willing to tackle such exotica as differentiable manifolds, fiber bundles, and global topology, if that is the price to pay for saying something significant about space-time theories. It must be noted that, so far, it has been primarily philosophers of science—including such historically minded ones as John Earman, Clark Glymour, and John Norton, for example—who have taken the technical plunge. One of our hopes for this volume is that it will inspire more historians of science to occupy themselves with the subject. However much physicists and philosophers are able to accomplish in this area, the skills of trained historians of science are badly needed, not least in the formulation of the right questions to be answered by future research.

The Einstein Archive at the Hebrew University in Jerusalem, and the duplicates of it in the Mudd Manuscript Library, Princeton University, and Mugar Memorial Library, Boston University, are now readily accessible to interested scholars. However, there is still an urgent need for an organized effort to collect and preserve manuscript materials on the history of the general theory of relativity, and to create an archive of oral interviews with relativists. Hopefully, copies of all such material would then become available at one or more central repositories.

**********

A few areas of research on the general theory have already been fairly well charted. While we are far from having the last word on the origins of the general theory, the stages of Einstein's work on it have been well delineated, and a number of major issues have been clarified and explored in some detail. The papers by Norton and Stachel in this volume discuss this topic, and provide references to a number of other important contributions. The papers of Cattani and De Maria, focusing on the roles of Levi-Civita and Abraham, and Kox's paper on Lorentz, also explore aspects of the period through 1916, the year in which Einstein gave the theory its classic formulation.

The development of the subject after 1916, on the other hand, is largely unexplored. This volume provides no more than a small sample of the fascinating problems that await further exploration. Eisenstaedt offers an English summary of his two extensive papers in French on the history of that most important of all solutions to the Einstein field equations, the Schwarzschild solution, as well as a provocative discussion of the reasons for the neglect of the general theory among physicists for about thirty years. Havas provides a short, tantalizing introduction to his extensive, largely unpublished work on the history of the discovery of a unique feature of the general theory: Its field equations delimit and in certain cases completely determine the motions of sources of the gravitational field. Bergmann provides an account of the early history of attempts to quantize the general theory, an area of work in which he played—and continues to play—an active role.

Two papers concern the history of early attempts at unified field theories. Vizgin discusses the origins of the geometrical unification program, while Biezunski discusses the exchange of letters between Einstein and Cartan on the attempt to unify gravitation and electromagnetism using the concept of distant parallelism.

Ellis gives an impressive classification and survey of the first forty years of work on relativistic cosmology, and provides an invaluable bibliography, while Kerszberg studies in detail the origins of that subject in an early dispute between Einstein and de Sitter.

The participants in the Osgood Hill meeting were all encouraged to persevere in their hitherto rather solitary efforts by the opportunity to meet each other, often for the first time, and to discuss together, both formally and informally, many problems of common interest. They resolved to plan a second conference, which was held in France in September 1988 under the auspices of the Centre Nationale de la Recherche Scientifique. It is hoped that this meeting will institute regular meetings on the subject. May this volume serve to convey to a wider circle of physicists, historians of science, and philosophers of science at least part of the sense of enthusiasm and the challenge to do further work felt by the participants in the first meeting.

# What was Einstein's Principle of Equivalence?*

JOHN NORTON

## 1. Introduction

In October and November 1907, just over two years after the completion of his special theory of relativity, Einstein made the breakthrough that set him on the path to the general theory of relativity. While preparing a review article on his new special theory of relativity, he became convinced that the key to the extension of the principle of relativity to accelerated motion lay in the remarkable and unexplained empirical coincidence of the equality of inertial and gravitational masses. To interpret and exploit this coincidence, he introduced a new and powerful physical principle, soon to be called the "principle of equivalence," upon which his search for a general theory of relativity would be based. Moreover, with the completion of the theory and throughout the remainder of his life, Einstein insisted on the fundamental importance of the principle to his general theory of relativity.

Einstein's insistence on this point has created a puzzle for philosophers and historians of science. It has been argued vigorously that the principle in its traditional formulation does not hold in the general theory of relativity. Consider, for example, a traditional formulation such as Pauli's in his 1921 *Encyklopädie* Article. For Pauli the principle asserts that one can always transform away an arbitrary gravitational field in an infinitely small region of space-time, by transforming to an appropriate coordinate system (Pauli 1921, p. 145).

In response, such eminent relativists as Synge (1960, p. ix), and even Eddington before him (1924, pp. 39–41), have objected that a coordinate transformation or change of state of motion of the observer can have no effect on the presence or absence of a gravitational field. The presence of a "true" gravitational field is determined by an invariant criterion, the curvature of the metric. The gravitation-free case of special relativity is just the case in which this curvature vanishes, whereas the true gravitational fields of general relativity are distinguished by the nonvanishing of this curvature.

This objection has immediate ramifications for the "Einstein elevator" thought experiment, which is commonly used in the formulation of the principle of equivalence. In this thought experiment, a small chamber, such as an elevator, is accelerated in order to transform away a gravitational field present within it or, depending on the version at hand, to produce a gravitational field in an initially gravitation-free chamber. Now in general relativity, nonvanishing metrical curvature is responsible for tidal gravitational forces. Their effects can be used by an observer within the chamber to decide whether the gravitational field present is a true gravitational field or is due to the acceleration of the chamber in gravitation-free space. Alternatively, they can be used to determine whether an apparently gravitation-free chamber is in free fall in a gravitational field or moving uniformly in gravitation-free space. It is significant that the effects of these tidal forces do not vanish as the box becomes arbitrarily small. For example, the tidal bulges arising in a freely falling liquid droplet do not vanish as the droplet in made arbitrarily small, ignoring such effects as surface tension (Ohanian 1977).

Of course it has proved possible to retain a principle of equivalence in general relativity. But to do this, the principle might be given quite new formulations, which seem to carry us far from Einstein's original intentions. For example, in its "weak" form the principle merely asserts the equality of inertial and gravitational mass.[1] Or in another form, it asserts that all phenomena distinguish a unique affine structure for space-time (Anderson 1967, pp. 334–338). Alternatively, we can retain a traditional formulation of the principle, such as Pauli's, by reading the restriction to infinitely small regions of space-time as denying access to certain quantities such as curvature, which are constructed from the higher derivatives of the metric tensor. But then the principle is reduced to a simple and, as far as questions of foundations are concerned, not especially interesting theorem in general relativity. Certainly Einstein could not represent such a result as a fundamental principle of his theory.

My purpose in this paper is to determine precisely what Einstein took his principle of equivalence to be, to show how it figured historically in his discovery of the general theory of relativity, and to show the sense in which he took it to be fundamental to that theory. In particular I will seek to demonstrate that Einstein's version of the principle and the way he sought to use it are essentially different from the many later versions and applications of the principle. As a result, we shall see that the objections rehearsed earlier from the later debate over the principle of equivalence are peripheral to the concerns of Einstein's version of the principle and that this version does find completely satisfactory and uncontroversial expression in the general theory of relativity.

In the following section, as a focus for the remainder of the paper, I will present one of the clearest and most cautious of Einstein's formulations of the principle of equivalence and in Section 3, I will develop sufficient formal apparatus to negotiate certain ambiguities in it. In particular, I will introduce the concept of a three-dimensional relative space of a frame of reference, which is essential to the understanding of Einstein's principle and much of his early work on his general theory of relativity.

In Sections 4 and 5, I will review the role the principle played in the 1907 to 1912 period of Einstein's search for his general theory of relativity. In Section 4, I will outline how the principle enabled Einstein to construct a novel relativistic theory of static gravitational fields and, in Section 5, I will outline the sense in which he believed the principle would enable an extension of the principle of relativity to accelerated motion.

In Sections 6, 7, and 8, I will examine the principle of equivalence within Einstein's general theory of relativity, whose basic formal structure was laid down by Einstein and Marcel Grossmann in 1912 and 1913 and which achieved its final form in November 1915. In Section 6, I will review aspects of Einstein's transition from a three- to a four-dimensional formalism, and, in Sections 7 and 8, I will review the status of the principle in the theory. In particular, we shall see its crucial heuristic role in the transition from the special to the general theory.

In Sections 9 and 10, I will relate Einstein's version of the principle and the results he drew from it to the "infinitesimal" principle of equivalence, such as that formulated by Pauli, and which is now commonly but mistakenly regarded as Einstein's version of the principle. In particular, I will analyze in some detail a devastating objection Einstein had to this version of the principle. It follows from the objection that, insofar as it can be precisely formulated, the infinitesimal principle is trivial. In Section 11, I will review Einstein's attitude to Synge's now popular identification of "true" gravitational fields with metrical curvature.

Finally, in Section 12, I will draw together the threads of my story and answer the question posed in the title of this paper.

## 2. Einstein's Formulation of the Principle of Equivalence

Einstein has given us many statements of the principle of equivalence in his treatments and discussions of the general theory of relativity. But none is clearer or more cautious than the formulation he gives in a 1916 reply to Kottler's claim that Einstein had given up the principle of equivalence in the general theory of relativity (Einstein 1916b). Einstein began by introducing the limiting case of special relativity in which he defined a "Galilean system." I quote this here for later reference:

1. *The Limiting Case of the Special Theory of Relativity.* Let a finite space-time region be free from a gravitational field, i.e., it is possible to set up a reference system $K$ ("Galilean system"), relative to which the following holds for the region considered. Coordinates are measured directly in the well-known way with unit measuring rods, times with unit clocks, as is customarily assumed in the special theory of relativity. In relation to this system an isolated material point moves uniformly and in a straight line, as was assumed by Galileo.

He then proceeded to his statement of the principle:

2. *Principle of Equivalence.* Starting from this limiting case of the special theory of relativity, one can ask oneself whether an observer, uniformly accelerated relative to $K$ in the region considered, must understand his condition as accelerated, or whether there remains a point of view for him, in accord with the (approximately) known laws of nature, by which he can interpret his condition as "rest." Expressed more precisely: do the laws of nature, known to a certain approximation, allow us to consider a reference system $K'$ as at rest, if it is accelerated uniformly with respect to $K$? Or somewhat more generally: Can the principle of relativity be extended also to reference systems, which are (uniformly) accelerated relative to one another? The answer runs: As far as we really know the laws of nature, nothing stops us from considering the system $K'$ as at rest, if we assume the presence of a gravitational field (homogeneous in the first approximation) relative to $K'$; for all bodies fall with the same acceleration independent of their physical nature in a homogeneous gravitational field as well as with respect to our system $K'$. The assumption that one may treat $K'$ as at rest in all strictness without any laws of nature not being fulfilled with respect to $K'$, I call the "principle of equivalence."

For Einstein, the basic assertion of the principle of equivalence is that "one may treat $K'$ as at rest...." I will defer discussion of exactly what he intended with this assertion until Section 5. The assumption upon which this assertion is based—that acceleration can produce a gravitational field—is at present more commonly associated with the principle of equivalence. The way in which it is used, however, is distinct from its use in "traditional" formulations of the principle such as Pauli's. In the latter, by reversing Einstein's argument, one assumes that one can always transform away an arbitrary gravitational field in general relativity within an infinitesimal region of space-time. Einstein however considers only the homogeneous gravitational field produced by uniform, nonrotating acceleration in the Minkowski space-time of special relativity. In addition, there is clearly no restriction to infinitesimal regions.

These last features are typical characteristics of Einstein's preferred formulation of the principle and appear in many of the statements of the principle that Einstein gave throughout the half century of his working life. These include his first published formulation of the principle in 1907, some five years prior to the completion of the general theory of relativity (Einstein 1907, p. 454), his well-known 1911 communication on gravitation (Einstein 1911, pp. 898–899), and his 1916 review of the just-completed theory (Einstein 1916a, pp. 772–773).[2] The principle is defined in these terms in *The Meaning of Relativity*, the work which came closest to his "textbook" on relativity (Einstein 1922, pp. 57–58). Finally, it appears again in this form in one of his last discussions of the question, the 1952 appendix to his popular book, *Relativity* (Einstein 1952, pp. 151–152).

Einstein's next step in his reply to Kottler was to insist pointedly that his principle did not allow one to transform away arbitrary gravitational fields. Rather it dealt only with those gravitational fields that could be transformed away and which we would now identify as associated with Minkowski space-time.

3. *Gravitational Fields not only Kinematically Conditioned.* One can also invert the previous consideration. Let the system $K'$, formed with the gravitational field considered above, be the original one. Then one can introduce a new reference system $K$, accelerated with respect to $K'$, with respect to which (isolated) masses (in the region considered) move uniformly in a straight line. But one may *not* go on and say: if $K'$ is a reference system provided with an arbitrary gravitational field, then it is always possible to find a reference system $K$, in relation to which isolated bodies move uniformly in a straight line, i.e., in relation to which no gravitational field exists. The absurdity of such an assumption is quite obvious. If the gravitational field with respect to $K'$, for example, is that of a stationary mass point, then this field certainly cannot be transformed away for the entire neighborhood of the mass point, no matter how refined the transformation artifice. Therefore, one may in no way assert that gravitational fields should be explained so to speak purely kinematically; a "kinematic, not dynamic understanding of gravitation" is not possible. Merely by means of acceleration transformations from a Galilean system into another, we do not become acquainted with *arbitrary* gravitational fields, but those of a quite special kind, which, however, must still satisfy the same laws as all other gravitational fields. This is only again another formulation of the principle of equivalence (in particular in its application to gravitation).

In short, he rules out an extension of the principle to arbitrary gravitational fields on the grounds that an acceleration of the reference system can only produce gravitational fields of a quite special kind. Such comments appear quite frequently in Einstein's writings, throughout his life. They appear in his publications[3] and in his correspondence, right up to the last years of his life.[4]

What might seem striking to the modern reader here is Einstein's failure to consider the possibility of transforming away arbitrary gravitational fields *in infinitesimal regions* of space-time. The omission was not a peculiarity of this particular discussion of the principle, for I have been unable to find any sustained treatment by Einstein of such an extension of the principle.[5] Nevertheless we can readily infer Einstein's attitude to this possibility. In Section 9, we shall see that he believed that one cannot distinguish the motion of a point-mass uninfluenced by a gravitational field from other motions if one considers only infinitesimal regions of the manifold. It follows immediately from Einstein's comments above that it is meaningless to talk in any thoroughgoing sense of transforming away a gravitational field in such infinitesimal regions.

The task of explicating Einstein's formulation of the principle of equivalence and even some of the preceding discussion is by no means straightforward. To begin, we must deal with Einstein's failure to maintain such distinctions as those between frames of reference and coordinate systems and between three-dimensional and four-dimensional concepts.[6] For example, we shall see that when Einstein speaks of a four-dimensional coordinate system, he may be referring to a four-dimensional coordinate system *simpliciter*, a frame of reference, or even a three-dimensional space associated with the frame. In the following section, I will introduce sufficient formal apparatus to deal with this problem, and then with it, we shall find that there is little difficulty in understanding Einstein's intentions. Then we can turn to ask precisely what Einstein means when he talks of a gravitational field produced by acceleration and in what sense the associated states of acceleration can be regarded as being "at rest."

## 3. On Reference Systems and Relative Spaces

In this section, I will deal with structures associated with the semi-Riemannian manifolds of special and general relativity.

In such manifolds, it is now customary to represent the intuitive notion of a physical frame of reference as a congruence of time-like curves. Each curve represents the world line of a reference point of the frame. The velocity of these points is given by the tangent vectors to the curves, where defined. We shall usually deal with frames of reference in rigid-body motion and we can readily nominate the state of motion of such frames because of the limited number of degrees of freedom associated with them.[7] In particular, an inertial frame of reference in a Minkowski space-time is a congruence of time-like geodesics in rigid-body motion, and therefore its reference points move with constant velocity.

A coordinate system $\{x^i\}$ ($i = 1, 2, 3, 4$) is said to be "adapted" to a given frame of reference just in case the curves of constant $x^1$, $x^2$, and $x^3$ are the curves of the frame. These three coordinates are "spatial" coordinates and the $x^4$ coordinate a "time" coordinate.

With these definitions, Einstein's talk of "accelerated coordinate systems" can be made precise. A coordinate system is "accelerated" just in case it is adapted to an accelerating frame of reference. In this manner of speaking, a transformation from one frame of reference to another can be represented at least locally by a transformation between coordinate systems adapted to each frame.

Similarly we can represent the "Galilean" reference system mentioned in the last section as a coordinate system in Minkowski space-time, adapted to an inertial frame of reference and chosen so that the metric has components diag($-1, -1, -1, c^2$), where $c$ is a positive constant—the coordinate speed of light. In such a coordinate system, differences of coordinates along curves, for which all but one coordinate is held fixed, are equal to the proper time or proper length of that segment of the curve, according to whether the curve is space-like or time-like. This implements Einstein's requirement that the coordinates be given directly by clock readings and measuring operations with rigid rods.

Presumably Einstein required the coordinates of his accelerated coordinate systems to have as much of a similar direct metrical significance as was possible. Methods and scope for constructing analogous coordinate systems in the context of Newtonian theory and special and general relativity are well known (see, for example, Friedman 1983, pp. 79–84, 129–135, 181–183).

However this discussion of Galilean and other systems in four-dimensional space-time does not entirely capture Einstein's intentions. He was also concerned with certain three-dimensional spaces, which are alluded to throughout his discussion of the principle of equivalence. It is appropriate to call these spaces "relative spaces," because of their similarity to the "relative space" Newton defined to contrast with his absolute space (Newton 1729, p. 6).[8] Einstein himself introduces the concept of this space in the introductions to his accounts of relativity theory, where it is presented as our most primitive notion of space (Einstein 1922, pp. 3–4; 1954a, pp. 5–8). It arises through our experience that a given physical body can be extended by bringing other bodies into contact with it. The space of all such possible extension is the relative space of the body.

If we think of the time-like curves of a frame of reference as the world lines of physical bodies, then these bodies define a single relative space, insofar as each of the bodies can be extended to contact any other body of the frame. The geometric properties of this space can be investigated in the familiar manner by laying out infinitesimal rigid rods, which are at rest in the frame. An example of this, which Einstein discussed frequently, is the relative space of a uniformly and rigidly rotating frame of reference in Minkowski space-time. In particular on finds there that the geometry of the relative space is non-Euclidean.[9]

The properties of the relative space defined by a given frame of reference can be precisely specified, although not in general by isomorphism with a three-dimensional hypersurface in the space-time manifold with the associated induced geometrical structure. The nature candidates for such hypersurfaces—the three-dimensional hypersurfaces orthogonal to the curves of the frame of reference—simply fail to exist if the frame of reference is rotating even in Minkowski space-time, for example.

Rather, we formally define the relative space $R_F$ of a frame of reference $F$ in a four-dimensional manifold $M$ as follows. $F$ defines an equivalence relation $f$ under which points $p$ and $p'$ of $M$ are equivalent if and only if they lie on the same curve $c$ of $F$. The relative space $R_F$ is the quotient manifold $M/f$ and has the curves of $F$ as elements. Coordinate charts of $R_F$ are inherited directly from the coordinate charts of $M$, which are adapted to the frame, ensuring that $R_F$ has a well-defined local topology. That is, if $\{x^i\}$ ($i = 1, 2, 3, 4$) is a chart in a neighborhood of $M$ adapted to $F$, then there will be a chart $\{y^i\}$ ($i = 1, 2, 3$) in the corresponding neighborhood of $R_F$ for which $y^i(c) = x^i(p)$ ($i = 1, 2, 3$) whenever $p$ lies on $c$.

A positive-definite metric $g_r$ is induced on $R_F$ as follows. At any point $p$ on $c$ we define the (unique) orthogonal metric $g_{\text{orth}}$ as the restriction of the space-time metric $g$ to any three-dimensional hypersurface $H_c(p)$ orthogonal to $c$ at $p$. A diffeomorphism $h$, which maps points of $H_c(p)$ in a neighborhood of $p$ to points in a neighborhood of $c$ in $R_F$, is such that, if $p'$ lies on the curve $c'$ of $F$, then $h(p') = c' \cdot g_r$ at $c$ is defined as the image of $g_{\text{orth}}$ at $p$ under $h$.[10] (Intuitively, we take $g_r$ to be the three-dimensional spatial metric revealed to an observer co-moving with the frame through the laying out of infinitesimal rods.)

Since point $p$ of $c$ here is chosen arbitrarily, it is clear that the resulting induced metric will only be uniquely defined in certain special cases. These special cases turn out to be just those in which the frame of reference is in rigid-body motion, for the requirement of rigid-body motion can be expressed as the requirement of constancy of the orthogonal metric along the world lines of the body. More specifically, what is required is the vanishing of the Lie derivative of $g_{\text{orth}}$, that is, $L_V g_{\text{orth}} = 0$, where $V$ is the tangent vector field of $F$.[11] General relativity deals with space-times that do not always admit rigid-body motions. Obviously, in these cases we will be unable to construct a relative space with a well-defined metric.

To deal with the phenomena Einstein considers, we need to define a few more structures in these relative spaces. A gravitational field will be represented by a scalar field in nearly all the cases we need consider. A moving point-mass $M$ will be represented by a scalar, its rest mass, and an appropriately parameterized curve $C$, its trajectory in the relative space $R_F$. $C$ can be inferred readily from the points of intersection of $M$'s world line with the time-like curves of the frame. That is, if $M$'s world line $c$ at parameter value $x$ intersects the curve $c'$ of frame $F$, then $C$ is the map that takes $x$ to $c'$. The velocity and acceleration vectors of $C$ can now be defined in the usual way. If $c$ is parameterized by proper time, we would then arrive at the point-mass's proper velocity and proper acceleration.

In certain important special cases, it is possible to introduce a "frame time" into the relative space $R_F$ of a frame $F$. These cases are those in which the relevant neighborhood of the manifold can be foliated by a family of hypersurfaces, orthogonal to the curves of the frame $F$. Pick any curve $c$ of $F$, parameterized by proper time. Informally, we shall think of this curve as the frame clock of $F$ and its relative space $R_F$. Disseminate the time it marks by the following procedure. Define a scalar field $T$ on the space-time manifold whose constant-value hypersurfaces coincide with the hypersurfaces of the foliation and whose value agrees with the proper-time parameterization of $c$. Of course $T$ will only be defined up to an additive constant.

This frame-time can now be transferred to the structures defined in $R_F$ by obvious means. For example the trajectory $C$ of a moving point-mass $M$ in $R_F$ can be parameterized by $T$, if $T$ is also used to parameterize $M$'s world line in the procedure for constructing $C$. From this parameterization, we would then arrive at $M$'s frame velocity and frame acceleration. Through a similar procedure, a time–varying field in $R_F$, induced by a field defined in the space-time manifold, can be represented by a family of fields indexed by $T$. The parameterization and indexing of structures in $R_F$ by $T$ gives a criterion of simultaneity.[12]

Clearly, in general we shall not be able to define a frame time. A rotating frame, for example, has no orthogonal hypersurfaces. Even if there are such hypersurfaces, the frame time may not be unique. A rigid, uniformly accelerating frame in Minkowski space-time admits orthogonal hypersurfaces; but the frame times defined by each of its curves differ by a multiplicative constant, although they yield the same simultaneity criterion. However, if the frame is an inertial frame in Minkowski space-time then the same frame time is defined by all curves of the frame, up to an additive constant.

We can recover a "standard formulation" of special relativity—corresponding to the original three-dimensional formulation of the theory introduced by Einstein in 1905—by writing the laws that govern physical processes in Minkowski space-time in terms of structures defined within the relative space of an inertial frame, using the relative space's frame time. This formulation will hold just in any relative space of an inertial frame. Quantities describing the same process viewed from two different inertial relative spaces will be related by the Lorentz transformation in the familiar manner.

Generalizing, we construct a standard formulation of a four-dimensional space-time theory, in any given relative space that admits a frame time, by re-expressing its laws in terms of structures defined in the relative space, parameterized where necessary by the frame time. Thus we can construct a standard formulation of special relativity in the relative space of a rigid uniformly accelerating frame—and it will look quite different from the standard formulation associated with an inertial frame.

Einstein commenced his description of the principle of equivalence in his reply to Kottler by mention of space-time. It is now clear, however, that the phenomena he proceeded to describe are considered in relation to the relative spaces of the frames of reference. An isolated material point in a Galilean system can only be properly described as "mov[ing] uniformly and in a straight line" in the relative space. There it is represented by a geodesic of the relative space ("straight line"); its proper time and its frame time parameterization are directly proportional to the metrical distance along the curve ("move uniformly"). Use of either parameterization in this way also gives two general definitions of "uniform straight-line motion" in relative spaces, which agree in this case.

Similarly, it is more natural to understand Einstein's requirement that the coordinates of the Galilean system be "measured directly in the well-known way" with rods and clocks as referring to operations described in the relative space and out of which the Galilean space-time coordinate sytem is constructed.

But most important of all, when Einstein speaks of "the presence of a gravitational field" in his reply to Kottler, clearly we should understand it to be present in the relative space of the frame of reference in question. In Minkowski space-time, there is a gravitational field in the relative space of the accelerated reference system but not in the relative space of the Galilean system. This is certainly more satisfactory than trying to speak of the presence of a gravitational field in space-time in this context. For then we would have to assume that a change of frame of reference can "produce" a gravitational field in space-time even though it does not change the world line of the point-mass on which the newly produced field is supposed to act.

This somewhat cumbersome mixture of three- and four-dimensional concepts in Einstein's formulation of the principle of equivalence derives directly from the fact that, for the first five years of its life, the principle and the gravitation theory associated with it were treated entirely within the same three-dimensional formalism Einstein had used in his 1905 special relativity paper. In particular, the spaces Einstein dealt with in this period were invariably the relative spaces of frames of reference. Nevertheless, Einstein's 1916 formulation and his original 1907 formulation of the principle read almost identically, even though the former was associated with a theory that could not readily be written in a three-dimensional formalism. In the following section I turn to examine this early period of Einstein's work. I will be concerned with showing precisely which structures Einstein chose to represent the gravitational field in the relative spaces he dealt with.

## 4. A New Theory of Gravitation

### 4.1. A New Concept of Gravitational Field

Einstein made clear from the inception of the principle of equivalence in 1907 that its main purpose was to enable the extension of the principle of relativity to accelerated motion.[13] But for the five years following 1907, his actual use of the principle involved the development of a novel relativistic theory of static gravitational fields out of which his general theory of relativity would emerge in 1912 and 1913. The principle assured him that a certain structure ("inertial field") arising in the relative space of a uniformly accelerated frame of reference in Minkowski space-time was just one special type of gravitational field. The properties of this structure could be examined minutely using the known results of special relativity and the properties of other types of gravitational fields could then be inferred.

That this structure (whose properties will be developed and outlined in Section 4.2) could be regarded as a gravitational field requires a change in our understanding of what a gravitational field is. We must now accept that gravitational fields can have an existence dependent on the relative space considered and that the choice of relative space may decide whether or not a single given process is regarded as acted on by a gravitational field. The obvious objection, which was put by Laue to Einstein in 1911, is that this type of gravitational field cannot be "real" since it has no source masses.[14] Einstein's later response to this objection was that it is essential to field theory to be able to conceive of fields, such as gravitational fields, as existing independently of their sources.[15]

In effect, Einstein asks us to give up the familiar concept of gravitational field as that which mediates the gravitational interaction of bodies. In its place in the relative space of frames of reference, regardless of whether they are accelerated or not, we infer the existence of a structure that is responsible for the deviations from uniform straight-line motion of a free point-mass, without concerning ourselves with what generates that structure. Following Einstein's lead, we would take such a structure to be a gravitational field by definition, if the deviations associated with it are independent of the point's mass.

Using this definition, we could now describe as gravitational fields the inertial fields arising in relative spaces of rigid frames of reference in *arbitrary* states of acceleration in Minkowski space-time. It is difficult to imagine that Einstein would contradict this result. Nevertheless, as I have pointed out, he formulated his principle of equivalence only for the case of *uniform* acceleration.

There were most probably several reasons for this additional restriction. In the early years of the principle of equivalence, in order to convince skeptical contemporaries that inertial fields could be regarded as gravitational fields, he had to show that they behaved *exactly* like known gravitational fields—that is, like Newtonian gravitational fields—aside of course from the question of source masses. If the principle of equivalence is formulated in a Newtonian space-time, as Einstein did sometimes in these earlier years,[16] the requirement that the inertial field behave exactly like a Newtonian field places severe restrictions on the allowed states of motion of the frame of reference.

In Newtonian mechanics, the inertial field induced on the relative space of a rotating frame of reference contains a Coriolis field, which exerts a force on a body dependent on its velocity. A structure representing such a field will contain vector potentials, such as those arising in electromagnetic theory, rather than the familiar scalar potential of the Newtonian gravitational field.[17] The inertial field induced on the relative space of a frame of reference in rectilinear acceleration can be represented by a scalar potential satisfying Laplace's equation. But if the acceleration is not uniform the resulting field will be nonconservative due to the explicit time dependence of the potential.

In this case of a Newtonian space-time, we are led directly to Einstein's choice of a uniformly accelerated frame of reference for the formulation of the principle of equivalence. For only in this case will the structure concerned in the relative space behave exactly like a Newtonian gravitational field. It will be a scalar field, it will satisfy Laplace's equation, and its gradient will be equal to the acceleration of otherwise free point-masses in the space.

It would be natural for Einstein to continue to formulate the principle of equivalence in terms of the special case of uniform accleration in Minkowski space-time as well, if only in the interests of continuity. In addition, we can identify at least three complexities arising with the use of rotating frames of reference or those in nonuniform acceleration in Minkowski space-time.

First, the associated relative spaces would have non-Euclidean geometries, if they were well defined. This was a problem Einstein was well aware of from a very early stage. But he treated it as a separate issue from his principle of equivalence, usually by consideration of a rotating frame of reference.

Second, he would be unable to introduce a frame time into the relative space, making very difficult the description of phenomena in the space by a standard formulation of a theory such as he used in 1907–1912.

Third, the trajectory of a light signal exchanged between two points in the relative space would differ on the forward and return journeys. In a letter of June 1912 to Ehrenfest, in which Einstein discussed the failure of his 1912 gravitation theory to deal with the fields associated with rotating frames of reference, he mentioned this failure of the "reversibility of light paths" in such fields and described how dealing with them would be the next step (EA 9-333).

In any case, after the completion of the general theory of relativity, when the difficulties of the earlier gravitation theory had been resolved, there is a suggestion in one or two places in Einstein's writings that he was prepared to extend the formulation of the principle to the case of frames of reference in rotation or nonuniform acceleration (for example, Einstein 1922, p. 59; 1952, pp. 151–154).

4.2. THE 1907–1912 THEORY

Einstein's 1907–1912 theory of static gravitational fields achieved its most developed form in two consecutive papers in the latter year (Einstein 1912a; 1912b). The theory may be represented most precisely in four-dimensional terms, although Einstein had not yet begun to use them. It was based on exploiting certain especially simple properties of uniformly accelerating frames of reference in Minkowski space-time.

These special properties can be derived from the result that one can always find a coordinate system $\{x^i\}$ ($i = 1, 2, 3, 4$) adapted to a uniformly accelerating frame in Minkowski space-time in which the metric has the form

$$\text{diag}(-1, -1, -1, c^2)$$

where $c = 1 + bx^1$ and $b$ is a constant. It follows immediately that the geometry of the relative space is Euclidean, inheriting the coordinates $\{x^i\}$ ($i = 1, 2, 3$) as Cartesian coordinates. Further, the space-time can be foliated by a family of hypersurfaces orthogonal to the frame, the hypersurfaces of constant $x^4$. Therefore we can introduce a frame time.

For convenience, select the world line of the frame for which $x^1 = x^2 = x^3 = 0$ as the frame clock and call $t$ the frame time disseminated by it. The choice as frame clock of any of the other world lines of the frame would alter $t$ by a constant multiplicative factor and thus not materially affect the results.

Thus Einstein could introduce a standard formulation of special relativity in the relative space. In particular, it followed in this standard formulation that the motion of a free point-mass, whose world line was a geodesic in the space-time, was governed by the equation

$$d/dt(\beta v^i/c) = -\beta \partial c/\partial x^i,$$

where $\beta = 1/(1 - v^2/c^2)^{1/2}$, $v^i = d/dt(x^i)$ is the three-velocity of the point-mass, and $v$ is its magnitude.

This relation closely parallels the relation

$$\text{acceleration} = -\text{gradient of scalar field}$$

governing the motion of a freely falling point-mass in traditional Newtonian gravitation theory and in which the point's mass also does not appear. Thus in accord with the discussion of Section 4.1, Einstein could view the motion of the point-mass in the relative space as under the influence of a gravitational field whose scalar potential was $c$ and which was responsible for the deviations from uniform straight-line motion.

Note that while the scalar field $c$ was introduced earlier via the $g_{44}$ component of the Minkowski metric in a particular coordinate system, it can be described in coordinate-free terms: $c$ is just the Minkowski norm of the tangent four-vector of the curves of the frame, when parameterized by the frame time. It can be seen that $c$ will have a constant value along each of these curves and therefore a unique, well-defined value at each point of the relative space.

Recalling that the coordinates $\{x^i\}$ ($i = 1, 2, 3$) are inherited as Cartesian coordinates by the Euclidean relative space, the relation $c = 1 + bx^1$ now can be seen to assert that the gravitational potential $c$ varies linearly with (Euclidean) distance in one direction in the relative space. This is exactly the way a traditional Newtonian potential behaves in the case of a homogeneous gravitational field.

There were some complications however, in addition to the usual relativistic corrections; $c$ turned out to be the isotropic speed of light in the relative space, measured with frame time, which it now followed must also vary with position in the relative space. It could be shown that the rates of clocks at rest in the relative space would vary with $c$ and, therefore, with position.

Now that Einstein had a firm grasp on relativistic gravitational fields in the one special case of homogeneous fields, it was a simple matter to infer the properties of arbitrary static gravitational fields by a natural and hopefully unproblematic generalization. To do this, Einstein left the standard formulation of the theory unchanged, except for relaxing the condition that $c$ vary linearly with distance in the direction of acceleration. Following the model of Newtonian theory, he now required that $c$ satisfy a weaker condition, the field equation

$$\Delta c = \kappa c \sigma,$$

where $\sigma$ is the mass density and $k$ a constant.

This step amounted to the transition to the relative spaces of more general semi-Riemannian manifolds with static space-time metrics of Lorentz signature. The relative spaces are those of frames of reference whose velocity vectors are Killing vector fields. The metric must be static rather than just stationary, since the space-time must admit a foliation by a family of hyper-surfaces orthogonal to these frames, in order for a frame time to be defined for use in the standard formulation. The requirement that the relative spaces still be Euclidean further restricts the space-time metric to those whose orthogonal metrics are Euclidean.

It follows that there always exists a coordinate system $\{x, y, z, t\}$ adapted to the frame in which the space-time metric has the form $\text{diag}(-1, -1, -1, c^2)$ and the relative space inherits the coordinates $\{x, y, z\}$ as Cartesian coordinates. As a result, Einstein's 1912 theory is sometimes described as a theory of space-times with the line element

$$ds^2 = -dx^2 - dy^2 - dz^2 + c^2 dt^2,$$

where $c = c(x, y, z)$, although his theory actually deals with the relative spaces of such space-times.

It is interesting that the field equation chosen here for the relative space corresponds to the field equation for the space-time metric

$$R = k'T$$

where $R$ is the Riemann curvature scalar, $T$ is the trace of the stress-energy tensor of a dust cloud, and $k'$ is a constant, although when Einstein formulated his theory he could not have known this.

In the second of the 1912 papers cited, Einstein described the difficulties his bold new theory soon encountered. In order to retain the equality of action and reaction of forces, that is, to retain a law of momentum conservation, Einstein found himself forced to a modified field equation

$$\Delta\sqrt{c} = (k/2)\sqrt{c}\sigma.$$

This new field equation no longer admitted the homogeneous field associated with uniform acceleration in Minkowski space-time as a solution, unless one considered only infinitely small regions of the relative space. Einstein confessed that he had resisted this development, since it now meant that his principle of equivalence could only be formulated in infinitely small regions of the relative space, even though it still dealt only with the simplest case of uniform acceleration in Minkowski space-time.[18]

4.3. THE TEMPORARY LIMITATION TO INFINITESIMAL REGIONS

Because of the superficial similarity between this version of the principle and the infinitesimal principle of equivalence now common in the context of arbitrary gravitational fields in general relativity, some writers have regarded this development as, for example, "the dawn of the correct formulation of the principle of equivalence as a principle that holds only locally" (Pais 1982, p. 205). It certainly was not as far as Einstein was concerned. The limitation to infinitesimal regions of the relative space was not introduced to homogenize inhomogeneous fields, as it is in the modern infinitesimal principle. His principle still dealt only with homogeneous fields produced by uniform acceleration. (Note that the inhomogenous fields of his 1912 theory were not produced by acceleration but by generalizing the properties of homogeneous fields.) Therefore, the need for such a limitation, in the case of fields that were *already* homogeneous, was a source of some puzzlement to him and he dispensed with it as soon as he could. But before he could, there were yet more problematic developments concerning the principle of equivalence. I relate them here in the hope of nipping in the bud the myth of Einstein's 1912 introduction of the modern infinitesimal principle of equivalence.

In late 1912 and early 1913, in this climate of uncertainty about the principle, Einstein made his major breakthrough to the *Entwurf* theory with the mathematical assistance of his friend Marcel Grossmann (Einstein and Grossmann 1913). The new theory contained virtually all the essential features of the final general theory of relativity. However, they were unable to incorporate generally covariant gravitational field equations in it. Einstein was able to remove this defect only after nearly three years of intense work and thereby arrived at his final general theory of relativity (see Norton 1984).

During this period, Einstein omitted to mention the catastrophe that had befallen the principle of equivalence. Because of their restricted covariance, it can be shown that the field equations of the *Entwurf* theory do not hold in coordinate systems adapted to uniformly accelerating frames of reference in Minkowski space-time, even allowing restrictions to infinitely small regions of space-time. In the language of Einstein's 1916 formulation of the principle in his reply to Kottler, this meant that he could not regard such coordinate systems as "at rest." That is, according to his new theory, the principle of equivalence was false if formulated for this standard and simple case.

Therefore, in the introduction to the *Entwurf* paper, Einstein had to present the principle of equivalence as a result drawn from his earlier theory of static fields; for he still based the principle on the assumption that a uniform acceleration of the reference system in Minkowski space-time produced a homogeneous gravitational field even if only in an infinitely small region of the relative space. Presumably because of this problem, Einstein avoided the detailed discussion of the equivalence of the inertial field of uniform acceleration and homogeneous gravitational fields in the three years in which he held to the *Entwurf* theory, for this theory entailed no such equivalence. But he retained the principle of equivalence, for it was essential to the conceptual development of his theory. In addition, the notion of the equivalence of inertial and gravitational fields was central to the theory. However, the extent to which his *Entwurf* theory admitted this equivalence was not entirely clear.

This difficulty was resolved dramatically and completely with Einstein's November 1915 adoption of the generally covariant field equations of his completed general theory of relativity. The restriction of the principle of equivalence to infinitely small regions of space disappeared from his writings.

## 5. Extending the Principle of Relativity

Einstein's early success in constructing a new gravitation theory from his principle of equivalence is partly responsible for the still prevalent misconception that this was its essential purpose. To combat this, he frequently stressed that the principle did not provide a recipe for producing arbitrary gravitational fields by acceleration. The real point of the principle, as he had made clear in 1907, was that it enabled an extension of the principle of relativity to accelerated motion. Thus in the 1916 formulation of the principle quoted in Section 2, the principle itself is "the assumption that one may treat [the uniformly accelerated reference system] $K'$ as at rest in all strictness without any laws of nature not being fulfilled with respect to $K'$."

Prior to 1913 and the development of the basic formal structure of the general theory of relativity, Einstein gave no sustained discussion of precisely what he required in an extension of the principle of relativity and how the principle of equivalence was to help bring it about. However, we can reconstruct Einstein's position on these questions in this early period by considering the discussion he gave in an introductory section of his 1916 review of the general theory of relativity, called "On the grounds which suggest an extension of the postulate of relativity" (Einstein 1916a, pp. 771–773). This section concluded with a formulation of the principle of equivalence. Further, it dealt only with concepts that would have arisen in the pre-1913 period, suggesting that he was rehearsing arguments essentially from this period of his work. In particular, the discussion focused exclusively on the relative spaces of frames of reference.

Einstein began by pointing out an "epistemological defect" of classical mechanics and special relativity, enabling us to locate his arguments in Newtonian and Minkowski space-times. In a celebrated thought experiment, he considered two fluid spheres in relative rotation and noted that only one of them can be free of centrifugal distortion. But there is no observable difference between the relative spaces of the rest frames of each sphere, other than the state of motion of the distant masses of the universe, in which, he concluded, the cause of the centrifugal distortion is to be sought. This led to the following requirement for relative spaces

> Of all imaginable spaces $R_1$, $R_2$, etc., in any kind of motion relatively to one another, there is none which we may look upon as priviledged *a priori* without reviving the above-mentioned epistemological objection. *The laws of physics must be of such a nature that they apply to systems of reference in any kind of motion.* (Einstein 1916a, p. 772)

Einstein then proceeded to formulate the principle of equivalence that enables a *uniformly* accelerated observer to avoid inferring that he is "really" accelerated and enables us to regard the uniformly accelerated reference system $K'$ as just as "privileged" or "stationary" as the unaccelerated system $K$.

Since Einstein's discussion was in terms of relative spaces, it is clear that the "laws of physics" were being considered in their "standard formulations," described in Section 3. The standard formulations of classical mechanics and special relativity in question would be those then generally available, that is, those defined in the relative spaces of inertial frames (henceforth "inertial spaces"). These standard formulations would hold *only* in inertial spaces and therefore fail to satisfy Einstein's requirement that they "apply to [the relative spaces of] systems of reference in any kind of motion." Thus they would single out inertial spaces and their associated inertial frames as privileged.

In response, Einstein used the principle of equivalence to propose a more general theory, a theory of homogeneous gravitational fields, whose standard formulation will hold not only in inertial spaces but in uniformly accelerated spaces as well. The relativistic version of this theory is quite familiar to us now from Section 4 and presumably also to Einstein's readers of 1916. It is just his 1907–1912 gravitation theory, restricted to the case of a homogeneous gravitational field. In this way, Einstein broadened the set of privileged frames and relative spaces to include those in uniform acceleration.

Precisely what Einstein achieved with this result has not always been properly understood. His point can be made more clearly by avoiding reference to the standard formulation of theories, which has proven to be confusing to modern readers steeped in the four-dimensional formulation of these theories.

The focus of Einstein's concern is the necessity in special relativity and classical mechanics of presuming an immutable division of relative spaces and frames of reference into the privileged inertial and the noninertial. The principle of equivalence enabled him to eliminate the immutability of this division, by reinterpreting the nature of the inertial effects which distinguish the privileged inertial spaces and frames from all others. He explained this to a correspondent in a letter of July 12, 1953, reminding him that the principle could not be used to generate arbitrary gravitational fields by acceleration:

> The equivalence principle does not assert that every gravitational field (e.g., the one associated with the Earth) can be produced by acceleration of the coordinate system. It only asserts that the qualities of physical space, as they present themselves from an accelerated coordinate system, represent a special case of the gravitational field. It is the same in the case of the rotation of the coordinate system: there is *de facto* no reason to trace centrifugal effects back to a 'real' rotation.[19]

Through the principle of equivalence, Einstein proposed that we do not regard these distinguishing inertial effects as depending on an immutable property of the accelerating relative space, but as arising from the presence of a field in the relative space, which was to be seen as a special case of the gravitational field. This view could be extended beyond the case of uniform acceleration of the principle. Within this view, relative spaces would have no intrinsic states of motion—none would be "really" rotating for example—and in this sense they would all be indistinguishable. However, any relative space could become inertial according to the particular instances of the gravitational field defined on the relative spaces. Simiarly, all frames of reference would be indistinguishable, until the introduction of any particular instance of the gravitational field made some inertial and others not.

This crucial aspect of Einstein's account has been commonly misunderstood. The fact that an accelerated frame remains distinguishable from an unaccelerated frame in both special and general relativity is irrelevant to the extension of the principle of relativity. Einstein's account *requires* that each instance of the gravitational field distinguish certain frames as inertial and others as accelerating. The decision as to which frames will be inertial and which accelerated, however, must depend only on the particular instance of the gravitational field at hand and not on any intrinsic property of the frames.[20]

At this stage of his development of general relativity, Einstein's important innovation did not yet lie in the introduction of an empirically new theory. According to the principle of equivalence, his theory of static gravitational fields was predictively identical to special relativity in the case of homogeneous gravitational fields. Rather, it lay in a new way of looking at the division of structures between space and the fields it contains in the context of special relativity. Specifically, he no longer regarded the structures accounting for inertial effects as a part of space. Rather he now looked upon them as associated with the fields defined in space and, in particular, intimately related to gravitation. This move stripped space of the privileged frames to which he objected.

Einstein's "*Gestalt* switch" can be described more precisely if we present it more explicitly in four-dimensional terms. Of course, Einstein himself did not begin to work explicitly in such terms until five years after his original 1907 formulation of the principle of equivalence.

In the old view of special relativity, the background arena of space and time, against which physical processes unfold, is a Minkowski space-time, that is, a pair: $\langle M, g \rangle$, where $M$ is a four-dimensional manifold and $g$ a Minkowski metric. This background arena admits certain privileged structures: inertial frames of reference and their associated inertial spaces.

In the new view of special relativity, we are informed by the principle of equivalence that the structure responsible for inertial effects, the Minkowski metric $g$, is not an intrinsic part of the background arena of space and time. Rather, it is a field defined against that background and actually a special case of the field structure that also accounts for gravitational effects. The background arena of space and time is now just the bare space-time manifold $M$. In $M$ in the absence of a metric, we can still introduce frames of reference as congruences of curves, although we cannot require them to be time-like, and we can still define their relative space, although they will have no induced metric. Clearly in terms of $M$ alone, all such frames and correspondingly all relative spaces will be indistinguishable and therefore none will be privileged.

Following the model of classical gravitation theory, special relativity in this new view circumscribes the metric fields allowed on the manifold by a differential field equation. It requires a metric of Lorentz signature and with a vanishing Riemann curvature tensor

$$R_{iklm} = 0.$$

This requirement does not specify a unique Minkowski metric, but a large set of Minkowski metrics. Because of this, the *theory* does not single out any frame of reference as privileged in a particular "background space" (*i.e.*, manifold), even though each metric allowed by the theory will single out certain frames as inertial and others as noninertial. For, speaking informally, it can be shown that there is always a Minkowski metric allowed by the theory in which any well-behaved noninertial frame would become inertial. This result, given more precisely later, rests entirely on an active interpretation of the general covariance of the preceding field equation.

In a space-time manifold $M$, let $g$ be a Minkowski metric and $F$ an inertial frame of reference, that is, one whose time-like curves are geodesics in rigid-body motion. Let $F'$ be *any* frame of reference in the neighborhood $U'$ of $M$ (or even any congruence of curves which need not be all time-like), for which there exists a coordinate sytem $\{x'^i\}$ with domain $U'$ adapted to $F'$. (Such a frame is "well behaved".) Now in some neighborhood $U$ of $M$ there exists a coordinate system $\{x^i\}$ adapted to $F$ whose range coincides with that of $\{x'^i\}$. $h$ is a diffeomorphism that maps $p$ to $hp$ such that $x^i(p) = x'^i(hp)$. Then it follows that $F'$ is an inertial frame of reference, with respect to the Minkowski metric $g'$, which is the image of $g$ under $h$.[21]

The essential features of the old and new way of viewing special relativity are summarized in Table 1.

TABLE 1. Comparison of old view of special relativity with new view informed by principle of equivalence.

|  | Old view | New view |
|---|---|---|
| Background arena of space and time | Minkowski space-time $= \langle M, g \rangle$ where $M$ = four dimensional manifold $g$ = Minkowski metric | Four-dimensional manifold $M$ only |
| Examples of contents/ processes in space and time | Electromagnetic fields, matter in dust clouds, etc. | Electromagnetic fields, matter in dust clouds, etc. Any Minkowski metric = special case of structure inducing gravitational fields |
| Privileged frames of reference in background of space and time? | Yes, each $\langle M, g \rangle$ has a unique set of inertial frames. | No, bare manifold $M$ has no privileged frames of reference. Any well-behaved frame can be made inertial by defining an appropriate Minkowski metric on $M$. |

The equivalence of all frames embodied in this new view goes well beyond the result that Einstein himself claimed in 1916 from the principle of equivalence. He claimed only an equivalence of inertial and uniformly accelerated relative spaces, that is, of inertial and uniformly accelerated frames. The establishment of a wider equivalence would have been straightforward, even if inessential in view of the fact that he had the general theory of relativity in hand by then. But he most likely chose to avoid this extension because it would have required him to find standard formulations of a gravitation theory, similar to his 1907–1912 theory, which would hold in relative spaces of frames in rotation or nonuniform acceleration. I listed some of the difficulties Einstein would face in this task in the last section.

In any case, Einstein could not simply take special relativity, viewed in the new way, as a theory extending the principle of relativity in the way required for two reasons. First, the principle of equivalence clearly indicated that the theory was not complete. The structure accounting for inertia must also account for *all* gravitational effects. The Minkowski metric of special relativity, however, could only account for effects due to gravitational fields which could be transformed away over some neighborhood of a relative space by transforming to a new relative space. So Einstein immediately continued from his statement of the principle of equivalence, quoted earlier from his 1916 review article, by observing that "in pursuing the general theory of relativity, we shall be led to a theory of gravitation...." We shall see that it was the completion of this task that yielded the general theory of relativity.

The second reason was more subtle but far more important and can only be touched on informally here. The theory was also causally incomplete. As we have seen, Einstein required a complete theory of inertia to account for the disposition of inertial frames in space-time in terms of the only available *observable* cause, the distribution and motion of the masses of the universe. Special relativity in any of the forms described cannot be that theory. The disposition of inertial frames and the Minkowski metric which determines them is completely unaffected by any change in these masses. In some large neighborhood of space-time, such changes might include the setting of all masses into rotation about a central axis or even the conversion of all their energy into radiation and its resulting dissipation.

However it was natural for Einstein to expect that the extended theory, which dealt with general gravitational effects, would explain the observed disposition of inertial frames of reference in terms of the matter distribution of the universe. For the structure that determined this disposition would behave in many aspects like a traditional gravitational field and therefore be strongly influenced by any motion of its sources, the masses of the universe.

Although Einstein's hopes were not borne out by later developments, he made clear in his earliest relevant publications that he expected his new general theory of relativity to implement a "hypothesis of the relativity of inertia," which required inertia to be nothing other than the resistance of a body to acceleration with respect to other bodies (Einstein 1913b, pp. 1260–

1262). This, of course, would forbid universes, all of whose masses were rotating about a local inertial compass. He had already sought and found small effects he felt were consistent with this hypothesis. They included the dragging of the inertial frames of reference inside a rotating shell of matter and were similar to those discussed in his *Meaning of Relativity* (Einstein 1922, pp. 100–103). Clearly he also related this hypothesis to his 1907–1912 theory of static gravitational fields, for in 1912 he had published a paper which demonstrated the existence of similar such effects in that theory too (Einstein 1912d).

## 6. The Breakdown of Relative Spaces

It was inevitable that Einstein would give up the use of standard formulations of theories in his search for a general theory of relativity. For the relative spaces used by these formulations would only have well-defined geometries if the associated frame is in rigid motion, which is by no means generally the case. Even in Minkowski space-time, no *non*uniformly rotating frame can move rigidly. Worse, the relative space will only have the frame time required by standard formulations if the space-time admits a foliation by hypersurfaces orthogonal to the frame. Even uniformly rotating frames in Minkowski space-time do not admit such a foliation.

In his general theory of relativity, Einstein turned to the four-dimensional space-time formulation of theories. As indicated in the last section, he now also came to regard the four-dimensional space-time manifold without further structure as the background of space and time against which physical processes unfold.

One can define very few reference structures in such a manifold. Frames of reference as congruences of world lines can be defined. But without further structure, such as a metric, they cannot be described as time-like or have an overall state of motion assigned to them. The richest reference structure available is the arbitrary space-time coordinate system, whose coordinate values can have no metrical significance, such as Einstein had required in his Galilean reference systems.

So in the general theory of relativity, Einstein proceeded to use arbitrary space-time coordinate systems as the reference structures from which to view physical processes and formulate physical principles. In his expositions of general relativity, Einstein typically made this transition from frame of reference and relative space to arbitrary space-time coordinate system by considering the relative space of a frame of reference in uniform rigid rotation in Minkowski space-time (for example, Einstein 1916a, pp. 773–776; 1922, pp. 59–62). He would show that the spatial geometry is non-Euclidean and conclude that the coordinate system used there could not have the same direct metrical significance of spatial coordinates in his Galilean reference systems. Similar results followed from attempts to retain a time coordinate, presumably for space-time, whose value would coincide with the readings of clocks at rest in the frame. Einstein then introduced the use of arbitrary space-time coordinate systems as a natural extension of the methods developed in the nineteenth century for dealing with non-Euclidean spatial geometries.

This argument gave psychologically natural grounds for introducing the methods of differential geometry into relativity theory. However, it failed to demonstrate the completeness of the demise of relative spaces in general relativity. The relative space of the argument's uniformly rotating frame of reference still has a well-defined geometry, unlike the relative spaces of other frames of reference in space-times with more general semi-Riemannian metrics. Einstein turned to this problem in his popularization *Relativity* (1954a), most of whose discussion is set in terms of the relative spaces of "reference bodies" (=frames of reference). In chapter 28 he points out that rigid reference bodies will in general no longer be available in general relativity and that "the Gauss coordinate system has to take the place of the body of reference." He then proceeds to describe the difficulties and artificiality of retaining the use of nonrigid reference bodies (and by implication their associated relative spaces with ill-defined geometries) through the discussion of what he calls "reference molluscs."

In the same chapter, Einstein gave his well-known reformulation of the extended principle of relativity—"All Gaussian co-ordinate systems are essentially equivalent for the formulation of the general laws of nature"— and proceeded to explain that this requirement was satisfied by a theory if its laws were written in a generally covariant form. Naturally, this meant that his generally covariant general theory of relativity realized the extended principle of relativity.

Einstein has taken the principle of equivalence to assert the equivalence of inertial and uniformly accelerated relative spaces, an assertion that is subsumed by the extended principle of relativity. So it was easy for Einstein to conclude, in continuing his reply to Kottler, that the principle of equivalence was automatically satisfied by his general theory of relativity:

A gravitation theory violates the principle of equivalence, in the sense which I understand it, only then, if the equations of gravitation are satisfied in *no* reference system $K'$, which is moving non-uniformly relative to a Galilean reference system. That this reproach cannot be raised against my theory with *generally* covariant equations is evident; for here the equations are satisfied with respect to each reference system. *The requirement of general covariance of equations embraces the principle of equivalence as a quite special case.* (Einstein 1916b, p. 641)[22]

Einstein's reformulation of the extended principle of relativity as the requirement of general covariance is unproblematic in so far as it is based on the fact that the space-time manifold without any additional structure has no privileged coordinate systems. This fact immediately entails that there are no privileged frames of reference and, therefore, no privileged relative spaces. For were any frames privileged, the coordinate systems adapted to them would also be privileged.

However, as has been frequently objected, it is hard to see how this requirement could capture all that Einstein required in an extension of the principle of relativity, when there are simple generally covariant formulations of many other theories apart from general relativity. These include special relativity, Nordström's theory of gravitation, and Newtonian gravitation theory. Of course Einstein was aware of this at least in the case of the first two theories.

A thorough analysis of Einstein's intentions here and their refinement in his later work is a complex task that goes well beyond this paper. Nevertheless, I will make a few tentative comments concerning Einstein's early view of the question to make his remarks more plausible.

For Einstein, violations of the extended principle of relativity need not be limited to the laws of a theory. They could also arise in its solutions, that is, in models or classes of models of the theory. For example he pointed out in a 1917 paper on the cosmological problem that it was "contrary to the spirit of the relativity principle" to introduce solutions of the field equations of general relativity by imposing a boundary condition of a Minkowski metric at matter-free spatial infinity (Einstein 1917, p. 147). This introduces privileged coordinate systems in which the metric approaches the form diag($-1$, $-1$, $-1$, $1$) as the limit to spatial infinity is taken. In addition, these privileged coordinate systems were objectionable since there was no observable cause for their special status, contradicting the hypothesis of the relativity of inertia.

Clearly, solutions of generally covariant formulations of special relativity and Newtonian theory would necessarily involve the introduction of similarly objectionable privileged coordinate systems in one form or other. Minkowski space-time, even regarded as a model of general relativity, would be objectionable for the same reason. However, Einstein believed that the introduction of these boundary conditions would not always be needed in the case of his general theory of relativity. In his 1917 paper, he continued to demonstrate how the field equations of general relativity, augmented with the cosmological term, admitted solutions without the use of boundary conditions at spatial infinity. To arrive at these solutions, one needed only to specify the mass and world lines of the universe's smoothed-out dust cloud of matter on the manifold and invoke other natural requirements, such as the symmetry of the metric with respect to these world lines, and its isotropy about them.

In 1918, Einstein described a solution generated in this way as satisfying "Mach's Principle" (Einstein 1918a, p. 241). This principle required that the metric tensor be determined completely by the matter of the universe and was taken to be the natural generalization of the hypothesis of the relativity of inertia. In a footnote, he pointed out that he had not previously distinguished this principle from the (extended) principle of relativity and that this had caused confusion. So, at least at this time, the general theory of relativity seemed to be the only viable theory satisfying all his requirements concerning the relativity of motion. It was clearly impossible for special relativity or Nordström's theory to exhibit such Machian behavior, irrespective of the covariance of their formulations.

## 7. Generating General Relativity

Einstein had come to recognize that a general theory of relativity was to be found as a four-dimensional theory of gravitation. The principle of equivalence provided the crucial starting point: the identification of the Minkowski metric as an *instance* of the four-dimensional space-time structure representing gravitational fields. For Einstein had found that the Minkowski metric can induce gravitational fields on the relative spaces of a Minkowski space-time.

Einstein's discovery of the gravitational properties of the Minkowski metric was a remarkable feat. Unlike so many other discoveries in physics, it seems to have been almost totally unanticipated by his contemporaries.

The role of the principle of equivalence in Einstein's development of his new gravitation theory remained essentially the same as in his earlier 1912 theory of gravitation. The principle yields a special case of the gravitational field, whose properties are then generalized in a natural way to arrive at a general theory of gravitation.

However, from the perspective of the general theory of relativity, Einstein had no prospect of arriving at the correct laws of a general theory of the gravitational fields of relative spaces, as long as he worked within the framework of his 1912 theory. This follows immediately if we recall that Einstein sought to characterize arbitrary static gravitational fields as structures induced onto relative spaces by the special type of static space-times I described in Section 4.2.

In these space-times, in the source-free case, one can readily demonstrate that the field equations of general relativity, that is, the requirement of the vanishing of the Ricci tensor

$$R_{im} = 0,$$

entails the vanishing of the Riemann–Christoffel curvature tensor

$$R_{iklm} = 0.$$

This in turn entails that the only source-free gravitational fields in relative spaces which the theory can deal with correctly, from the perspective of the general theory of relativity, are those induced by acceleration in Minkowski space-time. In addition, it follows from an evaluation of the components of the curvature tensor in a coordinate system adapted to the accelerating frame that this acceleration must be a uniform rectilinear acceleration.[23]

Unfortunately, in the period 1912 to 1915, Einstein believed that the arbitrary static space-times associated with his 1912 theory ought also to be solutions of the field equations of his new general theory of relativity. I have argued elsewhere in detail that this played a major role in his failure to adopt the generally covariant field equations of his final theory in this period. (see Norton 1984).

Nevertheless, Einstein commonly used the principle of equivalence to recover and motivate the basic formal structure of his general theory of relativity in an argument whose strategy was essentially the same as that used in 1912. Einstein presents the argument in a compact and well-developed form in a 1951 letter to Becquerel, in which the role of the principle of equivalence is made especially clear.[24] He begins by using the equality of inertial and gravitational mass to justify introduction of the principle, which is formulated in terms of relative spaces: "An inertial space without gravitational field is physically equivalent to a uniformly accelerated space, in which there is a (homogeneous) gravitational field. (Equivalence hypothesis.)" Then after introducing the requirement of general covariance, he proceeds with the steps he numbers as the third and fourth of his argument:

(3) *One* kind of space is completely known to us, that is empty Minkowski-space, in which the interval $ds$, as given by

$$ds^2 = -dx_1^2 - dx_2^2 - dx_3^2 + dx_4^2$$

can be measured immediately by resting clocks and measuring rods. Through a nonlinear transformation, this becomes

$$ds^2 = g_{ik} dx_i dx_k,$$

where $ds$ has the same *value* as a Minkowski system. The $g_{ik}$ depend on the coordinates and, according to the equivalence hypothesis, describe a gravitational field (of a more special kind).

(4) In general coordinates, a gravitational field of the more special kind satisfies the differential equations

$$R^i{}_{klm} = 0$$

from the loosening of which the field law of an *arbitrary* pure gravitational field must follow. For this, only

$$R_{kl} = R^s{}_{kls}$$

comes into consideration. It is natural to assume that $ds$ expresses the naturally measured interval also in the case of a *general* pure gravitational field.

Because of its extreme brevity, Einstein's argument requires some explication. In his step 3, he appears to identify a coordinate effect, the nonconstancy of the components $g_{ik}$, with the presence of a gravitational field. His real intention emerges, however, if we recall his practice of tacitly associating changes of frame of reference with coordinate transformations. In particular, a nonlinear coordinate transformation can represent the change from an inertial frame of reference to a rigidly and uniformly accelerated frame of reference, which is precisely the case considered in the statement of the principle of equivalence just given. In this case, the nonconstancy of the $g_{ik}$ is now associated with the presence of a homogeneous gravitational field in the relative space of the accelerated frame, for as we have seen in Section 4, the potential of such a field is given by $g_{44}$ in a coordinate system adapted to the frame.

Thus Einstein's step 3 is multifaceted. The introduction of an arbitrary coordinate system makes the presence of a metric tensor in Minkowski space-time formally explicit as a matrix of components $g_{ik}$. At the same time Einstein uses the principle of equivalence to point out that this metric induces a gravitational field of a special type in the relative space of an accelerated frame of reference. This justifies interpreting the Minkowski metric as a particular instance of the four-dimensional generalization of such gravitational fields.

Interpreting the Minkowski metric in this way indicates that Einstein can arrive at a four-dimensional theory of arbitrary gravitational fields, which will also be his general theory of relativity, by generalizing the properties of the Minkowski metric in a manner analogous to the way that uniform gravitational fields can be generalized to nonuniform fields in Newtonian theory. He finds that the way to proceed is straightforward. The general theory will deal not only with Minkowski metrics, but also others of Lorentz signature.

This argument appears throughout Einstein's earlier work, but in a slightly less-developed form.[25] For it was only in his later years that he explicitly renounced the use of a separate stress-energy tensor as the source term in the field equations and used these equations only in their source-free form. This source-free form of the field equations can be arrived at readily in the argument, as Einstein shows earlier, by merely contracting the flat space-time condition of special relativity. The argument appears commonly in this more complete form in his later writings.[26]

The earlier examples of the argument also contained an important addition to the example quoted earlier. Einstein would note that in the Galilean reference system of special relativity, a free point mass moves uniformly in a straight line. Such motion is represented in Minkowski space-time by a time-like geodesic, which satisfies the condition that the interval be extremal along the curve:

$$\delta \int ds = 0.$$

It was natural to *assume*, the argument continued, that this requirement would also be satisfied by the world line of a free point-mass in the more general case of the general theory of relativity. I will return to the importance of this point in Section 9.

In short, we have seen in this section that the principle of equivalence enabled Einstein to see that one structure was responsible for inducing both inertial and gravitational fields and that the Minkowski metric was a special case of it. Einstein summarized this insight in a compact 1918 statement of the principle:

> *Principle of Equivalence*: inertia and gravity are *wesensgleich* [identical in essence]. From this and from the results of the special theory of relativity it necessarily follows that the symmetrical "fundamental tensor" ($g_{\mu\nu}$) determines the metrical properties of space, the inertial behavior of bodies in it, as well as gravitational action. (Einstein 1918a, p. 241)[27]

## 8. A Manner of Speaking

It was not uncommon for Einstein to associate the nonconstancy of the components of the metric tensor, or, equivalently, the nonvanishing of the Christoffel symbols in a given coordinate system with the presence of a gravitational field. In particular, he would describe the Christoffel symbols as the "gravitational field strengths" or "components of the gravitational field," for in a coordinate system in which these symbols vanished, free point-masses move "uniformly in a straight line." Therefore, these components "condition the deviation of the motion from uniformity" (Einstein 1916a, p. 802).

As in the last section, this association of the Christoffel symbols with gravitational field strengths can be explicated by recalling that Einstein often tacitly referred to frames of reference and their relative spaces when he talked explicitly only of a coordinate system adapted to them. If a coordinate system adapted to a uniformly accelerating frame of reference in Minkowski space-time is chosen so that its spatial coordinates are Cartesian, then the Christoffel symbols will contain only the spatial derivatives of the $g_{44}$. However, these derivatives together form a field strength, the three-vector gradient of the potential of the homogeneous gravitational field in the associated relative space.

The connection made here between the Christoffel symbols and the field strengths of the gravitational fields in relative spaces depends on a careful choice of space-time and coordinate system. Einstein, however, did not make this clear in his work and rarely qualified the identification of nonvanishing Christoffel symbol and gravitational field strength.

This practice has undoubtedly caused confusion. In a letter of January 1951, Laue challenged Einstein on this point.[28] He gave the example in Minkowski space-time of the transformation to curvilinear *spatial* coordinates from a Galilean coordinate system with no alteration in the time coordinate. Since this transformation is not associated with a change of state of motion, the resulting nonvanishing of "field strengths" is physically counterintuitive.

Einstein began his response by stressing that the Newtonian concept of gravitational field ("all the expressions obtained from the potential") is different from the concept of the relativistic gravitational field ("everything formed out of the symmetrical $g_{ik}$").[29] This corresponds to the distinction made here between the gravitational fields of relative spaces, which are usually represented by scalar fields, and their four-dimensional generalization, the metric field. Nevertheless, as he continued to explain, it was possible to forge a heuristic link between these two concepts and this link was the principle of equivalence:

> Heuristically, the interpretation of the field existing relative to a system, parallelly accelerated [*parallel beschleunigten*] against an inertial system (equivalence principle) was naturally of decisive importance, since this field is equivalent to a Newtonian gravitational field with parallel lines of force. In this case, the Newtonian field strengths are equal to the spatial derivatives of the $g_{44}$. Correspondingly, if one wants to, one can designate the first derivatives of the $g_{ik}$ or the displacement quantities Γ [affine connection] as gravitational field strengths, which certainly have no tensor character. In this manner of speaking, the introduction of cylindrical coordinates leads to the appearance of field strengths in a Galilean space. With this it is only a question of a manner of speaking.

Here Einstein uses the special case described earlier to justify speaking of the first derivatives of the $g_{ik}$ (which determine the Christoffel symbols and the affine connection in these space-times) as gravitational field strengths. One can continue to use this manner of speaking in other cases, but, as Einstein's response indicates, it should be used with some caution.

This attitude to the description of the Christoffel symbols as gravitational field strengths was not a later development in Einstein's thought. It is also clearly evident in his 1916 reply to Kottler. There he says of this nomenclature, referring also to the nongenerally covariant stress-energy pseudo-tensor of the gravitational field, that "it is meaningless in principle and only intended to make concessions to our physical thought habits," but that it "appears to me, at least provisionally, not without value to maintain the continuity of thought" (Einstein 1916b, p. 641).

Today, some fifty years later, we insist that coordinate effects be carefully distinguished from physical effects. Examples such as Laue's show the confusion that would otherwise arise. Therefore, the provisional value of Einstein's manner of speaking is no longer evident. Einstein continued his response to Laue by stressing the important point beneath his manner of speaking, which involved no equivocation about coordinate effects:

> It is essential however, that a gravitational field exists in the sense of general relativity also in the case of a Galilei or a Minkowski space, even if the field strengths in the sense defined above vanish. In the theory of relativity, just the dimensionality of the field is the only thing that remains of the earlier physically independent (absolute) space.

In a given space-time, the nature, and even existence, of a gravitational field in a relative space will depend on the choice of frame of reference defining the relative space. But this relative-space dependence of these gravitational fields does not extend to their four-dimensional generalization, the space-time metric. All space-times of general relativity contain such a metric field—a gravitational field "in the sense of general relativity"—regardless of the frame of reference or relative space under consideration. This holds equally for Minkowski space-times, even though we can always find relative spaces in them that are gravitation-free in the older sense. In short, in general relativity a Minkowski space-time is *not* the gravitation-free special case.

## 9. The Infinitesimal Principle of Equivalence

Einstein's contemporaries of the early 1920s regarded the relative-space dependence of the gravitational field as the basic assertion of the principle of equivalence, rather than the occasion for inference to a more fundamental structure. Naturally, they were dissatisfied that Einstein dealt only with this relative-space dependence in the very simple case of the homogeneous gravitational fields of uniformly accelerated reference systems in Minkowski space-time. They sought an extended statement of this dependence that would apply directly to arbitrary gravitational fields (Pauli 1921, pp. 145–147; Silberstein 1922, pp. 10–13). They believed that this could be achieved in general relativity on the basis of the notion that special relativity holds in infinitesimally small regions of the space-time manifold, tacitly assuming that special relativity is a gravitation-free special case. As a result, their construal of the principle was very different from Einstein's and lays stress on the notion that a gravitational field can always be transformed away.[30] Pauli's classic formulation of the resulting principle reads:

> For every infinitely small world region (i.e., a world region which is so small that the space- and time-variation of gravity can be neglected in it) there always exists a coordinate system $K_0(X_1, X_2, X_3, X_4)$ in which gravitation has no influence either on the motion of particles or any other physical processes (Pauli 1921, p. 145).[31]

Pauli continued to explain a little later that

> The special theory of relativity should be valid in $K_0$. All its theorems have thus to be retained, except that we have put the system $K_0$, defined for an infinitely small region, in place of the Galilean coordinate system.

In particular, this meant that the metric adopted the form diag$(1, 1, 1, -1)$ in $K_0$.

This "infinitesimal principle of equivalence" can be connected to Einstein's version at least superficially by noting that classical gravitational fields become homogeneous in infinitesimal regions of the relative space. Inverting Einstein's usual argument, they can then be transformed away at least infinitesimally by an appropriate acceleration of the reference system. One then regards the Pauli version of the principle as a four-dimensional restatement of these two results.

Of course this infinitesimal principle and the discussion of its connection to Einstein's version is beset with a number of serious technical difficulties.

The notion of both three- and four-dimensional "infinitesimal regions" and the sense in which special relativity holds in such regions are unclear. Further, the actual statement of the principle makes it look as though it deals solely with a coordinate effect. These problems will be addressed shortly.

The popularity of the infinitesimal principle derives at least in part from its leading to a particularly attractive result: that it is possible to reconstruct much of the space-time manifold of general relativity as a patchwork of infinitesimal pieces in which special relativity holds.

Moritz Schlick, in his influential two-part article on space and time in the March 1917 issues of *Die Naturwissenschaften*, attempted just such a reconstruction (Schlick 1917). "We stipulate," he wrote, "that in an infinitely small region and in a reference system in which the bodies considered have no acceleration the special theory of relativity holds." It followed that in a "local" coordinate system, such as Pauli's $K_0$, the interval between two infinitesimally separated events is given by

$$ds^2 = (dX_1)^2 + (dX_2)^2 + (dX_3)^2 - (dX_4)^2.$$

Transforming to an arbitrary space-time coordinate system $\{x_i\}$ ($i = 1, 2, 3, 4$), the expression for the interval became

$$ds^2 = g_{11}(dx_1)^2 + 2g_{12}dx_1 dx_2 + \cdots + g_{44}(dx_4)^2,$$

where the symmetric coefficients $g_{ik}$ ($i, k = 1, 2, 3, 4$) represent the components of the metric tensor in the new coordinate system. Schlick was thus able to infer that the new theory would involve a metric tensor and to arrive at many of its properties by considering the properties of the interval as given in special relativity.

In addition, Schlick considered the motion of a free material point. By reviewing its motion in the relative spaces of both local and accelerated coordinate systems and invoking the principle of equivalence, he concluded that the components of the metric tensor in the new coordinate system determine the gravitational field in the latter space. It also followed from special relativity that the world line of such a particle in the local coordinate system ($X_i$) would be a geodesic. Since this was an invariant property, it would also be true of the world line in all coordinate systems, such as ($x_i$). He then invoked the "principle of continuity" to justify the important conclusion that the world line of a free material point would be a geodesic in finite regions of the manifold as well.

Einstein has used arguments very similar to those just described. In particular, he used the assumption that special relativity holds in infinitesimal regions of the space-time manifold of general relativity in a manner close to that of Schlick, to introduce the metric tensor and some of its properties, especially those relating to the behavior of infinitesimal rods and clocks (Einstein 1916a, pp. 777–778; 1922, pp. 62–64).[32] However, this assumption was never related to the principle of equivalence, which was always formulated in Minkowski space-times. In addition, he was cautious in his use of this assumption, since he held that it was only true to a limited extent. This emerged in the correspondence between Einstein and Schlick following Schlick's article.

We know from this correspondence that Einstein had seen Schlick's article prior to its publication and that he approved of it wholeheartedly.[33] Six weeks after their initial exchange, however, Einstein wrote to Schlick to point out an error in one of the arguments sketched out here:

> The derivation of the law of motion of a point mass given on page 184 proceeds from the motion of a point being a straight line, when considered in the local coordinate system. But from this nothing can be derived. In general, the local coordinate system has a meaning only in the infinitely small and in the infinitely small every continuous line is a straight line. The correct derivation runs as follows: in principle there can exist finite (matter-free) parts of the world for which
>
> $$ds^2 = dX_1^2 + \cdots - dX_4^2$$
>
> with an appropriate choice of the reference system. (If this were not the case, then the Galilean law of inertia and the special theory of rel. could not have held good.) In such a part of the world, the Galilean law of inertia holds with this choice of reference system; and the world line is a straight line, and therefore a geodesic, with an arbitrary choice of coordinates.
>
> That the world line of a point is a geodesic in other cases too (if none other than gravitational forces act) is an hypothesis, even if a very obvious one.[34]

Einstein's objection bears directly on the assumption that special relativity does hold in an infinitesimal region of the space-time manifold of general relativity. He claims that it can only hold in a limited sense, for in such regions we cannot formulate the requirement that the world line of a free point-mass be a geodesic. (Note that Einstein called such lines "straight" in a Galilean reference system, since their spatial coordinates are linear functions of the time coordinate.)

Rather, as Einstein indicates here and as was his own practice elsewhere, when one discusses the motion of free point-masses, one must consider finite regions of the manifold in both special and general relativity. From the assumption that special relativity holds infinitesimally in general relativity, it does not follow that the world line of a free point-mass will be a geodesic in general relativity. Einstein's approach here and throughout his early work was to take this result in general relativity as strongly suggested by the corresponding result in special relativity, but in the last analysis still an independent assumption. (Of course, later he sought to derive this result in general relativity from the gravitational field equations.)

Finally, Einstein's comments here provide one more reason for his failure to retain an infinitesimal principle of equivalence after he briefly entertained one in 1912. As he came to realize, such a principle could not deal with the motion of bodies, the consideration of which formed the core of his principle. In the next section, I turn to examine whether Einstein's objection to Schlick holds. If it does, then he has pointed out a rarely acknowledged, but nevertheless devastating, difficulty for the traditional infinitesimal principle of equivalence.[35] If he is correct, then the restriction to infinitesimal regions makes it impossible to distinguish the geodesic world lines of free point-masses from other world lines and thus it is impossible to judge whether—in the words of Pauli's formulation—"gravitation has no influence on ... the motion of particles."

## 10. The Problem of Infinitesimal Regions

When Pauli and Schlick wrote of special relativity holding in infinitely small regions of the space-time manifold of general relativity, they could not have meant that special relativity holds in its usual sense. For whatever an infinitesimal or infinitely small region is, it must contain at least one point. Special relativity requires the vanishing of the Riemann–Christoffel curvature tensor. This requirement is well defined at every point of the manifold and is typically not satisfied in general relativity.

Rather they referred to a coordinate-dependent result, as is suggested by their qualification that special relativity hold in the region of an appropriately defined coordinate system. In a neighborhood of any given point $p$ in the space-time manifold in general relativity, it is possible to introduce a "local" corrdinate system $K_0$ so that at $p$: the components of the metric $g_{ik}$ have the values $\text{diag}(1,1,1,-1)$; the first (coordinate) derivatives of the components of the metric tensor $g_{ik,m}$ and thus also the Christoffel symbols vanish; but, in general, the second derivatives $g_{ik,mn}$ will not vanish.

When special relativity is said to hold in $K_0$ in an infinitesimal region around $p$, what is meant is the following. In $K_0$ at $p$, structures defined on the manifold, which do not deal with second and higher (coordinate) derivatives of the metric tensor, behave identically to their special relativistic counterparts at any point of a Minkowski space-time in a Galilean coordinate system. The criterion of identical behavior is equality of components of the quantities concerned. For example, in both cases the metric has components $\text{diag}(1,1,1,-1)$, which means that the coordinate velocity of light will be unity. Both cases are commonly regarded as gravitation free insofar as the Christoffel symbols, the "gravitational field strengths," vanish. And the world line of a free point-mass is a "straight" line, in the sense that it satisfies the condition $d^2 X^i/ds^2 = 0$ at $p$, where $s$ is the interval. The two cases differ, however, when quantities containing $g_{ik,mn}$ are considered. Most notably the curvature tensor vanishes only in the case of Minkowski space-time.

The ignoring of second and higher derivatives of the metric tensor is usually justified by the introduction of a hierarchy of nested orders of quantities. Examples of first-order quantities contain the $g_{ik}$ alone; of second-order quantities, the $g_{ik}$ and $g_{ik,m}$; of third-order quantities, the $g_{ik}$, $g_{ik,m}$, and $g_{ik,mn}$; and so on. One must now imagine that the $g_{ik}$ are given at $p$ alone; the $g_{ik,m}$ are given by comparing the $g_{ik}$ at $p$ and at an infinitesimally close point; and the $g_{ik,mn}$ by comparing the $g_{ik}$ at two points infinitesimally close to $p$, the second more removed than the first. Then, finally, we imagine that access to quantities higher than any designated order can be denied by restricting consideration to sufficiently small infinitesimal regions around $p$.

It is now clear that the notion of these infinitesimal regions is problematic in differential geometry, since such regions cannot be equated with neighborhoods in their usual sense or any other structure commonly employed.

If we are to make a consistent evaluation of Einstein's objection to Schlick, the foregoing discussion must be made more precise. First, ambiguous restrictions concerning infinitesimal regions will be replaced by restrictions concerning orders of quantities. The assertion that special relativity holds infinitesimally in general relativity will be taken to mean only that special relativity holds at a point in the space-time manifold when quantities up to second order only are considered.

Second, we can eliminate the dependence on the coordinate system $K_0$ and on Galilean coordinate systems in Minkowski space-time by replacing the quantities $g_{ik}, g_{ik,m}$, and $g_{ik,mn}$ in the examples of first-, second-, and third-order quantites mentioned earlier, by the covariant quantities $g_{ik}$, $D_i$, and $D_iD_k$, respectively. $D_i$ is the unique covariant derivative operator compatible with the metric $g_{ik}$. The coordinate-dependent notion of identity of quantities in the space-time manifold of general relativity with corresponding quantities in a Minkowski space-time is also naturally replaced by a requirement of diffeomorphic equivalence at the two corresponding points of each manifold.

Finally, we can extend the hierarchical ordering of quantities to those not constructed solely out of the metric and its derivatives by a technique based on one outlined by Geroch.[36] We generate subsets of the set of all diffeomorphisms $\{h\}$ whose domain is some neighborhood of $p$ and which map $p$ back onto itself. Let $g'$ be the image of $g$ under such a diffeomorphism and $D'_i$ the derivative operator constructed from $g'$. $\{h_1\}$ are all those diffeomorphisms for which $g' = g$ at $p$. $\{h_2\}$ are all those diffeomorphisms for which $D'_i = D_i$ at $p$. $\{h_3\}$ are all those for which $D'_iD'_k = D_iD_k$ and so on. We find[37]

$$\{h_1\} \supset \{h_2\} \supset \{h_3\} \supset \cdots .$$

We can think of the members of $\{h_n\}$ as disturbing the manifold about $p$ in a way that will not affect the particular $n$th order quantity used at $p$ to define them. More figuratively, they leave undisturbed the infinitesimal region about $p$ needed to determine that quantity. Hence it is natural to use these sets of diffeomorphisms to define the hierarchy of orders of other quantities defined on the manifold. If $Q$ is a quantity defined at $p$, then the order of any quantity $F(Q)$ derived from it in the hierarchy of orders engendered by $Q$ is the smallest value of $n$ for which we always have $F(Q') = F(Q)$, where $Q'$ is the image of $Q$ under any member of $\{h_n\}$.

Let $c$ be a curve through $p$ differentiable to all orders with a tangent vector $X$. We can also classify the hierarchy of quantities generated by $c$ at $p$ by considering the images of $c$ under members of $\{h\}$. If an image curve $c'$ has the tangent vector $X'$, then we find that $X'$ is first order since $X' = X$ only under any member of $\{h_1\}$. Writing $D_X = X^i D_i$, we find $D_{X'} X' = D_X X$ only under the members of $\{h_2\}$. Hence $D_X X$ is a second-order quantity. Similarly $(D_X)^n X$ is of order $n+1$ for all positive integers $n$.[38]

Now let the curve $c$ passing through $p$ be a geodesic parameterized by the interval $s$ and have tangent vector $X = d/ds$. By definition, at every point of $c$ in some neighborhood of $p$, $X$ will satisfy the condition

$$D_X X = 0.$$

It necessarily follows that at $p$

$$D_X D_X X = 0 \quad D_X D_X D_X X = 0 \ldots (D_X)^n X = 0 \ldots$$

for all positive integers $n$.[39]

Einstein's objection that "in the infinitely small every continuous line is a straight line" can now be made more precise. If we restrict ourselves to quantities of first order, then at $p$ we can only characterize curves through $p$ by their tangent vectors, if defined. But if $c^*$ is *any* curve through $p$ with tangent vector $X^*$, then there will always be a geodesic $c$ through $p$ with tangent vector $X$ equal to $X^*$. That is, as far as first-order quantities are concerned one cannot distinguish smooth curves from geodesics. If we read Einstein's "continuous line" as "smooth curve," then this first-order indistinguishability seems to express his point more precisely.

In the context of the infinitesimal principle of equivalence however, access to first- and second-order quantities is allowed. It follows that a geodesic $c$ with tangent vector $X$ will be indistinguishable from any sufficiently smooth curve $c^*$ with tangent vector $X^*$, provided $X^* = X$ and $D_{X^*} X^* = D_X X = 0$. Of course, the higher derivatives of $X^*$ along $c^*$ will not vanish in general. So $c^*$ need not be a geodesic. Since Einstein's objection was concerned in effect with this second-order case, it would have been better stated as "the world lines of any particles *unaccelerated* at $p$ (i.e., $D_X X = 0$) are indistinguishable from geodesics."

It is now also clear that any restriction on the order of quantities accessible at $p$ will make it impossible to distinguish geodesics from other curves. If quantities to order $n$ are allowed, then we cannot distinguish a geodesic $c$ from any other sufficiently smooth curve $c^*$ if they agree on quantities up to order $n$. Nevertheless, $c^*$ need not be a geodesic since any of the $(D_{X^*})^m X^*$ may fail to vanish for $m > n - 1$.

Another way to arrive at similar results is to consider $c'$, the image of $c$ under any member of $\{h_n\}$. By definition, $c'$ will be indistinguishable from $c$ to order $n$ at $p$. That is, they will agree on *any* quantity up to order $n$ that characterizes them. For example, $X' = X$, $D_{X'} X' = D_X X = 0$, ... $(D_{X'})^{n-1} X' = (D_X)^{n-1} X = 0$. But as before, $c'$ will not be a geodesic in general since its derivatives of order greater than $n-1$ need not vanish.

The results of this section vindicate Einstein's objection to Schlick. If we understand the infinitesimal principle of equivalence to assert that special relativity holds at a point in the space-time manifold of general relativity up to second-order quantities only, then it follows that we cannot formulate special relativity's requirement that the world line of a free point-mass be a geodesic.

In the terminology used by Pauli, Schlick, and Einstein, we would say that in the infinitesimal region concerned in the "local" coordinate sytem $K_0$, the fact that a world line satisfies the condition $d^2 X^i/ds^2 = 0$ does not mean that it is a geodesic. This much is obvious once we realize that the restriction to infinitesimal regions effectively involves a restriction to the consideration of quantities at a single point in the manifold. However, we now also see that, under a consistent treatment of this restriction, the higher derivative terms, which might enable us to distinguish other curves satisfying this condition from geodesics, are not accessible from within these infinitesimal regions.

## 11. Real and Fictitious Gravitational Fields

The infinitesimal principle of equivalence tells us that the space-time manifolds of special and general relativity share the same first- and second-order structure at a point. For example, it tells us that metric $g$ and compatible derivative operator $D_i$ at a single point in each manifold are diffeomorphically equivalent. This result is not deep—it really only depends on the fact that both metrics have the same signature.

Presumably, this result is what Synge had in mind when he lamented in the introduction to his well-known text on general relativity that he never understood what I assume to be the infinitesimal principle of equivalence.

> Does it mean that the signature of the space-time metric is $+2$ (or $-2$ if you prefer the other convention)? If so, it is important, but hardly a Principle. Does it mean that the effects of a gravitational field are indistinguishable from the effects of an observer's acceleration? If so, it is false. In Einstein's theory, either there is a gravitational field or there is none according as the Riemann tensor does not or does vanish. (Synge 1960, p. ix)

Synge's response to this difficulty is to insist that the effects of a true gravitational field are distinguishable from those of a fictious field produced by the acceleration of the observer, through an invariant criterion based on the Riemann-Christoffel curvature tensor.

It should now be clear that Einstein would not endorse this response to the difficulties of the infinitesimal principle of equivalence. For here Synge is proposing to resurrect precisely the distinction whose breakdown was crucial to Einstein's discovery of the general theory of relativity. Einstein explained his attitude to this question in correspondence with Laue, after Laue had pointed out that the Riemann-Christoffel curvature tensor vanishes in the context of the rotating disk problem:

> It is true that in that case the $R_{iklm}$ vanish, so that one could say: "There is no gravitational field present." However, what characterizes the existence of a gravitational field from the empirical standpoint is the non-vanishing of the $\Gamma_{ik}^l$ [coefficients of

the affine connection], not the non-vanishing of the $R_{iklm}$. If one does not think intuitively in such a way, one cannot grasp why something like a curvature should have anything at all to do with gravitation. In any case, no reasonable person would have hit upon such a thing. The key for the understanding of the equality of inertial and gravitational mass is missing.[40]

Here Einstein reminds Laue that he had been able to recognize that the relativistic theory of gravitational fields should be a theory dealing with metrics of *nonvanishing* curvature, precisely because he was able to recognize that special relativity, the theory which dealt with a metric of *vanishing* curvature, was really also the theory of a special type of gravitational field. He could see this because, in turn, the Minkowski metric induced a structure identical to a classical gravitational field on the relative spaces of accelerating frames of reference and, unlike Synge, he had resisted the temptation of regarding this structure as somehow fictitious or different from "real" gravitational fields. (We have seen earlier how the $\Gamma^l_{ik}$ can appear as the field strengths of this structure in the relative spaces concerned.)

In the last analysis, over a half century after Einstein found and used this key, it matters little to one's application of the theory if one follows Synge and says that "the Riemann tensor ... *is* the gravitational field" (Synge 1960, p. viii) or if one follows Einstein and calls the metric tensor the gravitational field. For the connection between these structures and the gravitational fields of relative spaces which they generalize is essentially only a heuristic one. Perhaps Synge's approach is more comfortable for those who wish to continue thinking of special relativity as a gravitation-free case. For them, the presence of a gravitational field is the intrusion of some kind of perturbation into the Minkowski metric, in the same way as classical gravitational fields arise as anisotropies in otherwise constant scalar fields. If the curvature of a metric field is nonvanishing, then even a freely falling observer can detect this perturbation through the presence of tidal gravitational forces and he may well also be able to identify some nearby massive body that is largely responsible for it.[41]

Personally however, I find Einstein's attitude more comfortable and the association of gravitational fields only with metrics of nonvanishing curvature an arbitrary and unnecessary distinction. For such a distinction masks one of the most beautiful of Einstein's insights, that there is no essential difference between inertia and gravity. According to general relativity, the same structure—the metric—governs the motion of a body in free-fall in the "gravitation-free" case of special relativity or in free-fall in a classically recognizable gravitational field. If we are to call any structure "gravitational field" in relativity theory, then it should be the metric.

## 12. What was Einstein's Principle of Equivalence?

Einstein's principle of equivalence asserted that the properties of space that manifest themselves in inertial effects are really the properties of a field structure *in* space; moreover this same structure also governs gravitational effects. As a result, the privileged inertial states of motion defined by inertial effects are not properties of space but of this structure and the various possible dispositions of inertial motions in space are determined completely by it. Space of itself is to be expected to designate no states of motion as privileged.

This principle guided Einstein to seek his general theory of relativity as a gravitation theory of which special relativity was a special case. There the principle found precise theoretical expression. The structure responsible for inertial and gravitational effects is the metric tensor. The space-time manifold itself has no properties that would enable us to designate the motion associated with any given world line as privileged, that is as "inertial" or "unaccelerated." This designation depends entirely on the metric and the affine structure for space-time that it determines.

The purpose of the "Einstein elevator" thought experiment was to show that the structures associated with supposedly gravitation-free special relativity were already intimately connected with gravitation. To demonstrate this, he transformed from an inertial frame of reference to a uniformly accelerated frame and showed that a structure indistinguishable from a classical homogeneous gravitational field was induced by the Minkowski metric on the associated relative space.

This property of the Minkowski metric enabled Einstein to identify it as an *instance* of the four-dimensional generalization of classical gravitational fields. This identification set Einstein on a royal road to his general theory of relativity. For it effectively reduced his task to that of finding a theory that generalized the properties of the Minkowski metric in a way enabling treatment of arbitrary gravitational fields.

Unfortunately, Einstein's contemporaries seized upon one of Einstein's intermediate results, that in certain cases the gravitational fields of relative spaces have a relative existence, dependent on the choice of frame of reference. They sought to generalize this result from the simple cases in Minkowski space-time that Einstein considered to arbitrary gravitational fields. It has rarely been acknowledged that Einstein never endorsed the principle that results, here called the "infinitesimal principle of equivalence." Moreover, his early correspondence contains a devastating objection to this principle: in infinitesimal regions of the space-time manifold it is impossible to distinguish geodesics from many other curves and therefore impossible to decide whether a point-mass is in free fall.

Some readers may feel dissatisfied that Einstein's principle of equivalence finds the uncontroversial expression indicated above in the general theory of relativity. On the contrary, I find it a source of great satisfaction and a testament to the coherence and clarity of Einstein's vision. For it shows that Einstein has been completely successful in taking an idea, which was quite extraordinary when conceived in 1907, and incorporating it completely into the body of a now universally accepted physical theory. In recent decades there has been much criticism of "the" principle of equivalence. But the principle under cogent attack has rarely been Einstein's version. For, to paraphrase Einstein's 1916 reflection on the critics of Mach, "even those who regard themselves as Einstein's opponents barely known how much of Einstein's views they have imbibed, so to speak, with their mother's milk" (Einstein 1916c, p. 102).[42]

*Acknowledgments.* I am most grateful for the generous hospitality of the Einstein Project, Princeton, and the Center for the Philosophy of Science, University of Pittsburgh, and for the sponsorship of the Fulbright Exchange Program during the researching and writing of this paper. I wish to thank Michael Friedman, Al Janis, David Malament, John Stachel, and Roberto Torretti for helpful discussion and comments on earlier drafts. I also wish to thank the Hebrew University of Jerusalem for its kind permission to quote the material in this paper from Einstein's unpublished writings, to which it holds copyright.

NOTES

* Reprinted from *Studies in History and Philosophy of Science*, vol. 16. © 1985 by Pergamon Press, Ltd. with permission.

[1] For a compact discussion of some principles of equivalence, see Thorne, Lee, and Lightman 1973, pp. 3570–3572.

[2] This hypothesis is not labeled as the "principle of equivalence" in this article—the term does not appear anywhere in the article.

[3] For example, Einstein 1911, p. 899; 1954a, pp. 77–78.

[4] For example, Einstein to T. Levi-Civita, March 20, 1915, EA 16-233; to E. Klug, February 13, 1929, EA 25-126; to L.R. and H.G. Lieber, November 20, 1940, EA 15-135; to J. Reyntjens, August 26, 1950, EA 27-144; to A. Rehtz, July 12, 1953, EA 27-134.

[5] In all the places cited in this section, the only weak exception to this is in the letter to the Liebers where he allows that the gravitational field at a point is "in a certain way fictitious," because it can be transformed away.

[6] Earman and Glymour have also remarked on this (1978, p. 254).

[7] Specifically, six degrees of freedom in Newtonian space-times, three in Minkowski space-time and three or less (if any) in an arbitrary semi-Riemannian manifold. See Pauli 1921, pp. 130–132. So a "(rigid) uniformly accelerated frame of reference" in Minkowski space-time is specified by requiring the reference points to be in rigid motion and *one* of them to be uniformly accelerated. I shall always read "uniform (rectilinear) acceleration" in Minkowski space-time as referring to hyperbolic motion (Pauli 1921, pp. 74–76).

[8] Torretti 1983, pp. 14–15, 28, defines a similar "relative space."

[9] Stachel (1980) has discussed Einstein's use of this example in detail.

[10] If $F$ is rotating, $H_c(p)$ will be orthogonal to $c$ only. So in general this mapping procedure must be repeated with a new orthogonal hypersurface for each $c$ in $R_F$. Most of the discussion of this section can be transferred to Newtonian space-times with little modification. Similar induced metrics could be defined in the relative spaces of Newtonian space-times by deriving them from the three-dimensional metrics of hypersurfaces of simultaneity.

[11] Pauli 1921, p. 131 writes this as the requirement of the constancy along $c$ of the components of $g_{\text{orth}}$ in an adapted coordinate system. This condition is equivalent to the vanishing of the frame's expansion tensor, as defined in Hawking and Ellis 1973,

p. 82. Informally, the condition ensures constancy of the orthogonal interval between $c$ and an infinitesimally close curve $c'$ of $F$ in the hypersurfaces $H_c$.

[12] In Newtonian space-times, the scalar field $T$ is already given for all frames by the absolute time field. Therefore every relative space will have a frame time.

[13] Einstein 1907, pp. 414, 454. Then he wrote (p. 454): "This assumption extends the principle of relativity to the case of uniformly accelerated translational motion of the reference system." Einstein did not begin to describe his hypothesis with the compact labels "equivalence principle" and "equivalence hypothesis" until 1912 and 1913.

[14] Laue to Einstein, December 27, 1911, EA 16-008.

[15] See Einstein 1918b, p. 700; 1950, p. 347; 1955b, p. 140.

[16] In his early (1911) version, Einstein notes that he will "disregard the theory of relativity" and confine himself to "customary" kinematics and "ordinary" mechanics.

[17] Einstein briefly rehearses the problem of characterizing such fields as Newtonian gravitational fields in 1920a.

[18] Einstein relayed his puzzlement at this result to Ehrenfest in a letter of June 1912, EA 9-333. See also Einstein 1912c.

[19] Einstein to A. Rehtz, July 12, 1953, EA 27-134. In his 1920b, Einstein summarizes the principle in similar terms: "... the physical properties of space prevailing relative to $K'$ are completely equivalent to a gravitational field." $K'$ is a reference system in uniform rectilinear acceleration with respect to a Galilean system.

[20] Friedman 1983, pp. 191–195, has given a lucid analysis of the limited prospects of using a principle of equivalence to yield a generalized principle of relativity, if the latter is understood to require this type of indistinguishability.

[21] $g'$ must be a Minkowski metric, since if $g$ has the form $\mathrm{diag}(-1, -1, -1, 1)$ in a coordinate system $\{y^i\}$, then $g'$ will have the same form in $\{y'^i\}$, the image of $\{y^i\}$ under $h$. Similarly the components of $g$ in $\{x^i\}$ at $p$ will equal the components of $g'$ in $\{x'^i\}$ at $hp$; therefore: (a) since the curves of constant $x^i$ ($i = 1, 2, 3$) are geodesics of $g$, the curves of constant $x'^i$ ($i = 1, 2, 3$) will be geodesics of $g'$; and (b) since the orthogonal metric of $g$ in the frame $F$ satisfies the rigid-body motion condition, the same will be true of the orthogonal metric of $g'$ in $F'$. From (a) and (b) it follows that $F'$ will be an inertial frame of $g'$.

[22] In his correspondence about his early work on the general theory, Einstein commented briefly that he saw the principle of equivalence incorporated into the new theory through its covariance properties; Einstein to P. Ehrenfest, Winter 1913–1914?, EA 9-347; Einstein to M. Besso, March 1914 (Speziali 1972, p. 53).

[23] These results also make plausible the failure of Einstein's first 1912 field equation to yield a conservation law, in spite of its similarity to the field equations of general relativity. From the perspective of general relativity, we would only expect his first field equation to yield consistent results in the trivial case of Minkowski space-time.

[24] Einstein to Becquerel, August 16, 1951, EA 6-074 and 6-075. Einstein's argument is especially interesting and important, since it is intended to take a skeptic who accepts special but not general relativity step by step from the former to the latter, carefully delineating the assumptions of each step.

[25] See Einstein 1913a, pp. 285–286; 1913b, pp. 1255–1256; 1914a, p. 177; 1914b, pp. 1032–1033. See also Einstein 1954a, pp. 100–101, for a very clear exposition without formalism.

[26] See Einstein 1936, pp. 308–309; 1949, pp. 70–73; 1950, pp. 350–351; 1952, pp. 153–154; 1955a, pp. 14–15.

[27] Einstein used this same notion of identity of essence elsewhere in Einstein 1912c, p. 1063; Einstein and Grossmann 1913, p. 226; and Einstein 1922, p. 58.

[28] Laue to Einstein, January 8, 1951, EA 16-152.

[29] Einstein to Laue, January 16, 1951, EA 16-154.

[30] Compare with Einstein's: "There is no space without gravitational or inertial field. What one calls empty space in the sense of classical or Maxwell's theory, is a gravitational field of a special kind, that is one in which the gravitational potentials are constant with an appropriate choice of coordinates." Einstein to H. Titze, January 16, 1954, EA 23-026/027.

[31] See also Silberstein 1922, p. 12.

[32] In a letter to P. Painlevé, December 7, 1921, EA 19-003, Einstein stresses that the general theory rests completely on the assumption that space-time behaves as it does in special relativity in infinitely small elements of the space-time manifold.

[33] Schlick to Einstein, February 4, 1917, EA 21-568; Einstein to Schlick, February 6, 1917, EA 21-612.

[34] Einstein to Schlick, March 21, 1917, EA 21-614. Schlick corrected the argument in accord with Einstein's remarks in the republication of the article in monograph form. See Schlick 1920, pp. 60–62.

[35] Torretti 1983, pp. 150–151, 316, has made the same objection in this context using virtually the same words as Einstein, but independently of him. Torretti writes: "In a Riemannian manifold, *every* curve is 'straight in the infinitesimal'." He illustrates his point vividly by pointing out that the streets which run along both parallels of latitude and meridians on the earth's surface are straight in the infinitesimal of such cities as Chicago, but only the meridians are geodesics.

[36] I am grateful to David Malament for making available to me mimeographed lecture notes of Robert Geroch, in which the technique is outlined.

[37] If members of $\{h\}$ map a point with coordinates $x^k$ to one with $y^i$, then at $p$ members of $\{h_1\}$ satisfy $y^i{}_{,k} = \delta^i_k$; members of $\{h_2\}$ satisfy the additional condition $y^i{}_{,km} = 0$; members of $\{h_3\}$ satisfy the additional condition $y^i{}_{,kmn} = 0$ and so on. Commas denote differentiation with respect to $x^k$.

[38] It is important to note that one can only consistently compare orders of quantities if their orders are assigned within a hierarchy generated by the same structure. Any tensor will generate a hierarchy of quantities in which that tensor is of first order, since all tensors are invariant under the members of $\{h_1\}$. For example, the curvature tensor will be of first order in a hierarchy it generates, whereas it is of third order in the hierarchy generated by the metric tensor. In the text I tacitly assume that one can compare the orders of quantities in the metric tensor hierarchy with the orders of quantities in the hierarchy engendered by a geodesic through $p$. This is justified by the fact that these two hierarchies can be combined as follows. Each member of the set of geodesics $\{c\}$ through $p$ has a parametrization by the interval $s$ induced upon it by the metric tensor $g$. Conversely, given this same parametrization we can recover the original $g$, through the condition $g(X, X) = 1$ for all tangent vectors $X = d/ds$. Therefore, for the present purpose, we can consider $g$ and associated quantities as well as the set of tangent vectors $\{X\}$ and associated quantities as dependent on $\{c\}$ and its parametrization. In particular, the image of $\{c\}$ and its parametrization under a member of $\{h\}$ will generate a new metric tensor $g'$ and a new set of tangent vectors $\{X'\}$. We can now determine the orders of these and related quantities in the manner outlined earlier. The expected results do obtain. For example, both $g$ and $X$ are first order in this hierarchy.

[39] This argument establishes the necessity of these additional conditions. Their necessity can be illustrated in the example of a two-dimensional Euclidean space. In

the usual Cartesian coordinate system, geodesics passing through the origin are $y = mx$, for $m$ a constant. However, the curves $y = x^n$ for all $n > 2$ satisfy the condition $D_X X = 0$ at the origin. The conditions $(D_X)^n X = 0$ for all positive integers $n$ are *not* sufficient. In the Euclidean space they are satisfied at the origin by the smooth curve $y = 0$ when $x = 0$; $y = \exp(-1/x^2)$ for all other $x$, but this curve is not a geodesic. (I am grateful to Al Janis for this last point.)

[40] Einstein to Laue, September 12, 1950, EA 16-148.

[41] Einstein and Rosen 1935 have added a curious twist to the standard objection that the gravitational fields produced by acceleration cannot be "true" gravitational fields since they have no sources. Recalling the principle of equivalence by name, they consider a coordinate system $\{x_i\}$ adapted to a uniformly accelerated frame of reference in Minkowski space-time and, in the now familiar manner, associate a homogeneous gravitational field with it. This accelerated frame cannot fill all of Minkowski space-time. In the case they consider, their frame fills the submanifold given by $(y_1)^2 \geq (y_4)^2$, where $\{y^i\}$ is the Galilean coordinate system used to define the frame (see their footnote, p. 74). They note that the Minkowski metric is a solution of the usual gravitational field equations of general relativity in the coordinate system $\{x_i\}$, but that certain components ($T_{22}$ and $T_{23}$) of the otherwise everywhere vanishing source stress-energy tensor become singular along the hypersurface $x_1 = 0$, which is a boundary of the submanifold containing the accelerated frame. This represents a kind of source mass or energy distribution. They introduce the example so they can proceed to illustrate how such singularities can be removed. For further details see Einstein and Rosen 1935, p. 74.

[42] Of course, the original quotation is recovered by replacing "Einstein" by "Mach." This image may complement Synge's memorable image of the principle of equivalence as a midwife at the birth of general relativity who is now to suffer burial, but at least with appropriate honors. (Synge 1960, pp. ix–x).

REFERENCES

Anderson, James L. (1967). *Principles of Relativity Physics*. New York: Academic Press.

Earman, John, and Glymour, Clark (1978). "Lost in the Tensors: Einstein's Struggles with Covariance Principles 1912–1916." *Studies in History and Philosophy of Science* 9: 251–278.

Eddington, Arthur Stanley (1924). *The Mathematical Theory of Relativity*, 2nd ed. Cambridge: Cambridge University Press.

Einstein, Albert (1907). "Über das Relativitätsprinzip und die aus demselben gezogenen Folgerungen." *Jahrbuch der Radioaktivität und Elektronik* 4: 411–462.

——— (1911). "Über den Einfluss der Schwerkraft auf die Ausbreitung des Lichtes." *Annalen der Physik* 35: 898–908. Translated as "On the Influence of Gravitation on the Propagation of Light." In Lorentz et al. 1923, pp. 97–108.

——— (1912a). "Lichtgeschwindigkeit und Statik des Gravitationsfeldes." *Annalen der Physik* 38: 355–369.

——— (1912b). "Zur Theorie des statischen Gravitationsfeldes." *Annalen der Physik* 38: 443–458.

——— (1912c). "Relativität und Gravitation. Erwiderung auf eine Bemerkung von M. Abraham." *Annalen der Physik* 38: 1059–1064.

——— (1912d). "Gibt es eine Gravitationswirkung die der elektrodynamischen

Induktionswirkung analog ist?" *Vierteljahrsschrift für gerichtliche Medizin und öffentliches Sanitätswesen* 44: 37–40.

———— (1913a). "Physikalische Grundlagen einer Gravitationstheorie." *Naturforschende Gesellschaft Zürich. Vierteljahrsschrift* 58: 284–290.

———— (1913b). "Zum gegenwärtigen Stande des Gravitationsproblems." *Physikalische Zeitschrift* 14: 1249–1266.

———— (1914a). "Prinzipielles zur verallgemeinerten Relativitätstheorie und Gravitationstheorie." *Physikalische Zeitschrift* 15: 176–180.

———— (1914b). "Die formale Grundlage der allgemeinen Relativitätstheorie." *Königlich Preussische Akademie der Wissenschaften* (Berlin). *Sitzungsberichte*: 1030–1085.

———— (1916a). "Die Grundlage der allgemeinen Relativätstheorie." *Annalen der Physik* 49: 769–822. Translated as "The Foundation of the General Theory of Relativity." In Lorentz et al. 1923, pp. 109–164.

———— (1916b). "Über Friedrich Kottlers Abhandlung 'Über Einstein's Äquivalenzhypothese und die Gravitation.'" *Annalen der Physik* 51: 639–642.

———— (1916c). "Ernst Mach." *Physikalische Zeitschrift* 17: 101–104.

———— (1917). "Kosmologische Betrachtungen zur allgemeinen Relativitätstheorie." *Königlich Preussische Akademie der Wissenschaften* (Berlin). *Sitzungsberichte*: 142–152. Translated as "Cosmological Considerations on the General Theory of Relativity." In Lorentz et al. 1923, pp. 175–188.

———— (1918a). "Prinzipielles zur allgemeinen Relativitätstheorie." *Annalen der Physik* 55: 241–244.

———— (1918b). "Dialog über Einwände gegen die Relativitätstheorie." *Die Naturwissenschaften* 6: 697–702.

———— (1920a). "Inwiefern lässt sich die moderne Gravitationstheorie ohne die Relativität begründen?" *Die Naturwissenschaften* 8: 1010–1011.

———— (1920b). "Grundgedanken und Methoden der Relativitätstheorie in ihrer Entwicklung dargestellt." Unpublished manuscript EA 2-070. This document can be dated to 1920, in part by Einstein's mention of the "English [eclipse] expedition of the previous year" on p. 32.

———— (1922). *The Meaning of Relativity*. Princeton: Princeton Univesity Press. Page numbers are cited from the 5th ed. Princeton: Princeton University Press, 1955.

———— (1936). "Physics and Reality." *Journal of the Franklin Institute* 221: 349–382. Page numbers are cited from the reprint in Einstein 1954b, pp. 290–323.

———— (1949). "Autobiographical Notes." In *Albert Einstein: Philosopher-Scientist*. Paul Arthur Schilpp, ed. Evanston, Illinois: The Library of Living Philosophers, pp. 1–95.

———— (1950). "On the Generalized Theory of Gravitation." *Scientific American* 182: 13–17. Page numbers are cited from the reprint in Einstein 1954b, pp. 341–356.

———— (1952). "Relativity and the Problem of Space." Appendix 5 in Einstein 1954a, pp. 135–157.

———— (1954a). *Relativity, The Special and the General Theory: A Popular Exposition*, 15th ed. Robert W. Lawson, trans. London: Methuen, 1954.

———— (1954b). *Ideas and Opinions*. Carl Seelig, ed. Sonja Bargmann, trans. New York: Crown; reprint London: Souvenir Press, 1973.

———— (1955a). "Erinnerungen—Souvenirs." *Schweizerische Hochschulzeitung* 28 (*Sonderheft*): 145–148, 151–153. Reprinted as "Autobiographische Skizze." In *Helle Zeit-Dunkle Zeit*. Carl Seelig, ed. Zurich: Europa Verlag, 1956, pp. 9–17.

———— (1955b). "Relativistic Theory of the Non-Symmetric Field." Appendix 2 in *The Meaning of Relativity*, 5th ed. Princeton: Princeton University Press, pp. 133–166.

Einstein, Albert, and Grossmann, Marcel (1913). *Entwurf einer verallgemeinerten Relativitätstheorie und einer Theorie der Gravitation. I. Physikalischer Teil von Albert Einstein. II. Mathematischer Teil von Marcel Grossmann.* Leipzig and Berlin: B.G. Teubner.

Einstein, Albert, and Rosen, Nathan (1935). "The Particle Problem in the General Theory of Relativity." *Physical Review* 48: 73–77.

Friedman, Michael (1983). *Foundations of Space-Time Theories.* Princeton: Princeton University Press.

Hawking, Stephen W., and Ellis, George F.R. (1973). *The Large-Scale Structure of Space-Time.* Cambridge: Cambridge University Press.

Lorentz, Hendrik Antoon, et al. (1923). *The Principle of Relativity: A Collection of Original Memoirs on the Special and General Theory of Relativity.* W. Perrett and G.B. Jeffery, trans. London: Methuen; reprint New York: Dover, 1952.

Newton, Isaac (1729). *Sir Isaac Newton's Mathematical Principles of Natural Philosophy & His System of the World.* Andrew Motte, trans. London. Reprint New York: Greenwood, 1969.

Norton, John (1984). "How Einstein Found His Field Equations: 1912–1915." *Historical Studies in the Physical Sciences* 14: 253–316. See this volume, pp. 101–159.

Ohanian, Hans C. (1977). "What is the Principle of Equivalence?" *American Journal of Physics* 45: 903–909.

Pais, Abraham (1982). *"Subtle is the Lord ...": The Science and the Life of Albert Einstein.* Oxford: Clarendon Press.

Pauli, Wolfgang (1921). "Relativitätstheorie." In *Encyklopädie der mathematischen Wissenschaften, mit Einschluss ihrer Anwendungen.* Vol. 5, *Physik*, part 2. Arnold Sommerfeld, ed. Leipzig: B.G. Teubner, 1904–1922, pp. 539–775. [Issued November 15, 1921.] Page numbers cited from the English translation *Theory of Relativity.* With supplementary notes by the author. G. Field, trans. London: Pergamon, 1958; reprint New York: Dover, 1981.

Schlick, Moritz (1917). "Raum und Zeit in der gegenwärtigen Physik. Zur Einführung in das Verständnis der allgemeinen Relativitätstheorie." *Die Naturwissenschaften* 5: 161–167 (March 16), 177–186 (March 23).

——— (1920). *Raum und Zeit in der gegenwärtigen Physik. Zur Einführung in das Verständnis der Relativitäts- und Gravitationstheorie*, 3rd ed. Berlin: Julius Springer, 1920. English translation *Space and Time in Contemporary Physics: An Introduction to the Theory of Relativity and Gravitation.* Henry Brose, trans. New York: Oxford University Press, 1920.

Silberstein, Ludwik (1922). *The Theory of General Relativity and Gravitation.* Toronto: University of Toronto Press.

Speziali, Pierre, ed. (1972). *Albert Einstein–Michele Besso. Correspondance, 1903–1955.* Paris: Hermann.

Stachel, John (1980). "Einstein and the Rigidly Rotating disk." In *General Relativity and Gravitation: A Hundred Years after the Birth of Einstein.* A. Held, ed. New York: Plenum, pp. 1–15. See this volume, pp. 48–62.

Synge, John Lighton (1960). *Relativity: The General Theory.* Amsterdam: North-Holland.

Thorne, Kip S., Lee, David L., and Lightman, Alan P. (1973). "Foundations for a Theory of Gravitation Theories." *Physical Review D* 7: 3563–3578.

Torretti, Roberto (1983). *Relativity and Geometry.* Oxford: Pergamon Press.

# The Rigidly Rotating Disk as the "Missing Link" in the History of General Relativity*

JOHN STACHEL

## 1. Introduction

Working with the Einstein Archive at the Institute for Advanced Study has given me a chance to become familiar with some of the material in this most extensive repository of documents on the life and activities of Albert Einstein, collected indefatigably over the last quarter of a century by the Trustees of his Estate, Dr. Otto Nathan and Miss Helen Dukas; and organized by Miss Dukas, the Archivist of the collection, whose memory is undoubtedly its most important single resource. It has also made me realize how much the material held in the Archive can contribute towards the study of many problems in the history of modern physics, to say nothing of many cultural, social, and political topics.

As a small example, I shall discuss the question of the relativistic rigidly rotating disk, a topic that has been the subject of extensive—and intensive— discussion from the early days of the special theory of relativity to the present.[1] An examination of Einstein's treatment of this problem is of interest not only because it shows his way of treating the issues involved, but because it seems to provide a "missing link" in the chain of reasoning that led him to the crucial idea that a nonflat metric was needed for a relativistic treatment of the gravitational field.

## 2. Einstein's Treatment of the Rotating Disk

Einstein's first mention of rigidly rotating bodies that I have located in the Archive is in a letter of September 29, 1909, to Sommerfeld:

> The treatment of the uniformly rotating rigid body seems to me to be of great importance on account of an extension of the relativity principle to uniformly rotating systems along analogous lines of thought to those that I tried to carry out for uniformly accelerated translation in the last section of my paper published in the *Zeitschrift für Radioaktivität* (EA 21–377).[2]

Einstein is referring here to his first published attempt in 1907 to develop a relativistic treatment of gravitation based on the equivalence principle applied to a spatially uniform static gravitational field (Einstein 1907).

The occasion for Einstein's comment was probably the discussion of Born's paper (1909) at the Salzburg meeting of the German Society of Scientists and Physicians, which took place September 21–25, 1909. Born had presented his definition of rigid motions (and thus of rigid bodies, insofar as they are capable of existing) in special relativity. Sommerfeld (1909) had commented on Born's talk, and Born in a later paper (1910) noted that he and Einstein had discussed the rigid-body problem at Salzburg, and were puzzled "that a [rigid] body at rest can never be brought into uniform rotation"; this problem was discussed almost simultaneously by Paul Ehrenfest in a paper (1909) submitted September 29—the same day as Einstein's letter to Sommerfeld—and soon became known as "Ehrenfest's paradox." In spite of the importance he attached to the problem, and the intense discussion occasioned by Ehrenfest's paper,[3] Einstein published nothing directly on the question during the next years. His only contribution to the discussion was an answer to one of the points raised by Varičak (1911) in a comment on Ehrenfest's paradox.[4] Einstein's note (1911) made no reference to the rotating-disk problem but confined itself to rebutting Varičak's aspersions on the "reality" of the Lorentz contraction.

Einstein's first published reference to the rigidly rotating disk is a hesitant one. It occurs in the first of two papers on static gravitational fields written in 1912 during his stay in Prague (Einstein 1912a, 1912b). It dates from February 1912 and begins by reviewing his previous work on the uniformly accelerating coordinate system, pointing out that

> Such a system $K$, according to the equivalence principle, is strictly equivalent to a system at rest in which a matter-free static gravitational field of a certain kind exists. Let spatial measurements in $K$ be made with measuring rods which—when compared with each other at rest at some point of $K$—have the same length; assume that the theorems of [Euclidean] geometry are valid for lengths measured in this way, and thus also for the relationship between the coordinates $x$, $y$, $z$ and other lengths. This stipulation is not automatically permissible, but contains physical assumptions which ultimately could prove to be invalid. For example, they most probably do not hold in a uniformly rotating system, in which, on account of the Lorentz contraction, the ratio of the circumference of a circle to its diameter would have to differ from $\pi$ using our definition of length. The measuring rod, as well as the coordinate axes, are to be treated as rigid bodies. This is permissible in spite of the fact that, according to [special] relativity theory, rigid bodies cannot really exist. For one can imagine the rigid measuring body replaced by a large number of small non-rigid bodies so aligned alongside each other that they do not exert any pressure forces on each other since each is separately held in place.

The tentative nature of his conclusions reflects Einstein's puzzlement during this period over the problem of the relationship between coordinates and measurements with rods and clocks, a point to which we shall later return.[5]

There are references to rotating frames of reference in several of Einstein's papers on the developing general theory of relativity,[6] as well as in the correspondence;[7] but the context is Einstein's interpretation of the equivalence principle: the explanation of the inertial forces occurring in such frames as equivalent to gravitational forces, and there is no direct reference to the rotating-disk problem. The next time that the rotating-disk argument occurs in Einstein's writings—this time without the tentative note—is in the 1916 review paper in which he presented the final version of the general theory, together with various arguments for it (Einstein 1916). Although the paper is well known and easily accessible, I quote the paragraph in full for the sake of completeness:

> In a space which is free of gravitational fields we introduce a Galilean system of reference $K(x, y, z, t)$, and also a system of co-ordinates $K'(x', y', z', t')$ in uniform rotation relatively to $K$. Let the origins of both systems, as well as their axes of $Z$, permanently coincide. We shall show that for a space-time measurement in the system $K'$ the above [special relativistic] definition of the physical meaning of lengths and times cannot be maintained. For reasons of symmetry it is clear that a circle around the origin in the $X$, $Y$ plane of $K$ may at the same time be regarded as a circle in the $X'$, $Y'$ plane of $K'$. We suppose that the circumference and diameter of this circle have been measured with a unit measure infinitely small compared with the radius, and that we have the quotient of the two results. If this experiment were performed with a measuring-rod at rest relatively to the Galilean system $K$, the quotient would be $\pi$. With a measuring rod at rest relatively to $K'$, the quotient would be greater than $\pi$. This is readily understood if we envisage the whole process of measuring from the "stationary" system $K$, and take into consideration that the measuring rod applied to the periphery undergoes a Lorentzian contraction, while the one applied along the radius does not. Hence Euclidean geometry does not apply to $K'$. The notion of co-ordinates defined above, which presupposes the validity of Euclidean geometry, therefore breaks down in relation to the system $K'$. So, too, we are unable to introduce a time corresponding to physical requirements in $K'$; indicated by clocks at rest relatively to $K'$. To convince ourselves of this impossibility, let us imagine two clocks of identical constitution placed, one at the origin of co-ordinates, and the other at the circumference of the circle, and both envisaged from the "stationary" system $K$. By a familiar result of the special theory of relativity, the clock at the circumference—judged from $K$—goes more slowly than the other, because the former is in motion and the latter at rest. An observer at the common origin of co-ordinates, capable of observing the clock at the circumference by means of light, would therefore see it lagging behind the clock beside him. As he will not make up his mind to let the velocity of light along the path in question depend explicitly on the time, he will interpret his observations as showing that the clock at the circumference "really" goes more slowly than the clock at the origin. So he will be obliged to define time in such a way that the rate of a clock depends upon where the clock may be.
>
> We therefore reach this result:—In the general theory of relativity, space and time cannot be defined in such a way that the differences of the spatial co-ordinates be directly measured by the unit measuring-rod, or differences in the time co-ordinate by a standard clock. (Einstein 1916, pp. 115–117)

Note that the "rigidly rotating disk" is not actually referred to in these considerations. However, that Einstein had it in mind—or at least was not averse to its consideration in this context—is made clear by the expanded discussion of the topic that he gave in his *Relativity* book of 1916 (Einstein 1917), chapter 23 of which is devoted to the topic "Behavior of Clocks and Measuring-Rods on a Rotating Body of Reference." In this amplified discussion, he adds: "In order to fix our ideas, we shall imagine $K'$ to be in the form of a plane circular disk, which rotates uniformly in its own plane about its centre," and thereafter phrases his discussion in terms of the disk. He also adds an element not made explicit in his previous discussion, asserting that an observer at rest on the disk is entitled to regard "the force acting on himself, and in fact on all other bodies which are at rest relative to the disk ... as the effect of a gravitational field." Thus, he links up his treatment of the rotating disk with his earlier treatment of rotating reference frames; but we shall continue to confine ourselves to the disk aspect of his discussion.

He draws the conclusion "that the propositions of Euclidean geometry cannot hold exactly on the rotating disk, nor in general in a gravitational field, at least if we attribute the length 1 to the [measuring] rod in all positions and in every orientation." He immediately follows this discussion with a discussion in chapter 24 of the "Euclidean and Non-Euclidean Continuum," and in chapter 25 with a discussion of the use of "Gaussian Co-ordinates" to treat non-Euclidean continua mathematically. Since the book is still in print and easily accessible, I shall omit this long discussion.

In *The Meaning of Relativity*, based upon his 1921 Princeton lectures, Einstein again gives a similar discussion of the rotating disk. This time I shall quote his conclusions, since they briefly summarize the material in chapters 23, 24, and 25 just mentioned:

> Space and time, therefore, cannot be defined with respect to $K'$ as they were in the special theory of relativity with respect to inertial systems. But, according to the principle of equivalence, $K'$ may also be considered as a system at rest, with respect to which there is a gravitational field (field of centrifugal force, and force of Coriolis). We therefore arrive at the result: the gravitational field influences and even determines the metrical laws of the space-time continuum. If the laws of configuration of ideal rigid bodies are to be expressed geometrically, then in the presence of a gravitational field the geometry is not Euclidean.
> 
> The case that we have been considering is analogous to that which is presented in the two-dimensional treatment of surfaces. It is impossible in the latter case also to introduce co-ordinates on a surface (e.g., the surface of an ellipsoid) which have a simple metrical significance, while on a plane the Cartesian co-ordinates, $x_1$, $x_2$, signify directly lengths measured by a unit measuring rod. Gauss overcame this difficulty, in his theory of surfaces, by introducing curvilinear co-ordinates which, apart from satisfying conditions of continuity, were wholly arbitrary, and only afterwards these co-ordinates were related to the metrical properties of the surface. In an analogous way we shall introduce in the general theory of relativity arbitrary co-ordinates, $x_1$, $x_2$, $x_3$, $x_4$, which shall number uniquely the space-time points, so

that neighboring events are associated with neighboring values of the co-ordinates; otherwise, the choice of co-ordinates is arbitrary. We shall be true to the principle of relativity in its broadest sense if we give such a form to the laws that they are valid in every such four-dimensional system of co-ordinates, that is, if the equations expressing the laws are co-variant with respect to arbitrary transformations.

The most important point of contact between Gauss's theory of surfaces and the general theory of relativity lies in the metrical properties upon which the concepts of both theories, in the main, are based. (Einstein 1922a, pp. 60–61)

The topics are presented in the same order as in the *Relativity* book (Einstein 1917): first the discussion of the disk, leading to the conclusion that a non-Euclidean geometry holds on the disk, and therefore space (and time) coordinates cannot be given a direct physical meaning as in the special theory. The two-dimensional Gaussian theory of curved surfaces is recalled, based upon the possibility of introducing entirely arbitrary coordinate systems, and then using the metric tensor to describe the metrical properties of the surface. The analogy is then drawn with the use of arbitrary space-time coordinates and the metric tensor to characterize the gravitational field mathematically. Later on, I shall comment upon the possible historical significance of this order of presentation. But first I shall finish the account of Einstein's discussions of the disk.

In *The Evolution of Physics*, his popular book written with Leopold Infeld, Einstein again reverts to the example of the rotating disk to show the necessity to introduce non-Euclidean geometry if one wants to generalize the principle of relativity so that it applies to noninertial frames of reference (Einstein and Infeld 1938, pp. 226–234).

I have not found any other discussion of the rotating disk in Einstein's published writings (in a far from exhaustive search, I must add); but there are a number of letters in which Einstein refers to the subject, giving a much more detailed discussion than in any of the printed sources, and explicitly replying to some of the objections that were offered to his treatment. He also makes several comments on the importance of the problem for his development of the general theory of relativity, which will be of great value for my later historical discussion.

The first such letter I have come across is one to Joseph Petzoldt, the well-known positivist philosopher, and the author of several early essays claiming special relativity theory as a triumph of the positivistic approach (see, for example Petzoldt 1912). On July 26, 1919, Petzoldt wrote Einstein a letter (EA 19-055) in which he raised the objection to Einstein's treatment of the rotating disk (an objection that will not be unknown to the connoisseur of the literature on this subject) that the Lorentz contraction of the rotating rods on the circumference implies that the circumference of the rotating disk should be *shorter* than $2\pi$ times the radius. Einstein replied at length in a letter of August 19, 1919, which has been published in German; I translate it here:

As concerns the rotating disk, I cannot agree with you at all. It is well to remark that a rigid circular disk at rest must break up if it is set into rotation, on account of the Lorentz contraction of the tangential fibers and the noncontraction of the radial ones. Similarly, a rigid disk in rotation (produced by casting) must explode as a consequence of the inverse changes in length, if one attempts to bring it to the rest state. If you fully take into account this state of affairs, your paradox vanishes.

Now you believe that a rigidly rotating circular line must have a circumference that is less than $2r\pi$ because of the Lorentz contraction. The basic error here is that you instinctively set the radius $r$ of the rotating circular line equal to the radius $r_0$ that the circular line has in the case when it is at rest. This however, is not correct; because of the Lorentz contraction rather $2\pi r = 2\pi r_0 \sqrt{1 - (v^2/c^2)}$.

The treatment of the metric of the circular disk runs as follows in detail. Let $U_0$ be the circumference, $r_0$ the radius of the rotating disk, considered from the standpoint of $K_0$ [that is, the rest frame]; then, on account of ordinary Euclidean geometry,

$$U_0 = 2\pi r_0 \tag{1}$$

$U_0$ and $r_0$ naturally are to be thought of as measured with nonrotating measuring rods, i.e., at rest relative to $K_0$.

Now let me imagine corotating measuring rods of rest length 1 laid out on the rotating disk, both along a radius as well as the circumference. How long are these, considered from $K_0$? Let us imagine, in order to make this clearer to ourselves, a "snapshot" taken from $K_0$ (definite time $t_0$). On this snapshot the radial measuring rods have the length 1, the tangential ones, however, the length $\sqrt{1 - (v^2/c^2)}$. The "circumference" of the circular disk (considered from $K$) is nothing but the number of tangential measuring rods that are present in the snapshot along the circumference, whose length considered from $K_0$ is $U_0$. Therefore,

$$U = U_0/\sqrt{1 - (v^2/c^2)}. \tag{2}$$

On the other hand, obviously

$$r = r_0 \tag{3}$$

(since the snapshot of the radial unit measuring rod is just as long as that of a measuring rod at rest relative to $K_0$).

Therefore, from (2), (3), $U/r = U_0/r_0(1/\sqrt{1 - (v^2/c^2)})$, or on account of (1) $= 2\pi/\sqrt{1 - (v^2/c^2)}$.

You make the analogous error for clocks as for measuring rods. The *rotating observer notes very well that, of his two equivalent clocks, that placed on the circumference runs slower than that placed at the center.* We again prove this by considering the entire process from $K_0$. Let $U_z$ be the clock at the center, $U_p$ the one at the periphery. Considered from $K_0$, $U_p$ goes slower than $U_z$; a corotating observer placed next to $U_z$ therefore also sees $U_p$ as going slower than $U_z$. For it is clear that—judged from $K_0$—the time between the occurrence of a position of the hands of the clock and its perception by our observer is constant (independent of the time). I hope this explanation will suffice. (EA 19-069; Thiele 1971, pp. 71–73)

Apparently, Petzoldt did not find this explanation fully satisfactory, and in a missing letter must have objected to the introduction of rigid bodies into the argument (again, an objection not unknown to the connoisseur). Einstein replied in a second letter, of August 23, 1919, which has also been published in the original German:

> I also think that only a personal discussion can produce real clarity. I request that you therefore visit me soon (after making an appointment by telephone). In the meantime, the following on the matter: I know quite well, naturally, that rigid bodies cannot exist according to relativity theory. But one can proceed with advantage as if such did exist; i.e., it is a question of an idealization that can be applied in certain considerations without any contradiction. The considerations of my letter are to be understood in this sense.
>
> You have incorrectly set the radius of the rotating "rigid" circular line equal to $r_0$. Because the circumference, *thought of as materialized by itself*, contracts because of the Lorentz transformation. It would be otherwise, if only the *radii* were thought of as materialized, *but not the tangential connections of their endpoints*.
>
> What you say about peripheral measuring rods and clocks is quite untenable. It is a question of the unjustified taking over of results of special relativity to accelerated reference systems (relative to the inertial systems). Freundlich and Schlick are absolutely correct here. By your sort of reasoning one could just as well conclude that every light ray must propagate rectilinearly with respect to an arbitrary rotating system, etc. Your misunderstanding is quite fundamental. (EA 19-072; Thiele 1971, p. 73)

As late as 1951 Einstein again gave a detailed discussion of the disk, in his draft reply (EA 25-482) to a letter from an Australian medical student named Leonard Champion, who had been teaching himself general relativity but could not find anyone on the staff of Melbourne University able to answer all of his questions.[8] In particular, he had run across the account of the rotating disk by Whittaker (1949), who mentioned that Lorentz and Eddington regarded the geometry of the disk as Euclidean, while others, including Einstein, took it as non-Euclidean. Attempting to resolve the conflict between the sources, Champion made a calculation that amounted essentially to taking the metric of the disk to be given by the line element orthogonal to the world lines of the disk; i.e.,

$$d\sigma^2 = g_{ij} - g_{0i}g_{0j}/g_{00},$$

where the world lines of the points of the disk are given by $x^i = \text{const}$ ($i = 1, 2, 3$). Einstein's reply agreed fully with Champion's calculation. He stated that one had to assume the existence of rigid infinitesimal rods, which implies that if two such rods once agreed in length, when compared, they would always do so, no matter what sort of gravitational field each might afterwards have been subjected to; and a similar assumption must be made for clocks. That is, physical objects that measure the metrical interval are assumed to exist. (Einstein points out that his assumption could be wrong, even though the gravitational field equations were correct.) It follows that the length of an elementary measuring rod is the orthogonal interval between the world lines of the endpoints of the rod.

He then states that he does not know what Eddington meant by claiming the geometry on the rotating disk is flat. While four-dimensional Minkowski space is naturally flat, no matter what coordinates are used, this is not the case for the geometry of the disk as measured with measuring rods rotating with the disk.

He points out that to set up a rigidly rotating disk one would first have to melt a disk at rest, then set the molten disk into rotation and solidify it while it rotates. He admits that there are not really any completely rigid bodies, since if there were one could signal with superluminal velocities; but he maintains that the use made of rigid bodies in his argument seems justified. He states that this example of the disk was of "decisive importance" to him in setting up the general theory of relativity because it showed that a gravitational field (here equivalent to the centrifugal field) causes non-Euclidean arrangements of measuring rods, and thus compelled a generalization of Euclidean space. He emphasizes that the behavior of the rotating measuring rods can be obtained from special-relativistic considerations, since everything is considered from the nonrotating frame of reference.

In this letter Einstein thus brings together in summary form all of his considerations on the rigidly rotating disk, together with his answers to many objections to his treatment.

Although this paper is primarily historical in its aim, it is perhaps worth noting one epistemological feature of Einstein's argument, because it is of some importance for the historical discussion. Einstein sees the argument for the necessity of non-Euclidean metrical relations on the rigidly rotating disk to be based upon three premises:

1. Special relativity holds in a global inertial frame, in which no gravitational field is present.
2. Any coordinate system may be used, and indeed not only mathematically, but may be interpreted as a physical frame of reference, provided that the appropriate gravitational (cum-inertial) field is introduced.
3. A small measuring rod does not change its length in any gravitational field.

The first assumption represents Einstein's conviction that special relativity theory retained its validity within any gravitational theory as the important limiting case in which no noninertial gravitational field occurs. The second assumption is finally embodied in general relativity in the postulate that the metric tensor is the appropriate mathematical representation of the gravitational field potentials. The third assumption, in the context of the metric interpretation of the second, gives physical significance to the metrical interval. That the third assumption really is an independent one is a point that Einstein emphasized a number of times.

In addition, he makes a most significant remark for the history of the development of general relativity, about the importance of his considerations in convincing him of the need to go over to non-Euclidean geometries in his treatment of the gravitational field. While I have not found any discussion of such a role for the rotating disk in Einstein's published writings on the origins of the general theory of relativity (Einstein 1921, 1933, 1949), it is not the first time that the claim is made in his correspondence. In a letter (EA 26-351) presumed to date from the winter of 1939–40 to Hyman Levy, the English Marxist mathematician and author of a number of books on modern science and philosophy for popular audiences, Einstein comments on Levy's latest book (Levy 1939). After stating how pleased he was with much of the book, he recommends that Levy correct one glaring error in later editions. Levy had stated on page 595 that observers on the disk would verify Euclidean geometry. Einstein points out that just the opposite is the case, and adds that it was just the recognition that non-Euclidean geometry holds on the rotating disk which convinced him, at the time he was working on his gravitation theory, that Euclidean geometry could not hold for rigid bodies in the presence of a gravitational field.

There are other references to the rotating disk problem in the correspondence; but we now have essentially all of Einstein's basic ideas connected with the problem. I will try to use these ideas to help solve a problem in the history of general relativity.

## 3. The Rotating Disk as a "Missing Link"

In his discussion of the development of general relativity in the "Autobiographical Notes," Einstein points out that the significance of the equivalence principle in requiring a generalization of the special theory was clear to him in 1908. He then adds

> Why were another seven years required for the construction of the general theory of relativity? The main reason lies in the fact that it is not so easy to free oneself from the idea that co-ordinates must have an immediate physical meaning. (Einstein 1949, p. 67)

In the context of the development of the general theory of relativity as a theory of gravitation (leaving aside the question of possible generalized unified field theories), I think it is clear that what is meant is that only the coordinates-cum-metric tensor in some coordinate system have a physical meaning.

In trying to trace Einstein's journey from the special to the general theory, the following difficulty presents itself.[9] In the papers up to and including those published in 1912, there is no mention of the need for a nonflat space-time, much less of the metric tensor as mathematical representation of the gravitational field. Yet the first paper of 1913 presents us with a full-fledged argument for the representation of the gravitational field by $g_{\mu\nu}$, together with the development of four-dimensional tensor analysis on a Riemannian manifold, the Riemann tensor, etc. (Einstein and Grossmann 1913). Of course, the problem of the correct field equations for the metric tensor was not resolved until late in 1915; but once the crucial step of the correct mathematical description of the gravitational field had been taken, it was only a matter of time until the right field equations were found. I shall argue that the consideration of the rotating disk is a "missing link" in the crucial developments which must have taken place in late 1912.

By the end of March 1912, Einstein had completed his work on the static gravitational field, which he treated by introducing the concept of a variable speed of light, which played the role of gravitational potential (Einstein 1912a, 1912b). At this point he even felt compelled to give up the symmetry between space and time that had characterized the special theory of relativity, especially in Minkowski's formulation of space-time.[10] He had also arrived at the conclusion that his previous formulations of the equivalence principle only held locally. Rather than quote his papers on this question, we shall summarize his account in a letter to Ehrenfest, since this letter also takes us an important step forward beyond the papers. In this letter (EA 9-333; undated, but marked by Ehrenfest as received July 7, 1912), he states that his papers on the static gravitational field (Einstein 1912a, 1912b) indicate that the equivalence hypothesis can only hold for infinitesimally small fields. He notes that his discussion of static gravitational fields corresponds to the electrostatic case in electromagnetic theory; while what he calls "the general static case" would include the analogue of magnetostatic fields. He mentions the "rotating ring" as an example of a system that will generate such a nonstatic but time-independent field.

Thus, by July 1912 Einstein, pursuing his step-by-step approach to the problem, was ready to attack what he called the "general static case"—what we would today call the case of stationary gravitational fields.[11] This presumably led him to look again at the rotating-disk problem—the simplest case of a stationary gravitational field—which he had already tentatively discussed early in 1912, as mentioned earlier. There exist some notes, unfortunately undated, in Einstein's notebooks from roughly this period that may preserve evidence of this study. At this point the argument quoted herein could have occurred to him: from special-relativistic considerations (which he certainly did not doubt hold in the absence of a gravitational field) plus the hypothesis that any coordinate system may be used, provided it is treated as a frame of reference with a corresponding gravitational field (which was a leitmotiv of his search for a general theory of relativity), and the assumption that a unit measuring rod always measures the same length in any gravitational field, he would have concluded that "in a gravitational field Euclidean geometry could not hold with respect to the arrangement of rigid bodies," as he put it in the letter to Levy. The results obtained could also have helped to shake him free from any lingering idea that "co-ordinates must have an immediate metrical meaning." As we have seen, this was one of the points implicit in his earliest printed discussion of the rotating disk. Together with the idea that the equivalence principle only holds infinitesimally, this may have reminded him of the use of Gaussian coordinates to describe the line element of curved surfaces, in which the idea that Euclidean geometry holds infinitesimally plays such a major role. We have evidence that Einstein, in his studies at the ETH, had become familiar with the Gaussian formula for the line element through the lectures on infinitesimal geometry given by Professor Geiser,[12] lectures which stood out in Einstein's memory fifty years later, when he described them as "true masterpieces of pedagogical art that later helped me very much in wrestling with general relativity" (Einstein 1955). At any rate, we have Einstein's words to assure us that

> I first had the decisive idea of the analogy of mathematical problems connected with the theory and Gauss's theory of surfaces in 1912 after my return to Zurich [which took place in August 1912] without knowing at that time Riemann's and Ricci's or Levi-Civita's work. (Einstein 1922b)[13]

Minkowski's four-dimensional formulation played an important role in Einstein's considerations at this point, as he tells us (Einstein 1955), and he soon saw that what was needed was a four-dimensional generalization of Gauss's two-dimensional surface theory, and that the flat metric tensor of Minkowski's formulation of special relativity had to be generalized to a nonflat metric. At this point, with the mathematical problem already well formulated, he approached Marcel Grossmann for help,[14] with the well-known results, which must have been largely attained by October 29, 1912, when he wrote Sommerfeld

> I am now occupying myself exclusively with the problem of gravitation and believe that, with the aid of a local mathematician who is a friend of mine [Grossmann], I'll now be able to master all difficulties. But one thing is certain, that in all my life I have never struggled as hard and that I have been infused with great respect for mathematics, the subtler parts of which, in my simple-mindedness, I had considered pure luxury up to now! Compared to this problem the original relativity theory [i.e., special relativity] is child's play (EA 13-082; Hermann 1968, p. 26)[15]

Three years were actually to elapse before Einstein was truly able to "master all the difficulties" of the general theory—but that is another story! Meanwhile, if my reconstruction is approximately accurate, we can see that Einstein's recapitulations of the rotating-disk story in the 1916 paper and the popular book (Einstein 1916, 1917), as well as in the Princeton lectures (Einstein 1922a) and the Einstein-Infeld book (Einstein and Infeld 1938), would not only represent a certain logical order of presentation of the material leading up to the recognition of the need for a nonflat space-time structure to describe the gravitational field, but would also represent a fairly accurate historical reconstruction of Einstein's own journey. Since Einstein notes in the preface of the 1916 book that he has attempted to present his ideas "on the whole, in the sequence and connection in which they actually originated" (Einstein 1917), this is perhaps not so surprising.[16] What is more surprising, if our reconstruction is more or less correct, is the lack of mention of the rotating-disk problem in any of his papers on gravitational theory from 1907 through 1915. We can hazard the guess that the reason for this is the amazingly brief period—some time between mid-July and mid-October 1912—when the problem played its role; and the fact that the next critical step of generalizing the result to the four-dimensional metric tensor was taken almost immediately afterwards. It was this latter great leap forward which provided the starting point for the Einstein-Grossmann investigations, and Einstein started his section of their paper (1913) with an account of this step. Thus the disk problem, having played its role of a "missing link," modestly stayed in the background; but that role was never forgotten by Einstein, as his later letters show.

NOTES

* © 1980 by John Stachel. Quotations from Einstein's published writings reproduced with the kind permission of the Trustees of the Einstein Estate.

[1] Bibliographies of the literature on the rotating disk may be found in Arzeliès 1966 and Grøn 1975. A recent paper by Grünbaum and Janis (1977) gives additional references.

[2] "Die Behandlung des gleichförmig rotierenden starren Körpers scheint mir von grosser Wichtigkeit wegen einer Ausdehnung des Relativitätsprinzips auf gleichförmig rotierende Systeme nach analogen Gedankengängen, wie ich sie im letzten § meiner in der *Zeitschr. f. Radioaktivit.* publizierten Abhandlung für gleichförmig beschleunigte Translation durchzuführen versucht habe." This letter (EA 21-377) is not included in the published volume of the Einstein-Sommerfeld correspondence (Hermann 1968).

[3] See Klein 1970, pp. 152–154, for an account of this discussion, with references.

[4] In a letter of April 12, 1911, to Ehrenfest (EA 9-316), Einstein suggested that Ehrenfest reply, noting that a short reply was needed to avoid confusion; however, he took on the job himself.

[5] In a reply to an attack by Abraham a few months later, Einstein notes "One sees already from the previously treated highly special case of the gravitation of masses at rest that the space-time coordinates lose their simple physical interpretation; and it still cannot be foreseen what form the general space-time transformation equations may take. I should like to ask all colleagues to have a try at this important problem!" (Einstein 1912c, pp. 1063–1064). Curiously enough, a paper using a definition of spatial distances on the disk equivalent to Einstein's and actually deriving the metric of the rotating disk had been published two years earlier by Theodor Kaluza (1910). The paper was to have been delivered by Kaluza at the 1910 Naturforscherversammlung in Königsberg, where he was then working; but he took sick and only the published version appeared under the title "Zur Relativitätstheorie," which gave no idea of its contents. I have found no evidence that Einstein—or anyone else in the long history of the rotating-disk problem for that matter—was aware of the existence of Kaluza's work.

[6] See, for example, pp. 1031–1032 of Einstein's summary survey of the state of general relativity theory (Einstein 1914). The fact that rotating reference frames did not satisfy the field equations of the Einstein-Grossmann theory, while they satisfy the generally covariant field equations of general relativity, played an important role in motivating Einstein's abandonment of the former in favor of the latter, when he discovered this in 1915, as has been pointed out in Earman and Glymour 1978b.

[7] See, for example, Einstein's letter to Mach (EA 17-454), thought to date from late 1913, published in Herneck 1966.

[8] No record exists in the Archive to indicate that a reply was actually sent.

⁹ I have consulted on this topic Illy 1977; Lanczos 1972; Guth 1970; and Earman and Glymour 1978a, 1978b. I am extremely grateful to Dr. Illy, of the Institute of Isotopes of the Hungarian Academy of Sciences, for making his work available to me.

¹⁰ For example, he states in a letter to Smoluchowski of March 24, 1912; "The simple schema of the equivalence of the four dimensions does not hold here in the way it does with Minkowski" (EA 20-597; Teske 1969).

¹¹ In another letter to Ehrenfest, undated but surely from a little earlier in 1912, Einstein speaks of his work on the static case as finished, and states that he is considering the "dynamic case" now, "again proceeding from the special to the more general case" (EA 9-321).

¹² Marcel Grossmann's notes of Geiser's lectures have been preserved and are now in the ETH Library HS 421:15. They may have been used by Einstein to study for his examinations (see Einstein 1955). They contain a discussion of curvilinear coordinates and the Gaussian line element for the plane (private communication from Professor Res Jost).

¹³ "Den entscheidenden Gedanken von der Analogie des mit der Theorie verbundenen mathematischen Problems mit der Gauss'schen Flächentheorie hatte ich allerdings erst 1912 nach meiner Rückkehr nach Zurich, ohne zunächst Riemanns und Riccis, sowie Levi-Civitas Forschungen zu kennen."

¹⁴ Both Einstein 1955 and Einstein 1922b state explicitly that the mathematical problems to be solved were formulated by Einstein before he approached Grossmann for help in their solution. On the other hand, neither of them makes any reference to Georg Pick or any mathematical help received from him in Prague in the formulation of the problem. Indeed, I have quoted the passage from the Preface to the Czech edition of Einstein 1917 to the effect that the "decisive idea" occurred to Einstein after his return to Zurich. This stands in contrast to Philip Frank's account in his biography of Einstein (Frank 1947). Since Frank is usually very careful in his account, and moreover was Einstein's successor in Prague and presumably had a chance to speak to Pick, the discrepancy is puzzling.

¹⁵ For all but the last sentence, I have used the translation in McCormmach 1976, p. xxviii.

¹⁶ Max Wertheimer, in his discussion of the origins of special relativity based on discussions with Einstein, states "In the course of one of his books he did report some steps in the process" (Wertheimer 1945, p. 168), and later makes it clear that the book was Einstein 1917. On page 174 he says: "For what now followed in Einstein's thinking we can fortunately report paragraphs from his own writing [then follows a reference to pages 14 ff. of the German edition of Einstein 1917]. He wrote them in the form of a discussion with the reader. What Einstein says here is similar to the way his thinking proceeded...." Wertheimer's account of the development of the special theory has been attacked recently as unreliable, but not on this point; see Miller 1975.

REFERENCES

Arzeliès, Henri (1966). *Relativistic Kinematics.* Oxford and New York: Pergamon.
Born, Max (1909). "Über die Dynamik des Elektrons in der Kinematik des Relativitätsprinzips." *Physikalische Zeitschrift* 10: 814–817.
——— (1910). "Über die Definition des starren Körpers in der Kinematik des Relativitätsprinzips." *Physikalische Zeitschrift* 11: 233–234.

Earman, John, and Glymour, Clark (1978a). "Lost in the Tensors: Einstein's Struggles with Covariance Principles 1912–1916." *Studies in History and Philosophy of Science* 9: 251–278.

——— (1978b). "Einstein and Hilbert: Two Months in the History of General Relativity." *Archive for History of Exact Sciences* 19: 291–308.

Ehrenfest, Paul (1909). "Gleichförmige Rotation starrer Körper und Relativitätstheorie." *Physikalische Zeitschrift*. 10: 918.

Einstein, Albert (1907). "Über das Relativitätsprinzip und die aus demselben gezogenen Folgerungen." *Jahrbuch der Radioaktivität und Elektronik* 4: 411–462.

——— (1911). "Zum Ehrenfestschen Paradoxon. Bemerkung zu V. Varičaks Aufsatz." *Physikalische Zeitschrift* 12: 509–510.

——— (1912a). "Lichtgeschwindigkeit und Statik des Gravitationsfeldes." *Annalen der Physik* 38: 355–369.

——— (1912b). "Zur Theorie des statischen Gravitationsfeldes." *Annalen der Physik* 38: 443–458.

——— (1912c). "Relativität und Gravitation. Erwiderung auf eine Bemerkung von M. Abraham." *Annalen der Physik* 38: 1059–1064.

——— (1914). "Die formale Grundlage der allgemeinen Relativitätstheorie." *Königlich Preussische Akademie der Wissenschaften* (Berlin). *Sitzungsberichte*: 1030–1085.

——— (1916). "Die Grundlage der allgemeinen Relativitätstheorie." *Annalen der Physik* 49: 769–822. Reprinted in translation in Lorentz, Hendrik Antoon, et al. *The Principle of Relativity: A Collection of Original Memoirs on the Special and General Theory of Relativity*. W. Perrett and G.B. Jeffery, trans. London: Methuen, 1923; reprint New York: Dover, 1952, pp. 109–164.

——— (1917). *Über die spezielle und die allgemeine Relativitätstheorie. (Gemeinverständlich)*. Braunschweig: Friedrich Vieweg & Sohn. The preface is dated December 1916. Page numbers are cited from the English translation *Relativity, the Special and General Theory: A Popular Exposition*. Robert W. Lawson, trans. London: Methuen, 1920; New York: Holt, 1921.

——— (1921). "A Brief Outline of the Development of the Theory of Relativity." *Nature* 106: 782–784.

——— (1922a). *The Meaning of Relativity: Four Lectures Delivered at Princeton University*. Princeton: Princeton University Press.

——— (1922b). "Vorwort des Autors zur Tschechischen Ausgabe." In *Einstein a Praha*. J. Bičak, ed. Prague: Prometheus, 1979, p. 42; written in 1922 for a Czech edition of Einstein 1917.

——— (1933). "Notes on the Origin of the General Theory of Relativity." In *Ideas and Opinions*, pp. 285–290. New York: Crown, 1954.

——— (1949). "Autobiographical Notes." In *Albert Einstein: Philosopher-Scientist*. Paul Arthur Schilpp, ed. Evanston, Illinois: The Library of Living Philosophers, pp. 1–95. A corrected text and translation has been issued as *Albert Einstein: Autobiographical Notes*. Paul Arthur Schilpp, ed. La Salle, Illinois and Chicago: Open Court, 1979. Page numbers are cited from the 1949 edition.

——— (1955). "Erinnerungen—Souvenirs." *Schweizerische Hochschulzeitung* 28 (*Sonderheft*): 145–148, 151–153. Reprinted as "Autobiographische Skizze." In *Helle Zeit-Dunkle Zeit*. Carl Seelig, ed. Zurich: Europa Verlag, 1956, pp. 9–17.

Einstein, Albert, and Grossmann, Marcel (1913). *Entwurf einer verallgemeinerten Relativitätstheorie und einer Theorie der Gravitation. I. Physikalischer Teil von Albert Einstein. II. Mathematischer Teil von Marcel Grossmann*. Leipzig and Berlin:

B.G. Teubner. Reprinted with added "Bemerkungen," *Zeitschrift für Mathematik und Physik* 62 (1914): 225–261.

Einstein, Albert, and Infeld, Leopold (1938). *The Evolution of Physics: The Growth of Ideas from Early Concepts to Relativity and Quanta.* New York: Simon & Schuster.

Frank, Philipp (1947). *Albert Einstein: His Life and Times.* George Rosen, trans. Shuichi Kusaka, ed. New York: Alfred Knopf.

Grøn, Ø. (1975). "Relativistic Description on a Rotating Disk." *American Journal of Physics* 43: 869–876.

Grünbaum, Adolf, and Janis, Allen I. (1977). "The Geometry of the Rotating Disk in the Special Theory of Relativity." *Synthese* 34: 281–299.

Guth, Eugene (1970). "Contribution to the History of Einstein's Geometry as a Branch of Physics." In *Relativity. Proceedings of the Relativity Conference in the Midwest, held at Cincinnati, Ohio, June 2–6, 1969.* Moshe Carmeli, Stuart I. Fickler, and Louis Witten, eds. New York and London: Plenum Press, pp. 161–207.

Hermann, Armin, ed. (1968). *Albert Einstein–Arnold Sommerfeld. Briefwechsel.* Basel and Stuttgart: Schwabe.

Herneck, Friedrich (1966). "Die Beziehungen zwischen Einstein und Mach, dokumentarisch dargestellt." *Wissenschaftliche Zeitschrift der Friedrich-Schiller-Universität Jena. Mathematisch-Naturwissenschaftliche Reihe* 15: 1–14.

Illy, József (1977). "The Birth of Einstein's Theory of Relativity." Typescript.

Kaluza, Theodor (1910). "Zur Relativitätstheorie." *Physikalische Zeitschrift* 11: 977–978.

Klein, Martin (1970). *Paul Ehrenfest: The Making of a Theoretical Physicist.* Vol. 1. Amsterdam: North-Holland.

Lanczos, Cornelius (1972). "Einstein's Path From Special to General Relativity." In *General Relativity: Papers in Honour of J.L. Synge.* L. O'Raifeartaigh, ed. Oxford: Clarendon Press, pp. 5–19.

Levy, Hyman (1939). *Modern Science: A Study of Physical Science in the World Today.* New York: Alfred Knopf.

McCormmach, Russell (1976). "Editor's Forward." *Historical Studies in the Physical Sciences* 7: xi–xxxv.

Miller, Arthur I. (1975). "Albert Einstein and Max Wertheimer: A Gestalt Psychologist's View of the Genesis of Special Relativity Theory." *History of Science* 13: 75–103.

Petzoldt, Joseph (1912). "Die Relativitätstheorie im erkenntnistheoretischen Zusammenhange des relativistischen Positivismus." *Deutsche Physikalische Gesellschaft. Verhandlungen* 14: 1055–1064.

Sommerfeld, Arnold (1909). "Diskussion." Following Born 1909. *Physikalische Zeitschrift* 10: 826–829.

Teske, Armin (1969). "Einstein und Smoluchowski. Zur Geschichte der Brownschen Bewegung." *Sudhoffs Archiv. Zeitschrift für Wissenschaftsgeschichte* 53: 292–305.

Thiele, Joachim (1971). "Briefe Albert Einsteins an Joseph Petzoldt." *NTM—Schriftenreihe für Geschichte der Naturwissenschaften, Technik und Medizin* 8: 70–74.

Varičak, V. (1911). "Zum Ehrenfestschen Paradoxon." *Physikalische Zeitschrift* 12: 169–170.

Wertheimer, Max (1945). *Productive Thinking.* New York: Harper Brothers.

Whittaker, Edmund (1949). *From Euclid to Eddington: A Study of Conceptions of the External World.* Cambridge: Cambridge University Press; reprint New York: Dover, 1958.

# Einstein's Search for General Covariance, 1912–1915*

JOHN STACHEL

## 1. Introduction

Einstein listed the stages of his search for a generally covariant theory of gravitation in a biographical note written in 1916:

- 1907  Basic idea for the general theory of relativity.
- 1912  Recognition of the non-Euclidean nature of the metric and of the physical determination of the latter by gravitation.
- 1915  Field equations of gravitation. Explanation of the perihelion motion of Mercury. (EA 11-196)[1]

There were thus three key moments in Einstein's development of the general theory of relativity:

1. Adoption of the principle of equivalence as the crucial element in a relativistic theory of gravitation (1907).
2. Recognition that the gravitational field must be characterized mathematically via a four-dimensional (pseudo-) Riemannian metric tensor (1912).
3. Discovery of the final form of the field equations relating the metric tensor to the sources of the gravitational field (1915).

Elsewhere, I have discussed this story in broad outlines (Stachel 1979a, 1979b), and have tried to contribute to a more detailed understanding of the decisive second step (Stachel 1980). Most discussions of the development of general relativity have focused on the third step, in particular on the puzzling question: Why—once the decisive step of representing the gravitational field by the metric tensor had been taken—did Einstein take so long to arrive at the final form of the field equations? (See Earman and Glymour 1978a, 1978b; Hoffmann 1972; Lanczos 1972; Mehra 1973; Vizgin and Smorodinskiî 1979.)

I shall suggest an answer that is (no doubt) still incomplete, but that differs from the existing accounts in several respects. In particular, I shall try to explain why it took Einstein over two years to return to general covariance after rejecting it in 1913:

1. without convicting Einstein or Marcel Grossmann of an elementary mathematical error in their original argument rejecting covariance (Einstein and Grossmann 1913) [see Section 1];
2. showing that Einstein's subsequent "hole" argument against general covariance involves a nontrivial *physical* question about the nature of space-time [see Sections 2 and 3]; and
3. giving a new interpretation of what seem to be some of Einstein's most "operationalistic" declarations in his 1916 exposition of the general theory (Einstein 1916) [see Section 5].

Circumstances outside of Einstein's purely scientific activities may have played some role in the length of the delay. Nineteen-fourteen was the year of his move to Berlin and his separation from his first wife and their children, who soon returned to Zurich. It was also, of course, the year of the outbreak of the first World War. It is clear that Einstein was affected deeply by these personal and political events. His turn towards political activity dates from his disillusionment with the chauvinistic reaction of most German intellectuals to the war.[2] It is also clear, however, that he found solace for his personal and political worries in his work. Leaving these issues aside, I shall confine myself here to the intellectual aspects of the problem.

In his "Autobiographical Notes," Einstein himself tried to answer the question of why it took so long from the time of his initial insight into the significance of the equivalence of inertial and gravitational mass until the general theory was completed:

This happened in 1908 [actually it was 1907, as he noted elsewhere—*JS*]. Why were another seven years required for the construction of the general theory of relativity? The main reason lies in the fact that it is not so easy to free oneself from the idea that co-ordinates must have an immediate metrical significance. (Einstein 1949, p. 67)

Einstein's comment is often interpreted as a reference to the need to recognize the role of the metric tensor in the mathematical representation of the gravitational field. But Einstein had taken this step by the end of 1912, as mentioned earlier. If this interpretation were complete, why the further delay?

I shall try to show that it was because Einstein still had to free himself from the idea that points of space-time (events) are physically individuated apart from their metrical properties; or, more accurately, from the idea that the points of a matter-free portion of a four-dimensional manifold are individuated as spatio-temporal events in some way that is independent of the properties they inherit, so to speak, from the presence of a metric tensor field on the manifold.

If the latter assertion is included in the interpretation of Einstein's statement about coordinates, then I believed that Einstein's answer is essentially correct. Be this as it may, my thesis is that the problem of the individuation of space-time points was the basic reason for Einstein's long delay in accepting general covariance *after* he had recognized the role of the metric tensor.

## 2. The Einstein-Grossmann Paper

In the fall of 1912, Einstein returned to Zurich from Prague to accept a professorship in physics at the Eidgenössische Technische Hochschule (ETH). He later recalled:

> I had the decisive idea of the analogy between the mathematical problems connected with the theory [of gravitation—*JS*] and the Gaussian theory of surfaces only after my return to Zurich in 1912, without then knowing Riemann's and Ricci's or Levi-Civita's investigations. I was made aware of these only by my friend Grossmann in Zurich, when I posed him the problem of finding generally covariant tensors whose components depend only on derivatives of the coefficients of the quadratic fundamental invariant [components of the metric tensor—*JS*]. (Einstein 1922; EA 1-014)[3]

Other reminiscences by Einstein,[4] as well as a study of Marcel Grossmann's ETH notebooks,[5] demonstrate that the lectures of Professor Carl Friedrich Geiser, which Einstein heard as a student at the ETH, had familiarized him with the Gaussian theory of two-dimensional surfaces. Thus, aside from probable anachronism in his use of the words, "generally covariant tensors," to describe what he wanted from Grossmann, there is no reason to doubt Einstein's account of what he knew, and what he wanted to learn, when in the fall of 1912 he turned for help to his old school friend, now professor of mathematics at the ETH. Grossmann "at once caught fire" ("fing sofort Feuer"), as Einstein recounts (Einstein 1955, p. 151), and promised to help Einstein with the mathematical problems involved in working out his gravitational theory, but on the condition that Grossmann bear no responsibility for the physics.

By late 1912, Einstein had thus achieved the fundamental breakthrough in his search for a relativistic theory of gravitation based upon the equivalence principle: adoption of a locally Minkowskian metric tensor on a four-dimensional manifold as a mathematical representation of the gravitational field. The preceding quotation indicates that Einstein was looking for generally covariant field equations from the start. The reason seems clear: for Einstein, a mathematical coordinate system is associated with a physical frame of reference; general covariance thus corresponds to the admissibility of arbitrary reference frames. If the entire theory of gravitation could be formulated in terms of generally covariant equations, this would constitute for Einstein an extension of the (special-relativistic) relativity principle from inertial frames of reference to arbitrary frames—a truly *general* relativity (note that the German word "allgemein," as in "allgemeine Relativität," has the connotation of "universal in application"). Any frame of reference could be interpreted as an inertial frame with a certain gravitational field present in it. Thus, the distinction between gravitational and inertial effects would not be absolute, but would depend upon the frame of reference adopted.[6]

With Grossmann's help, Einstein soon succeeded in mathematically representing the effect of gravitation on all other physical processes in a generally covariant form. But there still remained the question of the equations determining the gravitational field itself. The sources of this field are represented by the rank-two stress-energy tensor (SET), the relativistic generalization of the Newtonian mass density. So the gravitational field equations must be a set of differential equations that generalize the Newtonian equation for the gravitational potential (Poisson's equation) by relating the metric tensor and its first two derivatives to the stress-energy tensor. Grossmann found that the Ricci tensor is almost the only rank-two generally covariant tensor of second differential order that can be formed from the metric tensor and its derivatives.[7] However, Einstein and Grossmann became convinced that field equations setting the Ricci tensor equal to the SET do not lead to the correct Newtonian limit for weak static gravitational fields. They gave the reasons for this belief in a joint paper on gravitation, the "Entwurf" paper (Einstein and Grossmann 1913), that was *first* published early in 1913 as a separate booklet by B.G. Teubner Verlag, and then again early in 1914 in the *Zeitschrift für Mathematik und Physik*, with an addendum signed by Einstein. (The fact that this addendum is *not* in the original edition has caused some later confusion.)

The second, "mathematical," part of the paper was written by Grossmann. After defining the Riemann tensor Grossmann states:

> The outstanding significance of these concepts for the *differential geometry* of a manifold characterized by a line element makes it *a priori* probable that these general differential tensors would also be of significance for the problem of the differential equations of a gravitational field. Indeed one can give a covariant differential tensor of second rank and second order $G_{im}$ [the Ricci tensor, now more commonly denoted by $R_{im}$—JS], which could enter into those equations, namely:
> 
> $$G_{im} = \sum_{kl} \gamma_{kl}(ik, lm) = \sum_k \{ik, km\}.$$
> 
> [In notation more common today, $R_{im} = g^{kl} R_{ik\,lm} = R_i{}^k{}_{km}$, where $R_{ik\,lm}$ are the covariant components of the Riemann tensor—JS.] (Einstein and Grossmann 1913, p. 257)[8]

Then comes a fateful sentence:

> But it turns out that this tensor [the Ricci tensor—JS] does *not* reduce, in the special case of an infinitely weak static gravitational field, to the expression $\Delta\varphi$ [the Laplacian of $\varphi$—JS]. (Einstein and Grossmann 1913, p. 257)[9]

What did this mean? The usual interpretation (see, e.g., Vizgin and Smorodinskiĭ 1979) is that Grossmann failed to see how to extract a d'Alembertian operator from the linearized approximation to the Ricci tensor because of his inability to use suitably the general covariance of the Ricci tensor— for example, by imposing a suitable coordinate condition. Since neither Grossmann nor Einstein gave any further details of the argument, it is impossible to rule out such an interpretation, and just as impossible to prove it. However, there is a simpler interpretation of the sentence that is entirely consistent with the rest of the Einstein–Grossmann paper, and that leaves Grossmann innocent of such a gross mathematical error. This interpretation does depend on the existence of a physical misconception on the part of Einstein. However, its existence is not based on conjecture. The misconception is clearly stated in part I of the paper, written by Einstein, and it explains why Grossmann arrived at a result that led Einstein to reject the Ricci tensor. The argument hinges on the word "static" in the sentence by Grossmann quoted above. At the time he was working on the Einstein–Grossmann paper, Einstein thought that, even before having the correct field equations, he knew the correct form of the metric tensor for a static gravitational field based on his earlier work on the subject (Einstein 1912a, 1912b). On page 229 of part I he wrote down the special-relativistic Minkowski metric, with $g_{44} = c^2$ ($c =$ the velocity of light), $g_{11} = g_{22} = g_{33} = -1$, and all other components zero; and he immediately added: "The same type of degeneration is shown by static gravitational fields of the type previously considered, except that for the latter $g_{44} = c^2$ is a function of $x_1, x_2, x_3$ [the spatial coordinates—$JS$]" (Einstein and Grossmann 1913, p. 229).[10] (The words "previously considered" refer to pages 227–228, where Einstein summarized his earlier work on static gravitational fields, in which the gravitational field is represented by a time-independent but spatially variable speed of light.)

We may well imagine Grossmann substituting this *Ansatz* for the metric tensor of a static gravitational field into the formula for the Ricci tensor, neglecting all but linear terms, since he was dealing with an "infinitely weak" field. If that is what he did, he found (in linearized approximation):[11]

$$R_{44} = \tfrac{1}{2}\Delta g_{44}, \qquad R_{4n} = 0$$

$$R_{mn} = \frac{1}{2}\frac{\partial 2g_{44}}{\partial x_m \partial x_n} \qquad (m, n = 1, 2, 3).$$

At first sight this result seems promising, since $R_{44}$ does reduce to the Laplacian of $g_{44}$ (as it does for *any* static metric, without any coordinate conditions). A moment's thought shows, however, that the result is really disastrous. For if the Ricci tensor vanishes (as it must outside of the sources of the gravitational field), it follows from $R_{mn} = 0$ that $g_{44}$ will depend at most *linearly* on the coordinates. Thus, it cannot possibly represent the gravitational potential of any (finite) distribution of matter, static or otherwise. If Einstein's *Ansatz* for the static metric tensor had been correct, the vanishing of the Ricci tensor *could not* provide gravitational field equations outside of matter. Einstein's response to this dilemma (if I have correctly described it) was to keep his *Ansatz* for static metrics, and drop the Ricci tensor field equations.

The root of the trouble is Einstein's belief that the spatial metric of a static gravitational field (in coordinates adapted to the static nature of the field — in coordinates adapted to the static Killing vector, as we say now) must be flat. This belief, which is implicit in his geometrical reinterpretation of his 1912 papers on static gravitational fields (Einstein 1912a, 1912b), may well have been reinforced by the circumstance that it was the move beyond static fields to consideration of a particular stationary gravitational field—the gravitational field on a rotating disk—that first led Einstein to introduce a nonflat spatial geometry (see Stachel 1980).

He apparently remained convinced that static gravitational fields have spatially flat cross sections—at least to "first order," i.e., in a linearized approximation (see the following discussion)—until late 1915, when a study of approximate solutions to a set of generally covariant field equations finally showed that this is not the case. Indeed, it is this spatial nonflatness of the static, spherically symmetric solution that yields the additional corrections to the precession of the perihelion of Mercury, whose agreement with the observed anomalous precession helped to demonstrate the validity of the covariant equations (see Einstein 1915c). In a letter to Besso, dated December 21, 1915, Einstein remarked:

> Most gratifying [about the new theory—*JS*] is the agreement of the perihelion motion [of Mercury—*JS*] and the general covariance; most curious, however, the circumstance that Newton's theory of the field, even for terms of the 1st order, is incorrect for the field (occurrence of [nonflat—*JS*] $g_{11}$-$g_{33}$). Only the circumstance that $g_{11}$-$g_{33}$ do not occur in the first approximation of the eq[uations] of mot[ion] of a point[-particle] causes the simplicity of Newton's theory. (Speziali 1972, p. 61)[12]

Interestingly enough, in connection with the attempt to reconstruct the Einstein–Grossmann argument of 1913, this is not the first letter in which Einstein told Besso about this result. In his previous letter of December 10, 1915, Einstein told Besso about the success of his generally covariant equations, and immediately added:

> This time the most obvious thing was the correct one; but Grossmann and I believed that the conservation equations were not fulfilled, and that Newton's law did not come out in first approximation. You will be astonished by the occurrence of $g_{11}...g_{33}$. (Speziali 1972, p. 60)[13]

It is not difficult to surmise that Einstein was indicating his own astonishment by this comment. It is also not hard to see more than a fortuitous conjunction in the juxtaposition of the remark about the Einstein–Grossmann errors of 1913 and the result about the nonflat values for $g_{11}...g_{33}$. It is plausible to conjecture that Einstein thought: "If only I had realized that the static spatial metric was nonflat, I could have avoided the retreat from general covariance in 1913." I shall now return to this topic.

Faced with the check to their hopes for a generally covariant theory, Einstein and Grossmann still left open the possibility of finding, ultimately, a generally covariant set of gravitational field equations. Immediately after his previously quoted remark Grossmann continued:

> Such a connection [between the general theory of tensors and the differential equations for the gravitational field—*JS*] would have to be present, insofar as the gravitational equations were to permit *arbitrary* substitutions [of coordinates—*JS*]; but in this case finding differential equations of the *second* order seems to be excluded. On the other hand, if it were determined that the gravitational equations only allow a certain group of transformations [i.e., a more limited one—*JS*], then

it would be understandable if one could not manage with the differential tensors offered by the general theory. As explained in the Physical Part [by Einstein—*JS*], we are not able to take a position on these questions. (Einstein and Grossmann 1913, p. 257)[14]

In part I of the paper, after noting that it might be possible to retain general covariance without using the Ricci tensor by going to differential equations of order higher than second, Einstein stated:

The attempt to discuss such possibilities would be premature in the present state of our knowledge of the physical properties of the gravitational field. Therefore, the restriction to second order [field equations—*JS*] is dictated to us and we must therefore renounce attempting to set up gravitational equations that are covariant under arbitrary transformations [of coordinates—*JS*]. Besides, it is to be emphasized that we have no sort of justification for general covariance of the gravitational equations. (Einstein and Grossmann 1913, p. 234)[15]

After then finding what he considered satisfactory equations for the gravitational field, Einstein returned to this equation:

We have only been able to prove for the gravitational equations that they are covariant with respect to arbitrary *linear* transformations; however, we do not know if there is a general group of transformations, with respect to which the equations are covariant. The question of the existence of such a group ... is the most important one connected with the considerations given here. (Einstein and Grossmann 1913, p. 240)[16]

He again states that,

we are not justified in the present state of the theory in demanding the covariance of physical equations with respect to arbitrary substitutions [of coordinates—*JS*].
   On the other hand, we have seen that an energy-momentum balance equation for material processes can be set up ... that allows arbitrary transformations. Therefore, it seems rather natural if we assume that all physical systems of equations with the exception of the gravitational equations are to be so formulated that they are covariant with respect to arbitrary substitutions. This exceptional position of the gravitational equations with respect to all other systems is connected, in my opinion, with their having to contain only the first two derivatives of the components of the fundamental [metric—*JS*] tensor. (Einstein and Grossmann 1913, p. 240)[17]

How did Einstein find nongenerally covariant gravitational field equations? He based his search on the criteria that the field equations should:

1. generalize Poisson's equation for the Newtonian gravitational potential (obey a correspondence principle, in modern terms);
2. be invariant at least under linear transformations (manifest no *less* relativity than in the special theory);
3. include a gravitational stress-energy complex, built from the metric tensor and its first derivatives (by analogy with Maxwell's electromagnetic theory), that is a tensor under linear transformations and enters the field equations as a source term in the same way as does the stress-energy tensor of ordinary

(nongravitational) matter or fields. (See Einstein and Grossmann 1913, pp. 233–234, 239 for these criteria.)

The third requirement serves to ensure that conservation laws for the *total* gravitational and nongravitational energy and momentum follow from the gravitational field equations. Einstein was able to derive a set of field equations satisfying these three requirements in part I of the Einstein–Grossmann paper.

He also gave a sketch of how Newton's law of gravitation follows from the linear approximation to the Einstein–Grossmann field equations for the static case, a result that he later demonstrated in detail in his 1913 lecture to the Vienna meeting of the Gesellschaft Deutscher Naturforscher and Ärzte (Einstein 1913a). As mentioned earlier, he explicitly noted that the spatial metric remains flat in this approximation: "One recognizes that coordinate lengths are at the same time natural lengths ($dt = 0$). Measuring rods thus undergo no deformation due to the 'Newtonian' gravitational field" (Einstein 1913a, p. 1260).[18]

At the time he finished the paper with Grossmann, Einstein seems to have felt that its negative results regarding general covariance could be only provisional. For example, he wrote to Paul Ehrenfest on May 28, 1913:

> I am now inwardly convinced that I have found that which is correct [the solution to the gravitational problem—*JS*], and, at the same time, that a murmur of disappointment will, of course, go through the ranks of our colleagues when the work appears, which will be in a few weeks.... The conviction to which I have slowly struggled through is that *there are no preferred coordinate systems of any kind*. However, I have only partially succeeded, even formally, in reaching this standpoint. (EA 9-340)[19]

However, by November 1913, Einstein had developed a simple "meta-argument" against all generally covariant field equations. He wrote to Ludwig Hopf on November 2:

> I am now very content with the gravitation theory. The fact that the gravitational equations are not generally covariant, which a short time ago still disturbed me so much, has proved to be unavoidable; it is easily proved that a theory with generally covariant equations cannot exist if one demands that the field be mathematically completely determined by matter. (EA 13-290)[20]

The proof in question is the notorious "hole" ("Loch") argument, alluded to in the introduction to this paper, and discussed in detail in the next section. Einstein now wrote to Ehrenfest, in an undated letter of later 1913:

> The gravitational affair has been clarified to my *complete satisfaction* (the circumstance, namely, that the equations for the gr[avitational] field are only covariant with respect to *linear* transformations). It can be proven, namely, that *generally covariant* equations that determine the field *completely* from the matter tensor cannot exist at all. What can be prettier than this, that the necessary specialization [of the coordinate system—*JS*] follows from the conservation laws? The conservation laws now determine, out of all surfaces, those which are to be chosen as preferred

coordinate surfaces. We can designate these preferred surfaces as planes, since the only remaining allowed substitutions are *linear* ones. (EA 9-342)[21]

He wrote to Ernst Mach in a similar vein, in late 1913 or early 1914:

> For me it is absurd to attribute physical properties to "space." The totality of masses generates a $g_{\mu\nu}$-field (gravitational field), which in turn regulates the course of all processes, including the propagation of light rays and the behavior of measuring rods and clocks. Events are first referred to four *quite arbitrary* space-time variables. These must then, if the conservation laws of energy and momentum are to be satisfied, be specialized in such a way that only (completely) *linear* substitutions lead from one justified reference system to another. The reference system is, so to speak, adapted to the existing world with the help of the law of conservation of energy, and loses its nebulous *a priori* existence. (Herneck 1963; EA 17-454)[22]

At this point, then, Einstein believed that the conservation laws for the total energy and momentum, including the gravitational contribution, are valid only in a preferred set of coordinate systems, related to each other by linear transformations. He soon gave up the restriction to linearly related coordinate systems, as we shall see; but the "hole" argument still prevented him from accepting general covariance. Let us turn to this argument.

## 3. The "Hole" Argument, or Covariance Lost

The argument was first published in the "Bemerkungen" ("Remarks"), which form the previously mentioned addendum to the "Entwurf" paper (Einstein and Grossmann 1913), signed by Einstein alone and not published in the original (separatum) printing of the paper. We can date the origin of the argument fairly closely. The letter to Hopf quoted earlier shows that Einstein had it by November 1913. In a lecture given to the Annual Meeting of the Swiss Naturforschende Gesellschaft on September 9, 1913, Einstein stated: "It has been possible to demonstrate by a general argument that equations that completely determine the gravitational field cannot be generally covariant with respect to arbitrary substitutions" (Einstein 1913b, p. 289).[23] And in his Vienna lecture of September 23, 1913, mentioned in the previous section, he stated: "In the last few days I have found the proof that such a generally covariant solution cannot exist at all" (Einstein 1913a, p. 1257, footnote 2).[24] Since it is always possible that both of these remarks were added when the texts were printed, all we can assert is that the "hole" argument probably first occurred to Einstein in September 1913, but surely by November of that year.

The first version of the argument runs as follows:

> Let there be a portion $L$ [for "Loch"—*JS*] in our four-dimensional manifold in which a "material process" is not occurring, in which therefore the $\Theta_{\mu\nu}$ [the components of the nongravitational stress-energy tensor—*JS*] vanish. In accord with our assumptions, the $\gamma_{\mu\nu}$ [components of the metric tensor—*JS*] are completely deter-

mined everywhere, therefore also inside $L$, by the $\Theta_{\mu\nu}$ given outside $L$. We now imagine that, instead of the original coordinates $x_\nu$, new coordinates $x'_\nu$ are introduced of the following type. Outside of $L$ let $x_\nu = x'_\nu$ everywhere; inside $L$, however, let $x_\nu \neq x'_\nu$ for at least a part of $L$ and for at least one index $\nu$. It is clear that by means of such a substitution it can be arranged that, at least for a part of $L$, $\gamma'_{\mu\nu} \neq \gamma_{\mu\nu}$. On the other hand, $\Theta'_{\mu\nu} = \Theta_{\mu\nu}$ everywhere; outside of $L$ because, in this region, $x'_\nu = x_\nu$; inside of $L$, however, because, for this region, $\Theta_{\mu\nu} = 0 = \Theta'_{\mu\nu}$. It follows from this that, in the case considered, if all [coordinate—JS] substitutions are allowed as justified, more than one system of $\gamma_{\mu\nu}$ belongs to the given system of $\Theta_{\mu\nu}$.

If therefore—as is the case in this paper—one maintains the requirement that the $\gamma_{\mu\nu}$ should be completely determined by the $\Theta_{\mu\nu}$, then one is forced to restrict the choice of coordinate system. (Einstein and Grossmann 1913, p. 260)[25]

Substantially the same argument is repeated in two subsequent papers, Einstein 1914a and Einstein and Grossmann 1914, both dating from early 1914. In a footnote to his paper, however, Einstein added a significant explanatory comment. To the equations $g'_{\mu\nu} = g_{\mu\nu}$ for the two metric tensors, which he notes, "surely do not all hold" ("sicherlich nicht alle erfüllt sein"), he appends a note: "The equations are to be thus understood, that on the left sides each of the independent variables $x'_\nu$ are assigned the same numerical values as the variables $x_\nu$ on the right sides" (Einstein 1914a, p. 178, column 2, footnote 1).[26]

A second version of the argument is presented in his definitive exposition of the Einstein–Grossmann theory, presented to the Berlin Academy in October, 1914 (Einstein 1914b). This version is much more explicit than the earlier one about the coordinate system used at each point in the argument. It may represent a significant evolution in Einstein's thinking about the "hole" argument, as Earman and Glymour (1978a) argue. At any rate, I shall base my discussion of the argument in the next section on this final version. After developing generally covariant formulations for the behavior of physical systems in the gravitational field, Einstein turned to the laws of the gravitational field itself. He wrote:

The epistemologically satisfying thing about the previously developed theory lies in the fact that it fulfills the relativity principle in its most far reaching sense. Formally considered, this rests on the fact that the systems of equations are *generally* covariant, i.e., with respect to arbitrary substitutions of the $x_\nu$.

Accordingly, this suggests the requirement that the differential laws for the $g_{\mu\nu}$ must also be *generally* covariant. However, we shall show that we must restrict this requirement, if we want the law of causality to be completely valid. Namely, we prove that the laws that determine the course of events for the gravitational field cannot be *generally* covariant.

§12. *Proof of the Necessity of a Restriction of the Choice of Coordinates.*

We consider a finite portion $\Sigma$ of the continuum, in which no material process occurs. What happens physically in $\Sigma$ is then completely determined if the quantities $g_{\mu\nu}$ are given as functions of the coordinates $x_\nu$ with respect to a coordinate system

$K$ used for the description. The totality of these functions will be symbolically designated by $G(x)$.

Let a new coordinate system $K'$ be introduced that, outside of $\Sigma$, coincides with $K$, within $\Sigma$, however, diverges from $K$ in such a way that the $g'_{\mu\nu}$ referred to $K'$, like the $g_{\mu\nu}$ (and their derivatives), are everywhere continuous. The totality of the $g'_{\mu\nu}$ will be symbolically designated by $G'(x')$. $G'(x')$ and $G(x)$ describe the same gravitational field. If we replace the coordinates $x'_\nu$ by the coordinates $x_\nu$ in the functions $g'_{\mu\nu}$, i.e., if we form $G'(x)$, then $G'(x)$ also describes a gravitational field with respect to $K$, which however does not correspond to the actual (i.e., originally given) gravitational field.

Now if we assume that the differential equations of the gravitational field are generally covariant, then they are satisfied by $G'(x')$ (with respect to $K'$) if they are satisfied by $G(x)$ with respect to $K$. They are then also satisfied with respect to $K$ by $G'(x)$. Relative to $K$ there then exist the solutions $G(x)$ and $G'(x)$, which are different from each other, in spite of the fact that at the boundary of the region both solutions coincide; i.e., *what happens in the gravitational field cannot be uniquely determined by generally covariant differential equations for the gravitational field.*

Therefore, if we demand that the course of events in the gravitational field be completely determined by the laws to be set up, then we are forced so to restrict the choice of coordinate systems that it is impossible, without abandonment of the restrictive conditions, to introduce a new coordinate system $K'$ of the type previously characterized. The continuation of the coordinate system into the interior of a region $\Sigma$ must not be arbitrary. (Einstein 1914b, pp. 1066–1067)[27]

## 4. The Meaning of the Argument

In trying to understand these passages, we must remember that the modern terminology of differential geometry, which expresses geometrical concepts in coordinate-free language and carefully distinguishes between point and coordinate transformations, for example, was not available to Einstein (I do not know when these distinctions were first clearly enunciated by mathematicians working in differential geometry, but I have the impression it was much later).[28]

On the other hand, Einstein had an astounding geometrical intuition that usually led him, using coordinate language, to conclusions that make excellent geometrical sense. In trying to interpret these passages, then, I have proceeded on the assumption that he was trying to express something nontrivial. I also presume that, if the "hole" argument *had* been based on a trivial mathematical error, Marcel Grossmann—with whom he was still in close contact, as their second joint paper (Einstein and Grossmann 1914) proves—would have set him right. Yet, as noted in the previous section, this paper repeats the hole argument.

Thus, I am led to reject one common interpretation of the "hole" argument, which assumes that Einstein did not realize that the transformation of the components of the metric tensor under a coordinate transformation results in a redescription of the *same* gravitational field in a *different* coordinate system.

This *might* hold as a critique of the first (1913) version of the argument (Einstein and Grossmann 1913, p. 260; see Section 2 of this paper).[29] But the second (1914) version *explicitly affirms* that "$G'(x')$ and $G(x)$ describe the same gravitational field" (Einstein 1914b, pp. 1066–1067; see Section 2 of this paper). So Einstein's point is *not* that $G'(x')$ and $G(x)$ are different. His point is that $G'(x)$ and $G(x)$—same $x$—are different: "$G'(x)$ also describes a gravitational field *with respect to K* which however *does not correspond* to the actual (i.e., originally given) gravitational field" [emphasis added—*JS*].

What he is here asserting, I believe, is that (in more modern language) in a region of purely gravitational field the *dragged-along field* under a point transformation results in a gravitational field physically distinct from the original one.[30] And I further assert that the physical equivalence or nonequivalence of these two fields is *not* a purely mathematical question. Or, more accurately, it does not *become* one unless and until one introduces the additional, *nonmathematical* assumption or postulate that, in regions where no matter is present, the points of a manifold are physically differentiated only by the properties that they inherit from the metric field.

Since the explanation of this assertion does not really depend on the fact that it involves the gravitational field, nor on the fact that the latter is represented mathematically by a metric tensor field, I shall discuss the question in a wider context. I shall carry out much of the discussion in coordinate-free language, but a reader not used to dispensing with coordinates need merely interpret all statements as references to the components of the objects in question with respect to a single coordinate system.

Consider a class of theories in which space-time is represented mathematically by a four-dimensional manifold, the points of which are associated with physical events (just *how* they represent such events will be the major question). Physical processes are represented by some set of geometrical object fields on the manifold, whose exact transformation properties under coordinate transformations need not concern us.[31] Now consider a point transformation that takes each point of the manifold (or some region of it) into another point (the transformation is assumed to be invertible). We may drag the coordinates along with the points, thus defining a new coordinate system, called the dragged-along coordinate system. We may also drag along a geometrical object field with such a point transformation; this produces a new geometrical object field, the components of which have values at each dragged-along point in the dragged-along coordinate system that are numerically equal to those of the original field at the original (undragged) point in the original coordinate system. This new field is called the dragged-along field (with respect to the point transformation). Since there is a (nondenumerable) infinity of possible point transformations on a manifold, there is a (nondenumerable) infinity of such dragged-along fields. Although all are examples of the same type of geometrical object, each is a geometrical object field mathematically distinct from the others. If there are two geometrical object fields on a manifold, one may be dragged along with a point transformation, while the other is left undragged, or left behind, as we shall say. Then we may refer to one field as dragged along relative to the other. As far as the relative effect goes, it does not really matter which field is regarded as dragged along and which as left behind. Clearly, if more than two fields are given, additional possibilities exist for dragging some and leaving others behind.

Suppose a set of geometrical object fields in some region of the manifold completely describes the physical situation in that region. Now drag along *all* of these fields with some point transformation restricted to this region. Should we regard the original and dragged-along fields as representing *distinct* physical situations in this region of space-time? The answer depends on whether we regard the points of the manifold as *physically* individuated apart from the properties they inherit from the presence of the geometrical object fields. If these fields really include *all* the information about the physical situation in the region, then such properties, which would enable us to distinguish physically one point of the region from another, cannot exist. On the other hand, if the points of the region *are* physically individuated in some independent fashion, the assertion that the original fields exhaust the physical properties of the region is incorrect. In the latter case, the point transformation dragged along all the fields, *but* left behind the distinguishing properties—and such a "relative dragging" is certainly of physical significance.

It may be the case that it is not the points directly, but other structures on the manifold that are regarded as labeled independently of all other physical processes. We shall refer to such structures as individuating structures or fields. For example, Newton's assertion of an absolute space amounts to the postulation of a preferred fibration of space-time. His assertion of an absolute time amounts to postulation of a preferred foliation of space-time. Together, they give an "absolute" individuation to the points of space-time (they provide much more of course). Relative dragging of other fields with respect to such individuating structures does have physical significance. But even the most confirmed Newtonian will have to admit that he or she does not wish to distinguish *physically* between mathematical situations in which *all* of the fields—the individuating fields as well as the other fields—are dragged along by the same point transformation.

One might be tempted to assert that coordinate-labeling fields provide such individuating fields for the points of the manifold. And so they do for *mathematical* properties, such as the differential-topological structure of the manifold. But the point is (no pun intended!) that no mathematical coordinate system is *physically* distinguished per se; and without such a distinction there is no justification for physically identifying the points of a manifold—which are mathematically homogeneous, anyway—as physical events in space-time. Thus, the mathematician will always correctly regard the original and the dragged-along fields as distinct from each other. But the physicist (or indeed anyone applying differentiable manifolds) must examine this question in a different light; and the answer will depend upon the means available for making extra-mathematical distinctions between the points of the manifold—or, more concisely upon the presence of individuating structures.

A look at the question of how the abstract or generic mathematical concept of a differentiable manifold can be reached by starting from concrete examples sheds a little more light on the problem of the (nonmathematical) individuation of the points of such a manifold. Riemann, in his *Habilitationsschrift* (Riemann 1854), introduced the concept of an "$n$-fold extended magnitude," which he also refers to as a "manifold" ("Mannigfaltigkeit"). He cited two examples of a "continuous manifold" that arise from the experiences of everyday life: "the positions of sense-objects and the colors [are—*JS*] indeed the only simple concepts whose modes of determination form a multiply extended manifold" (Riemann 1854, p. 3). Of course, Riemann was not referring to "the manifold of physical colors [that] has infinitely many dimensions," as Weyl put it, but to the three-dimensional "space of the perceptively given *color-qualities* of colored light" (Weyl 1927, pp. 70–71).

Russell pointed out an important difference between these two examples: "There is no inherent quality in a single point [of physical space—*JS*], as there is in a single color, by which it can be quantitatively distinguished from another" (Russell 1897, pp. 66–67). Let us explore this difference. Each element within the set of colors is individuated by its qualities as a color. Three quantities can be attached to the colors in such a way that each color is uniquely characterized by a particular set of values of the three quantities. These values are taken to vary continuously, so that we may think of the colors as forming a space having three dimensions.[32] One may analyze the properties of this color space, introduce a metric on it, and so forth.[33] Color space is often referred to as the color manifold, following Riemann. As colors, the elements of this manifold are obviously not homogeneous. Only if we abstract its quality as a color from each of these elements, while retaining the continuity relations between these elements, are we are left with a space of homogeneous elements, each of which may be identified with a point of a generic three-dimensional mathematical manifold.[34] Conversely, if we start from such a generic manifold and wish to turn it into a color manifold, we must introduce what I have previously called individuating structures or fields in order to restore the qualitative (and quantitative) distinctions between the elements—that is, just the qualities that make them colors and nothing else.

Now let us turn to physical space and time. Any sufficiently rich theoretical model of the world around us must include enough detail to individuate each event taking place in the model. Suppose the model can be represented mathematically as a four-dimensional differentiable manifold with certain additional structures over it.[35] In the full model, the points of the manifold represent events. It is obvious that if *all* the additional structures over the manifold are abstracted, the points of the manifold lose their character as events. How many of the additional structures can be abstracted without the points of the manifold entirely losing their individual spatio-temporal qualities? The answer to this question depends on the theoretical model. If one starts from Newton's model of absolute space and time, for example, all other physical processes may be abstracted, and one will still be left with physically individuated points of space and time. (As noted previously, absolute space and time constitute individuating structures for Newton's model of space-time.) One would have to abstract absolute space and time while retaining the continuity and differentiability properties they induce, in order to get a generic four-dimensional manifold, the elements of which are completely homogeneous.

Suppose one takes a Galilean or special-relativistic model; that is, a model that postulates the existence of a privileged group of nonaccelerated inertial frames of reference, each in uniform relative motion with respect to the others. One may still individuate the points of space and time, for example by introducing a framework of unaccelerated rigid rods (or better, rods in rigid, unaccelerated motion), together with a set of clocks distributed at each point of the framework and then suitably synchronized.[36] Such a materialization of the abstract concept of inertial frame forms a significant part of Einstein's conception of the special theory of relativity. In 1907 he wrote:

> In order to be able to describe any physical process, we must be in a position to evaluate spatially and temporally the changes that take place at the individual points of space.
>
> For the spatial evaluation of a process of infinitely short duration taking place in an element of space (point-event) we use a Cartesian coordinate system, i.e., three perpendicular rigid rods, rigidly bound together, as well as a rigid unit rod. [In a footnote at this point, Einstein adds: "Instead of 'rigid' bodies, one ... can just as well speak of solid bodies not subject to deforming forces"—*JS*] ... For the temporal evaluation of a point-event we use a clock, which is at rest relative to the coordinate system and in the immediate neighborhood of which the point-event takes place....
>
> We now assume that *the clocks can be set so that the speed of propagation of any light signal in vacuum—measured with the aid of these clocks—is always equal to a universal constant c*, supposing that the coordinate system is not accelerated. (Einstein 1907, p. 415)[37]

Thus, in abstracting physical structures from a Galilean or special-relativistic model of the world, one may begin by abstracting all dynamical processes and leaving this framework of rods and clocks, which constitutes an inertial reference system in Einstein's sense. The points of space and time are still physically individuated by this reference system. One would have to abstract the inertial reference system, while preserving the continuity and differentiability properties that are associated with such a system, in order to arrive at a generic four-dimensional differentiable manifold. Conversely, if we start with such a generic manifold and want to arrive at a Galilean or special-relativistic model of the world, there are now two steps involved. We first introduce a frame of reference (or some equivalent set of individuating structures or fields) in such a way that the individuation of each point of space-time is fixed. Then, all the dynamical physical processes in the model are introduced.

Einstein was well aware of the artificiality of the distinction thus set up between rods and clocks on the one hand, and all other physical processes on the other; yet he emphasized the need for such a division, given the current state of physics:

> It is striking that the theory [of special relativity—*JS*] (except for the four-dimensional space) introduces two kinds of physical things, i.e., (1) measuring rods and clocks, (2) all other things, e.g., the electromagnetic field, the material point, etc. This, in a certain sense, is inconsistent; strictly speaking, measuring rods and clocks should emerge as solutions of the basic equations ... not, as it were, as theoretically self-sufficient entities. The procedure justifies itself, however, because it was clear from the very beginning that the postulates of the theory are not strong enough to deduce from them equations for physical events sufficiently complete and sufficiently free from arbitrariness in order to base upon such a foundation a theory of measuring rods and clocks. If one did not wish to forego a physical interpretation of the coordinates in general (something that, in itself, would be possible), it was better to permit such inconsistency—with the obligation, however, of eliminating it at a later stage of the theory. (Einstein 1949, pp. 59, 61)[38]

So far Russell's distinction between the color manifold and physical space—which we have replaced by space-time—appears valid. Now consider a general-relativistic model. In this case, there are no nondynamical physical processes in the theory. In particular, for regions from which matter or nongravitational fields are absent, the only additional structure on the manifold is the metric tensor field. This case thus resembles the color manifold case. If we abstract the metric field while retaining the continuity and differentiability properties that it induces, we are left with a generic four-dimensional manifold that has no spatial and temporal properties associated with its points. The metric tensor field plays the roles of both the dynamical and the individuating fields in the previous two examples.[39] One may still introduce a physical frame of reference, but only when the metric tensor field is present. More generally, spatio-temporal individuation of the points of the manifold in a general-relativistic model is possible only after the specification of a particular metric field, i.e., only after the field equations of the theory (which constitute its dynamical problem) have been solved. Once this is done, the points of the manifold-with-metric become full-fledged physical events, endowed with gravitational as well as spatio-temporal properties. There is no half-way house, so to speak, as there is in absolute, Galilean, or special-relativistic theories, where some independent individuating structure gives the points of the manifold spatio-temporal properties (a certain point in space at a certain moment in time), before a particular solution to some set of dynamical equations (Maxwell's equations, for example) imparts additional properties to these points, making them into full-fledged physical events (the point in space where, and the moment in time when, the electric and magnetic field vectors assume certain values).

One way to obtain this individuation of the points of the manifold once the metric tensor is present is to imagine the space-time to be traversed by a number of trajectories of test particles, that is, particles of mass so small that their effect on the gravitational field in which they move may be neglected. Then the intersections of the trajectories of two such particles will constitute an event that serves to individuate physically a point of the manifold. As we shall see in the final section, this is the method that Einstein adopted at the end of 1915, once he had rejected the "hole" argument and adopted generally covariant field equations. But now let us return to the general discussion of the argument.

Suppose that the geometrical object fields are subject to generally covariant field equations involving no additional fields. That is, no additional fields, given a priori, occur in the formulation of these equations. It follows that, if a particular solution to these field equations exists in some region of the manifold, then all the dragged-along field generated from this solution by point transformations in the region are also solutions to the field equations. Are these to be regarded as physically distinct solutions? Clearly, we are again faced with our previous question, and must give the same answer: Only if there exist some individuating fields for the points of the region. If such fields exist, then all of the solutions to the generally covariant field equations generated from the initial one by dragging must be regarded as physically distinct, and the following version of the "hole" argument is valid.

Suppose we try to fix a unique solution to the generally covariant field equations inside the region in question. Let us specify the sources outside the region, the values of the fields outside and on the boundary of the region, together with any finite number of normal derivatives on the boundary. We still cannot specify a unique solution inside the region, because there will always be point transformations inside the region that reduce to the identity on the boundary, together with any finite number of derivatives of the transformation, but which do not reduce to the identity inside the region. These point transformations give rise to a family of dragged-along solution fields that are identical to each other to the required differential order on the boundary, but differ inside the region.

Thus, if one requires field equations for which well-posed initial or boundary value problems can be set, with or without sources, there is only one way to escape the "hole" argument: to *identify* the class of mathematically distinct dragged-along solutions with *one* physical solution. This identification does not guarantee that a given set of generally covariant field equations has a well-posed initial or boundary value problem; but at least it allows for the possibility. As we have seen, however, the only warrant for such a physical identification of mathematically distinct fields is the denial of any a priori physical labeling of the points of the manifold. The points must inherit all of their distinguishing physical properties and relations from the set of fields that is subject to the generally covariant equations. No individuating fields can be allowed in such a theory.

We can now apply the results of this discussion to Einstein's "hole" argument. Einstein considered a region in which the metric tensor is the only field present. It obeys generally covariant field equations that do not involve any other fields in this region. Our general discussion of the "hole" argument, when applied to this situation, really has two aspects. The first is the problem of the physical identity of points in the manifold, and the consequent question of whether $G(x)$ and $G'(x)$ represent distinct gravitational fields, regardless of whether or not they satisfy certain field equations. This question is relevant whether the metric tensor is specified a priori, as, for example, in the special theory of relativity, or whether it is determined by field equations relating it to the sources of the gravitational field—the sort of theory for which Einstein was searching. The second aspect of the hole argument relates exclusively to the latter case, in which the components of the metric tensor are dynamical variables to be determined from a set of partial differential equations relating them to the components of the stress-energy tensor. (The argument is complicated, but not in any essential way, by the fact that, in many cases, the expressions for the stress-energy tensor components themselves involve the metric tensor.) Then it is indeed the case that any set of generally covariant field equations that has $G(x)$ as a solution in some empty region of space-time will also have $G'(x)$ as a solution in that region. Nothing can prevent $G'(x)$ from being a solution to generally covariant field equations for the metric tensor field if $G(x)$ is a solution. If the points of the manifold were in some way distinguished physically as events in space-time *before* the metric tensor field were determined, this would indeed imply that no criteria could be given to single out a *physically* unique solution to the field equations. This would be a serious problem, which might well be called a violation of the law of causality, as Einstein termed it in 1914. However, once it is stipulated that the points of the manifold have no physical properties except those that they inherit from the metric field, the problem disappears. $G(x)$ and $G'(x)$, together with all other mathematically distinct metric tensor fields that can be transformed into each other by being dragged along with a point transformation, form an *equivalence class* of solutions. This equivalence class of mathematically distinct metric tensor fields corresponds to *one* physical solution to the field equations—that is, to one gravitational field. There is no further problem of uniqueness or causality—or, more correctly, none that arises from general covariance.

Some modern treatments of general relativity discuss at least one aspect of this problem: the need to formulate the theory in such a way as to allow a unique solution to the Cauchy problem for the vacuum Einstein equations.[40] This is done by introducing the concept of isometric equivalence. Two differentiable manifolds with metric tensor fields are regarded as isometrically equivalent if there exists a sufficiently smooth invertible mapping (a diffeomorphism) between the manifolds that carries the metric tensor field of one into that of the other. Two metric fields on the *same* manifold are then isometrically equivalent if and only if one can be dragged into the other. A space-time may then be defined as a four-dimensional differentiable manifold with metric tensor field of Lorentz signature. (Additional global topological restrictions may be imposed, and the differentiability structure must be explicitly stated. but we forego discussion of such details.) A gravitational field is then defined as an equivalence class of isometrically equivalent space-times—isometric equivalence being an equivalence relation. Then the set of solutions to the field equations generated by continuing properly specified Cauchy or initial-value data off a space-like hypersurface will all be isometrically equivalent, and thus correspond to a unique gravitational field.[41] The physical presupposition of this mathematical treatment—that the lack of independent physical individuation of the points of a manifold permits identification of isometrically equivalent space-times as one gravitational field—is not usually made explicit.

Einstein, of course, started from this second aspect of the "hole" argument: the lack of mathematically unique solutions to generally covariant field equations for the metric tensor field. But I maintain that he was initially unable to accept the idea that the points of an empty region of a manifold do not have any preexistent spatio-temporal individuality, apart from that which they inherit from the metric tensor field. It was this last remnant of his belief in the physical significance of coordinates that led Einstein to regard an important *mathematical* property of generally covariant field equations as the source of nonuniqueness in their *physical* solutions, making such equations physically unacceptable to him. He was able to accept generally covariant equations for the gravitational field only after he realized that mathematically distinct solutions to such equations could be regarded as physically equivalent if one gave up the idea of any inherent physical distinction between the points of a manifold.

Einstein had worked hard to give a physical significance to the coordinates in the special theory of relativity, and the existence of an individuation of the points of space-time by means of rods and clocks formed an integral part of his interpretation of the special theory, as we noted earlier in this paper. One presumes that it was difficult for him to unlearn this lesson, which had served him so well.

Letters that Einstein wrote to Besso and Ehrenfest after he had abandoned the "hole" argument, in which he explained to them just what was wrong with it, provide strong evidence in support of my interpretation. They are cited and discussed in the final section, to which the reader may now turn, if so minded.

## 5. Einstein and Fokker on Nordström's Theory

After the Einstein–Grossmann (1913) paper, Einstein's next significant references to the curvature tensor and general covariance occur in a paper that Einstein wrote early in 1914 together with a young Dutch physicist, Adriaan D. Fokker (Einstein and Fokker 1914). This paper is a discussion of a gravitational theory proposed by Gunnar Nordström, a Finnish physicist then working on the problem of finding a gravitational theory compatible with the principles of special relativity (Nordström 1913). Einstein and Fokker showed that Nordström's scalar theory can be rewritten as a metric theory of gravitation, using only generally covariant tensor equations, if one assumes that the metric tensor is conformally related to the Minkowski metric tensor of special relativity. They phrased the latter requirement as follows: "Nordström's theory ... is based on the assumption that, by proper choice of the reference system, it is possible to satisfy the principle of the constancy of the velocity of light" (Einstein and Fokker 1914, p. 324).[42] Thus, they not only assumed the metric tensor to be conformally Minkowskian, but also adopted conformally (quasi-)Cartesian coordinates. With a few additional requirements, such as that the gravitational field equations depend only linearly on the stress-energy tensor and the second derivatives of the metric tensor, they were led to field equations of the form (in modern notation):

$$R = \kappa T,$$

where $R$ is the Ricci scalar (the trace of the Ricci tensor, formed by contracting it with the metric), $\kappa$ is the gravitational coupling constant, and $T$ is the trace of the stress-energy tensor. They showed that this equation reduces to Nordström's gravitational field equation in conformally Cartesian coordinates.

The main interest of the paper for present purposes lies in some remarks in the concluding section. Einstein and Fokker first note that they have

> shown that, basing oneself upon the principle of the constancy of the velocity of light [i.e., the hypothesis of the conformally Minkowskian nature of the metric tensor—JS], one can arrive at the Nordström theory by purely formal considerations, i.e., without having recourse to further physical hypotheses. It seems to us, therefore, that this theory merits preference over all others retaining this principle. (Einstein and Fokker 1914, p. 328)[43]

What is responsible for this success? "We note that only the application of the theory of invariants of the absolute differential calculus [i.e., tensor analysis—JS] allowed us to give a clear insight into the formal content of Nordström's theory" (Einstein and Fokker 1914, p. 328).[44] This seems to be Einstein's first recorded acknowledgment of the impression made upon him by the power of formal mathematical methods in the search for a relativistic theory of gravitation, and later in the search for a unified field theory. (Another is cited in the final section.)

But of even more direct relevance is the final paragraph of the paper, and its closing footnote:

> Finally, the role which the Riemann–Christoffel [curvature—JS] tensor plays in the present investigation, suggests that it should also open the way to a derivation of the Einstein–Grossmann gravitational equations independent of physical assumptions. The proof of the existence or non-existence of such a connection would constitute important theoretical progress. (Einstein and Fokker 1914, p. 328)[45]

This sentence, the last in the paper, is footnoted: "The argument given in §4, p. 36 [of Einstein and Grossmann 1913—*JS*] ... for the non-existence of such a connection does not withstand a closer examination" (Einstein and Fokker 1914, p. 328).[46]

The reference is to the separatum version of the Einstein–Grossmann paper; it corresponds to the Ricci tensor argument on page 257 of the version in the *Zeitschrift für Mathematik und Physik*, translated in Section 2 herein. Neither here, nor anywhere else, have I found any explanation by Einstein of exactly why he and Fokker believed this argument to be invalid. I can offer an explanation consistent with my earlier interpretation of Einstein's and Grossmann's comments on the Ricci tensor, but I cannot cite any further evidence that it was Einstein's explanation. The Einstein–Fokker reformulation of the Nordström theory is based on the Ricci scalar for a conformally Minkowskian metric. It is obvious that Nordström's theory has the correct static Newtonian limit. Einstein or Fokker may have noticed that this is also true of the Ricci scalar in the case of Einstein's *Ansatz* for the static metric tensor. Thus, a theory based on the trace of the Ricci tensor does yield the correct Newtonian limit, even with Einstein's *Ansatz*.

At any rate, Einstein did not follow up the question raised in this paper in the period immediately after his work on Nordström's theory. By November 1914 Einstein thought that he could derive the Einstein–Grossmann field equations uniquely by "purely formal considerations"; but these considerations did not involve the Riemann–Christoffel tensor. It was only the recognition of the fallaciousness of this derivation, coupled with other difficulties encountered by the Einstein–Grossmann theory, that finally led Einstein in late 1915 to reexamine the entire basis of that theory.

But before turning to the final chapter of the story, I shall discuss briefly a question that may already have occurred to the reader. If Einstein and Fokker were able to give a generally covariant formulation of Nordström's theory in 1914, why did Einstein not immediately drop his "hole" argument? This argument purports to prove the impossibility of a generally covariant theory of gravitation based exclusively on the metric tensor and satisfying Einstein's causality criterion; yet Nordström's theory appears to satisfy both of these conditions. Is there not a palpable contradiction here?

Actually, there is not. The reason is that, as Misner, Thorne, and Wheeler put it, Nordström's theory is based upon a "prior geometry," that is, to quote them, "an aspect of the geometry of space that is fixed immutably, i.e., that cannot be changed by changing the distribution of gravitating sources" (Misner et al. 1973, p. 429). In Nordström's theory, the conformal metric structure is fixed a priori to be *Minkowskian*; only the conformal factor is dynamically determined by the (trace of) the stress-energy tensor. Thus the space-time structure has a built-in "rigidity," which allows Einstein and Fokker to choose a family of quasi-Cartesian coordinates. So it is plausible to assume that Einstein realized that the "hole" argument does not apply to the Nordström theory. The existence of an additional element of "prior geometry" on the space-time manifold causes the argument to fail.

## 6. Covariance Regained

Einstein and Grossmann soon returned to the problem of finding all transformations under which their field equations are invariant.[47] They realized that the components of their gravitational stress-energy complex only behave like the components of a tensor under linear transformations; but they now considered the requirement that it be a tensor "unjustified" for unspecified reasons (Einstein and Grossmann 1914, p. 218, footnote 1). Evidently, they regarded the possibility of invariance under nonlinear coordinate transformations as much more important for the physical interpretation of the theory (see the quotation from Einstein and Grossmann 1914, p. 216, in endnote 6). They showed that their field equations could be derived from a variational principle, and investigated the maximal group of transformations under which this principle remains invariant. They found a condition on the coordinates that they took as a criterion for membership in the set of "adapted coordinates" ("angepasste Koordinaten"), i.e., coordinate systems in which the Einstein–Grossmann field equations hold. Transformations between such adapted coordinate sytems were called "justified transformations" ("berechtigte Transformationen"). They showed that, while this group includes more than the linear transformations, it does not include all possible transformations of coordinates. (We shall refer to this set of transformations as the restricted covariance group.)

In March 1914 Einstein wrote to Besso triumphantly, explaining this new development, and adding:

> Now I am completely satisfied and no longer doubt the correctness of the entire system.... The reasonableness of the thing is too obvious.... The general theory of invariants only acted as a hindrance. The direct route proved to be the only accessible one. What is inconceivable is only that I had to feel my way so long before I found what lay at hand. (Speziali 1972, p. 53)[48]

The problem of demonstrating the uniqueness of the field equations remained, that is, the problem of demonstrating that they are the only equations that are invariant under the restricted covariance group. Einstein thought that he had provided such a proof in his next major paper on the subject, a complete review of his gravitational theory, written soon after his move to Berlin (Einstein 1914b, presented to the Prussian Academy of Sciences on October 29—the paper in which he presented the final version of the "hole" argument). He gave an argument to demonstrate that the Lagrangian $H$, variation of which leads to the Einstein–Grossmann field equations, is uniquely fixed by the requirement that it be invariant under the restricted covariance group. He was evidently quite pleased with this accomplishment: "We have now arrived in a purely formal way, i.e., without direct appeal to our physical knowledge of gravitation, at completely determinate field equations" (Einstein 1914b, p. 1076).[49]

His triumph did not last very long. Difficulties in connection with the field equations and the variational principle began to accumulate. About a year later, shortly after these difficulties had been resolved, Einstein summarized some of them in a letter to Sommerfeld of November 28, 1915:

> I realized ... that my previous gravitational field equations were completely untenable! This was indicated by the following points:
>
> 1) I proved that the gravitational field on a uniformly rotating system does not satisfy the field equations.
> 2) The motion of the perihelion of Mercury came out as 18″ instead of 45″ per century.
> 3) The covariance argument in my paper of last year [Einstein 1914b—*JS*] does not give the Hamiltonian function $H$ [today usually called the Lagrangian function—*JS*]. When suitably generalized, it allows an arbitrary $H$. It followed that covariance relative to "adapted" coordinate systems was a wild goose chase.
>
> After all confidence thus had been lost in the results and methods of the earlier theory, I saw clearly that only in connection with the general theory of covariants, i.e., with Riemann's covariant [tensor—*JS*], could a satisfactory solution be found. I have unfortunately immortalized the last errors in this battle in the papers of the Academy. (Hermann 1968, pp. 32–33)[50]

The last sentence is a reference to problems that Einstein encountered in the course of arriving at the generally covariant form of the field equations, first announced to the Prussian Academy of Sciences at its session of November 25, 1915 (see Einstein 1915d). This paper had been preceded during November by two papers that chronicle the tortuous path by which Einstein reached the field equations, including his errors.[51] We shall not go into the details of this last phase of Einstein's peregrinations here.

Einstein again wrote to Besso triumphantly, but in a rather different vein than in his previously quoted letter of March 1914: "My wildest dreams have been fulfilled. *General* covariance. The perihelion motion of Mercury wonderfully exact.... This time the most obvious thing was the correct one" (Einstein to Besso, December 10, 1915; Speziali 1972, p. 60).[52]

Starting with his paper of November 18, devoted to explaining the precession of Mercury's perihelion, a new phrase appears in Einstein's discussions of general covariance: "though which time and space are deprived of the last trace of objective reality" (Einstein 1915c, p. 831).[53] In his paper of November 25 he used a rather different formulation: "The relativity postulate in its most universal formulation, which makes the spatio-temporal coordinates into physically meaningless parameters" (Einstein 1915d, p. 847).[54] But in a letter to Schlick, he returned to words closer to his original phrasing:

> Thereby [through the general covariance of the field equations—*JS*] time and space lose the last remnant of physical reality. All that remains is that the world is to be

conceived as a four-dimensional (hyperbolic) continuum of 4 dimensions. (Einstein to Moritz Schlick, December 14, 1915; EA 21-610)[55]

Similar wording occurs in Einstein's now-classic review and exposition of the final version of the theory, written early in 1916 (Einstein 1916). The phrase is quoted and the significance of his choice of words is discussed below.

Einstein still had to explain just what is wrong with the "hole" argument against general covariance, in which, as we have seen, he believed so strongly for over two years. He discussed this question in letters to Besso and Ehrenfest. (The reader may refer back to the translation of section 12 of the Einstein-Grossmann "Entwurf" paper (Einstein and Grossmann 1913) in Section 2 for the version of the "hole" argument referred to in the two letters.) On January 3, 1916, he wrote to Besso:

> Everything in the hole argument was correct up to the final conclusion. It has no physical content if, with respect to *the same* coordinate system $K$, two different solutions $G(x)$ and $G'(x)$ exist. To imagine two solutions simultaneously on the same manifold has no meaning and indeed the system $K$ has no physical reality. The hole argument is replaced by the following consideration. Nothing is physically *real* but the totality of space-time point coincidences. If, for example, all physical happenings were to be built up from the motions of material points alone, then the meetings of these points, i.e., the points of intersection of their world lines, would be the only real things, i.e., observable in principle. These points of intersection naturally are preserved during all transformations [of coordinates—*JS*] (and no new ones occur) if only certain uniqueness conditions are observed. It is therefore most natural to demand of the laws that they determine no *more* than the totality of the space-time coincidences. From what has been said, this is already attained through the use of generally covariant equations. (Speziali 1972, pp. 63–64)[56]

In a letter to Ehrenfest a week earlier, Einstein had given a more detailed explanation of his new argument:

> In §12 of my paper of last year, everything is correct (in the first three paragraphs) up to the italicized part at the end of the third paragraph. Absolutely no contradiction to the uniqueness of events follows from the fact that both systems $G(x)$ and $G'(x)$ satisfy the conditions for the gravitational field with respect to the same reference system. The apparent compulsion of this argument disappears at once if one considers that 1) the reference system signifies nothing real, 2) that the (simultaneous) realization of two different $g$-systems (better said, of two different grav[itational] fields) in the same region of the continuum is impossible by the nature of the theory.
> 
> The following consideration should replace §12. The physically real in what happens in the world (as opposed to what depends on the choice of the reference system) consists of *spatio-temporal coincidences*. [In a footnote, Einstein adds "and of nothing else!"—*JS*] For example, the points of intersection of two world lines, or the assertion that they *do not* intersect, are real. Such assertions referring to the physically real are thus not lost because of any (single-valued) coordinate transformations. If two systems of $g_{\mu\nu}$ (or generally, of any variables applied in the description of the world) are so constituted that the second can be obtained from the first by a pure space-time transformation, then they are fully equivalent. For they have all

spatio-temporal point coincidences in common, that is, everything that is observable. This argument shows at the same time how natural the demand for general covariance is. (Einstein to Ehrenfest, December 26, 1915; EA 9-363)[57]

In Einstein's classic exposition of the general theory of relativity, written early in 1916, the "hole" argument naturally does not appear (Einstein 1916, submitted March 20, 1916). However the coincidence argument, developed in the two letters just quoted to replace the "hole" argument, is given early in the paper as the principal argument for general covariance:

*The general laws of nature are to be expressed by equations which hold good for all systems of coordinates, that is, are covariant with respect to any substitutions whatever (generally covariant).*

It is clear that a physical theory that satisfies this postulate will also be suitable for the general postulate of relativity. For the set of *all* substitutions in any case includes those which correspond to all relative motions of three-dimensional systems of co-ordinates. That this requirement of general covariance, which takes away from space and time the last remnant of physical objectivity, is a natural one, will be seen from the following reflexion. All our space-time verifications invariably amount to a determination of space-time coincidences. If, for example, events consisted merely in the motion of material points, then ultimately nothing would be observable but the meetings of two or more of these points. Moreover, the results of our measurings are nothing but verifications of such meetings of the material points of our measuring instruments with other material points, coincidences between the hands of a clock and points on the clock dial, and observed point-events happening at the same place at the same time.

The introduction of a system of reference serves no other purpose than to facilitate the description of the totality of such coincidences. We allot to the universe four space-time variables $x_1, x_2, x_3, x_4$ in such a way that for every point-event there is a corresponding system of values of the variables $x_1 \ldots x_4$. To two coincident point-events there corresponds one system of values of the variables $x_1 \ldots x_4$, i.e., coincidence is characterized by the identity of the co-ordinates. If, in place of the variables $x_1 \ldots x_4$, we introduce functions of them $x'_1, x'_2, x'_3, x'_4$, as a new system of co-ordinates, so that the systems of values are made to correspond to one another without ambiguity, the equality of all four co-ordinates in the new system will also serve as an expression for the space-time coincidence of the two point-events. As all our physical experience can be ultimately reduced to such coincidences, there is no immediate reason for preferring certain systems of co-ordinates to others, that is to say, we arrive at the requirement of general covariance. (Einstein 1916, pp. 776–777)[58]

The context in which Einstein presented the coincidence argument and referred to space and time losing "their last remnant of physical reality" should now be clear. We see here the traces of his rejection of the "hole" argument and his concomitant acceptance of general covariance. They serve to emphasize his recognition that only a physical process can individuate the events that make up space-time, or "the world," as it was then often called, following Minkowski (1909). Removed from the context of the "hole" argument, Einstein's words seem to invite a purely operationalistic interpretation. The emphasis on observability, and even the phrase "quantities observable in principle," used in his letter to Besso, were later employed in a totally different context in discussions of the quantum theory.[59] But for Einstein, I suggest, the coincidence argument was not primarily an element of a positivistic philosophical credo. Rather, it was an important link in the chain of arguments by which he convinced himself that a manifold only becomes a space-time with a certain gravitational field after the specification of the metric tensor field, and that, prior to such a specification, there is no physical distinction between the elements of the manifold. Similarly, his comments about space and time losing their last vestiges of objective reality were not meant, I believe, to indicate that space and time have *no* physical reality, but that they no longer have any *independent* reality, apart from their significance as the spatial and temporal aspects of the metrical field. The language in which Einstein expresses his thoughts is not very precise, and—it must be admitted— certainly influenced current positivistic terminology. But as my earlier discussion indicates, his words were primarily intended to clarify physical problems connected with his work on a relativistic theory of gravitation.

NOTES

* This paper is based on an invited talk given July 17, 1980, at the Ninth International Conference on General Relativity and Gravitation, Jena, Germany (DDR). A written version of this talk has circulated privately since the fall of 1980. Since that version has been cited in print several times since then, I have decided to publish it with only minor revisions to improve the exposition, but without attempting to update it. For a more recent account of my views see Stachel 1986b.

[1] "1907 Grundgedanke für die allgemeine Relativitätstheorie
 1912 Erkenntnis der nicht-euklidischen Natur der Metrik und der physikalischen Bedingtheit derselben durch die Gravitation
 1915 Feldgleichungen der Gravitation. Erklärung der Perihelbewegung des Merkur."

The note was prepared for Erwin Freundlich's use in his book on Einstein's theory (Freundlich 1916).

[2] See Nathan and Norden 1960, section I, "The Reality of War (1914–1918)," for documentation of this period of Einstein's life.

[3] "Den entscheidenden Gedanken von der Analogie des mit der Theorie verbundenen mathematischen Problems mit der Gauss'schen Flächentheorie hatte ich allerdings erst 1912 nach meiner Rückkehr nach Zürich, ohne zunächst Riemanns und Riccis sowie Levi-Civitas Forschungen zu kennen. Auf diese wurde ich erst durch meinen Freund Grossmann in Zürich aufmerksam, als ich ihm das Problem stellte, allgemein kovariante Tensoren aufzusuchen, deren Komponenten nur von Ableitungen der Koeffizienten der quadratischen Fundamental-Invariante abhängen."

[4] See Einstein 1955, and Einstein to Walter Leich, April 24, 1950 (EA 60-252).

[5] See Reich 1979. I am also indebted to Dr. Alvin Jaeggli and Professor Res Jost, both of the ETH, for private communications about the contents of the Grossmann notebooks, which Einstein used to prepare for his examinations.

[6] A clear statement of this view occurs in Einstein and Grossmann 1914, p. 216:

"The whole theory [of gravitation—*JS*] developed from the conviction that all physical processes in a gravitational field proceed in exactly the same way as the corresponding processes proceed without a gravitational field, if one refers them to a suitably accelerated (three-dimensional) coordinate system ("equivalence hypothesis"). This hypothesis, based upon the experimental fact of the equality of gravitational and inertial mass, is particularly convincing if the 'apparent' gravitational field, which exists with respect to the accelerated (three-dimensional) coordinate system, can be considered as a 'real' gravitational field, that is if acceleration transformations (i.e., nonlinear transformations) are among the transformations permitted by the theory."
["Die ganze Theorie ist hervorgegangen aus der Überzeugung, daß alle physikalischen Vorgänge in einem Gravitationsfeld genau gleich ablaufen, wie die entsprechenden Vorgänge ohne Gravitationsfeld ablaufen, falls man sie auf ein passend beschleunigtes (dreidimensionales) Koordinatensystem bezieht ('Äquivalenzhypothese'). Diese auf die Erfahrungstatsache von der Gleichheit der schweren und der trägen Masse gegründete Hypothese erhält dann eine besondere Überzeugungskraft, wenn das 'scheinbare' Gravitationsfeld, welches in bezug auf das beschleunigte (dreidimensionale) Koordinatensystem existiert, als ein 'wirkliches' Gravitationsfeld aufgefaßt werden kann, wenn also Beschleunigungstransformationen (d.h. nichtlineare Transformationen) zu den berechtigten Transformationen der Theorie gehören."]

This point of view stands in contrast to the recent tendency to stress the nonvanishing of the Riemann tensor as the exclusive criterion for the presence of a "true" gravitational field. Einstein was aware of this viewpoint, but did not adopt it. In 1951, Max von Laue wrote to Einstein objecting to the latter's treatment of the rotating disc problem (see Stachel 1980). Von Laue noted that the Riemann tensor vanishes on the disc. Einstein replied: "It is true that in that case the $R_{iklm}$ [the components of the Riemann tensor—*JS*] vanish, so that one might say: 'There is no gravitational field present.' However, what characterizes the existence of a gravitational field from the empirical standpoint is the nonvanishing of the $\Gamma^l_{ik}$ [the components of the affine connection—*JS*], not the nonvanishing of the $R_{iklm}$. If one does not think in such intuitive ways, one cannot comprehend why something like curvature should have anything to do with gravitation at all. In any case, no reasonable person would have hit upon something in that way. The key to the understanding of the equality of inertial and gravitational mass would have been missing." ["Es ist wahr, dass in jenem Falle die $R_{iklm}$ verschwinden, sodass man sagen könnte: 'es ist kein Gravitationsfeld vorhanden'. Was aber die Existenz eines Gravitationsfeldes vom empirschen Standpunkt aus charakterisiert, ist das Nichtverschwinden der $\Gamma^l_{ik}$, nicht das Nichtverschwinden der $R_{iklm}$. Wenn man nicht in solcher Weise anschaulich denkt, begreift man nicht, warum überhaupt so etwas wie eine Krümmung etwas mit Gravitation zu thun haben soll. Jedenfalls würde kein vernünftiger Mensch auf so etwas verfallen. Es fehlt dann der Schlüssel für das Verständnis der Gleichheit der trägen und schweren Masse."] (Einstein to Max von Laue, September 1950; EA 16-148.)

[7] "Almost" because of the difference between the Ricci tensor and what we now call the Einstein tensor, a difference that was to cause Einstein further grief when he returned to the generally covariant approach in late 1915. For discussion of this episode, see Stachel 1979a, 1979b; Earman and Glymour 1978a, 1978b; Hoffmann 1972; Lanczos 1972; Mehra 1973; Vizgin and Smorodinskiĭ 1979.

[8] "Die hervorragende Bedeutung dieser Begriffsbildungen für die *Differentialgeometrie* einer durch ihr Linienelement gegebenen Mannigfaltigkeit macht es a priori

wahrscheinlich, daß diese allgemeinen Differentialtensoren auch für das Problem der Differentialgleichungen eines Gravitationsfeldes von Bedeutung sein dürften. Es gelingt in der Tat zunächst, einen kovarianten Differentialtensor zweiten Ranges und zweiter Ordnung $G_{im}$ anzugeben, der in jene Gleichungen eintreten könnte, nämlich

$$G_{im} = \sum_{kl} \gamma_{kl}(ik, lm) = \sum_{k} \{ik, km\}.$$"

⁹ This sentence, crucial for my interpretation: "Allein es zeigt sich, dass sich dieser Tensor im Spezialfall des unendlich schwachen statischen Schwerefeldes *nicht* auf den Ausdruck $\Delta\varphi$ reduziert," is mistranslated by Mehra (1973, p. 102) as: "This in itself shows that this tensor, in the special case of an infinitely weak gravitational field, would not reduce to the expression $\Delta\varphi$," and the crucial word "static" is also omitted. This mistranslation has apparently influenced the reading of Earman and Glymour (1978a, p. 256), who remark: "The fact that $R_{ij} = R^k{}_{ijk}$ in itself showed, according to Grossmann's reasoning, that $R_{ij}$ cannot have the right Newtonian limit."

¹⁰ "Dieselbe Art der Degeneration zeigt sich bei dem statischen Schwerefelde der vorhin betrachteten Art, nur daß bei diesem $g_{44} = c^2$ eine Funktion von $x_1, x_2, x_3$ ist."

¹¹ Hoffman 1972, p. 159, gives the form of the Riemann and Ricci tensors for Einstein's 1912 static metric without indicating its possible relevance to understanding the Einstein–Grossmann paper.

¹² "Das Erfreulichste ist das Stimmen der Perihelbewegung und die allgemeine Kovarianz, das Merkwürdigste aber der Umstand, dass Newtons Theorie des Feldes schon in Gl. 1. Ordnung unrichtig ist (Auftreten der $g_{11} - g_{33}$). Nur der Umstand, dass $g_{11} - g_{33}$ nicht in den ersten Näherungen der Bew. Gl. des Punktes auftreten, bedingt die Einfachheit von Newtons Theorie."

¹³ "Diesmal ist das Nächstliegende das Richtige gewesen; aber Grossmann u. ich glaubten, die Erhaltungssätze nicht erfüllt seien, und das Newton'sche Gesetz in erster Näherung nicht herauskomme. Du wirst über das Auftreten der $g_{11} \ldots g_{33}$ überrascht sein." The reference to the conservation equations concerns another argument against general covariance that Einstein developed in 1913; it is discussed at the end of this section. Other aspects of this letter are discussed in the final section of this paper.

¹⁴ "Ein solcher Zusammenhang müßte vorhanden sein, sofern die Gravitationsgleichungen *beliebige* Substitutionen zuzulassen hätten; allein in diesem Falle scheint es ausgeschlossen zu sein, Differentialgleichungen *zweiter* Ordnung aufzufinden. Würde dagegen feststehen, daß die Gravitationsgleichungen nur eine gewisse Gruppe von Transformationen gestatten, so wäre es verständlich, wenn man mit den von der allgemeinen Theorie gelieferten Differentialtensoren nicht auskommt. Wie im physikalischen Teile ausgeführt ist, sind wir nicht imstande, zu diesen Fragen Stellung zu nehmen."

¹⁵ "Der Versuch einer Diskussion derartiger Möglichkeiten wäre aber bei dem gegenwärtigen Stande unserer Kenntnis der physikalischen Eigenschaften des Gravitationsfeldes verfrüht. Deshalb ist für uns die Beschränkung auf die zweite Ordnung geboten und wir müssen daher darauf verzichten, Gravitationsgleichungen aufzustellen, die sich beliebigen Transformationen gegenüber als kovariant erweisen. Es ist übrigens hervorzuheben, daß wir keinerlei Anhaltspunkte für eine allgemeine Kovarianz der Gravitationsgleichungen haben."

¹⁶ "Wir haben für die Gravitationsgleichungen nur beweisen können, daß sie beliebigen *linearen* Transformationen gegenüber kovariant sind; wir wissen aber nicht,

ob es eine allgemeine Transformationsgruppe gibt, der genenüber die Gleichungen kovariant sind. Die Frage nach der Existenz einer derartigen Gruppe ... ist die wichtigste, welche sich an die hier gegebenen Ausführungen anknüpft."

[17] "[wir sind] bei dem gegenwärtigen Stande der Theorie nicht berechtigt, die Kovarianz physikalischer Gleichungen beliebigen Substitutionen gegenüber zu fordern.

"Anderseits aber haben wir gesehen, daß sich eine Energie-Impuls-Bilanzgleichung für materielle Vorgänge hat aufstellen lassen ... welche beliebige Transformationen gestattet. Es scheint deshalb doch natürlich, wenn wir voraussetzen, daß alle physikalischen Gleichungssysteme mit Ausschluß der Gravitationsgleichungen so zu formulieren sind, daß sie beliebigen Substitutionen gegenüber kovariant sind. Die diesbezügliche Ausnahmestellung der Gravitationsgleichungen gegenüber allen anderen Systemen hängt nach meiner Meinung damit zusammen, daß nur erstere zweite Ableitungen der Komponenten des Fundamentaltensors enthalten dürften."

[18] "Man erkennt, daß Koordinatenlängen zugleich natürliche Längen sind ($dt = 0$). Maßstäbe erleiden also durch das 'Newtonsche' Gravitationsfeld keine Verzerrung."

[19] "Ich bin nun innerlich überzeugt, das Richtige getroffen zu haben, zugleich freilich auch, dass ein Murmeln der Entrüstung durch die Reihe der Fachgenossen gehen wird, wenn die Arbeit erscheint, was in wenigen Wochen der Fall sein wird.... Die Überzeugung, zu der ich mich langsam durchgerungen habe, ist die, *dass es bevorzugte Koordinatensysteme überhaupt nicht gibt*. Doch ist es mir nur teilweise gelungen, auch formal bis zu diesem Standpunkte vorzudringen."

[20] "Mit der Gravitationstheorie bin ich nun sehr zufrieden. Die Thatsache, dass die Gravitationsgleichungen nicht allgemein kovariant sind, welche mich vor einiger Zeit noch ungemein störte, hat sich als unumgänglich herausgestellt; es lässt sich einfach beweisen, dass eine Theorie mit allgemein kovarianten Gleichungen nicht existieren kann, falls verlangt wird, dass das Feld durch die Materie mathematisch vollständig bestimmt wirt."

[21] "Die Gravitationsaffäre hat sich zu meiner *vollen Befriedigung* aufgeklärt (der Umstand nämlich, dass die Gleichungen des Gr. Feldes nur *linearen* Transformationen gegenüber kovariant sind). Es lässt sich nämlich beweisen, dass *allgemein kovariante* Gleichungen, die das Feld aus dem materiellen Tensor *vollständig* bestimmen, überhaupt nicht existieren können. Was kann es schöneres geben, als dies, dass jene nötige Spezialisierung aus den Erhaltungssätzen fliesst? Nun bestimmen die Erhaltungssätze unter allen Flächen diejenigen, welche als Koordinatenflächen mit Vorzug zu wählen sind. Diese bevorzugten Flächen können wir als Ebenen bezeichnen, weil ja nur *lineare* Substitutionen als berechtigt übrig bleiben."

[22] "Für mich ist es absurd, dem 'Raum' physikalische Eigenschaften zuzuschreiben. Die Gesamtheit der Massen erzeugt ein $g_{\mu\nu}$-Feld (Gravitationsfeld), das seinerseits den Ablauf aller Vorgänge, auch die Ausbreitung der Lichtstrahlen und das Verhalten der Massstäbe und Uhren regiert. Das Geschehen wird zunächst auf vier *ganz willkürliche* raumzeitliche Variable bezogen. Diese müssen dann, wenn die Erhaltungssätzen des Impulses und der Energie geleistet werden soll, derart spezialisiert werden, dass nur (ganz) *lineare* Substitutionen von einem berechtigten Bezugssystem zu einem andern führen. Das Bezugssystem ist der bestehenden Welt mit Hilfe des Energiesatzes sozusagen angemessen und verliert seine nebulose apriorische Existenz."

Herneck (1963) dated the letter to the turn of the years 1911–1912 or 1912–1913. The attribution to the turn of some year is made rather certain by the closing salutation, "With best wishes for the New Year;" but what year is left open. The years suggested by Herneck seem to be ruled out by the content of the letter, which clearly refers to the work contained in Einstein and Grossmann 1913. The point of view expressed concerning the conservation equations is closest in content to the remarks on p. 178 of Einstein 1914a, which is dated January 1914. This leads me to suggest the end of 1913 as the correct date for the letter.

[23] "Es hat sich durch eine allgemeine Überlegung zeigen lassen, dass Gleichungen, welche das Gravitationsfeld vollständig bestimmen, nicht beliebige Substitutionen gegenüber kovariant sein können."

[24] "In den letzten Tagen fand ich den Beweis dafür, daß ein deratige allgemein kovariante Lösung überhaupt nicht existieren kann."

[25] "Es gebe in unserer vierdimensionalen Mannigfaltigkeit einen Teil $L$, in welchem ein 'materieller Vorgang' nicht stattfinde, in welchem also die $\Theta_{\mu\nu}$ verschwinden. Durch die außerhalb $L$ gegebenen $\Theta_{\mu\nu}$ sind gemäß unserer Annahme überall, also auch im Innern von $L$ die $\gamma_{\mu\nu}$ vollkommen bestimmt. Wir denken uns nun statt der ursprünglichen Koordinaten $x_\nu$ neue Koordinaten $x'_\nu$ eingeführt von folgender Art. Außerhalb $L$ sei überall $x_\nu = x'_\nu$; innerhalb $L$ aber sei wenigstens für einen Teil von $L$ und wenigstens für einen Index $\nu$ $x_\nu \neq x'_\nu$. Es ist klar, daß durch eine derartige Substitution erreicht werden kann, daß wenigstens für einen Teil von $L$ $\gamma'_{\mu\nu} \neq \gamma_{\mu\nu}$ ist. Anderseits ist überall $\Theta'_{\mu\nu} = \Theta_{\mu\nu}$, nämlich außerhalb $L$, weil für dieses Gebiet $x'_\nu = x_\nu$ ist, innerhalb $L$ aber, weil für dies Gebiet $\Theta_{\mu\nu} = 0 = \Theta'_{\mu\nu}$ ist. Hieraus folgt, daß in dem betrachteten Falle, wenn alle Substitutionen als berechtigte zugelassen werden, zu dem nämlichen System der $\Theta_{\mu\nu}$ mehr als ein System der $\gamma_{\mu\nu}$ gehört.

"Wenn also—wie dies in der Arbeit geschehen ist—an der Forderung festgehalten wird, daß durch die $\Theta_{\mu\nu}$ die $\gamma_{\mu\nu}$ vollständig bestimmt sein sollen, so ist man genötigt, die Wahl des Bezugsystems einzuschränken."

[26] "Die Gleichungen sind so zu verstehen, daß auf den linken Seiten jeweilen den unabhängigen Variabeln $x'_\nu$, dieselben Zahlenwerten erteilt werden wie auf den rechten Seiten den Variabeln $x_\nu$."

[27] "Das erkenntnistheoretisch Befriedigende der bisher entwickelten Theorie liegt darin, daß dieselbe dem Relativitätsprinzip in dessen weitgehendster Bedeutung Genüge leistet. Dies beruht, formal betrachtet, darauf, daß die Gleichungssysteme *allgemein*, d.h. beliebigen Substitutionen der $x_\nu$ gegenüber, kovariant sind.

"Es scheint hiernach die Forderung geboten, daß auch die Differentialgesetze für die $g_{\mu\nu}$ *allgemein* kovariant sein müssen. Wir wollen aber zeigen, daß wir diese Forderung einschränken müssen, wenn wir dem Kausalgesetz vollständig Genüge leisten wollen. Wir beweisen nämlich, daß Gesetze, welche den Ablauf des Geschehens im Gravitationsfelde bestimmen, unmöglich *allgemein* kovariant sein können.

*§12. Beweis von der Notwendigkeit einer Einschränkung der Koordinatenwahl.*

"Wir betrachten einen endlichen Teil $\Sigma$ des Kontinuums, in welchem ein materieller Vorgang nicht stattfindet. Das physikalische Geschehen in $\Sigma$ ist dann vollständig bestimmt, wenn in bezug auf ein zur Beschreibung benutztes Koordinatensystem $K$ die Größen $g_{\mu\nu}$ als Funktion der $x_\nu$ gegeben werden. Die Gesamtheit dieser Funktionen werde symbolisch durch $G(x)$ bezeichnet.

"Es werde ein neues Koordinatensystem $K'$ eingeführt, welches außerhalb $\Sigma$ mit $K$ übereinstimme, innerhalb $\Sigma$ aber von $K$ abweiche, derart, daß die auf $K'$ bezogenen $g'_{\mu\nu}$ wie die $g_{\mu\nu}$ (nebst ihren Ableitungen) überall stetig sind. Die Gesamtheit der $g'_{\mu\nu}$ bezeichnen wir symbolisch durch $G'(x')$. $G(x')$ und $G(x)$ beschreiben das nämliche Gravitationsfeld. Ersetzen wir in den Funktionen $g'_{\mu\nu}$ die Koordinaten $x'_\nu$ durch die Koordinaten $x_\nu$, d.h. bilden wir $G'(x)$, so beschribt $G'(x)$ ebenfalls ein Gravitationsfeld bezüglich $K$; welches aber nicht übereinstimmt mit dem tatsächlichen (bzw. ursprünglich gegebenen) Gravitationsfelde.

"Setzen wir nun voraus, daß die Differentialgleichungen des Gravitationsfeldes allgemein kovariant sind, so sind sie für $G'(x')$ erfüllt (bezüglich $K'$), wenn sie bezüglich $K$ für $G(x)$ erfüllt sind. Sie sind dann also auch bezüglich $K$ für $G'(x)$ erfüllt. Bezüglich $K$ existierten dann die voneinander verschiedenen Lösungen $G(x)$ und $G'(x)$, trotzdem an den Gebietsgrenzen beide Lösungen übereinstimmten, d.h. *durch allgemein kovariante Differentialgleichungen für das Gravitationsfeld kann das Geschehen in demselben nicht eindeutig festgelegt werden.*

"Verlangen wir daher, daß der Ablauf des Geschehens im Gravitationsfelde durch die aufzustellenden Gesetze vollständig bestimmt sei, so sind wir genötigt, die Wahl des Koordinatensystems derart einzuschränken, daß es ohne Verletzung der einschränkenden Bedingungen unmöglich ist, ein neues Koordinatensystem $K'$ von der vorhin charakterisierten Art einzuführen. Die Fortsetzung des Koordinatensystems ins Innere eines Gebietes Σ hinein darf nicht willkürlich sein."

[28] For a history of tensor calculus, see Reich 1979. For a history of differential geometry, see Coolidge 1940, Book III. For a modern account of tensor analysis, see, for example, Bishop and Goldberg 1968. For textbooks of relativity based on the modern approach to differential geometry, see, for example, Hawking and Ellis 1973 and Sachs and Wu 1977. In this section, I have used many concepts and definitions from Schouten 1954, although it is not entirely modern in approach.

[29] However, a reading of the essential passage in this version of the argument in the light of Einstein's footnote to the "hole" argument in Einstein 1914a, p. 178 (also cited in Section 2) makes me dubious that Einstein ever committed this trivial error.

[30] For the definition of the dragged-along field of a geometrical object, see, for example, Schouten 1954, pp. 102–103. A brief discussion is given later in this section.

[31] A geometrical object, a generalization of a tensor, may be defined informally as an object whose components are determined in every coordinate system once its components are specified in one coordinate system. For example, an affine connection is a geometrical object but not a tensor. For a formal definition, see Schouten 1954, p. 57.

[32] Each color may be characterized by hue, brightness, and saturation, for example. More commonly, three primary colors are selected, and other colors are characterized by the amounts of each that must be mixed to produce a particular color.

[33] Schrödinger carried out an extensive analysis of color space and, following Helmholtz, introduced such a metric. See Schrödinger 1920, 1926. For a more recent discussion, see Weinberg 1976.

[34] I shall ignore as irrelevant for present purposes such questions as the global topology of the color manifold, the existence of a boundary, and so forth.

[35] In this case, as in the case of color space, we leave aside all questions as to whether the events in the world, or the colors we perceive, actually vary continuously and differentiably. We confine our attention to theoretical models that have such continuity and differentiability properties, without prejudice to the question of whether models without such properties may ultimately be required.

[36] Clearly, other methods exist for such an individuation, using various combinations of test particles, light rays, clocks, and other devices.

[37] "Um irgendeinen physikalischen Vorgang beschreiben zu können, müssen wir imstande sein, die in den einzelnen Punkten des Raumes stattfindenden Veränderungen örtlich und zeitlich zu werten.

"Zur örtlichen Wertung eines in einem Raumelement stattfindenden Vorganges von unendlich kurzer Dauer (Punktereignis) bedürfen wir eines Cartesischen Koordinatensystems, d.h. dreier aufeinander senkrecht stehender, starr miteinander verbundener, starrer Stäbe, sowie eines starren Einheitsmaßstabes. [In the footnote: "Statt von 'starren' Körpern, könnte ... ebenso gut von deformierenden Kräften nicht unterworfenen festen Körpern gesprochen werden."] ... Für die zeitliche Wertung eines Punktereignisses bedienen wir einer Uhr, die relativ zum Koordinatensystem ruht und in deren unmittelbarer Nähe das Punktereignis stattfindet....

"Wir nehmen nun an, *die Uhren können so gerichtet werden, daß die Fortpflanzungsgeschwindigkeit eines jeden Lichtstrahles in Vakuum—mit Hilfe dieser Uhren gemessen— allenthalben gleich einer universellen Konstante c wird*, vorausgesetzt, daß das Koordinatensystem nicht beschleunigt ist." See also Einstein 1905, part I, §1.

[38] "Es fällt auf, dass die Theorie (ausser dem vierdimensionalen Raum) zweierlei physikalische Dinge einführt, nämlich (1) Masstäbe und Uhren, (2) alle sonstigen Dinge, z.B. das elektromagnetische Feld, den materiellen Punkt, usw. Dies ist in gewissem Sinne inkonsequent; Masstäbe und Uhren müssten eigentlich als Lösungen der Grundgleichungen (Gegenstände bestehend aus bewegten atomistischen Gebilden) dargestellt werden, nicht als gewissermassen theoretisch selbständige Wesen. Das Vorgehen rechtfertigt sich aber dadurch, dass von Anfang an klar war, dass die Postulate der Theorie nicht stark genug sind, um aus ihr genügend vollständige Gleichungen für das physikalische Geschehen genügend frei von Willkür zu deduzieren, um auf eine solche Grundlage eine Theorie der Massstäbe und Uhren zu gründen. Wollte man nicht auf eine physikalische Deutung der Koordinaten überhaupt verzichten (was an sich möglich wäre), so war es besser, solche Inkonsequenz zuzulassen—allerdings mit der Verpflichtung, sie in einem späteren Stadium der Theorie zu eliminieren."

Einstein had given a less explicit formulation of the same distinction in 1921: "The concept of measuring body as well as the concept of measuring clock coordinated to it in the theory of relativity does not correspond exactly to any object in the real world. It is also clear that the rigid body and the clock do not play the role of irreducible elements in the conceptual structure of physics, but the role of composite structures, which should not play an independent role in the construction of theoretical physics. However, it is my conviction that these concepts, at the present stage of development of theoretical physics, still must be introduced as independent concepts" (Einstein 1921, p. 127). ["Der Begriff des Meßkörpers sowie auch der ihm in der Relativitätstheorie koordinierte Begriff der Meßuhr findet in der wirklichen Welt kein ihm exakt entsprechendes Objekt. Auch ist klar, daß der feste Körper und die Uhr nicht die Rolle von irreduzibeln Elementen im Begriffsgebäude der Physik spielen, sondern die Rolle von zusammengesetzten Gebilden, die im Aufbau der theoretischen Physik keine selbständige Rolle spielen dürfen. Aber es ist meine Überzeugung, daß diese Bergriffe beim heutigen Entwicklungsstadium der theoretischen Physik noch als selbständige Begriffe herangezogen werden müssen."]

[39] For example, in the generic case, the points of the manifold are completely individuated by the values of the four nonvanishing invariants of the Riemann tensor, as Kretschmann (1917) first pointed out. We omit consideration of complications that arise, for example, in case the metric tensor field has certain symmetries. In this case, additional structures are necessary to individuate physically points on an orbit of the symmetry group.

[40] See, for example, Hawking and Ellis 1973, p. 56 and chapter 7; or Sachs and Wu 1977, pp. 27, 262.

⁴¹ We omit numerous details and mathematical complications in this sketch. For a complete discussion see Hawking and Ellis 1973, chapter 7.

⁴² "Der Nordströmschen Theorie ... liegt die Annahme zugrunde, daß es möglich sei, durch passende Wahl des Bezugssystems dem Prinzip von der Konstanz der Lichtgeschwindigkeit zu genügen."

⁴³ "gezeigt ... daß man bei Zugrundelegung des Prinzips von der Konstanz der Lichtgeschwindigkeit durch rein formale Erwägungen, d.h. ohne Zuhilfenahme weiterer physikalischen Hypothesen zur Nordströmschen Theorie gelangen kann. Es scheint uns deshalb, daß diese Theorie allen anderen Gravitationstheorien gegenüber, die an diesem Prinzip festhalten, den Vorzug dient."

⁴⁴ "Wir bemerken, daß nur die Verwendung der Invariantentheorie des absoluten Differentialkalküls uns eine klare Einsicht in den formalen Inhalt der Nordstömschen Theorie zu geben vermag."

⁴⁵ "Endlich legt die Rolle, welche bei der vorliegenden Untersuchung der Riemann–Christoffelsche Differentialtensor spielt, den Gedanken nahe, daß er auch für eine von physikalischen Annahmen unabhängige Ableitung der Einstein–Großmannschen Gravitationsgleichungen einen Weg öffnen würde. Der Beweis der Existenz oder Nichtexistenz eines derartigen Zusammenhanges würde einen wichtigen theoretischen Fortschritt bedeuten."

⁴⁶ "Die in §4, s. 36, des "Entwurfs einer verallgemeinerten Relativitätstheorie" angegebene Begründung für die Nichtexistenz eines derartigen Zusammenhanges hält einer genaueren Überlegung nicht stand."

⁴⁷ Their work, published in Einstein and Grossmann 1914, was presumably well under way, if not completed, by the time Einstein left Zurich for Berlin in April 1914. Einstein explained the principal results of the paper in a letter written to Besso earlier in March, which is quoted below.

⁴⁸ "Nun bin ich vollkommen befriedigt und zweifle nicht mehr an der Richtigkeit des Systems.... Die Vernunft der Sache ist zu evident.... Die allgemeine Invariantentheorie wirkte nur als Hemmnis. Der direkte Weg erwies sich als der einzig gangbare. Unbegreiflich ist nur, dass ich solange tasten musste, bevor ich das Nächstliegende fand."

⁴⁹ "Wir sind nun auf rein formalem Wege, d.h. ohne direkte Heranziehung unserer physikalischen Kentnisse von der Gravitation, zu ganz bestimmten Feldgleichungen gelangt."

⁵⁰ "Ich erkannte ... dass meine bisherigen Feldgleichungen der Gravitation gänzlich haltlos waren! Dafür ergaben sich folgende Anhaltspunkte:
1) Ich bewies, dass das Gravitationsfeld auf einem gleichförmig rotierenden System den Feldgleichungen nicht genügt.
2) Die Bewegung des Merkur-Perihels ergab sich zu 18" statt 45" pro Jahrhundert.
3) Die Kovarianzbetrachtung in meiner Arbeit vom letzten Jahre liefert die Hamilton-Funktion $H$ nicht. Sie lässt, wenn sie sachgemäss verallgemeinert wird, ein beliebiges $H$ zu. Daraus ergab sich, dass die Kovarianz bezüglich 'angepasster' Koordinatensysteme ein Schlag ins Wasser war.

"Nachdem so jedes Vertrauen in Resultate und Methode der früheren Theorie gewichen war, sah ich klar, dass nur durch einen Anschluss an die allgemeine Kovariantentheorie, d.h. an Riemanns Kovariante, eine befriedigende Lösung gefunden werden konnte. Die letzten Irrtümer in diesem Kampfe habe ich leider in den Akademie-Arbeiten ... verewigt."

⁵¹ See Einstein 1915a, submitted on November 4, and Einstein 1915b, submitted on November 11. Einstein 1915c, submitted on November 18, is primarily devoted to an

explanation of the precession of Mercury's perihelion. In a footnote, it announces that a forthcoming communication will remove the last restriction on general covariance in his earlier papers. This is an allusion to Einstein 1915d, which was submitted a week later.

[52] "Die kühnste Träume sind nun in Erfüllung gegangen. *Allgemeine* Kovarianz. Perihelbewegung des Merkur wunderbar genau.... Diesmal ist das Nächstliegende das richtige gewesen." This quotation is continued in Section 1.

[53] "durch welche Zeit und Raum der letzten Spur objectiver Realität beraubt werden."

[54] "Das Relativitätspostulat in seiner allgemeinsten Fassung, welches die Raumzeit-koordinaten zu physikalisch bedeutungslosen Parametern macht." I thank Professor Roberto Torretti for pointing out that Einstein used words similar to these late in 1913, just before he formulated the "hole" argument (see Einstein 1913b, p. 1257). But he did so in the course of a discussion of the effect of the gravitational field on other physical processes, which he treated by generally covariant equations. He further noted that the gravitational problem would be satisfactorily solved if the gravitational equations were also generally covariant, but that this is not the case. In addition, there is a difference between stating that the coordinates are in themselves meaningless, and stating that space and time have lost the last trace of objective reality. The latter statement first occurs late in 1915, as noted in the text.

[55] "Dadurch verlieren Zeit und Raum den letzten Rest von physikalischer Realität. Es bleibt nur übrig, dass die Welt als vierdimensionales (hyperbolisches) Kontinuum von 4 Dimensionen aufzufassen ist."

[56] "An der Lochbetrachtung war alles richtig bis auf den letzten Schluss. Es hat keinen physikalischen Inhalt, wenn inbezug auf *dasselbe* Koordinatensystem $K$ zwei verschiedene Lösungen $G(x)$ und $G'(x)$ existieren. Gleichzeitig zwei Lösungen in dieselbe Mannigfaltigkeit hineinzudenken, hat keinen Sinn und das System $K$ hat ja keine physikalische Realität. Anstelle der Lochbetrachtung tritt folgende Überlegung. *Real* ist physikalisch nichts als die Gesamtheit der raum-zeitlichen Punktkoinzidenzen. Wäre z.B. das physikalische Geschehen aufzubauen aus Bewegungen materieller Punkte allein, so wären die Begegnungen der Punkte, d.h. die Schnittpunkte ihrer Weltlinien das einzig Reale, d.h. prinzipiell beobachtbare. Diese Schnittpunkte bleiben natürlich bei allen Transformationen erhalten (und es kommen keine neuen hinzu), wenn nur gewisse Eindeutigkeitsbedingungen gewahrt bleiben. Es ist also das natürlichste, von den Gesetzen zu verlangen, dass sie nicht *mehr* bestimmen als die Gesamtheit der zeiträumlichen Koinzidenzen. Das wird nach dem Gesagten bereits durch allgemein kovariante Gleichungen erreicht."

[57] "In §12 meiner Arbeit von letzten Jahre ist alles richtig (in der ersten 3 Absätzen) bis auf das am Ende des dritten Absatzes gesperrt Gedruckte. Daraus, dass die beide Systeme $G(x)$ und $G'(x)$, auf das gleiche Bezugssystem bezogen, den Bedingungen des Grav. Feldes genügen, folgt noch gar kein Widerspruch gegen die Eindeutigkeit des Geschehens. Das scheinbar Zwingende dieser Überlegung geht sofort verloren, wenn man bedenkt, dass 1) das Bezugssystem nichts Reales bedeutet 2) dass die (gleichzeitige) Realisierung zweier verschiedener $g$-Systeme (besser gesagt zweier verschiedener Grav. Felder), in demselben Bereich des Kontinuums der Natur der Theorie nach unmöglich ist.

"An Stelle des §12 hat folgende Überlegung zu treten. Das physikalisch Reale an dem Weltgeschehen (im Gegensatz zu dem von der Wahl des Bezugssystem Abhängigen) besteht *in raumzeitlichen Koinzidenzen*. [In the footnote: "und in nichts anderem!"]

Real sind z.B. die Schnittpunkte zweier verschiedener Weltinien, bezw. die Aussage, daß sie einander *nicht* schneiden. Diejenigen Aussagen, welche sich auf das physikalisch-Reale beziehen, gehen daher durch keine (eindeutige) Koordinatentransformationen verloren. Wenn zwei Systeme der $g_{\mu\nu}$ (bezw. allg. der zur Beschreibung der Welt verwandten Variabeln) so beschaffen sind, daß man das zweite aus dem ersten durch blosse Raum-Zeit-Transformation erhalten kann, so sind sie völlig gleichbedeutend. Denn sie haben alle zeiträumlichen Punktkoinzidenzen gemeinsam, d.h. alles Beobachtbare.

"Diese Überlegung zeigt zugleich wie natürlich die Forderung der allgemeinen Kovarianz ist."

[58] "*Die allgemeinen Naturgesetze sind durch Gleichungen auszudrücken, die für alle Koordinatensysteme gelten, d.h. die beliebigen Substitutionen gegenüber kovariant (allgemein kovariant) sind.*

"Est ist Klar, daß eine Physik, welche diesem Postulat genügt, dem allgemeinen Relativitätspostulat gerecht wird. Denn in *allen* Substitutionen sind jedenfalls auch diejenigen enthalten, welche allen Relativbewegungen der (dreidimensionalen) Koordinatensysteme entsprechen. Daß diese Forderung der allgemeinen Kovarianz, welche dem Raum und der Zeit den letzten Rest physikalischer Gegenständlichkeit nehmen, eine natürliche Forderung ist, geht aus folgender Überlegung hervor. Alle unsere zeiträumlichen Konstatierungen laufen stets auf die Bestimmung zeiträumlicher Koinzidenzen hinaus. Bestände beispielsweise das Geschehen nur in der Bewegung materieller Punkte, so wäre letzten Endes nichts beobachtbar als die Begegnungen zweier oder mehrerer dieser Punkte. Auch die Ergebnisse unserer Messungen sind nichts anderes als die Konstatierung derartiger Begegnungen materieller Punkte unserer Maßstäbe mit anderen materiellen Punkten bzw. Koinzidenzen zwischen Uhrzeigern, Zifferblattpunkten und ins Auge gefaßten, am gleichen Orte und zur gleichen Zeit stattfindenden Punktereignissen.

"Die Einführung eines Bezugssystems dient zu nichts anderem als zur leichteren Beschreibung der Gesamtheit solcher Koinzidenzen. Man ordnet der Welt vier zeiträumliche Variable $x_1$, $x_2$, $x_3$, $x_4$ zu, derart, daß jedem Punktereignis ein Wertesystem der Variablen $x_1 \ldots x_4$ entspricht. Zwei koinzidierenden Punktereignissen entspricht dasselbe Wertesystem der Variablen $x_1 \ldots x_4$; d.h. die Koinzidenz ist durch die Übereinstimmung der Koordinaten charakterisiert. Führt man statt der Variablen $x_1 \ldots x_4$ beliebige Funktionen derselben, $x'_1$, $x'_2$, $x'_3$, $x'_4$ als neues Koordinatensystem ein, so daß die Wertesysteme einander eindeutig zugeordnet sind, so ist die Gleichheit aller vier Koordinaten auch im neuen System der Ausdruck für die raumzeitliche Koinzidenz zweier Punktereignisse. Da sich alle unsere physikalischen Erfahrungen letzten Endes auf solche Koinzidenzen zurückführen lassen, ist zunächst kein Grund vorhanden, gewisse Koordinatensysteme vor anderen zu bevorzugen, d.h. wir gelangen zu der Forderung der allgemeinen Kovarianz."

[59] For a discussion of Einstein's use of the concept of observability in principle and its later use in quantum theory, see Stachel 1986a, pp. 369–374.

REFERENCES

Bishop, Richard L., and Goldberg, Samuel I. (1968). *Tensor Analysis on Manifolds*. New York and London: Macmillan.

Coolidge, Julian Lowell (1940). *A History of Geometrical Methods*. Oxford: Clarendon Press.

Earman, John, and Glymour, Clark (1978a). "Lost in the Tensors: Einstein's Struggle with Covariance Principles 1912–1916." *Studies in History and Philosophy of Science* 9: 251–278.

———  (1978b). "Einstein and Hilbert: Two Months in the History of General Relativity." *Archive for History of Exact Sciences* 19: 291–308.

Einstein, Albert (1905). "Zur Elektrodynamik bewegter Körper." *Annalen der Physik* 17: 891–921.

———  (1907). "Über das Relativitätsprinzip und die aus demselben gezogenen Folgerungen." *Jahrbuch der Radioaktivität und Elektronik* 4: 411–462.

———  (1912a). "Lichtgeschwindigkeit und Statik des Gravitationsfeldes." *Annalen der Physik* 38: 355–369.

———  (1912b). "Zur Theorie des statischen Gravitationsfeldes." *Annalen der Physik* 38: 443–458.

———  (1913a). "Zum gegenwärtigen Stande des Gravitationsproblems." *Physikalische Zeitschrift* 14: 1249–1260.

———  (1913b). "Physikalische Grundlagen einer Gravitationstheorie." *Naturforschende Gesellschaft Zürich. Vierteljahrsschrift* 58: 284–290.

———  (1914a). "Prinzipielles zur verallgemeinerten Relativitätstheorie und Gravitationstheorie." *Physikalische Zeitschrift* 15: 176–180.

———  (1914b). "Die formale Grundlage der allgemeinen Relativitätstheorie." *Königlich Preussische Akademie der Wissenschaften* (Berlin). *Sitzungsberichte*: 1030–1085.

———  (1915a). "Zur allgemeinen Relativitätstheorie." *Königlich Preussische Akademie der Wissenschaften* (Berlin). *Sitzungsberichte*: 778–786.

———  (1915b). "Zur allgemeinen Relativitätstheorie (Nachtrag)." *Königlich Preussische Akademie der Wissenschaften* (Berlin). *Sitzungsberichte*: 799–801.

———  (1915c). "Erklärung der Perihelbewegung des Merkur aus der allgemeinen Relativitätstheorie." *Königlich Preussische Akademie der Wissenschaften* (Berlin). *Sitzungsberichte*: 831–839.

———  (1915d). "Die Feldgleichungen der Gravitation." *Königlich Preussische Akademie der Wissenschaften* (Berlin). *Sitzungsberichte*: 844–847.

———  (1916). "Die Grundlage der allgemeinen Relativitätstheorie." *Annalen der Physik* 49: 769–822. Reprinted as a separatum Leipzig: Johann Ambrosius Barth, 1916. Quotations are from the English translation "The Foundation of the General Theory of Relativity." In Lorentz et al. *The Principle of Relativity: A Collection of Original Memoirs on the Special and General Theory of Relativity*. W. Perrett and G.B. Jeffery, trans. London: Methuen, 1923; reprint New York: Dover, 1952, pp. 109–164.

———  (1917). *Über die spezielle und die allgemeine Relativitätstheorie. (Gemeinverständlich)*. Braunschweig: Friedrich Vieweg und Sohn.

———  (1921). "Geometrie und Erfahrung." *Preussische Akademie der Wissenschaften* (Berlin). *Sitzungsberichte*: 123–130.

———  (1922). "Vorwort des Autors zur Tschechischen Ausgabe." In *Einstein a Praha*. J. Bičak, ed. Prague: Prometheus, 1979, p. 42; written in 1922 for a Czech edition of Einstein 1917.

———  (1949). "Autobiographical Notes." In *Albert Einstein: Philosopher-Scientist*. Paul Arthur Schilpp, ed. Evanston, Illinois: The Library of Living Philosophers, pp. 1–95. A corrected text and translation has been issued as *Albert Einstein: Autobiographical Notes*. Paul Arthur Schilpp, ed. La Salle, Illinois and Chicago:

Open Court, 1979. Page numbers are cited from the 1949 edition, with translations slightly modified using the 1979 edition.

——— (1955). "Erinnerungen—Souvenirs." *Schweizerische Hochschulzeitung* 28 (*Sonderheft*): 145–148, 151–153. Reprinted as "Autobiographische Skizze." In *Helle Zeit-Dunkle Zeit*. Carl Seelig, ed. Zurich: Europa Verlag, 1956, pp. 9–17.

Einstein, Albert, and Fokker, Adriaan D. (1914). "Die Nordströmsche Gravitationstheorie vom Standpunkt des absoluten Differentialkalküls." *Annalen der Physik* 44: 321–328.

Einstein, Albert, and Grossmann, Marcel (1913). *Entwurf einer verallgemeinerten Relativitätstheorie und einer Theorie der Gravitation. I. Physikalischer Teil von Albert Einstein. II. Mathematischer Teil von Marcel Grossmann*. Leipzig and Berlin: B.G. Teubner. Reprinted with added "Bemerkungen," *Zeitschrift für Mathematik und Physik* 62 (1914): 225–261. Page numbers are cited from the latter version.

——— (1914). "Kovarianzeigenschaften der Feldgleichungen der auf die verallgemeinerte Relativitätstheorie gegründeten Gravitationstheorie." *Zeitschrift für Mathematik und Physik* 63: 215–225.

Freundlich, Erwin (1916). *Die Grundlagen der Einsteinschen Gravitationstheorie*. Berlin: Julius Springer.

Hawking, Stephen W. and Ellis, George F.R. (1973). *The Large-Scale Structure of Space-Time*. Cambridge: Cambridge University Press.

Hermann, Armin, ed. (1968). *Albert Einstein–Arnold Sommerfeld. Briefwechsel*. Basel and Stuttgart: Schwabe.

Herneck, Friedrich (1963). "Zum Briefwechsel Albert Einsteins mit Ernst Mach (Mit zwei unveröffentlichten Einstein-Briefe)." *Forschungen und Fortschritte* 37: 239–243.

Hoffmann, Banesh (1972). "Einstein and Tensors." *Tensor* 26: 157–162.

Kretschmann, Erich (1917). "Über den physikalischen Sinn der Relativitätspostulate, A. Einsteins neue und seine ursprüngliche Relativitätstheorie." *Annalen der Physik* 53: 575–614.

Lanczos, Cornelius (1972). "Einstein's Path from Special to General Relativity." In *General Relativity: Papers in Honor of J.L. Synge*. L. O'Raifeartaigh, ed. Oxford: Clarendon Press, pp. 5–19.

Mehra, Jagdish (1973). "Einstein, Hilbert, and the Theory of Gravitation." In *The Physicist's Concept of Nature*. Jagdish Mehra, ed. Boston and Dordrecht: D. Reidel, pp. 92–178.

Minkowski, Hermann (1909). "Raum und Zeit." *Physikalische Zeitschrift* 10: 104–111.

Misner, Charles W., Thorne Kip S., and Wheeler, John Archibald (1973). *Gravitation*. San Francisco: W.H. Freeman.

Nathan, Otto and Norden, Heinz, eds. (1960). *Einstein on Peace*. New York: Simon and Schuster.

Nordström, Gunnar (1913). "Zur Theorie der Gravitation vom Standpunkt des Relativitätprinzips." *Annalen der Physik* 42: 533–554.

Reich, Karin (1979). "Einstein's mathematische Voraussetzungen," Section 4.1 of "Die Entwicklung des Tensorkalküls." Habilitationsschrift. Munich.

Riemann, Bernhard (1854). "Über die Hypothesen, welche der Geometrie zu Grunde liegen." *Königliche Gesellschaft der Wissenschaften zu Göttingen. Abhandlungen* 13 (1867): 133–152. Page numbers are cited from Bernhard Riemann. *Ueber die Hypothesen, welche der Geometrie zu Grunde liegen*. Hermann Weyl, ed. Berlin: Springer, 1919. An English translation by Henry S. White is found in *A Source Book in Mathematics*. Vol. 2. D.E. Smith, ed. New York and London: McGraw-Hill, 1929, pp. 411–425.

Russell, Bertrand (1897). *An Essay on the Foundations of Geometry*. Cambridge: Cambridge University Press; reprint New York: Dover, 1956.

Sachs, Rainer K. and Wu, Hung-Hsi (1977). *General Relativity for Mathematicians*. New York, Heidelberg, and Berlin: Springer-Verlag.

Schouten, Jan Arnoldus (1954). *Ricci-Calculus: An Introduction to Tensor Analysis*, 2nd. ed. Berlin, Göttingen, and Heidelberg: Springer-Verlag; London and New York: Lange, Maxwell and Springer.

Schrödinger, Erwin (1920). "Grundlinien einer Theorie der Farbenmetrik im Tagessehen." *Annalen der Physik* 63: 397–456, 481–520.

——— (1926). "Die Gesichtsempfindungen." In *Müller-Pouillets Lehrbuch der Physik*, 11th ed. O. Lummer, ed. Vol 2, *Lehre von der strahlende Energie (Optik)*. Braunschweig: Friedrich Vieweg & Sohn, pp. 456–560.

Speziali, Pierre, ed. (1972). *Albert Einstein–Michele Besso. Correspondance, 1903–1955*. Paris: Hermann.

Stachel, John (1979a). "Einstein's Odyssey." *The Sciences* 19: 14–15, 32–34.

——— (1979b). "The Genesis of General Relativity." In *Einstein Symposion Berlin*. H. Nelkowski et al., eds. Berlin, Heidelberg, and New York: Springer-Verlag, pp. 428–442.

——— (1980). "Einstein and the Rigidly Rotating Disk." In *General Relativity and Gravitation One Hundred Years After the Birth of Albert Einstein*. Vol. 1. A. Held, ed. New York: Plenum, pp. 1–15. See this volume, pp. 48–62.

——— (1986a). "Einstein and the Quantum: Fifty Years of Struggle." In *From Quarks to Quasars: Philosophical Problems of Modern Physics*. Robert G. Colodny, ed. Pittsburgh: University of Pittsburgh Press, pp. 349–385.

——— (1986b). "What a Physicist Can Learn from the History of Einstein's Discovery of General Relativity." In *Proceedings of the Fourth Marcel Grossmann Meeting on General Relativity*. Remo Ruffini, ed. Amsterdam: Elsevier, pp. 1857–1862.

Vizgin, Vladimir P., and Smorodinskiĭ, Y.A. (1979). "From the Equivalence Principle to the Equations of Gravitation." *Soviet Physics. Uspekhi* 22: 489–513.

Weinberg, Joseph (1976). "The Geometry of Color." *General Relativity and Gravitation* 7: 135–169.

Weyl, Hermann (1927). *Philosophie der Mathematik und Naturwissenschaft*. Handbuch der Philosophie. Munich: R. Oldenbourg. Quotations are from the revised and augmented English edition, based on a translation by Olaf Helmer, *Philosophy of Mathematics and Natural Science*. Princeton: Princeton University Press, 1949.

# How Einstein Found His Field Equations, 1912–1915*

JOHN NORTON

## 1. Introduction

By the middle of 1913, after less than nine months of collaboration with his mathematician friend Marcel Grossmann, Einstein had discovered virtually all the essential features of his general theory of relativity. For they had succeeded in constructing a gravitation theory in which the laws of nature could be written in a generally covariant form, that is, in a form which remained unchanged under all coordinate transformations (Einstein and Grossmann 1913). But, as Einstein confided to Lorentz, their new theory, the so-called *Entwurf* theory, was marred by an "ugly dark spot."[1] Its gravitational field equations, its most fundamental equations, were not generally covariant. It was not until November 1915 that Einstein could present the now familiar generally covariant field equations of the theory to the Prussian Academy of Sciences. In all, he had spent some three troubled years wrestling with the problem of these field equations.

Some of the highlights of this struggle are now well known. In the *Entwurf* paper, Einstein and Grossmann had come within a hair's breadth of the generally covariant field equations of the final theory. They had considered field equations based on the Ricci tensor—a choice virtually forced on them by the mathematical requirements of general covariance. But they discarded these equations on the ground that they failed to yield the correct Newtonian limit. Shortly after, Einstein came to believe that he had found two proofs for the physical unacceptability of all generally covariant field equations. The more notorious of these was his so-called "hole" argument.

Unfortunately, it has become common to dismiss these crucial turning points in Einstein's work in terms of barely excusable errors, even as simple mathematical slips by Grossmann or Einstein. The argument runs as follows. Generally covariant equations hold by definition in all coordinate systems, whereas the equations of Newtonian gravitation theory do not. So, in the process of recovering Newtonian theory as a limiting case from a generally covariant theory, it is necessary to restrict the set of coordinate systems under consideration. This is usually achieved through the explicit stipulation of a number of additional relations—called "coordinate conditions"—that must also be satisfied by the final solution. But—the argument continues—Einstein and Grossmann were ignorant of their freedom to apply such coordinate conditions and so failed to recover the correct Newtonian limit. Moreover, in working out his "hole" argument, Einstein is supposed not to have recognized the elementary fact that a given physical instance of a gravitational field will be represented by different mathematical functions in different coordinate systems.

My purpose in this paper is twofold. First I will seek to establish that Einstein was fully aware of his freedom to apply coordinate conditions to generally covariant field equations and knew how the process could help recover a Newtonian form from such equations. The evidence for this is contained primarily in one of Einstein's notebooks from this period and is, I think, irrefutable. Second, I will develop a more satisfactory account of Einstein's struggles with his field equations in these three troubled years. In particular, I will be concerned with showing that Einstein's difficulties were based on nontrivial misconceptions and that the path he followed was a thoroughly reasonable one. Stachel (1980b) was the first to try to approach the problem in this way.[2] He has conclued that:

- Einstein's understanding of the form of static gravitational fields in 1913 was inconsistent with his final general theory of relativity and, moreover, with the Ricci tensor as a gravitation tensor. This alone could account for Einstein and Grossman's rejection of the Ricci tensor.
- Einstein's "hole" argument admits a reading in which it focuses on a serious physical problem in the relationship between the space-time manifold and the gravitational field. This reading alone is consistent with Einstein's later resolution of the argument and, in this form, can be seen to contribute decisively to Einstein's understanding of space-time in his new theory.

These two points are essential to the account I offer here of Einstein's work on his field equations in the three years ending in November 1915. In outline, my account runs as follows.

The question whether Einstein was aware of his freedom to apply coordinate conditions to generally covariant field equations at the time of the *Entwurf* paper will be settled by examination of the contents of one of his notebooks from this period of his work in Zurich. It will be clear that Einstein was fully aware of this freedom and even knew of two different coordinate conditions that could be used to reduce the Ricci tensor to a Newtonian form. But, I shall argue, Einstein was not prepared to accept either condition because of a number of related misconceptions.

At the heart of these misconceptions lay the problem of the circumstances under which the ten gravitational potentials of the new theory would reduce to a more manageable single potential. On the basis of his earlier work on gravitation and the principle of equivalence, Einstein believed that there was such a reduction in the case of static fields. He chose the simplest and most natural weak-field equations and again found that they led to a similar reduction in the number of gravitational potentials. These and other signposts all pointed in the same direction. But, unfortunately for Einstein, it was the wrong direction. Both his assumptions about static fields and about the weak-field equations were inconsistent with his final theory. In addition he had one final and puzzling misconception about the form of these weak-field equations in rotating coordinate systems. Together, these were sufficient to thwart Einstein's attempts to construct acceptable generally covariant field equations from the Ricci tensor.

The suspension of the requirement of general covariance was soon to follow. Through these same misconceptions, Einstein convinced himself that if derivatives up to the second order only were considered, the conservation of energy and momentum led to a unique set of field equations which were not generally covariant. This formed the substance of his derivation of the *Entwurf* field equations and precluded any further search for generally covariant field equations.

Now convinced of the fruitlessness of this search, Einstein developed general arguments against the physical acceptability of all generally covariant field equations. The first of these was based on the impossibility of constructing a generally covariant conservation law in which the energy-momentum of the gravitational field and of other matter would each be represented by a generally covariant tensor. This argument was soon eclipsed by what appeared to be a far stronger one, the notorious "hole" argument. In it, he purported to show that generally covariant field equations could not uniquely determine the field generated by certain simple distributions of source masses, in contradiction with the requirement of physical causality.

Contrary to the usual account, the "hole" argument was not based on the naive misunderstanding that a given gravitational field is somehow physically changed by the transition to a new coordinate system simply because the mathematical functions that represent it have changed. But it still failed to establish the untenability of generally covariant field equations. Rather, the argument amounted to a demonstration that generally covariant field equations cannot uniquely determine the field as long as the point events of the space-time manifold are incorrectly thought of as individuated independently of the field itself. More figuratively, it showed that, if one could somehow take away the field, one would not be left with a bare space-time manifold replete with individual points. Nothing, not even this, would remain. Einstein did not interpret his "hole" argument in this way until his return to general covariance in November 1915.

Einstein's move from Zurich to Berlin in April 1914 ended his collaboration with Grossmann. But before this, Grossmann was still able to help Einstein with many of the preliminaries of the task of refining the mathematical formulation of the *Entwurf* theory and resolving the question of the theory's exact relation to the generalized principle of relativity. By means of variational techniques, Einstein developed a general way of formulating those field equations, which had exactly the maximum covariance permitted by his "hole" argument. He believed that this analysis led uniquely to his original *Entwurf* field equations, without any significant use of empirical knowledge of gravitation. Einstein was especially pleased with this outcome because it clearly demonstrated that the foundations of his new theory lay in covariance considerations. But he was unaware that his analysis by no means led uniquely to his *Entwurf* field equations. In the last step of his derivation he had made a mistake, for which I have been unable to find any explanation.

However, it has not generally been noted that Einstein's work on this question was not entirely in vain. He was able to use the mathematical machinery that he developed for it virtually unchanged in 1916 in his analysis of his final generally covariant field equations of November 1915. In particular, he developed a device for yielding four "adapted" coordinate conditions. These conditions had to be satisfied if the *Entwurf* field equations were to hold in a given coordinate system. With his final generally covariant field equations of November 1915, this device yielded an important set of identities, now known as the contracted Bianchi identities, from which the conservation laws could be derived.

Einstein's return to the search for generally covariant field equations towards the end of 1915 came after a period of growing dissatisfaction with his *Entwurf* theory. The theory did not account for the known anomaly in the motion of Mercury and he found, contrary to his earlier belief, that it was not covariant under transformations to rotating coordinate systems, which he felt was required by the generalized principle of relativity. His return to this search was precipitated by his discovery of the mistake in the final step of his derivation of the *Entwurf* field equations in 1914.

But Einstein could not yet proceed directly to his final generally covariant field equations, for he was still bedeviled by the same virtually untouched misconceptions about static fields and about the Newtonian limit as he had had three years earlier. The unraveling of these misconceptions can be traced in a dramatic series of weekly communications from Einstein to the Prussian Academy, beginning on November 4, 1915. Hitherto, the story behind Einstein's apparently erratic turns in this final month has remained untold and was, perhaps, untellable, without the clues from the Zurich notebook.

Einstein's first step was to return to a set of almost generally covariant field equations that he had considered with Grossmann three years earlier. Then, however, he had rejected them because he believed that they failed to yield the required weak-field equations in a rotating coordinate system. A week later he showed how the adoption of the hypothesis that all matter is electromagnetic in nature enabled a modification of these equations, which at last was generally covariant and which also satisfied the restrictive requirements of his enduring misconceptions about static fields and the weak-field equations. But these were still not the field equations of his final theory. Einstein was freed from the misconceptions that separated him from these equations through his successful calculation of the orbit of Mercury, which he reported the following week. There he was confronted with a static field in which the ten gravitational potentials did not reduce to a single potential in the way he had expected for so long.

This freed him to entertain a wide range of generally covariant field equations and the following week, with a weary tone of finality, he reported the generally covariant field equations of the final theory to the Prussian Academy. I argue that Hilbert's simultaneous discovery of these equations played little, if any, role in Einstein's final solution, despite the intense correspondence between them at this time. The delay in Einstein's discovery can be explained entirely in terms of the difficulties outlined here and a natural pathway to the final equations can be reconstructed for that final week. In addition, there is evidence that Einstein was unaware of the exact nature of Hilbert's discovery for several months.

## 2. Prelude: From 1905 to 1912

In June 1905, while still a patent examiner in Bern, Einstein submitted his famous work on the electrodynamics of moving bodies to the *Annalen der Physik*. This work contained his special theory of relativity, in which he asserted the equivalence of all inertial frames of reference as a fundamental postulate of physics. The question that then naturally arose, he recalled later, was whether it was possible to extend this principle of relativity to the more general case of frames of reference in arbitrary states of motion (Einstein, 1933, pp. 286–287). But he could find no workable basis for such an extension, until he tried to incorporate gravitation into his new special theory of relativity for a review article in 1907 (Einstein 1907, 1908). The difficulties of this task led him to a new principle, later to be called the "principle of equivalence."

On the basis of the fact that all bodies fall alike in a gravitational field, Einstein postulated the complete physical equivalence of a homogeneous gravitational field and a uniform acceleration of the frame of reference. This, Einstein noted in his review article, extended the principle of relativity to the case of uniform acceleration. It also foreshadowed the problem whose complete solution would lead him to his general theory of relativity: the construction of a relativistically acceptable theory of gravitation, based on the principle of equivalence.

Einstein did not publish any further on this question until 1911 and 1912, the years of his stay in Prague. Then he developed his speculations of 1907 into a substantial and innovative theory of static gravitational fields (Einstein 1911, 1912a, 1912b). His strategy was simple. He would consider a frame of reference in uniform acceleration. According to his principle of equivalence, the acceleration yielded a homogeneous gravitational field whose effect on a given phenomenon could be readily inferred. The result of this special case could be generalized easily to arbitrary static fields. He found that clocks were slowed by gravitational fields and that the now variable speed of light $c$ could stand for the single gravitational potential of static fields.

During this period, Einstein learned a lesson that would be important in his later search for the field equations of his general theory of relativity. He found that the conservation laws can circumscribe the range of admissible field equations very powerfully. In 1912 he sought a field equation that would describe how a given source distribution would generate the field. From a natural and simple generalization of Newtonian gravitation theory, he postulated

$$\Delta c = kc\sigma$$

for this equation, where $\Delta$ is the Laplacian operator, $\sigma$ the mass density, and $k$ a constant (Einstein 1912a). But he soon found to his dismay that this equation was inconsistent with the equality of action and reaction (Einstein 1912b). In effect, his field equation was inconsistent with the conservation of energy and momentum; it is impossible to construct a gravitational field stress tensor from this field equation and the associated force law. In a protracted discussion of several possible modifications to his theory, Einstein showed how he was forced to a specific and unpalatable resolution: His field equation had to be modified by the subtraction of a particular quantity, $(\text{grad } c)^2/2c$, from its left-hand side. He conceded that he resisted this modification to his field equation, for it required him to limit his principle of equivalence to infinitely small regions of space. Since his principle referred only to homogeneous gravitational fields and uniform acceleration, he found the need for such a limitation very puzzling.

With his return to Zurich in August 1912, Einstein took the major step towards his general theory of relativity. There, as he tells us in the foreword to the Czech edition of his popular book on relativity, he had the decisive idea of the analogy with Gauss's theory of surfaces.[3] Also, he began his collaboration with his mathematician friend, Marcel Grossmann, who assisted him with the unfamiliar mathematics required by the new theory. The first product of this collaboration was the *Entwurf* paper, which contained virtually all the essential features of the final general theory of relativity. It was the work of some months only, for Einstein could write to Paul Ehrenfest late in May 1913 that this new work was to appear within a few weeks.[4]

The essence of the new theory lay in the fusion of a number of earlier developments. Earlier in 1912, on the basis of the consideration of a rotating coordinate system, Einstein had argued that three-dimensional space need no longer remain Euclidean once frames of reference in arbitrary states of motion are introduced (Einstein 1912a, p. 356). Of course, such frames of reference must be introduced if the principle of relativity is to be extended. Also, it followed from Minkowski's work that one could treat the kinematics of Einstein's new special theory of relativity in terms of the geometry of a pseudo-Euclidean four-dimensional space-time. In particular, this meant that the familiar Pythagorean formula for the invariant length $l$ in space

$$dl^2 = dx^2 + dy^2 + dz^2$$

is extended to the pseudo-Euclidean formula for the invariant interval $s$ in space-time

$$ds^2 = -dx^2 - dy^2 - dz^2 + c^2 \, dt^2,$$

where $x$, $y$, and $z$ are the usual Cartesian spatial coordinates, and $t$ is the time coordinate. Just as straight lines in Euclidean space are those of minimal $l$, so the world lines of undeflected particles in space-time are geodesics, lines of extremal $s$.

Perhaps the juxtaposition of these two ideas was sufficient to lead Einstein to the central idea of his new theory, the consideration of space-times with a more general, non-Euclidean geometry than that of special relativity (see Stachel 1980a). Specifically, he considered those in which the interval $s$ is given in terms of the ten components of a symmetric metric tensor, $g_{\mu\nu}$

$$ds^2 = g_{11} \, dx_1^2 + 2g_{12} \, dx_1 \, dx_2 + \cdots + g_{44} \, dx_4^2.$$

The special attraction in this was that the four space-time coordinates, $x_1$, $x_2$, $x_3$, and $x_4$, could be selected arbitrarily, provided the values of the $g_{\mu\nu}$ were adjusted by the appropriate transformation. This associating of space-time coordinate systems with frames of reference suggests an equivalence of all frames of reference, as demanded by a generalized principle of relativity. Moreover, Einstein and Grossmann could turn to the absolute differential calculus of Christoffel, Ricci, and Levi-Civita, which enabled the writing of the basic laws of the new theory in a generally covariant form.

Finally, Einstein could interpret the physical significance of the metric tensor by examining certain special and limiting cases. From the requirement that special relativity be a limiting case, it followed that the metric tensor governed the behavior of rods and clocks in space. In addition, Einstein could compare the space-times of his new theory with the static gravitational fields of his earlier theory and conclude that nonconstancy of the components of the metric tensor corresponded to the presence of a gravitational field. This meant that the metric tensor could be regarded as the generalization of the Newtonian gravitational potential and, in particular, that this single potential was now to be replaced by ten gravitational potentials, the ten components of the metric tensor.

As a part of this reasoning, Einstein made an assumption about the form of static fields that was to cause him a great deal of trouble. In the first section of his part of the *Entwurf* paper, Einstein reviewed the results of his earlier theory of static gravitational fields. Within the second section, he translated some of these results into the formalism of his new theory. He noted that in special relativity—the "usual" theory of relativity, as he put it—the metric tensor degenerates to the simple form

$$\begin{pmatrix} -1 & 0 & 0 & 0 \\ 0 & -1 & 0 & 0 \\ 0 & 0 & -1 & 0 \\ 0 & 0 & 0 & c^2 \end{pmatrix} \qquad (1)$$

where $g_{44} = c^2$ is constant. Recalling his earlier theory, he wrote: "The same type of degeneration appears in static gravitational fields of the type considered just now, only that in these $g_{44} = c^2$ is a function of $x_1, x_2, x_3$" (Einstein and Grossmann 1913, p. 229). This formulation compactly expressed several results of his earlier theory that were now being transported intact to his new theory. According to his earlier theory, in static fields the speed of light $c$ is variable and the rate of clocks varies with it. This is encapsulated in the new theory by allowing $g_{44} = c^2$ to be a function of the three spatial coordinates. Einstein had also concluded that three-dimensional space would remain Euclidean in these static fields. This now meant that, for a proper choice of coordinate system, the remaining components of the metric tensor would retain their constant values as in (1).[5]

Einstein did not discover for three years that this last conclusion about static fields is incorrect. It follows from his final theory of November 1915 that static fields are not spatially flat in all but a very few special cases. But for the time being, Einstein had little reason to doubt this natural extension of the results of his earlier theory. In particular it provided a convenient special case in which the number of gravitational potentials was effectively reduced from ten to a more manageable single potential.

## 3. The Rejection of the Ricci Tensor

The saga of Einstein's search for his general theory of relativity should have ended rapidly and happily here, with the completion of the *Entwurf* paper. And it nearly did. All that he needed was to write the field equations. Einstein opened the critical section 5 in a familiar and promising way by writing the field equations in the general form

$$\kappa \Theta_{\mu\nu} = \Gamma_{\mu\nu}, \qquad (2)$$

where $\kappa$ is a constant, the source term $\Theta_{\mu\nu}$ is the contravariant stress-energy tensor and, the field term $\Gamma_{\mu\nu}$ is the as yet undetermined gravitation tensor, which is to be built up by differential operations out of the metric tensor, $g_{\mu\nu}$.[6] Since the gravitation tensor is the generalization of the corresponding Newtonian quantity, the Laplacian of the Newtonian gravitational potential $\Delta\varphi$, Einstein expected it to be of second order in the derivatives of the metric tensor.

Then Einstein quietly dropped his bombshell. It has proved impossible, he wrote, to find such a differential expression that is a generalization of $\Delta\varphi$ and a generally covariant tensor. Part of his justification is that covariant operations, corresponding to those that generate $\Delta\varphi$ out of $\varphi$ in Newtonian theory, yield degenerate results when applied to the metric tensor. The bulk of the justification, however, is a reference to a particular subsection of Grossmann's "Mathematical Part" of the paper.

Grossmann noted that the prominent position of the "Christoffel four index symbol," that is, the fourth-rank Riemann curvature tensor, would lead us to expect that its second-rank contraction, now called the "Ricci tensor," would be a natural candidate for the gravitation tensor. "However [he wrote] it turns out that this tensor does *not* reduce to the expression $\Delta\varphi$ in the special case of an infinitely weak, static gravitational field" (Einstein and Grossmann 1913, pp. 256–257). With this, both Grossmann and Einstein dropped the question of constructing generally covariant field equations out of the Riemann tensor and, apparently, out of any expression of second order in the derivatives of the metric tensor.

This was a catastrophe. A continued focus on the Riemann curvature tensor would have set them on a royal road to the generally covariant field equations of the final theory. The selection of the Ricci tensor as the gravitation tensor would have given them these equations in the source-free case. The discovery of the additional term necessary for the general case would have been but a small step, as it proved to be for Einstein in November 1915.[7]

It is clear from Grossmann's brief comment that the decision to turn away from the Ricci tensor resulted from a problem preventing recovery of a Newtonian limit from the field equations concerned. But he gives virtually no clues to the way they attempted this recovery or to the exact nature of the problem they encountered. Fortunately, in section 5 of their paper, Einstein went to some pains to establish the form that an appropriate generalization of $\Delta\varphi$ must have and, correspondingly, the form that the gravitation tensor must take if it is to reduce to this expression in appropriate cases. The required form $\Gamma_{\mu\nu}$ is given as

$$\Gamma_{\mu\nu} = \sum_{\alpha\beta} \frac{\partial}{\partial x_\alpha}\left(\gamma_{\alpha\beta}\frac{\partial\gamma_{\mu\nu}}{\partial x_\beta}\right) + \{\quad\}, \tag{3}$$

where the curly brackets indicate terms that drop out in first order when the derivatives of $g_{\mu\nu}$ are small. In brief, this is justified by noting that only the highest-order terms in equation (3) remain in the weak-field case and reduce to an expression of the form

$$\Gamma_{\mu\nu} = -\left(\frac{\partial^2\gamma_{\mu\nu}}{\partial x_1^2} + \frac{\partial^2\gamma_{\mu\nu}}{\partial x_2^2} + \frac{\partial^2\gamma_{\mu\nu}}{\partial x_3^2} - \frac{1}{c^2}\frac{\partial^2\gamma_{\mu\nu}}{\partial x_4^2}\right),$$

provided that the components of the weak field metric differ from those of the special relativistic metric of form (1) by infinitely small quantities. (The right-hand side of the preceding equation will be written as $\Box\gamma_{\mu\nu}$.) From this, $\Gamma_{\mu\nu}$ reduces to a satisfactory Newtonian form, with

$$\Gamma_{44} = -\Delta\gamma_{44}$$

as the only nonvanishing term, in the case of a static field, in which $g_{44}$ alone is variable.

This last case corresponds exactly to the special case considered by Grossmann, that of "an infinitely weak, static gravitational field." This promises to give us an immediate explanation of Einstein and Grossmann's rejection of the Ricci tensor as a gravitation tensor. For a direct inspection of the form of the Ricci tensor, written out explicitly in terms of derivatives of the metric tensor, shows that it fails to satisfy Einstein's condition (3). In fact the Ricci tensor contains four second-derivative terms in the metric tensor, which are of first order of smallness when the derivatives of the metric tensor are small. Only one of these corresponds to the one in condition (3).

However, the elimination of these three other second-derivative terms can be achieved readily in the process of recovery of the Newtonian limit. As I pointed out earlier, generally covariant field equations, such as those based on the Ricci tensor, hold in all coordinate systems, whereas the equations of Newtonian theory do not. So in the process of recovering Newtonian theory we must restrict the set of coordinate systems under consideration, by requiring, for example, the satisfaction of additional constraints. In particular, we could consider coordinate systems in which the "harmonic" coordinate conditions

$$g^{\mu\nu}\begin{Bmatrix}\alpha\\\mu\nu\end{Bmatrix}=0 \qquad (4)$$

are satisfied (the curly bracket is the Christoffel symbol of the second kind). In such coordinate systems, the three additional second derivative terms vanish and the Ricci tensor reduces to the form required by condition (3). The recovery of the expected Newtonian limit then follows readily from the consideration of weak fields in these coordinate systems.

Thus if we wish to explain Einstein and Grossmann's turning away from the Ricci tensor as a result of their inability to reduce it to the form required by condition (3), we must assume that neither had sufficient facility or familiarity with tensor calculus to be able to find such a condition as (4). Or worse, we might assume that neither was even aware of his freedom to apply coordinate conditions.

It is difficult to believe that both Einstein and Grossmann could have been ignorant of this freedom and that their ignorance should have persisted over several years of intense study and reflection. Nevertheless, several commentators believe it.[8] Their approach is made at least provisionally viable by the fact that neither Einstein nor Grossmann made any explicit acknowledgment of their freedom to use such coordinate conditions in their publications at the time of the *Entwurf* paper. The approach has significant attraction: It enables a very simple explanation of the apparent fallacy in Einstein's later "hole" argument against the physical admissibility of all generally covariant field equations. Einstein was unaware, it is supposed, that the imposition of coordinate conditions does not alter the physical content of the laws of his theory; rather it affects only their mathematical form in restricting them to certain coordinate systems. We shall see later that Einstein's "hole" argument can be approached by supposing that Einstein was unaware of a closely related result; that is, that a change in the coordinate system will affect only the mathematical functions representing the physical quantities of his theory, without actually changing the physical quantities themselves. That is, they will "look different" to us, but, of course, remain physically unchanged by our change of viewpoint.[9]

That such an account of Einstein's three-year struggle has become widespread is not surprising. Three out of four of Einstein's presentations of the "hole" argument are extremely brief and admit the standard interpretation. Moreover, the evidence necessary to acquit Einstein of the charge of ignorance of his freedom to apply coordinate conditions is neither as accessible nor, at present, as readily available as his publications from this period. Nevertheless, it seems that the simplifications of the standard account could have been avoided. I do not mean that we should assume that Einstein never made mistakes. In the course of this paper we shall see him make and correct quite a few mistakes. But we could have been more wary of accounts that try to explain the errors of three years of intense searching as beginners' blunders.

The first sustained attempt to write an account of these episodes that would not convict Einstein and Grossmann of such simple and fundamental errors was made by Stachel (1980b). He pointed out that the special case that Grossmann considered was not just that of an infinitely weak field, but that of an infinitely weak, *static* field. He also noted that Einstein's expectations for the form of static fields were inconsistent with his final theory, in the way we have seen here. If Grossmann or Einstein calculated the components of the Ricci tensor $G_{\mu\nu}$ in infinitely weak fields of this special type, it is easy to reconstruct what they found. If the metric of the field is of form (1) and the derivatives of $c$ are small, it follows that the Ricci tensor's only nonvanishing components of the first order are

$$G_{44} = \frac{1}{2}\Delta c^2 = \frac{1}{2}\left(\frac{\partial^2}{\partial x_1^2} + \frac{\partial^2}{\partial x_2^2} + \frac{\partial^2}{\partial x_3^2}\right)c^2$$

$$G_{ij} = -\frac{1}{2c^2}\frac{\partial^2 c^2}{\partial x_i \partial x_j}, \tag{5}$$

where $i$ and $j$ vary over 1, 2, and 3 only. The $G_{44}$ term looks promising, but the remaining terms are disastrous. In the source-free case, the field equations take the form

$$G_{\mu\nu} = 0.$$

With equations (5), this yields the unacceptable conclusion that $c^2$ can vary at most linearly with the spatial coordinates. Thus, if we assume that static gravitational fields have form (1), as Einstein believed, then it follows that the Ricci tensor is unacceptable as the gravitation tensor.

Through this conjecture, we can now see that Einstein and Grossmann's rejection of the Ricci tensor need not be explained in terms of a simple error, but that it may have resulted from a deep-seated misconception about the nature of static fields. Some problems remain, however. Beyond the brief remark of Grossmann's cited earlier, there is no direct evidence that Einstein or Grossmann ever actually performed the calculation described. Of greater importance, it still does not tell us whether Einstein and Grossmann were aware of their freedom to apply coordinate conditions. Of course the assumption that the static field in question has a metric with components of form (1) contains an *implicit* coordinate condition. For the coordinate system must be chosen in such a way that the components of the metric do in fact adopt the required form. But were they *explicitly* aware of this in 1912 and 1913?

## 4. The Zurich Notebook

In the Einstein Archive is a notebook originally cataloged as containing notes for Einstein's lectures at the University of Zurich in the period 1909 to 1911. However, the contents of the notebook, all written in Einstein's hand, are not lecture notes but scratch-pad calculations.[10] The subject matter includes statistical physics, thermodynamics, and basic principles of the four-dimensional representation of electrodynamics. The major part of the notebook, which extends from pages 5 to 29, belongs to 1912 to 1913, for it contains calculations made by Einstein during his work on the *Entwurf* paper. These calculations are accompanied by virtually no explanatory text. Fortunately their import can generally be deciphered.

The section in question begins with the heading "*Gravitation*" and contains various formulas, including a generally covariant expression of the equations of motion of a point mass and the laws of conservation of energy and momentum. Einstein's treatment here closely corresponds to that of the early sections of the *Entwurf* paper. He investigates various basic questions in his new theory, including the properties of general and rotating coordinate transformations. This part also contains a study of generalized d'Alembertian operators and the formation of associated expressions out of the metric tensor and its derivatives. These appear to be some of Einstein's earliest speculations on the problem of constructing field equations for his new theory. He writes next to one term "*vermutlicher Gravitationstensor*"—"presumed gravitation tensor."

After these preliminaries, which stop at page 14, Einstein began what seem to be his earliest attempts to construct a generally covariant gravitation tensor out of the Riemann curvature tensor. He wrote out this fourth-rank curvature tensor explicitly, with the notation "Grossmann/Tensor vierter/ Mannigfaltigkeit"—"Grossmann/tensor of fourth/rank." This, of course, suggests that this expression was provided for him by Grossmann, as we would expect at this early stage of their collaboration. So it is not surprising that

How Einstein Found His Field Equations    113

it matches Grossmann's formula (43) in the *Entwurf* paper exactly—even including the choice of letters used to label the indices. The four second derivative terms of the expression Einstein wrote down were:

$$\frac{1}{2}\left(\frac{\partial^2 g_{im}}{\partial x_\kappa \partial x_l} + \frac{\partial^2 g_{\kappa l}}{\partial x_i \partial x_m} - \frac{\partial^2 g_{il}}{\partial x_\kappa \partial x_m} - \frac{\partial^2 g_{\kappa m}}{\partial x_i \partial x_l}\right). \tag{6}$$

Einstein then multiplied by $\gamma_{\kappa l}$ and contracted over $\kappa$ and $l$ to form the Ricci tensor and proceeded to calculate its first derivative terms explicitly. Then the page was ruled off and Einstein wrote:

$$\text{``}\sum_\kappa \left(\frac{\partial^2 g_{\kappa\kappa}}{\partial x_i \partial x_m} - \frac{\partial^2 g_{i\kappa}}{\partial x_\kappa \partial x_m} - \frac{\partial^2 g_{m\kappa}}{\partial x_\kappa \partial x_i}\right) \text{ ought to vanish.''} \tag{7}$$

By comparison with equation (6), we can see that these three terms are taken from the explicit expression for the second derivative terms of the Ricci tensor.[11]

The point of Einstein's remark seems quite clear. The expression in (7) "ought to vanish" exactly in case Einstein were to require the Ricci tensor to have form (3). Since this expression does not vanish in general, Einstein had found that the Ricci tensor does not have form (3), which he required of a gravitation tensor. So far, what we have seen here is entirely in accord with Einstein's presumed ignorance of his freedom to apply coordinate conditions, for he had not mentioned the coordinate conditions that could be used to make these three terms vanish.

Einstein continued with an extended calculation of the explicit form of the Riemann curvature scalar. He had some difficulty working with all the terms since he dropped a factor of 1/4. But, under the constraint that the determinant of the metric tensor, written as $G$, should be equal to unity, he came to a final result on page 16. There his purpose becomes clear, for he tried to divide the scalar into the contraction of two tensors, $\Sigma_{ik} g_{ik} \mathcal{T}_{ik}$. This $\mathcal{T}_{ik}$ would presumably again be a candidate for the gravitation tensor. He then turned to a calculation of the contravariant components of the Ricci tensor. This calculation is broken off early as "*zu umständlich*," "too involved."

So far, Einstein seems to have made no real progress since he discovered the failure of the Ricci tensor to take on form (3). He introduced a new result of crucial importance, however, on page 19. This is the condition:

$$\gamma_{\kappa l}\begin{bmatrix} \kappa\ l \\ i \end{bmatrix} = 0. \tag{8}$$

Five pages earlier Einstein defined the term in square brackets to be the Christoffel symbol of the first kind. So, in assuming summation as implied over repeated indices, we recognize equation (8) as harmonic coordinate condition (4). Einstein proceeded to confirm that this condition reduces the Ricci tensor to the required form of equation (3). On pages 19–22, Einstein studied the behavior of the weak field that would follow from taking the Ricci

114   John Norton

tensor as a gravitation tensor in harmonic coordinates. He focused on the energy-momentum conservation law and the construction of a gravitational field stress-energy tensor.

Whatever he found there must not have pleased him greatly, for on page 22 he took a completely new approach to the problem of constructing a generally covariant gravitation tensor from the Riemann curvature tensor. Under the heading "Grossmann" he wrote an expression for the covariant Ricci tensor, denoted by $\mathcal{T}_{il}$, which exactly matches the contraction of Grossmann's expression (44) for the Riemann curvature tensor in the *Entwurf* paper. Einstein declared:

If $G$ is a scalar, then $\dfrac{\partial lg\sqrt{G}}{\partial x_i} = \mathcal{T}_i$ tensor of 1st rank;

$$\mathcal{T}_{il} = \underbrace{\left(\frac{\partial \mathcal{T}_i}{\partial x_l} - \sum \begin{Bmatrix} i & l \\ \lambda \end{Bmatrix} \mathcal{T}_\lambda \right)}_{\text{tensor of 2nd rank}} - \underbrace{\sum_{\kappa\lambda} \left( \frac{\partial \begin{Bmatrix} i & l \\ \kappa \end{Bmatrix}}{\partial x_\kappa} - \begin{Bmatrix} i & \kappa \\ \lambda \end{Bmatrix} \begin{Bmatrix} l & \lambda \\ \kappa \end{Bmatrix} \right)}_{\substack{\text{presumed gravitation} \\ \text{tensor. } \mathcal{T}_{il}^\times}} \quad (9)$$

The interpretation of this equation is made very easy by the fact that Einstein's procedure corresponds exactly to his first method of constructing a generally covariant gravitation tensor when he returned to the problem in late 1915 (Einstein 1915a).[12] There he showed that if one restricts oneself to coordinate transformations with a determinant of one, then $G$ becomes a scalar, the first term in the expansion of the Ricci tensor is itself a tensor as marked, and thus the second term, which is taken as the gravitation tensor, is also a tensor.

This close correspondence to the content of Einstein's later paper is in itself a fascinating discovery. It also provides an unexpected confirmation of Einstein's claim in the introduction to that paper that, three years earlier, he and Grossmann "had actually already come quite close to the solution of the problem given in the following" (Einstein 1915a, p. 778).[13] The reason that Einstein elected to split up the Ricci tensor in this way appears on page 22. There he showed that the application of the coordinate condition

$$\sum_\kappa \frac{\partial \gamma_{\kappa\alpha}}{\partial x_\kappa} = 0 \quad (10)$$

leads to the reduction of the new gravitation tensor to the form required in equation (3).

Einstein next investigated the behavior of his new gravitation tensor and coordinate condition particularly in connection with the energy-momentum conservation law. He constructed what I take to be the (coordinate) divergence of the gravitational field stress-energy tensor, which, he confirmed, vanishes in Minkowski space-time when viewed from rotating coordinates. Then, quite abruptly, he broke off the search for generally covariant gravitation tensors.

On page 26, under the heading "*Ableitung der Gravitationsgleichungen*"—
"derivation of the gravitation equations"—Einstein wrote out, or perhaps
transcribed, a tight summary of the derivation of identity (12) of his part of
the *Entwurf* paper. This comprises a major part of the derivation of the
*Entwurf* field equations. Its appearance here, and the want of these equations
earlier in the sequence of the bound pages of the notebook, indicate that the
*Entwurf* field equations and their derivation came after the attempts to construct an acceptable generally covariant gravitation tensor outlined earlier.[14]

4.1 THE PROBLEM OF COORDINATE CONDITIONS

From Einstein's notebook we learn that his search for a generally covariant
gravitation tensor was dominated by the requirement that it take the form
given in equation (3). We also see that he was already quite aware of his
freedom to apply coordinate conditions to achieve this form and that he
discovered *two* suitable gravitation tensors that reduced to the required form
with appropriate coordinate conditions.

In solving one problem, however, we have created another. Given that
Einstein had these results prior to the completion of the *Entwurf* paper, why
did he reject them and construct field equations of severely limited covariance?
Unfortunately, the question cannot be answered decisively on the basis of
the material at hand. However, we can note that there are two persistent
characteristics of the calculations in which Einstein appears to test out his
proposed gravitation tensors. First, they are concerned with the conservation
of energy-momentum and the construction of a gravitational field stress-
energy tensor. Second, they relate to a few special cases: weak and static fields
and Minkowski space-time viewed from rotating coordinates. These are sufficient clues to enable a reasonable conjecture why Einstein rejected these
gravitation tensors. I conjecture that his reasons reduce to two basic points:

- Both of the coordinate conditions that Einstein considered fail in at least
  one of the special cases in which he would have expected them to hold.
- Einstein believed that the derivation of the *Entwurf* field equations gave a
  unique result, which disagreed with both of the gravitation tensors proposed
  in the notebook, even after the application of the coordinate conditions.

I will return to the second point at the end of this section.

To begin analysis of the first point, note that Einstein's most favored special
case in this period is that of the static gravitational field, which he assumed
to have the form given in (1), where $c^2$ is a function of the spatial coordinates.
Einstein certainly considered this special case in the notebook (on pages 6 and
21) and after his examination of the Ricci tensor in harmonic coordinates.
Now such a field in form (1) does not satisfy the harmonic coordinate
condition. So the Ricci tensor would appear to Einstein not to reduce to the
required form of (3) in this most basic of cases. Einstein would have regarded
this failure as a major defect—perhaps in itself sufficient basis for Grossmann's

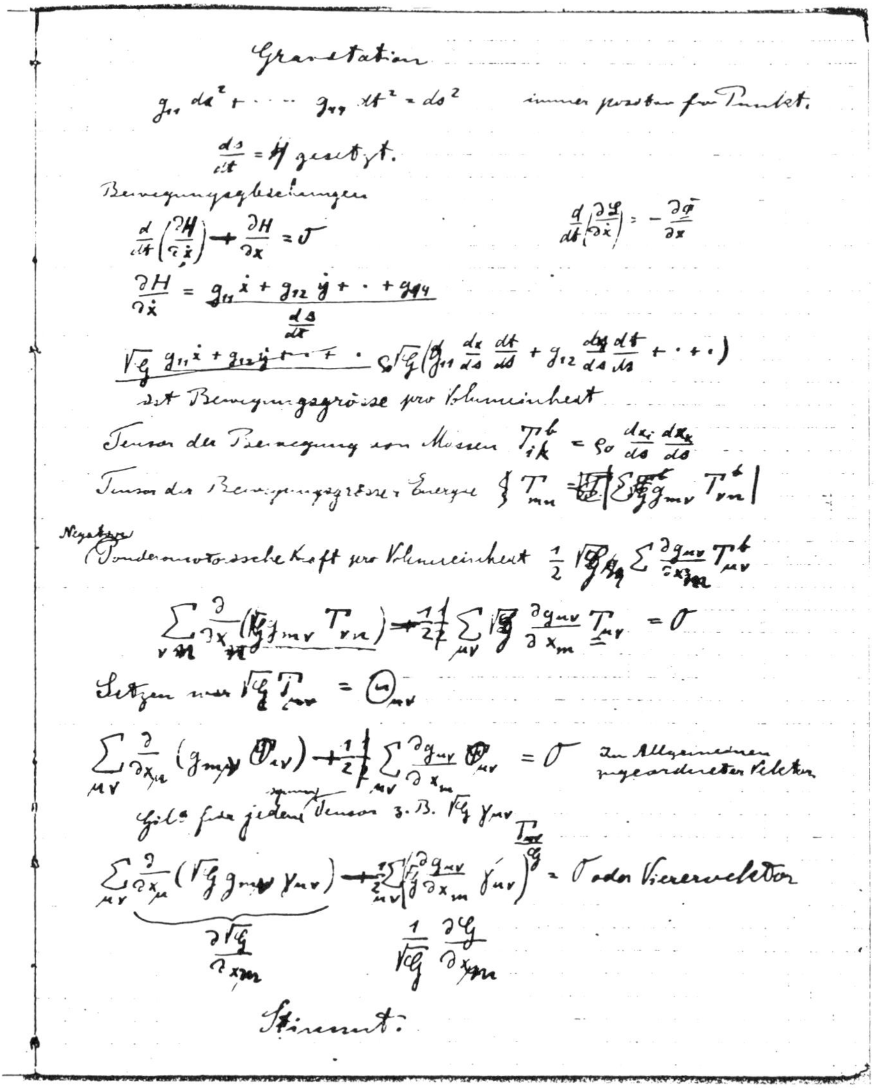

FIGURE 1. Zurich notebook, page 5 (right-hand side). The subject of gravitation is introduced with some elementary results from the *Entwurf* theory. Compare Einstein and Grossmann 1913, pp. 229, 232. These results include the equations of motion of a unit point-mass, written in a Hamiltonian form, and the identification of various "quantities of motion," which are combined to yield the law of conservation of energy and momentum, written as the vanishing of the covariant divergence of the stress-energy tensor. © Hebrew University of Jerusalem, reproduced by permission.

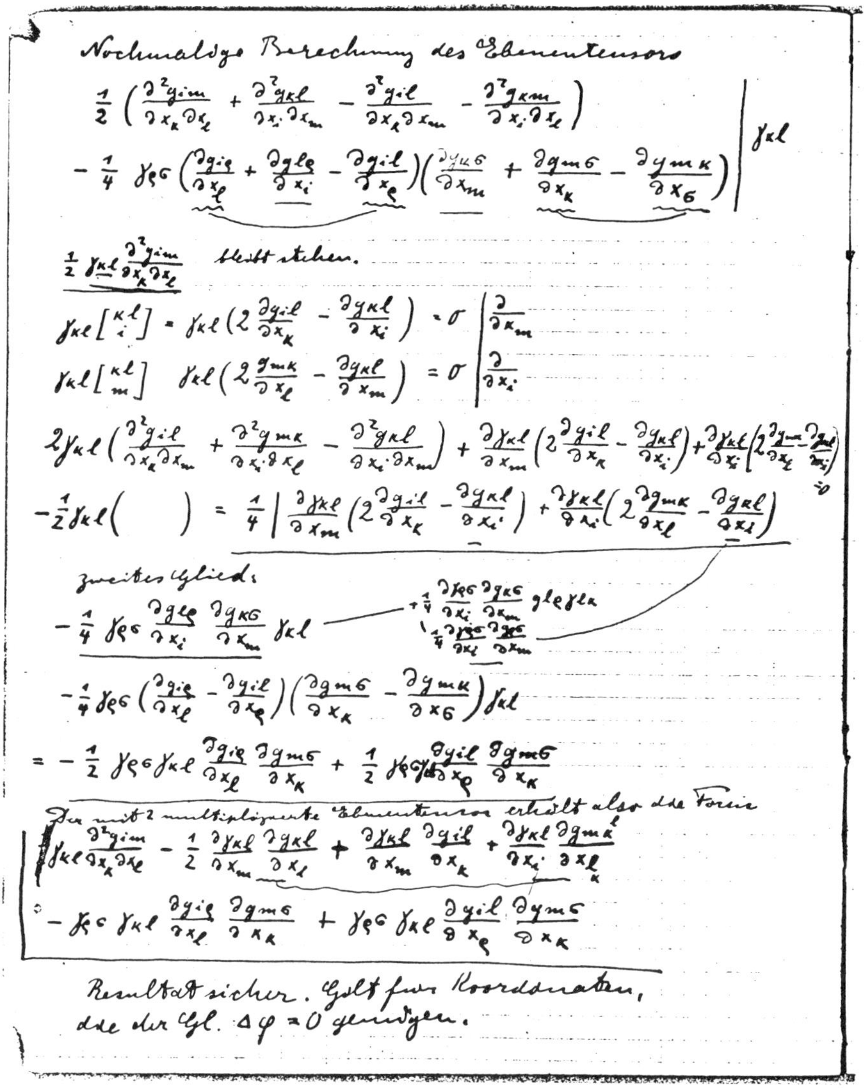

Nochmalige Berechnung des Elemententensors

$$\frac{1}{2}\left(\frac{\partial^2 g_{im}}{\partial x_\kappa \partial x_l} + \frac{\partial^2 g_{\kappa l}}{\partial x_i \partial x_m} - \frac{\partial^2 g_{il}}{\partial x_\kappa \partial x_m} - \frac{\partial^2 g_{\kappa m}}{\partial x_i \partial x_l}\right)$$

$$-\frac{1}{4}\gamma_{\varrho\sigma}\left(\frac{\partial g_{i\varrho}}{\partial x_l} + \frac{\partial g_{l\varrho}}{\partial x_i} - \frac{\partial g_{il}}{\partial x_\varrho}\right)\left(\frac{\partial g_{\kappa\sigma}}{\partial x_m} + \frac{\partial g_{m\sigma}}{\partial x_\kappa} - \frac{\partial g_{m\kappa}}{\partial x_\sigma}\right) \Bigg| \gamma_{\kappa l}$$

$\frac{1}{2}\gamma_{\kappa l}\frac{\partial^2 g_{im}}{\partial x_\kappa \partial x_l}$ bleibt stehen.

$$\gamma_{\kappa l}\begin{bmatrix}\kappa l\\ i\end{bmatrix} = \gamma_{\kappa l}\left(2\frac{\partial g_{il}}{\partial x_\kappa} - \frac{\partial g_{\kappa l}}{\partial x_i}\right) = 0 \quad \Big| \frac{\partial}{\partial x_m}$$

$$\gamma_{\kappa l}\begin{bmatrix}\kappa l\\ m\end{bmatrix} \quad \gamma_{\kappa l}\left(2\frac{\partial g_{m\kappa}}{\partial x_l} - \frac{\partial g_{\kappa l}}{\partial x_m}\right) = 0 \quad \Big| \frac{\partial}{\partial x_i}$$

$$2\gamma_{\kappa l}\left(\frac{\partial^2 g_{il}}{\partial x_\kappa \partial x_m} + \frac{\partial^2 g_{m\kappa}}{\partial x_i \partial x_l} - \frac{\partial^2 g_{\kappa l}}{\partial x_i \partial x_m}\right) + \frac{\partial \gamma_{\kappa l}}{\partial x_m}\left(2\frac{\partial g_{il}}{\partial x_\kappa} - \frac{\partial g_{\kappa l}}{\partial x_i}\right) + \frac{\partial \gamma_{\kappa l}}{\partial x_i}\left(2\frac{\partial g_{m\kappa}}{\partial x_l} - \frac{\partial g_{\kappa l}}{\partial x_m}\right) = 0$$

$$-\frac{1}{2}\gamma_{\kappa l}(\quad) = \frac{1}{4}\left|\frac{\partial \gamma_{\kappa l}}{\partial x_m}\left(2\frac{\partial g_{il}}{\partial x_\kappa} - \frac{\partial g_{\kappa l}}{\partial x_i}\right) + \frac{\partial \gamma_{\kappa l}}{\partial x_i}\left(2\frac{\partial g_{m\kappa}}{\partial x_l} - \frac{\partial g_{\kappa l}}{\partial x_m}\right)\right.$$

zweites Glied:

$$-\frac{1}{4}\gamma_{\varrho\sigma}\frac{\partial g_{l\varrho}}{\partial x_i}\frac{\partial g_{\kappa\sigma}}{\partial x_m}\gamma_{\kappa l} \quad + \frac{1}{4}\frac{\partial \gamma_{\varrho\sigma}}{\partial x_i}\frac{\partial g_{\kappa\sigma}}{\partial x_m} g_{l\varrho}\gamma_{l\kappa}$$

$$-\frac{1}{4}\gamma_{\varrho\sigma}\left(\frac{\partial g_{i\varrho}}{\partial x_l} - \frac{\partial g_{il}}{\partial x_\varrho}\right)\left(\frac{\partial g_{m\sigma}}{\partial x_\kappa} - \frac{\partial g_{m\kappa}}{\partial x_\sigma}\right)\gamma_{\kappa l}$$

$$= -\frac{1}{2}\gamma_{\varrho\sigma}\gamma_{\kappa l}\frac{\partial g_{i\varrho}}{\partial x_l}\frac{\partial g_{m\sigma}}{\partial x_\kappa} + \frac{1}{2}\gamma_{\varrho\sigma}\gamma_{\kappa l}\frac{\partial g_{il}}{\partial x_\varrho}\frac{\partial g_{m\sigma}}{\partial x_\kappa}$$

Der mit 2 multiplizierte Elemententensor erhält also die Form

$$\gamma_{\kappa l}\frac{\partial^2 g_{im}}{\partial x_\kappa \partial x_l} - \frac{1}{2}\frac{\partial \gamma_{\kappa l}}{\partial x_m}\frac{\partial g_{\kappa l}}{\partial x_i} + \frac{\partial \gamma_{\kappa l}}{\partial x_m}\frac{\partial g_{il}}{\partial x_\kappa} + \frac{\partial \gamma_{\kappa l}}{\partial x_i}\frac{\partial g_{m\kappa}}{\partial x_l}$$

$$-\gamma_{\varrho\sigma}\gamma_{\kappa l}\frac{\partial g_{i\varrho}}{\partial x_l}\frac{\partial g_{m\sigma}}{\partial x_\kappa} + \gamma_{\varrho\sigma}\gamma_{\kappa l}\frac{\partial g_{il}}{\partial x_\varrho}\frac{\partial g_{m\sigma}}{\partial x_\kappa}$$

Resultat sicher. Gilt für Koordinaten, die der Gl. $\Delta\varphi = 0$ genügen.

FIGURE 2. Zurich notebook, page 19 (left-hand side). In the first three lines, Einstein shows that he intends to contract the Riemann curvature tensor with $\gamma_{\kappa l}$ leaving only a d'Alembertian-like term in second derivatives of the metric. He succeeds easily with the harmonic condition $\gamma_{\kappa l}\begin{bmatrix}\kappa l\\ i\end{bmatrix} = 0$. Of the final result, given in the last two lines, he remarks, "Result certain. Holds for coordinates that satisfy the eq[uation] $\Delta\varphi = 0$." © Hebrew University of Jerusalem, reproduced by permission.

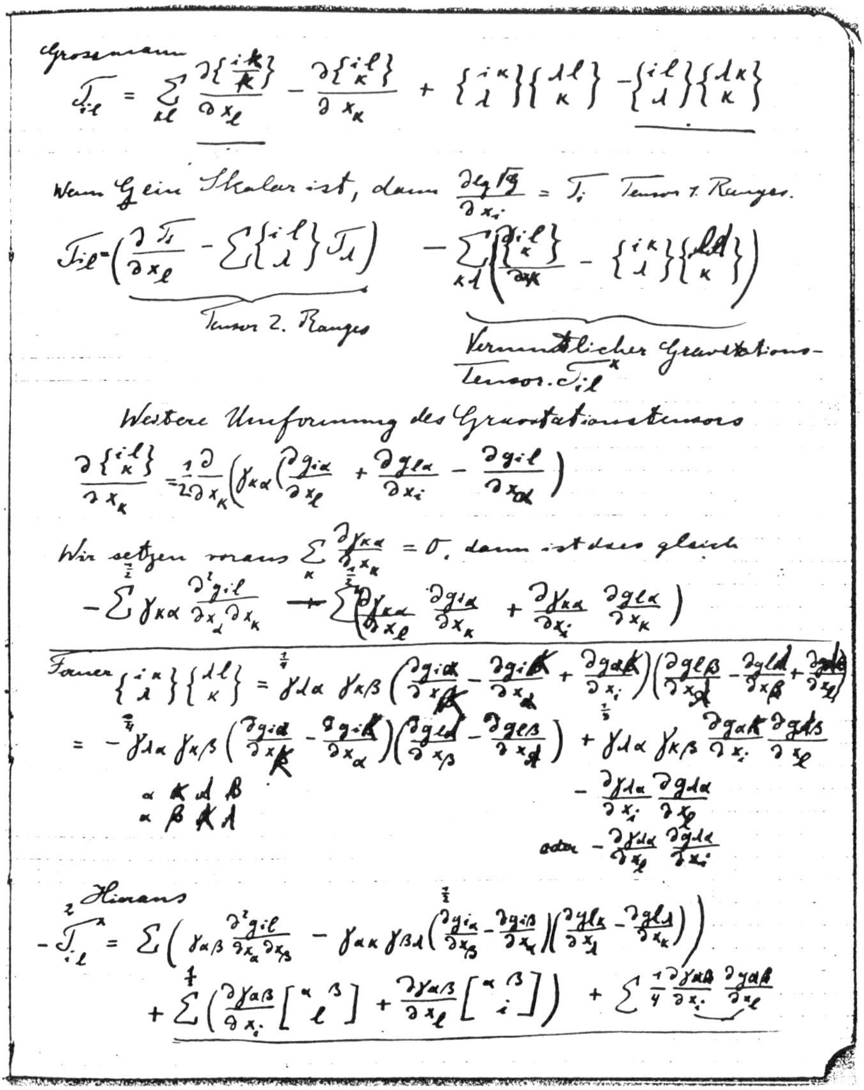

Grossmann

$$T_{il} = \sum_{kl} \frac{\partial \left\{ {ik \atop k} \right\}}{\partial x_l} - \frac{\partial \left\{ {il \atop k} \right\}}{\partial x_k} + \left\{ {ik \atop \lambda} \right\}\left\{ {\lambda l \atop k} \right\} - \left\{ {il \atop \lambda} \right\}\left\{ {\lambda k \atop k} \right\}$$

Wenn G ein Skalar ist, dann $\frac{\partial \lg \sqrt{G}}{\partial x_i} = T_i$ Tensor 1. Ranges.

$$T_{il} = \left(\frac{\partial T_i}{\partial x_l} - \sum \left\{ {il \atop \lambda} \right\} T_\lambda\right) - \sum_{k\lambda}\left(\frac{\partial \left\{ {il \atop k} \right\}}{\partial x_k} - \left\{ {ik \atop \lambda} \right\}\left\{ {\lambda l \atop k} \right\}\right)$$

Tensor 2. Ranges — Vermutlicher Gravitationstensor $T_{il}^x$

Weitere Umformung des Gravitationstensors

$$\frac{\partial \left\{ {il \atop k} \right\}}{\partial x_k} = \frac{1}{2}\frac{\partial}{\partial x_k}\left(\gamma_{k\alpha}\left(\frac{\partial g_{i\alpha}}{\partial x_l} + \frac{\partial g_{l\alpha}}{\partial x_i} - \frac{\partial g_{il}}{\partial x_\alpha}\right)\right)$$

Wir setzen voraus $\sum_k \frac{\partial \gamma_{k\alpha}}{\partial x_k} = 0$, dann ist dies gleich

$$-\sum \gamma_{k\alpha}\frac{\partial^2 g_{il}}{\partial x_\alpha \partial x_k} + \sum\left(\frac{\partial \gamma_{k\alpha}}{\partial x_l}\frac{\partial g_{i\alpha}}{\partial x_k} + \frac{\partial \gamma_{k\alpha}}{\partial x_i}\frac{\partial g_{l\alpha}}{\partial x_k}\right)$$

Ferner $\left\{ {ik \atop \lambda} \right\}\left\{ {\lambda l \atop k} \right\} = \frac{1}{4}\gamma_{\lambda\alpha}\gamma_{k\beta}\left(\ldots\right)\left(\ldots\right)$ …

Hieraus

$$-T_{il}^x = \sum\left(\gamma_{\alpha\beta}\frac{\partial^2 g_{il}}{\partial x_\alpha \partial x_\beta} - \gamma_{\alpha k}\gamma_{\beta\lambda}\left(\frac{\partial g_{i\alpha}}{\partial x_\beta} - \frac{\partial g_{i\beta}}{\partial x_\alpha}\right)\left(\frac{\partial g_{lk}}{\partial x_\lambda} - \frac{\partial g_{l\lambda}}{\partial x_k}\right)\right) + \sum\left(\frac{\partial \gamma_{\alpha\beta}}{\partial x_i}\begin{bmatrix}\alpha\beta\\ l\end{bmatrix} + \frac{\partial \gamma_{\alpha\beta}}{\partial x_l}\begin{bmatrix}\alpha\beta\\ i\end{bmatrix}\right) + \sum \frac{1}{4}\frac{\partial \gamma_{\alpha\beta}}{\partial x_i}\frac{\partial g_{\alpha\beta}}{\partial x_l}$$

FIGURE 3. Zurich notebook, page 22 (right-hand side). In the second equation, Einstein breaks up the Ricci tensor into two parts, as described in the text, the second being his presumed gravitation tensor. Application of the coordinate condition $\sum_\kappa \partial\gamma_{\kappa\alpha}/\partial x_\kappa = 0$ enables reduction of this tensor to a form whose only second-derivative term in the metric is of the required d'Alembertian form. The complete reduced form of the gravitation tensor is at the bottom of the page. © Hebrew University of Jerusalem, reproduced by permission.

claim that the Ricci tensor does not reduce to $\Delta\varphi$ in the case of a weak, static field. It is not surprising that Einstein should then have continued to search for another generally covariant gravitation tensor and proceeded to arrive (on page 22) at $\mathcal{T}_{il}^{\times}$. The coordinate condition associated with this tensor, equation (10), is satisfied in Einstein's static field, with the metric (1).

In the last relevant part of the notebook (page 24 ff.), Einstein reintroduced the case of a Minkowski space-time viewed from Cartesian coordinates in uniform rotation. We know that Einstein regarded this case as crucial to the general relativity of motion, so he expected his field equations to hold in such a case (Einstein, 1914b, pp. 1067–1068). We shall see that Einstein dealt with this case in a manner that suggests that he expected the gravitation tensor to retain form (3). This assumption was natural. For he readily, but mistakenly, came to believe that his *Entwurf* field equations held in such a rotating system—and that they have a gravitation tensor of form (3).[15] Now coordinate condition (10) does not hold in this case, nor does the harmonic condition, another argument for Einstein against the Ricci tensor. If Einstein approached the problem in the way set out here, he would inevitably be drawn to reject $\mathcal{T}_{il}^{\times}$ as a gravitation tensor.

This interpretation is consistent with a letter Einstein wrote to Sommerfeld late in 1915, in which he told him of his success in discovering generally covariant field equations:

> One can eminently simplify the whole theory by choosing the reference system so that $\sqrt{-g} = 1$. Then the equations take on the form
>
> $$-\sum_l \frac{\partial \begin{Bmatrix} i\ m \\ l \end{Bmatrix}}{\partial x_l} + \sum_{\alpha\beta} \begin{Bmatrix} i\ \alpha \\ \beta \end{Bmatrix} \begin{Bmatrix} m\beta \\ \alpha \end{Bmatrix} = -\kappa\left(T_{im} - \frac{1}{2}g_{im}T\right).$$
>
> I had already considered these equations three years ago with Grossmann (up to the second term on the right hand side). But then I had come to the result that they did not yield Newton's approximation, which was erroneous.[16]

This confirms the dating of the material in the Zürich notebook, for the field equation described here is exactly the one that would follow from placing $\mathcal{T}_{il}^{\times}$ in general field equation (2). The rejection of this equation, as Einstein tells us here, was due to a mistaken belief that it did not provide Newtonian theory as an approximation. We know that he was aware then that coordinate condition (10) led to a reduction of $\mathcal{T}_{il}^{\times}$ to the expected form in the weak-field case. We can only conclude that he had some objection to the coordinate condition itself.

In a letter to Paul Hertz, probably written in 1915, Einstein mentioned that he had had serious difficulties with coordinate conditions.[17] The general context is the problem of "adapted" coordinates, which he introduced with Grossmann in 1914. He singled out condition (10) as one coordinate condition that he had considered, but, unfortunately, he did not give further details on its use or the reasons for its rejection.

## 4.2 THE REDUCTION OF THE GRAVITATIONAL POTENTIALS

The last of the errors responsible for Einstein's rejection of the Ricci tensor as the gravitation tensor was an insistence that the gravitation tensor must adopt form (3) in rotating coordinate systems after the application of the appropriate coordinate condition. This was the least supportable of Einstein's misconceptions. Thus it is not surprising that its discovery came at an early stage in Einstein's 1915 return to general covariance.

The other errors were products of an extremely coherent but nonetheless mistaken view of static and weak fields. We have seen how Einstein came to believe that the metric degenerated in static fields to the form of (1), where $g_{44} = c^2$ is a function of the spatial coordinates and, due to the constancy of the remaining components of the metric, three-dimensional space is flat. This result seemed to be anchored firmly in the principle of equivalence. In his earlier theory, Einstein had invoked the principle to produce a homogeneous field by transforming to a coordinate system in uniform rectilinear acceleration. In terms of the *Entwurf* theory, the field that resulted had the degenerate form given earlier, with $g_{44}$ now a *linear* function of the spatial coordinates. It seemed entirely unremarkable to generalize from this very special case to the case of more general static fields through the relaxation of this condition, which now allowed $g_{44}$ to be an arbitrary function of the spatial coordinates. But this last generalization turned out to be inadmissible from the standpoint of the final general theory of relativity and led Einstein to a view of static fields inconsistent with that theory. For according to it, static fields are not spatially flat in all but a very few cases. At the time, however, the view that arbitrary static fields are spatially flat was simple and attractive, for it promised to reduce the number of gravitational potentials from the unwieldy ten of the general case to a more manageable and familiar single potential, $g_{44}$.

Einstein might well have been able to recover from this mistake quite rapidly were it not for an unfortunate coincidence. Einstein required that the gravitation tensor reduce to form (3) in appropriate coordinate systems. This amounts to requiring that the gravitation tensor reduce to the d'Alembertian of the metric tensor, $\Box g_{\mu\nu}$, in the weak-field case. The naturalness of this requirement is virtually unchallengeable. It guarantees at least Lorentz covariance for the theory in the weak-field case; it guarantees that this weak-field theory will generalize Newtonian theory in much the same way as electrodynamics generalizes electrostatics; and it does both in about the simplest way possible.

However, when combined with the assumption that the field equations have form (2), this natural requirement led to results that were inconsistent with Einstein's final general theory of relativity. For it led to the weak-field equations

$$\Box g_{\mu\nu} = \kappa T_{\mu\nu}, \tag{11}$$

whereas his final theory yielded

$$\Box g_{\mu\nu} = \kappa (T_{\mu\nu} - \tfrac{1}{2} g_{\mu\nu} T) \tag{11'}$$

in similar circumstances.[18] Unfortunately for Einstein, it turned out that his weak-field equation (11) could be solved in appropriate cases to yield exactly the static fields he expected from his earlier considerations. Equation (11'), however, does not yield such solutions.

As Einstein showed a little later in 1913, in the case of a static weak field, whose only source to first order is a pressureless, static dust cloud of density $\rho_0$, weak-field equation (11) reduces to

$$\Delta g_{44}^* = \kappa c^2 \rho_0$$
$$\Delta g_{\mu\nu}^* = 0 \quad \text{(all other } \mu, \nu\text{)}$$

where the asterisks denote deviations from Minkowskian values (Einstein 1913b, p. 1259). Provided the second equation holds everywhere in space-time, it can be solved to yield constant values for all the $g_{\mu\nu}$, except $g_{44}$. The first equation then solves to yield a $g_{44}$ that behaves exactly like the familiar Newtonian potential. If the background Minkowskian metric is taken to have the values of equation (1), then this solution amounts to the recovery of exactly the metric that Einstein expected in static fields.

Such a simple reduction to a single gravitational potential does not occur in the final theory. However, to Einstein at this early stage, it would have seemed quite natural, for some such reduction had to occur in the process of recovering Newtonian gravitation theory. And, of course, this simple reduction agreed exactly with the type of reduction Einstein believed would happen in static fields on the basis of a quite separate consideration, the principle of equivalence.

This reduction of the number of gravitational potentials in the weak-field case followed also from the form of equation (11) in weak fields that need not be static through an order-of-magnitude argument. If the source of the field is a dust cloud, which now need not be static, then the (4,4) term of the stress-energy tensor on the right-hand side of equation (11) is, typically, significantly larger than all its other components. Transferring this property to the field term on the left-hand side, it follows that the only first-order deviations from constant values in the components of the metric tensor are in the $g_{44}$ term.[19]

This argument cannot be used on the weak field equation (11') of Einstein's final theory. In this final theory, more components of the metric than $g_{44}$ are variable. However, only the $g_{44}$ component appears in the equations of motion of a slow-moving point-mass. From the point of view of the final general theory of relativity, this alone enables Newtonian theory to account successfully for the motion of a slow moving point-mass in terms of one gravitational potential only. Einstein commented to Besso late in 1915 on this remarkable feature of the equations of motion, which is common to both the *Entwurf* theory and the final theory. Perhaps this also helped to convince Einstein of his early misconceptions about the ease with which a reduction in the number of gravitational potentials could be achieved.

Thus, in 1912 and 1913 Einstein found himself driven to a single viewpoint by the interweaving of conclusions from quite disparate sources: in both weak and static fields the number of gravitational potentials effectively reduces from ten to one in the same way. This agreement underpinned his confidence in equation (11) as the weak-field equations and in metric (1) as the metric of static fields and helped make his passage to his final field equations such a difficult one.

The account being developed here is supported by comments Einstein made late in 1915, after he returned to seek generally covariant field equations and began to realize his earlier mistakes. "The difficulty was not finding generally covariant equations for the $g_{\mu\nu}$," he wrote Hilbert in mid-November, "for this worked out easily with the help of the Riemann tensor. But it was difficult to recognize that these equations formed a generalization and, indeed, a simple and natural generalization of Newton's law."[20] A month later, after all the difficulties were finally resolved, Einstein wrote to Besso in a more buoyant mood:

> The most delightful is the agreement of the perihelion motion and general covariance, the most remarkable however is the fact that Newton's theory *of the field* is already incorrect in equations of the 1st order (appearance of $g_{11}-g_{33}$). Only the fact that $g_{11}-g_{33}$ do not appear in the equations of motion effects the simplicity of Newton's theory.[21]

Max Born, who was at this time in Berlin and in close contact with Einstein, surely had it on the best authority when he wrote in 1916:

> What is remarkable about this [weak field] is that the $g_{23}, g_{31}\ldots$ in no way come out to be zero, so that there is more than one gravitational potential already in the first approximation; Einstein had first supposed the opposite and was forced into detours and incorrect assumptions before he found that his supposition was not confirmed. (Born 1916, p. 58)

### 4.3 THE DERIVATION OF THE *ENTWURF* FIELD EQUATIONS

The awkward episode of his 1912 field equation in his earlier theory of static fields seems to have convinced Einstein of the necessity of ensuring from the very beginning that any new field equation satisfy the conservation laws. This means that one should be able to construct a gravitational field stress tensor, or a stress-energy tensor in the four-dimensional case, using the field equations. This recognition provides the key to the understanding of some of the pages in the Zurich notebook and leads us directly to the derivation of the *Entwurf* field equations.

The law of conservation of energy-momentum, written as the vanishing of the covariant divergence of the stress-energy tensor $\Theta_{\mu\nu}$, takes the form

$$\sum_{\mu\nu} \frac{\partial}{\partial x_\nu}(\sqrt{-g}g_{\sigma\mu}\Theta_{\mu\nu}) - \frac{1}{2}\sum_{\mu\nu}\sqrt{-g}\frac{\partial g_{\mu\nu}}{\partial x_\sigma}\Theta_{\mu\nu} = 0 \qquad (12)$$

in equation (10) of Einstein's part of the *Entwurf* paper. The second term of this expression can be interpreted as the rate of transfer of energy-momentum out of the gravitational field. Thus, by analogy with the first term in equation (12), it should be possible to set it equal to the coordinate divergence of a tensor density corresponding to the gravitational field stress-energy tensor $\theta_{\mu\nu}$. If we write this equation and then, using equation (2), substitute the gravitation tensor $\Gamma_{\mu\nu}$ for the stress-energy tensor $\Theta_{\mu\nu}$, we then recover the equation

$$-\sum_{\mu\nu}\frac{\partial}{\partial x_\nu}(\sqrt{-g}g_{\sigma\mu}\kappa\theta_{\mu\nu}) = \frac{1}{2}\sum_{\mu\nu}\sqrt{-g}\frac{\partial g_{\mu\nu}}{\partial x_\sigma}\Gamma_{\mu\nu}. \quad (13)$$

This equation contains only the metric tensor and its derivatives and thus must be an identity. If we assume that the gravitation tensor has the form given in equation (3), then we can follow Einstein in taking the readily constructed

$$\sum_{\mu\nu\alpha\beta\tau\rho}\frac{\partial}{\partial x_\alpha}\left(\sqrt{-g}\gamma_{\alpha\beta}\left[\frac{\partial g_{\mu\nu}}{\partial x_\sigma}\frac{\partial \gamma_{\mu\nu}}{\partial x_\beta} - \frac{1}{2}g_{\beta\sigma}\gamma_{\tau\rho}\frac{\partial g_{\mu\nu}}{\partial x_\tau}\frac{\partial \gamma_{\mu\nu}}{\partial x_\rho}\right]\right)$$
$$= \sum_{\mu\nu\alpha\beta}\sqrt{-g}\frac{\partial g_{\mu\nu}}{\partial x_\sigma}\frac{\partial}{\partial x_\alpha}\left(\gamma_{\alpha\beta}\frac{\partial \gamma_{\mu\nu}}{\partial x_\beta}\right) \quad (14)$$

as approximating equation (13), where the equality in equation (14) is understood to hold only for quantities in the second order.

The fact that equation (13) should be an identity amounts to a simple test of whether the field equations in question satisfy the conservation equations and, conversely, provide a simple method of constructing the gravitational field stress-energy tensor. In the weak-field case considered here, we can read off an approximate expression for this tensor from the left-hand side of equation (14), by comparison with (13). On pages 19–21 of his notebook, immediately following the demonstration of how the harmonic coordinate condition reduces the Ricci tensor to form (3), Einstein probed the relationship between his new field equations and the conservation laws by using exactly this method in the weak-field case. Further, on pages 24–25, just after the construction of the gravitation tensor $\mathcal{T}_{il}^\times$, Einstein confirmed that the expression that he wrote as

$$\frac{d}{dx_i}\left(\gamma_{i\varepsilon}\frac{\partial g_{\alpha\beta}}{\partial x_\varepsilon}\frac{\partial \gamma_{\alpha\beta}}{\partial x_\sigma}\right) - \frac{1}{2}\frac{\partial}{\partial x_\sigma}\left(\gamma_{i\varepsilon}\frac{\partial g_{\alpha\beta}}{\partial x_\varepsilon}\frac{\partial \gamma_{\alpha\beta}}{\partial x_i}\right) \quad (15)$$

vanishes in Minkowski space-time as viewed from Cartesian coordinates rotating at uniform speed. With summation over repeated indices, this expression is equivalent to the coordinate divergence of the gravitation field stress-energy tensor recoverable from (14), although the $\sqrt{-g}$ factor is missing and the limitation to weak fields no longer seems to apply. Since this tensor is derived from a gravitation tensor of form (3), it seems that Einstein expected the gravitation tensor to have this form in the rotating-coordinate case considered here. Presumably his confirmation of the vanishing of expression

(15), which corresponds to the satisfaction of the conservation laws, would have confirmed for Einstein the correctness of this expectation.

We can now step directly to the method of deriving the field equations used in the *Entwurf* paper, which amount to a simple inversion of the method used to construct equation (13). As Einstein showed in section 5 of his part of the paper, once an identity has been decided upon to stand for equation (13), then one reads the gravitation tensor and gravitational field stress-energy tensor directly from it. The derivation of this identity,

$$\sum_{\alpha\beta\tau\rho} \frac{\partial}{\partial x_\alpha}\left(\sqrt{-g}\gamma_{\alpha\beta}\frac{\partial \gamma_{\tau\rho}}{\partial x_\beta}\frac{\partial g_{\tau\rho}}{\partial x_\sigma}\right) - \frac{1}{2}\sum_{\alpha\beta\tau\rho}\frac{\partial}{\partial x_\sigma}\left(\sqrt{-g}\gamma_{\alpha\beta}\frac{\partial \gamma_{\tau\rho}}{\partial x_\alpha}\frac{\partial g_{\tau\rho}}{\partial x_\beta}\right)$$

$$= \sum_{\mu\nu}\sqrt{-g}\frac{\partial g_{\mu\nu}}{\partial x_\sigma}\left\{\sum_{\alpha\beta}\frac{1}{\sqrt{-g}}\frac{\partial}{\partial x_\alpha}\left(\gamma_{\alpha\beta}\sqrt{-g}\frac{\partial \gamma_{\mu\nu}}{\partial x_\beta}\right)\right.$$

$$- \sum_{\alpha\beta\tau\rho}\gamma_{\alpha\beta}g_{\tau\rho}\frac{\partial \gamma_{\mu\tau}}{\partial x_\alpha}\frac{\partial \gamma_{\nu\rho}}{\partial x_\beta} + \frac{1}{2}\sum_{\alpha\beta\tau\rho}\gamma_{\alpha\mu}\gamma_{\beta\nu}\frac{\partial g_{\tau\rho}}{\partial x_\alpha}\frac{\partial \gamma_{\tau\rho}}{\partial x_\beta}$$

$$\left. - \frac{1}{4}\sum_{\alpha\beta\tau\rho}\gamma_{\mu\nu}\gamma_{\alpha\beta}\frac{\partial g_{\tau\rho}}{\partial x_\alpha}\frac{\partial \gamma_{\tau\rho}}{\partial x_\beta}\right\}, \tag{16}$$

is given by Grossmann in his section of the paper. It amounts to a generalization of equation (14) from the weak-field case to the general case. Equation (14) was constructed originally by expanding the terms on its right-hand side and retaining only quantities of second order to yield the left-hand side. The bulk of Grossmann's derivation of identity (16) is devoted to making this process exact. He took the terms of third order, which were dropped in constructing equation (14), and reworked and redistributed them until the identity had the form required by equation (13). This yielded identity (16) directly.

For our purposes, the important point is that Einstein introduced these identities as "eindeutig bestimmt," "uniquely determined" (Einstein and Grossmann 1913, p. 237). This, of course, suggests that he believed the *Entwurf* field equations to be uniquely determined, a belief that he stated explicitly a little later in 1913 (Einstein 1913a, p. 289). This conclusion seems to have put an end to his search for generally covariant field equations, so it is of great interest to us here. Neither Einstein nor Grossmann give any proof of their crucial result, the uniqueness of identity (16). On the basis of equations (13) and (3), Einstein required the identity to have the form

sum of differential quotients

$$= \frac{1}{2}\sum_{\mu\nu}\sqrt{-g}\frac{\partial g_{\mu\nu}}{\partial x_\sigma}\left\{\sum_{\alpha\beta}\frac{\partial}{\partial x_\alpha}\left(\gamma_{\alpha\beta}\frac{\partial \gamma_{\mu\nu}}{\partial x_\beta}\right)\right. \tag{17}$$

$$\left. + \text{ further terms which fall away in the first approximation}\right\}.$$

It is necessary to spell out the conditions that this identity had to satisfy more clearly; otherwise there is no possibility of developing a uniqueness proof. Most naturally, we would require the gravitational field stress-energy tensor to be a sum of terms quadratic in the first derivatives of the metric tensor. From comparison with equation (13), we see that this requirement specifies the form of the left-hand side of equation (17), which is just the usual divergence of this tensor. Following the form of Einstein's final result, we can also require that the "further terms" on the right-hand side can be constructed out of the metric tensor and its first derivatives only.

It is a little surprising to find that these natural, even somewhat restrictive, conditions still do not specify the identity uniquely. It is possible to add further terms to equation (16), which would still leave the equation an identity and that do not violate any of the conditions described herein.[22] This suggests that Einstein was mistaken in his claim that identity (16) is unique and thus mistaken in his belief in the uniqueness of the *Entwurf* field equations. Before we convict Einstein of this mistake—and perhaps also Grossmann, although the uniqueness claim does not appear in his part of the paper—we should consider the possibility that they placed an additional constraint on the form of the left-hand side of the identity.

In identity (16) the left-hand side contains only quantities built up out of the derivatives of the metric tensor in which the $\mu\nu$ indices of these derivatives $\partial g_{\mu\nu}/\partial x_\alpha$ are always summed as $\sum_{\mu\nu}(\partial g_{\mu\nu}/\partial x_\alpha)(\partial \gamma_{\mu\nu}/\partial x_\beta)$. Perhaps this was no accident. Einstein and Grossmann may have required that the terms on the left-hand side have this form only. They do not state this explicitly. However they may have tried to indicate it with their use of the term "sum of differential quotients." Their use of this term is sufficiently restricted in the *Entwurf* paper to allow it to refer to quantities that are not just the sum of differential quotients, but of differential quotients summed in the special way I have indicated here.

Why they would impose such a restriction is not entirely clear. The gravitational field stress-energy tensor derived in the weak-field case from equation (14) is built up out of terms of this form alone. Perhaps they wanted to retain this form in the *Entwurf* case, which was built up as a generalization of the weak-field case. Given the assumption that the gravitation field stress-energy tensor has to be made up out of terms quadratic in the derivatives of the metric tensor, we can add no further terms to the existing terms of the weak-field stress-energy tensor without interfering with the reduction to the weak-field case of (14). This would certainly restrict the summation to the way described herein, for it would require the gravitational field stress-energy tensor to be exactly the one derived from equation (14).

In any case, if Einstein and Grossmann did use this additional constraint, then it can be proved that their identity (16) is unique. (The proof is deceptively difficult.) The *Entwurf* field equations would be uniquely determined. Thus, with the conclusion of his work on the *Entwurf* paper, Einstein believed that he had found that the naturally suggested generally covariant tensors were inadmissible as gravitation tensors since they failed to yield the correct Newtonian limit and, furthermore, that he had been able to derive the only acceptable field equations, which did not turn out to be generally covariant.

It is clear that the problem in the derivation rests on Einstein's misconception of the Newtonian limit. The general method is sound. The familiar contracted Bianchi identities can be written in the form of equation (13) and from them Einstein's final generally covariant field equations can be recovered. But Einstein ruled out consideration of this identity by requiring the right-hand side of the identity to have form (17). This meant that he could only recover a gravitation tensor with form (3) from his identities: the use of the Bianchi identities would yield the Einstein tensor as the gravitation tensor and it does not have this form. Moreover, Einstein's method could not even yield the field equations of his final general theory of relativity after they had been reduced by some coordinate condition. Einstein's method assumes that the weak-field equations, in appropriate coordinate systems, take form (11), whereas his final theory yields the weak-field equation (11') in similar circumstances.

Finally, Einstein still admitted the possibility of generally covariant field equations if derivatives higher than the second in the metric tensor were allowed. Indeed, in 1914 he insisted that there must be a generally covariant generalization of the field equations if they have any physical content (Einstein 1914a, pp. 177–178). But this belief seems not to have had any practical effect on his work.

## 5. The Arguments against General Covariance

With the conclusion of the *Entwurf* paper, the problem of the field equations had altered radically. The question was no longer "What are the generally covariant field equations?" It had become "Why are there not any second-order generally covariant field equations?" and "How does the limited covariance of the field equations fit with the requirement of the general relativity of motion?" Once Einstein took this approach, answers to the first question came fairly fast.

We can date the discovery of the first answer to the first question quite exactly—August 15, 1913—from Einstein's correspondence with Lorentz.[23] Einstein's analysis focused on the law of conservation of energy-momentum, written as

$$\sum_v \frac{\partial}{\partial x_v}(\mathfrak{T}_{\sigma v} + \mathfrak{t}_{\sigma v}) = 0, \tag{18}$$

where $\mathfrak{T}_{\sigma v}$ and $\mathfrak{t}_{\sigma v}$ are the mixed stress-energy tensor densities for matter and the gravitational field, respectively (the $v$ index is contravariant). Einstein's argument has the general form of a *reductio ad absurdum*. If the theory were generally covariant, then we would expect the stress-energy tensor for the gravitational field to be generally covariant and to share the same transformation properties as the usual stress-energy tensor since they both enter into equations such as (18) in the same way. But if this were the case, equation (18) could not be generally covariant. Indeed, a "closer consideration" shows that such an equation (18) could only be covariant under linear coordinate transformations.[24] This means that the conservation laws and, as a result, the theory as a whole, can only hold in coordinate systems related by linear coordinate transformations, which contradicts the assumed general covariance.

Einstein was very pleased and, perhaps, even somewhat relieved with the discovery of this argument. In his letter to Lorentz he confided: "Only now does the theory please me, after this ugly dark spot seems to have been removed." We can understand Einstein's satisfaction, for there is a pleasing coherence in the way that the conservation laws first circumscribe powerfully the range of admissible field equations, as he found in the derivation of the *Entwurf* field equations, and then guarantee that the theory cannot be generally covariant. After Einstein felt that he had stronger reasons for rejecting the admissibility of generally covariant field equations, he was still pleased to note that the stronger restriction of the covariance of the theory to *linear* coordinate transformations should follow from the conservation laws. He wrote Ehrenfest: "What can be more beautiful than that necessary specialization flowing from the conservation laws."[25]

Of course what Einstein had not allowed for in his argument is that a generally covariant theory, with generally covariant field equations, might have a stress-energy tensor for the gravitational field that is itself not a generally covariant tensor, without compromising the general covariance of the theory. This is the case in the final general theory of relativity. That Einstein should miss this point is by no means a trivial oversight. The basic nature of the theory seems to demand that any physically meaningful quantity be represented by a generally covariant tensor. That this is not the case for the gravitational field stress-energy tensor of the final theory was to be a source of some confusion and is now explained in terms of the impossibility of localizing gravitational field energy and momentum.

Certainly we could not expect Einstein to anticipate this at a time when he still believed that there were no second-order, generally covariant field equations. There was, however, a second error in Einstein's use of this consideration. Einstein had concluded from it that the covariance of all versions of his new theory would be limited to linear transformations. This included the *Entwurf* theory. But, as we have seen, the restriction to linear coordinate transformations depended on the general covariance of the stress-energy tensor for the gravitational field and this tensor in the *Entwurf* theory was not generally covariant. So the argument from the conservation law did not entail a limitation of the covariance of the *Entwurf* theory to linear coordinate transformations. Einstein himself pointed out this error early in 1914 in a footnote to the paper in which he began concentrated study of the covariance properties of the *Entwurf* theory (Einstein and Grossmann 1914, p. 218). This was an important point, since Einstein came to regard it as essential that the *Entwurf* theory be covariant under more than just linear coordinate transformations if it was to realize any extension of the principle of relativity.

The argument from the conservation law then seems to have dropped from sight. It had already been eclipsed by what seemed to be a stronger argument against generally covariant field equations, the "hole" argument, discovered sometime late in 1913 or 1914.[26] The argument is intended to show that, if the field equations are generally covariant, then a given stress-energy tensor cannot uniquely determine the gravitational field through the field equations. The first three versions of the argument are virtually identical. I quote the second version:

> *If the reference system is chosen quite arbitrarily, then in general the $g_{\mu\nu}$ cannot be completely determined by the $\mathfrak{T}_{\sigma\nu}$* [stress-energy tensor density]. *For, think of the $\mathfrak{T}_{\sigma\nu}$ and $g_{\mu\nu}$ as given everywhere and let all $\mathfrak{T}_{\sigma\nu}$ vanish in a region of $\Phi$ of four-dimensional space. I can now introduce a new reference system, which coincides completely with the original outside $\Phi$, but is different from it inside $\Phi$ (without violation of continuity). One now relates everything to this new reference system, in which matter is represented by $\mathfrak{T}'_{\sigma\nu}$ and the gravitational field by $g'_{\mu\nu}$. Then it is certainly true that*
>
> $$\mathfrak{T}'_{\sigma\nu} = \mathfrak{T}_{\sigma\nu}$$
>
> *everywhere, but on the other hand the equations*
>
> $$g'_{\mu\nu} = g_{\mu\nu}$$
>
> *will definitely not all be satisfied inside $\Phi$. The assertion follows from this.*
>
> *If one wants a complete determination of the $g_{\mu\nu}$* (gravitational field) *by the $\mathfrak{T}_{\sigma\nu}$* (matter) *to be possible, then this can only be achieved by a limitation on the choice of reference systems.*

Einstein claimed to have shown with this argument that a single stress-energy tensor can determine two different gravitational fields, if the field equations are generally covariant. Einstein's argument seems to rest on a simple mistake: $g'_{\mu\nu}$ and $g_{\mu\nu}$ do not represent different gravitational fields. Rather they represent the same gravitational field, but as it appears in two different coordinate systems. All Einstein seems to have shown is that a given gravitational field will look different if viewed from different coordinate systems. On this basis, there seems no reason to doubt that the given stress-energy tensor does specify a unique gravitational field. This mistake is a trivial one and it has become customary in accounts of this argument to convict Einstein of making it.[27] Grossmann must then also have made the same mistake, for he was still collaborating with Einstein at this time and even coauthored one of the papers in which the argument appears.

That both Einstein and Grossmann could repeatedly make this same trivial mistake on such an important question is highly implausible, especially if we recall that Einstein was quite comfortable with the notion of applying coordinate conditions to generally covariant gravitation tensors prior to the completion of the *Entwurf* paper. Moreover, there is unequivocal evidence in both of the papers, in which the third and fourth versions of the "hole" argument appeared, that Einstein and Grossmann recognized that a change in coordinate system does not produce a new field, even though the components of the metric tensor may change (Einstein and Grossmann 1914, p. 223; Einstein 1914b, p. 1071). In both papers, an arbitrary infinitesimal change in the ten components of the metric tensor $g_{\mu\nu}$ is introduced. This is broken up into two parts, the first of which corresponds to a change in the gravitational field between "adapted" coordinates, the second to a change "that can be produced through mere variation of the coordinate system without a change of the gravitational field.... A variation of this kind is determined by four functions (variations of the coordinates), which are independent of one another" (Einstein 1914b, pp. 1071–1072). This shows a clear recognition of the fact that a change in the coordinate system does not alter the gravitational field, although the components of the metric tensor will change, and that this arbitrariness in the representation of the field is associated with four independent conditions.

Fortunately, it is possible to give a quite different account of the content and import of the "hole" argument, as it appears in its fourth version, which does not convict Einstein of a trivial mistake, and I will argue that this interpretation can also be used on the earlier three versions. The essential part of the text of this fourth version reads as follows:

> We consider a finite region of the continuum Σ, in which no material process takes place. Physical happenings in Σ are then fully determined, if the quantities $g_{\mu\nu}$ are given as functions of the $x_\nu$ in relation to the coordinate system $K$ used for description. The totality of these functions will be symbolically denoted by $G(x)$.
>
> Let a new coordinate system $K'$ be introduced, which coincides with $K$ outside Σ, but deviates from it inside Σ in such a way that the $g'_{\mu\nu}$ related to the $K'$ are continuous everywhere like the $g_{\mu\nu}$ (together with their derivatives). We denote the totality of the $g'_{\mu\nu}$ symbolically with $G'(x')$. $G'(x')$ and $G(x)$ describe the same gravitational field. In the functions $g'_{\mu\nu}$ we replace the coordinates $x'_\nu$ with the coordinates $x_\nu$, i.e., we form $G'(x)$. Then, likewise, $G'(x)$ describes a gravitational field with respect to $K$, which however does not correspond with the real (or originally given) gravitational field.
>
> We now assume that the differential equations of the gravitational field are generally covariant. Then they are satisfied by $G'(x')$ (relative to $K'$) if they are satisfied by $G(x)$ relative to $K$. Then they are also satisfied by $G'(x)$ relative to $K$. Then relative to $K$ there exist the solutions $G(x)$ and $G'(x)$, which are different from one another, in spite of the fact that both solutions coincide in the boundary region, i.e., *happenings in the gravitational field cannot be uniquely determined by generally covariant differential equations for the gravitational field.*

This version of the argument is identical to the three earlier versions, with the exception of the addition of a new and crucial step at the end. This step involves the construction of a new gravitational field that is also a solution of the field equations with the same stress-energy tensor and in the same coordinate system as the original field. It makes clear that the introduction of the alternate coordinte representation $G'(x')$ of the original field is only a device to enable construction of the new field. There is clearly no confusion over whether $G'(x')$ represents a new field. Einstein wrote: "$G'(x')$ and $G(x)$ describe the same gravitational field."

The way in which Einstein constructed this new field in the argument does bear some elucidation since Einstein's account of it is quite brief. The construction proceeds as follows. Consider a particular point of the space-time manifold, called $P_1$ for convenience, in $\Sigma$. It will have coordinates $x_v$ in $K$ and $x'_v$ in $K'$. There will be another point $P_2$ in $\Sigma$ whose coordinates *in $K'$* are numerically the same as $x_v$. The gravitational field at the point $P_2$ in coordinate system $K'$ will be described by the functions $G'(x'_v)$. Now consider the new field that would arise at the original point $P_1$ if the functions describing the field at $P_1$ were not $G(x_v)$ but $G'(x_v)$. Clearly the fields described by the functions $G(x_v)$ and $G'(x_v)$ are related to the same coordinate system $K$ since the arguments of both functions are the same numbers $x_v$, the coordinates of $P_1$ in $K$. But equally clearly they cannot describe the same field since $G$ and $G'$ are not the same functions. If a new field is constructed in this way for all points in the space-time manifold, then this new field will still satisfy the field equations. For we have done nothing to change the *mathematical* form of $G'(x'_v)$, which, of course, is a solution to the field equations, in constructing the new field. Rather, all that has been done occurs on the conceptual level. That is, we reassign the points in the space-time manifold which are thought of as belonging to a given set of coordinates—specifically, the point $P_1$ in the manifold is now assigned to the coordinates $x_v$ in the coordinate system $K'$. Since the stress-energy tensor vanishes everywhere in $\Sigma$, its new components, generated by exactly the same method as the new field, will still vanish everywhere in $\Sigma$. That is, its components will agree everywhere with those of the original stress-energy tensor in $K$. Thus both the new field and the old field are solutions to the field equations with the same stress-energy tensor in the same coordinate system $K$—and Einstein's result is established.

The case of the three earlier versions of the argument still remains. Were it not for one crucial piece of evidence, it would be difficult to escape the conclusion that Einstein and Grossmann were presenting a different argument from the fourth version and one in which the trivial mistake outlined earlier is committed. That crucial piece of evidence is a footnote appended to the sentence ending "the equations $g'_{\mu v} = g_{\mu v}$ will definitely not all be satisfied inside $\Phi$" in the second version of the argument quoted above:

> The equations are to be understood in such a way that each of the independent variables $x'_v$ on the left-hand side are to be given the same numerical values as the variables $x_v$ on the right-hand side.

In other words, Einstein required the inequality of $g'_{\mu v}$ and $g_{\mu v}$ to be read in a special way. In terms of points $P_1$ and $P_2$ defined earlier, the $g'_{\mu v}$ at $P_2$ are unequal to the $g_{\mu\nu}$ at $P_1$. This, of course, is the inequality crucial to the fourth version of the argument. If all Einstein were saying was that the different coordinate representations of the original field $g_{\mu\nu}$ and $g'_{\mu\nu}$ were actually different fields, then there would have been no reason to specify that the inequality be read in this special manner.

This suggests that all four versions of the argument were understood by Einstein to have the same content as the fourth and that his greatest mistake was only to present the first three versions in too compact a form to be readily understood. Presumably the crucial stipulation on how the inequality of the $g_{\mu\nu}$ and $g'_{\mu\nu}$ was to be read was obvious to Einstein, for it was appended only as an apparent afterthought to the second version of the argument in the footnote quoted. (It could equally have been added to the first and third versions and thus made their content clearer as well.) Perhaps if we were all Einsteins then such subtleties would be equally clear to us too!

Stachel has pointed out that, in effect, what Einstein did in the final step of the argument is to generate a new field from the old one by means of a point transformation (Stachel 1980b). Specifically, under a point transformation from $P_2$ to $P_1$, the image field represented in the image of coordinate system $K'$ amounts to the field with coordinate representation $G'(x)$. Further, Stachel has argued that, with the discovery of his generally covariant field equations in 1915, Einstein was able to draw a very significant physical conclusion about the relationship between the space-time manifold and the metric field from the machinery of the "hole" argument. This was that the individual points of the space-time manifold have no independent individuality and can only be distinguished with reference to the metric field (or perhaps some other material phenomena) in space-time. Thus it is the final step of the "hole" argument that is erroneous.

Within this understanding, it is impossible to drag the metric field away from a physical point in empty space-time and leave that physical point behind. For the physical individuation of the point only has meaning in terms of the metric field at that point. Or, in the terms Einstein used in the fourth version of the argument, it makes no sense to remove one field, $G(x)$, leave behind the bare space-time manifold, as represented by the coordinate system $K$, and then construct a new field, $G'(x)$, on this bare manifold. For this presupposes the concept of a space-time manifold, replete with points that have an existence independent of the metric field. Take away the metric field and one takes away the space-time points with it.[28]

This account is derived from—and indeed explains—Einstein's comments about the "hole" argument made in late 1915 and early 1916, after the discovery of the final generally covariant field equations. He wrote Besso:

> Everything was correct in the hole consideration up to the last conclusion. There is no physical content in two different solutions $G(x)$ and $G'(x)$ existing with respect to the *same* coordinate system $K$. To imagine two solutions simultaneously in the same manifold has no meaning and the system $K$ has no physical reality. In the place of the hole consideration we have the following. *Reality* is nothing but the

totality of space-time point coincidences. If, for example, physical happenings could be built up out of the motion of material points alone, then the meetings of the points, i.e., the points of intersection of their world lines, would be the only reality, i.e., in principle observable. Naturally these points of intersection remain unchanged in all transformations (and no new ones are added) if only certain uniqueness conditions are preserved. Therefore it is most natural to require of laws that they determine no more than the totality of time-space coincidences. Following what has been said before, this is already achieved with generally covariant equations.[29]

One of the important outcomes of Einstein's experience with the "hole" argument was the point coincidence argument for the need of generally covariant equations, which is sketched out in this letter to Besso. Einstein came to use this argument to good effect in his expositions of the general theory of relativity (for example, Einstein 1916a, pp. 776–777).

## 6. Covariance Properties of the *Entwurf* Field Equations

Now satisfied that there were good reasons to give up a search for a generally covariant theory, Einstein could devote his attention to the task of elucidating his *Entwurf* theory and, in particular, of determining the significance of its limited covariance. He began this work before his move to Berlin, while still collaborating with Grossmann, and the first product of their labor appeared early in 1914. In the introduction to this paper, they stressed that the field equations must be covariant under nonlinear coordinate transformations as well as linear if the theory was to contain an extension of the principle of relativity and satisfy the requirements of the principle of equivalence. They summarized the achievements of the paper:

> In the following it will be proved that the gravitation equations set up by us have that degree of general covariance which is conceivable under the condition that the fundamental tensor $g_{\mu\nu}$ should be completely determined by the gravitation equations; in particular, it turns out that the gravitation equations are covariant under acceleration transformations (i.e., nonlinear transformations) of many different kinds. (Einstein and Grossmann 1914, p. 216)

The "condition that the fundamental tensor $g_{\mu\nu}$ should be completely determined by the gravitation equations" refers, of course, to the "hole" argument. This passage therefore asserted that the *Entwurf* field equations have the maximum covariance consistent with the considerations of the "hole" argument. They proved this assertion simply and elegantly. However, there is no further detailed discussion of the second assertion, that this allowed covariance embraces a wide range of acceleration transformations, or any demonstration that some of these correspond to cases of special physical interest in the context of a generalized principle of relativity or the principle of equivalence. This is a curious and, as it turned out, serious omission and one that was maintained in Einstein's generalization of the work in this paper late in 1914.

The paper continued with a brief statment of the equations of the *Entwurf* theory and of the "hole" argument itself. Einstein and Grossmann then took the field equations, written in the compact form

$$\sum_{\alpha\beta\mu} \frac{\partial}{\partial x_\alpha}\left(\sqrt{-g}\gamma_{\alpha\beta}g_{\sigma\mu}\frac{\partial \gamma_{\mu\nu}}{\partial x_\beta}\right) = \kappa(\mathfrak{T}_{\sigma\nu} + t_{\sigma\nu}), \tag{19}$$

formed its coordinate divergence, and then applied the conservation law in the form of equation (18). This resulted in the condition

$$B_\sigma = \sum_{\alpha\beta\mu\nu} \frac{\partial^2}{\partial x_\nu \partial x_\alpha}\left(\sqrt{-g}\gamma_{\alpha\beta}g_{\sigma\mu}\frac{\partial \gamma_{\mu\nu}}{\partial x_\beta}\right) = 0, \tag{20}$$

which would clearly have to be satisfied in any coordinate system in which the *Entwurf* field equations held. They decided to call such coordinate systems "adapted."

The remainder of the paper was devoted to showing that this necessary condition is also a sufficient one. That is, if condition (20) was still satisfied after a coordinate transformation, then the field equations would still hold in the new coordinate system. In their proof, Einstein and Grossmann introduced mathematical techniques that would become of great importance to the development of the theory. They found a variational formulation of the field equations and studied their covariance properties by examining the behavior of the variational integral under an infinitesimal coordinate transformation. I pass over the details of their arguments now, for they are fully subsumed by the more general apparatus of Einstein's first major paper on the theory after his move to Berlin in April 1914. With that Einstein could guarantee that his field equations had the maximum covariance permitted by the "hole" argument.

We can approach condition (20) in terms of the problem of coordinate conditions discussed earlier. Einstein had made clear in a paper written earlier in 1914 that he believed that some generally covariant set of equations must correspond to the *Entwurf* equations (Einstein 1914a, p. 178). Condition (20) would then be the coordinate condition that would have to be applied to these equations in order to recover their *Entwurf* specialization.[30] Of course by this time Einstein had become convinced that the field equations in their generally covariant form were physically uninteresting as field equations. So there would have been little to gain from finding them.

The results of his and Grossmann's paper seemed at last to have reconciled Einstein to the limited covariance of his theory. He described these results to his friend Besso in a letter of March 1914, in which he claimed that the justified transformations included rotations transformations:

> Now I am completely satisfied and no longer doubt the correctness of the whole system, whether the observation of the solar eclipse works out or not. The sense [*Vernunft*] of the matter is too evident.[31]

Einstein's move to Berlin in April 1914 marked the end of his collaboration with Grossmann. Fortunately, by this time Einstein no longer seems to have needed Grossmann's mathematical guidance. By October 1914, he had completed a lengthy summary article on his new theory, whose form and detailed nature suggest that Einstein felt his theory had reached its final form. The article contained a review of the methods of tensor calculus used in the theory and, flexing his newfound mathematical muscles, Einstein could even promise to give new and simpler derivations of the basic laws of the "absolute differential calculus" (Einstein 1914b, p. 1030). Of great importance for us is the fact that Einstein had taken the new mathematical techniques of his last paper with Grossmann, generalized them, and found in them a quite new derivation of the field equations.

Einstein began his new derivation and treatment of the covariance properties of the field equations by introducing an undetermined action $H$ (for "Hamiltonian"?), which was to be some function of the metric tensor $g^{\mu\nu}$ and its first derivatives $g^{\mu\nu}_\sigma$ (Einstein 1914b, pp. 1066–1077).[32] From this the integral

$$J = \int H \sqrt{-g}\, d\tau \tag{21}$$

was formed, where $d\tau$ is an infinitesimally small element of space-time. The $g^{\mu\nu}$ were varied infinitesimally in such a way that their variation $\delta g^{\mu\nu}$ disappeared on the boundary of the region of space-time of the integration.

The variation produced in $J$ can be rewritten as

$$\delta J = \int \sum_{\mu\nu} \mathfrak{E}_{\mu\nu} \delta g^{\mu\nu}\, d\tau, \tag{22}$$

and the new quantity,

$$\mathfrak{E}_{\mu\nu} = \frac{\partial H \sqrt{-g}}{\partial g^{\mu\nu}} - \sum_\sigma \frac{\partial}{\partial x_\sigma}\left(\frac{\partial H \sqrt{-g}}{\partial g^{\mu\nu}_\sigma}\right), \tag{23}$$

was designated as the tensor density derived from the gravitation tensor for field equation (2). The covariance properties of the field equations that resulted were determined by examining the behavior of $J$ under infinitesimal coordinate transformations. Introducing such a transformation $\Delta$, Einstein found that

$$\frac{1}{2}\Delta H = \sum_{\mu\nu\sigma\alpha} g^{\nu\alpha} \frac{\partial H}{\partial g^{\mu\nu}_\sigma} \frac{\partial^2 \Delta x_\mu}{\partial x_\sigma \partial x_\alpha}, \tag{24}$$

given that $H$ is a function of $g^{\mu\nu}$ and $g^{\mu\nu}_\sigma$ alone and making the assumption that $H$ is invariant under linear transformations, which made it possible to neglect all terms in $\partial \Delta x_\mu / \partial x_\sigma$ in the general expression for $\Delta H$. From this assumption, it followed that

$$\frac{1}{2}\Delta J = \int d\tau \sum_\mu (\Delta x_\mu B_\mu) + F, \tag{25}$$

where

$$B_\mu = \sum_{\alpha\sigma\nu} \frac{\partial^2}{\partial x_\sigma \partial x_\alpha} \left( g^{\nu\alpha} \frac{\partial H \sqrt{-g}}{\partial g^{\mu\nu}_\sigma} \right), \tag{26}$$

and $F$ is a surface integral term that would vanish in case $\Delta x_\mu$ and $\partial \Delta x_\mu / \partial x_\sigma$ vanished on the boundary of the region of integration.

Einstein could then define "adapted" coordinates for a given field. He considered a series of infinitesimally separated coordinate systems $K$, $K'$, $K''$, ..., whose values agree on the boundary of the region of integration in such a way that if $\Delta$ represents the coordinate transformation between two adjacent systems, then both $\Delta x_\mu$ and $\partial \Delta x_\mu / \partial x_\sigma$ disappear on the boundary: $F$ vanishes for $\Delta$. Therefore we have

$$\frac{1}{2}\Delta J = \sum_\mu \int d\tau \, \Delta x_\mu B_\mu. \tag{27}$$

Einstein defined the adapted coordinate system to be that one for which $J$ is an extremum: in adapted coordinate systems,

$$B_\mu = 0. \tag{28}$$

Einstein proceeded to demonstrate his crucial result that

$$\Delta(\delta J) = 0, \tag{29}$$

provided that $\Delta$ relates adapted coordinate systems, or, using Einsteins term, is a "justified" coordinate transformation. In other words, $\delta J$ is a scalar under justified coordinate transformations. It follows directly from equation (22) that the gravitation tensor $1/\sqrt{-g}\,\mathfrak{E}_{\mu\nu}$ is a tensor under justified coordinate transformations.

These results have been derived so far with a largely undetermined $H$. If at the beginning $H$ had been set as

$$H = \frac{1}{4} \sum_{\alpha\beta\tau\rho} g^{\alpha\beta} \frac{\partial g_{\tau\rho}}{\partial x_\alpha} \frac{\partial g^{\tau\rho}}{\partial x_\beta}, \tag{30}$$

then these calculations would have corresponded to those of Einstein and Grossmann's earlier paper, in which they showed that the *Entwurf* field equations are covariant under justified coordinate transformations. The gravitation tensor resulting from this choice of $H$ is the *Entwurf* gravitation tensor, as Einstein and Grossmann showed, and condition (28) takes on the form of condition (20). Further, we can now see the basis of Einstein and Grossmann's claim that the *Entwurf* field equations have the maximum covariance allowed by the "hole" argument and can also note that this applies to the generalized field equations of Einstein's Berlin paper as well.

136   John Norton

The "hole" argument was built on the fact that generally covariant field equations in a given region of the space-time hold in any two coordinate systems whose values agree on the boundary of that region. Einstein's definition of an adapted coordinate system requires the selection of one of all those coordinate systems in a given region of space-time, whose values agree on the boundary, by means of the condition that $\Delta J$ in equation (27) be an extremum.[33] Such a restriction seems to be the minimum that the "hole" argument requires and this is the full extent of the limitation of the covariance of Einstein's field equations.[34]

In the Berlin paper, Einstein no longer felt that he had to stipulate the value of $H$ in order to recover his *Entwurf* field equations. He believed that he could derive equation (30) from the general formulation of the field equations. To do this, he substituted the gravitation tensor of equation (23) into the conservation law, written in the form of equation (12), which gave him

$$\sum_v \frac{\partial S_\sigma^v}{\partial x_v} - B_\sigma = 0, \tag{31}$$

where

$$S_\sigma^v = \sum_{\mu\tau} \left( g^{v\tau} \frac{\partial H \sqrt{-g}}{\partial g^{\sigma\tau}} + g_\mu^{v\tau} \frac{\partial H \sqrt{-g}}{\partial g_\mu^{\sigma\tau}} + \frac{1}{2} \delta_\sigma^v H \sqrt{-g} - \frac{1}{2} g_\sigma^{\mu\tau} \frac{\partial H \sqrt{-g}}{\partial g_v^{\mu\tau}} \right). \tag{32}$$

A count of the number of equations determining the field—ten field equations, four adapted coordinate conditions (equation (28)) and the four equations in (32)—showed that the field was overdetermined by four conditions. This could be resolved, Einstein concluded, if the $S_\sigma^v$ identically vanished:

$$S_\sigma^v \equiv 0. \tag{33}$$

Since this condition did not fully determine the field equations, Einstein stipulated that $H$ should be a homogeneous function of second order in the $g_\sigma^{\mu v}$. From this is followed unproblematically that $H$ would have to be equal to one of, or a linear combination of, the five linearly independent terms

$$\sum g_{\mu v} \frac{\partial g^{\mu v}}{\partial x_\sigma} \frac{\partial g^{\sigma\tau}}{\partial x_\tau}, \quad \sum g^{\sigma\sigma'} g_{\mu v} \frac{\partial g^{\mu v}}{\partial x_\sigma} g_{\mu' v'} \frac{\partial g^{\mu' v'}}{\partial x'_\sigma}, \quad \sum g_{\sigma\sigma'} \frac{\partial g^{\sigma\mu}}{\partial x_\mu} \frac{\partial g^{\sigma' v}}{\partial x_v},$$

$$\sum g_{\mu\mu'} g_{v v'} g^{\sigma\sigma'} \frac{\partial g^{\mu v}}{\partial x_\sigma} \frac{\partial g^{\mu' v'}}{\partial x'_\sigma}, \quad \text{and} \quad \sum g_{\alpha\beta} \frac{\partial g^{\alpha\sigma}}{\partial x_\tau} \frac{\partial g^{\beta\tau}}{\partial x_\sigma}. \tag{34}$$

Einstein then asserted that condition (33) eventually leads to the choice of the fourth of these terms, up to a constant factor. He gave no proof, but demonstrated that this choice of $H$ does indeed satisfy equation (33). Of course this choice of $H$ is equivalent to the selection of $H$ in equation (30). So Einstein's argument amounts to a new derivation of the *Entwurf* field equations.

Einstein had good reason to be pleased with this result. For it seemed to show that his theory was not just a theory of gravitation, but a generalized theory of relativity, insofar as it was concerned with establishing the widest covariance possible in its equations. His original derivation of the field equations had been based squarely on considerations in gravitation theory— that is, he sought tensor equations that would yield the correct Newtonian limit while remaining consistent with the conservation laws. The new derivation, however, focused on covariance considerations. He had found a simple way of formulating field equations that would have exactly the maximum covariance allowed by his "hole" argument, and they led him almost directly to his original *Entwurf* field equations. As a result, he could promise to "recover the equations of the gravitational field in a purely covariant-theoretical way" and to claim to "have arrived at quite definite field equations in a purely formal way, i.e., without directly drawing on our physical knowledge of gravitation" (Einstein 1914b, pp. 1030, 1076). Perhaps Einstein overstated the purity of his new derivation, but it certainly is far purer than the *Entwurf* derivation.[35]

Einstein's satisfaction with his new treatment of the field equations was short-lived. He soon found that the last step in his derivation was incorrect. Condition (33) in no way required that $H$ take on its *Entwurf* form. In effect, all that this condition required was that $H$ be a scalar under linear coordinate transformations: it just returned an assumption that Einstein had made earlier in the derivation. We can readily confirm that this is the import of condition (33) by writing out the general expression for $\Delta H$, which, by the usual methods, turns out to be

$$\frac{1}{2}\Delta H = \sum_{\mu\nu\sigma\alpha} \frac{1}{\sqrt{-g}} S_\mu^\alpha \frac{\partial \Delta x_\mu}{\partial x_\alpha} + g^{\nu\alpha} \frac{\partial H}{\partial g_\sigma^{\mu\nu}} \frac{\partial^2 \Delta x_\mu}{\partial x_\sigma \partial x_\alpha}. \tag{35}$$

If $H$ is a scalar under linear coordinate transformation, then $\Delta H$ must vanish in the case in which the $\partial \Delta x_\mu / \partial x_\alpha$ have arbitrary nonzero values, but all the $\partial^2 \Delta x_\mu / \partial x_\sigma \partial x_\alpha$ vanish. Clearly this will only be true if condition (33) is satisfied by $H$. Equation (35) is a generalization of equation (24). Presumably Einstein was unaware of the appearance of $S_\mu^\alpha$ in (35) since, in his derivation of (24), there would have been no need to collect terms in $\partial \Delta x_\mu / \partial x_\alpha$.

All five expressions in (34) and any of their linear combinations will satisfy condition (33) since they are all scalars under linear coordinate transformation. I have been unable to find any explanation why Einstein believed that condition (33) finally led to the result that $H$ had to take its *Entwurf* value. He asserted this result without proof in the paper and limited himself to confirming that this form of $H$ does in fact satisfy condition (33). His correspondence and later published discussion of his work of 1914 sheds only a little light on this episode and it remains an outstanding puzzle in the history of Einstein's theory.[36]

## 7. The Gradual Dawning

Einstein appears to have remained satisfied with the theory he developed in 1914 through the first half of 1915. In March, April, and early May, he defended the theory wholeheartedly in an intense correspondence with Levi-Civita, who challenged Einstein's derivation of the covariance properties of his gravitation tensor. But it seems that by mid-July he was less certain. He wrote enthusiastically to Sommerfeld about his visit to Göttingen of late June and early July, where he had lectured on his theory.[37] But he was less enthusiastic about Sommerfeld's proposal that one or two papers on general relativity be included in a new edition of *Das Relativitätsprinzip*, the well-known collection of original papers in the development of relativity theory. Einstein wrote that he would prefer to see the volume left unchanged since none of the current presentations of the theory was "complete."

By mid-October Einstein's points of dissatisfaction with his theory had grown in number and intensity. They soon culminated in some of the most agitated and strenuous weeks of his life, in which generally covariant field equations were discovered, or perhaps, rediscovered. In a letter of January 1, 1916, Einstein recounted the events of these months to Lorentz:

> The gradually dawning knowledge of the incorrectness of the old gravitational field equations gave me a rotten time last autumn. I had already found earlier that the perihelion motion of Mercury was too small. In addition, I found that the equations were not covariant for substitutions which corresponded to a uniform rotation of the (new) reference system. Finally I found that my approach of last year to the determination of Lagrange's function $H$ of the gravitational field was illusory throughout, since it could be easily modified so that one needed to apply no limiting condition at all to $H$, so that it could have been chosen quite freely. Thus I came to the conviction that the introduction of adapted systems was a false path and a more far-reaching covariance, where possible *general* covariance, must be demanded.[38]

Einstein gave a similar account of this dawning in an earlier letter to Sommerfeld.[39] There he noted that the old theory gave a figure of 18 minutes of arc per century rather than 45 minutes for the perihelion motion of Mercury.

The first result Einstein mentioned, the failure of his theory to account for the anomalous motion of Mercury's perihelion, might well have been known to him from the earliest days of the *Entwurf* theory. One of his earliest hopes for his new work on gravitation, as communicated in a letter of December 24, 1907 to Conrad Habicht, was that it might account for this anomaly.[40] The question seems to have arisen again early in 1915 in connection with the Berlin astronomer Freundlich, who had been the first to attempt astronomical tests of Einstein's new theory. In a postcard of March 1915 to Freundlich, Einstein confirmed that, according to his *Entwurf* theory, matter at rest can only yield a $g_{44}$ field, which proved that "a $g_{11}$ field cannot come into consideration in the problem of the planets."[41] This shows that Einstein still believed that a static field had to be spatially flat. But now he regarded it as a theorem of his *Entwurf* theory.

The second result that Einstein mentioned was his discovery that a uniform rotation of the coordinate axes does not belong to the justified coordinate transformations of his *Entwurf* field equations. We know that Einstein believed such transformations to be justified in 1914, although he never presented a proof. The Einstein Project has recently acquired a copy of a single page in Einstein's handwriting that bears on this matter.[42] Einstein here wrote out the *Entwurf* field equations. Beneath each of its terms, he put the values that their (4,4) components reduce to in a Minkowski metric viewed from uniformly rotating coordinates. In three-dimensional space-time, the transformation used seems to have been

$$t' = t, \quad r' = r, \quad \Theta' = \Theta - \omega t, \tag{37}$$

where the nonrotating coordinates are unprimed, rotating coordinates are primed, and the coordinates have their usual meaning.

Einstein's calculation is for the special weak-field case of small angular velocity $\omega$ and regions close to the axis of rotation. The results show that the field equations do not hold in the rotating case. Surrounding this calculation, in a way that indicates that it was added later, is the draft of a letter to Ministerial Director Naumann. This letter can be dated by content to late November or perhaps early December 1915. If the calculations are coeval with the letter, they were made around the time when Einstein returned to seek generally covariant field equations. The reverse side of the document contains some calculations on the form of the Minkowski metric in a uniformly rotating coordinate system. Einstein concentrated on the spatial part of the metric and deviated from the normal practice of equations (37) by leaving the radial coordinate in the rotating system $r'$ an undetermined function of the original nonrotating radial coordinate $r$.

The relationship—if any—between these calculations and those on the first side is unclear. Perhaps Einstein was investigating some problem in rotating coordinates, for example, the spatial geometry on a rotating disk, and perhaps this investigation led him to check whether the transformation to rotating coordinates is in fact justified. Or perhaps the discovery that the particular transformation (37) is not justified led him to try to find another transformation to rotating coordinates that is justified—hence the presumably unsuccessful examination of a more general transformation in which $r'$ is an undetermined function of $r$. What does remain a puzzle is how Einstein could have overlooked for so long the result that transformation (37) is not justified—if he in fact did. Or perhaps he knew that it does not hold, but was not disturbed to find that just one of many possible transformations to rotating coordinates is not justified.

The third result that Einstein described to Lorentz as precipitating his return to the search for generally covariant field equations was his discovery that his new derivation of 1914 did not actually determine $H$. He had found that an easy modification of his considerations no longer led to any restricting conditions on $H$. Lorentz would have been quite familiar with this last result, for in a letter of October 12, 1915, Einstein had described to him exactly what this modification was.[43] In his letter, Einstein recognized that condition (33) amounted only to the requirement that $H$ be a scalar under linear coordinate transformations. He described how he missed this in 1914 since he had assumed this property for $H$ at the beginning of the derivation.

In his modified derivation of the field equations in this letter of October 1915, he proceeded exactly as in 1914, but without this assumption. He found that the condition for adapted coordinate systems is equation (31) rather than (28). This immediately clarified the status of equation (33). Einstein continued to write the field equations that arose from the gravitation tensor in equation (23) as

$$-\sum \frac{\partial}{\partial x_\sigma}\left(g^{\nu\lambda}\frac{\partial Q}{\partial g_\sigma^{\mu\nu}}\right) = \kappa \mathfrak{T}_\mu^\lambda + \left(-\sum_\nu g^{\nu\lambda}\frac{\partial Q}{\partial g^{\mu\nu}} - \sum_{\nu\sigma} g_\sigma^{\nu\lambda}\frac{\partial Q}{\partial g_\sigma^{\mu\nu}}\right), \quad (38)$$

where $Q = H\sqrt{-g}$; and he expressed the conservation laws as

$$\sum_\lambda \frac{\partial}{\partial x_\lambda}(\mathfrak{T}_\mu^\lambda + t_\mu^\lambda) = 0, \quad (39)$$

where[44]

$$t_\mu^\lambda = \frac{1}{2\kappa}\sum_{\sigma\nu}\left(-g_\mu^{\nu\sigma}\frac{\partial Q}{\partial g_\lambda^{\nu\sigma}} + Q\delta_\mu^\lambda\right). \quad (40)$$

Einstein then required that the source term of the field equations, the right-hand side of (38), be equal to $\kappa(\mathfrak{T}_\mu^\lambda + t_\mu^\lambda)$. It follows directly from equation (40) that this will only be true if condition (33) is satisfied. (This result is interesting in itself. In effect, Einstein showed that the field equations can always be written with a source term of this form as long as $H$ is a scalar under linear coordinate transformations. So this result applies to his final generally covariant field equations as well.)

Einstein then completed what was to be his last derivation of the *Entwurf* field equations by noting that the choice of $H$ as its *Entwurf* form was dictated by the requirements of the Newtonian limit. This must have been quite a setback for Einstein, for, as we have seen, he had taken great pride in the fact that his earlier derivation of 1914 of the *Entwurf* field equations seemed to be based only on covariance arguments and did not need to draw directly on any physical knowledge of gravitation.

In his letter to Lorentz, Einstein did not explain how the requirement of the Newtonian limit was to be applied. Presumably he meant that the gravitation tensor had to be of form (3). If so, then he was wrong on two counts. First, equation (3) does not quite uniquely determine the form of $H$ as its *Entwurf* form, the fourth of those listed in (34). It admits a limited number of alternatives. For example, the gravitation tensor resulting from taking $H$ as the fourth plus an arbitrary constant times the difference of the third and the fifth also has this property. Second, as Einstein soon discovered, Newtonian theory can still be obtained as a limiting case if we dispense with restrictive condition (3).

But Einstein's work on this derivation had not been entirely in vain, for it had brought him both temporally and conceptually closer than ever before to a generally covariant theory. If we ignore Einstein's last fatal step, we find that the mathematical apparatus set up here by Einstein, and earlier by Grossmann, can be used almost unchanged in the final generally covariant theory. For if we make the now familiar selection

$$H = g^{\mu\nu}\left(\left\{\begin{matrix}\sigma\\ \mu\nu\end{matrix}\right\}\left\{\begin{matrix}\rho\\ \sigma\rho\end{matrix}\right\} - \left\{\begin{matrix}\rho\\ \mu\sigma\end{matrix}\right\}\left\{\begin{matrix}\sigma\\ \nu\rho\end{matrix}\right\}\right), \qquad (41)$$

then the field equations of the final theory follow.[45] The adapted coordinate condition still holds, but in a degenerate form, for now all coordinate systems are adapted. In fact the adapted coordinate condition, written as either equation (28) or (31) is none other than the contracted Bianchi identities.

Einstein clearly came to recognize these results, for they comprise the major part of his paper of late 1916 on a Hamiltonian formulation of the general theory of relativity (Einstein 1916b). By adopting a gravitational field action density based on the Riemann curvature scalar, he arrived at an $H$ of the form of (41). Since this $H$ is a scalar under linear coordinate transformations, he arrived at condition (33) and then directly at condition (28). With this choice of $H$, this last condition is equivalent to the contracted Bianchi identities, although Einstein was unaware presumably of the connection to the uncontracted Bianchi identities at this time.[46] He then proceeded to the field equations and conservation laws written in a form similar to equations (38), (39), and (40), but now using condition (28) to derive the conservation laws from the field equations and their covariance properties.

It is hard to imagine that Einstein was unprepared for the ease with which his formalism of 1914 could be applied to his final generally covariant theory. In 1914, in the paper in which he had first introduced the adapted coordinate condition, he remarked—prophetically—that, were this condition to be generally covariant, then all coordinate systems would be adapted and that this consequence would not compromise any step of his proof (Einstein and Grossmann 1914, pp. 224–225). But, he continued, these conditions were not generally covariant in the *Entwurf* theory; for if they were, the fully contracted gravitation tensor would have to be none other than the Riemann curvature scalar and it was not.

Hilbert, through his important paper of November 1915, is generally thought of as introducing the comprehensive use of these action principles to the theory (Hilbert 1915). My analysis shows that although Einstein might have drawn some of his work of 1916 in this area from Hilbert's, his basic mathematical apparatus and even the notation itself had its ancestry in his own work from 1914 and 1915.

## 8. "The Final Emergence into the Light"[47]

By mid-October 1915, Einstein was no longer satisfied with his theory. Presumably he had been disappointed that it had not accounted for the anomalous motion of Mercury. Perhaps this shortcoming had become all the more acute with the difficulties Freundlich faced in his attempts to set up and carry out astronomical tests of the theory. Then Einstein convinced himself that transformations to rotating coordinate systems were not "justified," which must have compromised his belief that his new theory extended the relativity of motion to accelerated motion. Finally, he found that all his elegant manipulation of covariance requirements and adapted coordinates did not even lead him to a definite set of field equations. With doubts accumulating about the empirical, physical, and formal foundations of his theory, Einstein took drastic action. He wrote to Sommerfeld that, at this point, he gave up the notion of requiring covariance with respect to adapted coordinate systems:

> After all trust in the results and methods of the earlier theory had thus given way, I saw clearly that only in a link to the general theory of covariants, i.e., to Riemann's covariant, could a satisfactory solution be found. Unfortunately I have immortalized the last errors in this struggle in the Academy papers, which I can send you soon.[48]

The first of these "last errors in this struggle" was presented to the Prussian Academy on November 4, 1915. He divided the Ricci tensor $G_{im}$ into the sum of two parts:[49]

$$G_{im} = R_{im} + S_{im}, \tag{42}$$

$$R_{im} = -\sum_l \frac{\partial \begin{Bmatrix} i\ m \\ l \end{Bmatrix}}{\partial x_l} + \sum_{\rho l} \begin{Bmatrix} i\ l \\ \rho \end{Bmatrix} \begin{Bmatrix} \rho m \\ l \end{Bmatrix}, \tag{43}$$

$$S_{im} = \sum_l \frac{\partial \begin{Bmatrix} il \\ l \end{Bmatrix}}{\partial x_m} - \sum_{\rho l} \begin{Bmatrix} i\ m \\ \rho \end{Bmatrix} \begin{Bmatrix} \rho\ l \\ l \end{Bmatrix}. \tag{44}$$

Einstein had shown that if we restrict ourselves to coordinate transformations of determinant one, then $\sqrt{-g}$ is a scalar. From this it followed easily that $S_{im}$ is a tensor under all such transformations and, since $G_{im}$ is a generally covariant tensor, then $R_{im}$ must also be a tensor under the restricted set of coordinate transformations. Einstein selected this as his gravitation tensor, and his field equations became

$$R_{\mu\nu} = -\kappa T_{\mu\nu}. \tag{45}$$

Why Einstein should choose this as his gravitation tensor rather than a generally covariant tensor, such as the Ricci tensor or even the Einstein tensor itself, has hitherto been a puzzle. It can now be solved by reference to my discussion of Einstein's original objections to the Ricci and related tensors as gravitation tensors. The results that led to his disillusionment with the *Entwurf* theory had left these original objections substantially intact. Einstein still expected his field equations to reduce to the form of (11) in the weak-field case. Einstein knew that the Ricci tensor reduced to the appropriate form with the application of the harmonic coordinate condition. But, as we have seen, this coordinate condition was unacceptable to him for it was inconsistent with the form of weak, static fields entailed by equation (11).

But there was a second possibility, the tensor $\mathcal{T}_{il}{}^{\times}$, which is the same as $R_{im}$. This reduces to the required form with coordinate condition (10). I argued that Einstein rejected this second tensor because he found that this coordinate condition was not satisfied in a Minkowski space-time viewed from rotating coordinates, a requirement that would ensure that the field equations retain the weak-field form of equation (11) in such rotating coordinates. But, as we have seen, by October 1915 Einstein had found that his *Entwurf* field equations, which had the required weak-field form, were not satisfied in such rotating coordinates. I conjecture that this discovery led him to reconsider the rather restrictive requirement that the field equations still have this weak-field form in such rotating systems, for he no longer had any objections to the tensor $R_{im}$ as a gravitation tensor in his paper of November 4.

In the concluding section of this paper, Einstein drew all these elements together. He stated that his new field equations reduce to the form of equation (11) in the weak-field case with the application of coordinate condition (10). This confirms my assertion that he still believed that his field equations must have this weak-field form and that his choice of $R_{im}$ as the gravitation tensor was based on the fact that they reduce to this form with the help of coordinate condition (10). Perhaps the juxtaposition is accidental, but Einstein completed the paper by noting that his new field equations are indeed covariant under transformation to rotating coordinate systems and to those in rectilinear acceleration, as required by the relativity of motion.

My account of Einstein's paper of November 4 has left a problem. The recovery of the weak-field equations described above involved the application of the four coordinate conditions (10) and the condition that only coordinate transformations of determinant one be admitted. This last condition amounts to one more coordinate condition than the four normally permitted. That Einstein did not in fact overdetermine his equations follows from his treatment of the single coordinate condition that arises from this last condition. First, he fully contracted field equation (45) to yield

$$\sum_{\alpha\beta}\frac{\partial^2 g^{\alpha\beta}}{\partial x_\alpha \partial x_\beta} - \sum_{\sigma\tau\alpha\beta} g^{\sigma\tau}\Gamma^\alpha_{\sigma\beta}\Gamma^\beta_{\tau\alpha} + \sum_{\alpha\beta}\frac{\partial}{\partial x_\alpha}\left(g^{\alpha\beta}\frac{\partial\lg\sqrt{-g}}{\partial x_\beta}\right) = -\kappa\sum_\sigma T^\sigma_\sigma. \quad (46)$$

Then, using familiar variational methods to define the stress-energy tensor of the gravitational field $t^\lambda_\mu$, he wrote the field equations in the mixed form[50]

$$\sum_{\alpha\nu}\frac{\partial}{\partial x_\alpha}(g^{\nu\lambda}\Gamma^\alpha_{\mu\nu}) - \frac{1}{2}\delta^\lambda_\mu \sum_{\mu\nu\alpha\beta} g^{\mu\nu}\Gamma^\alpha_{\mu\beta}\Gamma^\beta_{\nu\alpha} = -\kappa(T^\lambda_\mu + t^\lambda_\mu). \quad (47)$$

The coordinate divergence of the right-hand side of this equation vanishes as a result of the conservation laws. This yields the four conditions

$$\frac{\partial}{\partial x_\mu}\left(\sum_{\alpha\beta}\frac{\partial^2 g^{\alpha\beta}}{\partial x_\alpha \partial x_\beta} - \sum_{\sigma\tau\alpha\beta} g^{\sigma\tau}\Gamma^\alpha_{\sigma\beta}\Gamma^\beta_{\tau\alpha}\right) = 0, \qquad (48)$$

which correspond to the "adapted" coordinate conditions of the former theory. Equation (48) can be solved directly to yield a single condition—that the term in square brackets be a constant—which amounts to the coordinate condition imposed by the limitation of coordinate transformations to those with a determinant of one. Einstein set the value of this constant to zero, as one is free to do with such conditions, and thus arrived at

$$\sum_{\alpha\beta}\frac{\partial^2 g^{\alpha\beta}}{\partial x_\alpha \partial x_\beta} - \sum_{\sigma\tau\alpha\beta} g^{\sigma\tau}\Gamma^\alpha_{\sigma\beta}\Gamma^\beta_{\tau\alpha} = 0. \qquad (49)$$

These considerations resolve the problem of the overdetermination of the weak-field equations, for we can see immediately that in the weak-field case, in which terms quadratic in the derivatives of the metric tensor can be ignored. coordinate conditions (10) entail coordinate condition (49). Thus Einstein could introduce conditions (10), in the closing section of his paper, as a strengthening of condition (49).[51]

The conditions Einstein derived here led to one further result that was to be of great importance. Scalar conditions (46) and (49) could only both be true if the condition

$$\sum_{\alpha\beta}\frac{\partial}{\partial x_\alpha}\left(g^{\alpha\beta}\frac{\partial \lg\sqrt{-g}}{\partial x_\beta}\right) = -\kappa \sum_\sigma T^\sigma_\sigma \qquad (50)$$

held. Einstein noted that equation (50) meant that $\sqrt{-g}$ could not be set equal to unity, for the trace of the stress-energy tensor $T = \sum_\sigma T^\sigma_\sigma$ cannot be made zero.[52]

Einstein made one other point of special interest in his communication of November 4. He described "a fatal prejudice" in his earlier work: he had been induced to take the quantity

$$\frac{1}{2}\sum_\mu g^{\tau\mu}\frac{\partial g_{\mu\nu}}{\partial x_\sigma}$$

for the components of the gravitational field $\Gamma^\tau_{\nu\sigma}$ (Einstein 1915a, pp. 782–783). He now recognized that the Christoffel symbol of the second kind, or, to be exact, its negation, $-\left\{{\nu\sigma \atop \tau}\right\}$, was the quantity he should have selected and proceeded to argue for this new choice. Einstein did not explain why this prejudice was so fatal. A comparison of the equations of his *Entwurf* theory and the new theory may make it clearer. A weak correspondence between the forms of the equations of each theory appears if they are written in terms of

$\Gamma^\tau_{\nu\sigma}$, where these components have the appropriate forms as specified earlier. The action densities and gravitational field stress-energy tensors of both theories take on exactly the same form. Moreover, the second derivative terms of each of the two theories' gravitation tensors take on the same form, $\partial \Gamma^\tau_{\nu\sigma}/\partial x_\tau$. In the *Entwurf* case, this term is simply the d'Alembertian of the metric tensor. In the case of the new theory, however, this expression contains other second derivatives of the metric tensor, which Einstein had tried so hard to eliminate some three years earlier when trying to recover a Newtonian limit. So Einstein could write to Sommerfeld on November 28, 1915, that this final approach made his final field equations "the simplest conceivable since one is not tempted to transform them by multiplying out the symbols with the intention of more general interpretation."[53]

(It does not seem that pursuing this line of thought will help us further delineate the path Einstein took in 1912 and 1913. For Einstein's comments have all the flavor of an after-the-fact rationalization. Note in particular that Einstein did not introduce the notion of the "components of the gravitational field" into his *Entwurf* theory until 1914 [Einstein 1914b, p. 1058].)

Einstein did not remain satisfied with his theory of November 4 for very long. During the following week he found a simple modification to his theory that left its mathematical machinery essentially untouched but now brought field equations that were at last generally covariant. It seems likely that the modification occurred to him as a result of a reexamination of equation (50). From his standpoint on November 4, it followed that $\sqrt{-g}$ could not in general be a constant. Certainly there seemed to be no physical reason in the theory for such a limitation.

However, the equation can be read in a second way. We can regard it not as placing a limit on the form of $\sqrt{-g}$, but as restricting the value of $T$. Specifically, if $\sqrt{-g}$ has a constant value, then $T$ must vanish. Now this latter restriction does admit a simple physical interpretation, which Einstein was to seize with enthusiasm. If all matter were electromagnetic in nature, then this condition would be automatically satisfied. As is well known, the trace of the stress-energy tensor of an electromagnetic field is always zero.

On November 11 Einstein introduced his latest formulation of the theory with this hypothesis about the electromagnetic nature of all matter (Einstein 1915b).[54] He asserted that the hypothesis made possible the final step to generally covariant field equations. For these equations he wrote

$$G_{\mu\nu} = -\kappa T_{\mu\nu}. \tag{51}$$

From them he could recover the equations of November 4 and still use all their associated mathematical machinery by applying the coordinate condition

$$\sqrt{-g} = 1, \tag{52}$$

which in turn entailed the condition that only coordinate transformations of determinant one be admitted. Under condition (52), $S_{im}$ vanishes and his new field equation reduces to that of November 4 in form. He concluded his note of November 11 by noting that equation (50) entails the vanishing of $T$ if condition (52) holds.

In the unkind gaze of historical hindsight, Einstein's "mistake" seems simple. He had been forced to admit a dangerous conjecture about the nature of matter in order to conceal the fact that he had missed the now familiar trace term in the "correct" field equations. They are

$$G_{\mu\nu} - \tfrac{1}{2}g_{\mu\nu}G = -\kappa T_{\mu\nu}, \qquad (53)$$

where $G$ is the trace of $G_{\mu\nu}$, or, in an equivalent form,

$$G_{\mu\nu} = -\kappa(T_{\mu\nu} - \tfrac{1}{2}g_{\mu\nu}T). \qquad (54)$$

These field equations are consistent with the conservation laws without further hypothesis since the covariant divergence of the left-hand side of equation (53) vanishes identically. Transferring this property to the right-hand side gives the conservation laws, the vanishing of the covariant divergence of the stress-energy tensor. Now Einstein's field equations of November 11 do not have this property. But the assumption that $T = 0$ converts equations (53) and (54) into those field equations and brings them into accord with the conservation laws. That the assumption does this is not surprising since it came originally from equation (50), which in turn was derived with the help of the conservation laws.

This simple analysis completely misses what Einstein had achieved with his modification of November 11. He had finally succeeded in finding generally covariant field equations that reduced to the weak-field form of equation (11). This reduction could be effected by the application of coordinate conditions (10) and new condition (52). That these five conditions do not overdetermine the field followed almost immediately from his paper of November 4. There, as we have seen, Einstein showed the consistency in the weak-field case of using the four conditions (10) and limiting coordinate transformations to those with a determinant of one. This latter limitation, which amounted to the limitation to coordinate systems in which $\sqrt{-g}$ behaves like a scalar, was strengthened in the following note to the requirement that $\sqrt{-g}$ be a constant, specifically unity.

An examination of the relevant equations of Einstein's paper of November 4 (equations (46) to (50) here) shows that use of condition (10) strengthened with (52) does not compromise his consistency arguments, provided that we are willing to accept the new constraint on the field source term of $T = 0$. The adoption of the extra trace terms of equations (53) or (54) would have been out of the question, for they would have destroyed the hard-won and finely tuned agreement between the field equations and their weak-field limit. We can now appreciate why field equations (51) seemed the only possible generally covariant field equations to Einstein and thus why the adoption of the hypothesis $T = 0$ seemed a small price to pay for the final achievement of general covariance. Indeed, in the note of November 11, he seemed pleased to regard the information it contained as an unexpected bonus from the requirement of general covariance.

What still separated Einstein from his final field equations was not a simple oversight, but the same almost untouched misconceptions about the weak-field limit that he had had three years earlier. I argued that he expected the field equations to reduce to equation (11) in the weak-field case because of their formal simplicity and because they in turn enabled a simple reduction of the ten gravitational potentials of the full theory to a single Newtonian potential in a simple static-field case. The naturalness of such a reduction was corroborated by Einstein's belief that, on the basis of quite separate arguments, the number of gravitational potentials undergoes a similar reduction in the case of a general static field.

The final realization that his ideas on the behavior of weak and static fields were excessively restrictive and not justified by experience came over the two weeks following November 11. What seems to have catalyzed this realization was his calculation of the orbit of Mercury. He turned to this task immediately after he had arrived at the modified field equations and was able to present his results to the Prussian Academy just one week later, on November 18 (Einstein 1915c).[55] In this communication, Einstein used a method of successive approximations to solve his field equations for the gravitational field of the sun, that is, for the weak, static, spherically symmetric, source-free case, with Minkowskian values at spatial infinity. He presented a solution for his case that satisfies the condition $\sqrt{-g} = 1$, so that the field equations he was solving, those of November 11, reduce in form to those of November 4. Einstein had already shown that these latter field equations yield weak-field equations (11). Provided that the central mass is small, these equations solve to yield a spatially flat field. However, such a weak-field solution was no longer allowed to Einstein, for his solution of the field equations had to satisfy the condition $\sqrt{-g} = 1$. Thus Einstein was forced to a solution that had nonconstant $g_{11}$, $g_{12}, \ldots, g_{33}$ even in the first approximation. He commented on this crucial new development:

> From our theory it follows that, in the case of masses at rest, the components $g_{11}$ to $g_{33}$ are different from zero already in quantities of the first order. We shall see later that through this no contradiction arises with Newton's law (in the first approximation). (Einstein 1915c, p. 834)

This demonstration followed soon. Einstein was able to show that in quantities of the first order of smallness the equations of motion of a slow-moving mass-point reduce to those of Newtonian gravitation theory. Although it was not stated in this communication, this demonstration rested on the fact that only the $g_{44}$ component of the metric tensor was used in the construction of these equations in the first approximation. I have quoted Einstein's communication of this "most remarkable" result to Besso in December 1915.[56] Einstein seems to have remained impressed by it, for he still described this feature of the equations of motion as "remarkable" in his summary article on the theory written early the next year (Einstein 1916a, p. 817). Again, in his lectures of 1921 published as *The Meaning of Relativity*, he attributed the absence of an earlier recognition of the tensorial nature of the gravitational potential to this same feature of the equations of motion (Einstein 1921, p. 86). This comment can be applied to Einstein's own early treatment of weak and static fields.

Einstein's calculation of the orbit of Mercury was of crucial significance in the historical development of the theory. It gave the theory its first convincing empirical success. It was instrumental in freeing Einstein from his long-standing misconceptions about static and weak fields. It forced him to deal with a weak, static field, whose $g_{11}$ to $g_{33}$ components are not constant. In the communication of November 18, Einstein had already begun to tease out the implications of this revelation. He noted that his new theory predicts twice the deflection of a ray of starlight grazing the sun than that predicted by his earlier theory—including the field equations of November 4 (Einstein 1915c, p. 834).

Most significantly, Einstein was no longer constrained by the requirement that his field equations reduce to weak-field equation (11). It was now no longer necessary for the weak-field equations to give in the appropriate static cases a metric of form (1); field equations with additional terms to those in equation (11) could be contemplated. Einstein no longer had to adopt his field equations of November 11 and the associated $T = 0$ condition as the only possible generally covariant field equations. At last he was free to entertain field equations of the form of equations (53) and (54). Einstein may have realized this possibility very soon after completion of his paper on Mercury's motion. Although this communication still used the field equations of November 11 and the hypothesis $T = 0$, a footnote on its first page promised a new communication in which the hypothesis would be shown to be superfluous. The import of this footnote is not entirely clear since it makes no reference to new field equations. In any case, Einstein's next communication, presented to the Prussian Academy on November 25, gave as the results of nearly three years of labor Einstein's final field equations (54) (Einstein 1915d). He could also note—presumably with some relief—that this final modification does not affect the source-free form of the field equations and the resulting explanation of the anomalous motion of Mercury.

The question of the exact path that Einstein followed from November 11 to November 25 has become of some interest to historians of relativity. In his communication on the later date, Einstein dealt with what he called "the reasons that gave rise to my introduction of the second term on the right-hand side of the field equations [54]" (Einstein 1915d, p. 846). These, he tells us, arose in considerations analogous to those dealt with in equations (46) to (50). He noted that his new field equations, when fully contracted, become

$$\sum_{\alpha\beta} \frac{\partial^2 g^{\alpha\beta}}{\partial x_\alpha \partial x_\beta} - \kappa(T + t) = 0, \tag{55}$$

which corresponds to the earlier equation (46). However, he observed, in this new equation, both $T_\sigma^\lambda$ and $t_\sigma^\lambda$ appear in a fully symmetrical way, unlike the case of equation (46). Further reduction of the field equations with the conservation laws in the manner that produced equations (48) now yields

$$\frac{\partial}{\partial x_\mu}\left(\sum_{\alpha\beta}\frac{\partial^2 g^{\alpha\beta}}{\partial x_\alpha \partial x_\beta} - \kappa(T+t)\right) = 0. \tag{56}$$

In the context of the theory of November 11, Einstein found that the introduction of the hypothesis $T = 0$ was needed to bring equations (46) and (48) into accord. He noted two weeks later that this hypothesis was no longer necessary. The corresponding equation (55) actually entails the corresponding equation (56).

This necessity indicates that Einstein added the new trace term to his field equations of November 25 in an explicit attempt to modify conditions (46) and (48) so that the hypothesis $T = 0$ would no longer be necessary. Further, his observation on equation (55) suggests a natural way in which Einstein could have found exactly what this modification to the field equations should be: they should be modified so that both $T_\sigma^\lambda$ and $t_\sigma^\lambda$ appear in a symmetrical way. That this is not the case with the field equations of November 11 becomes especially clear if we write them in mixed form in a coordinate system in which $\sqrt{-g} = 1$, in which case they take the form of equation (47). The second term on the right-hand side is equal to

$$-\tfrac{1}{2}\delta_\mu^\lambda \kappa t,$$

and it is in this term that the asymmetry lies. If this term is replaced by

$$-\tfrac{1}{2}\delta_\mu^\lambda \kappa(T+t),$$

then the field equations become fully symmetrical in $T_\sigma^\lambda$ and $t_\sigma^\lambda$ and equivalent to the final equations of November 25.

The preceding considerations suggest a natural path for Einstein to have followed between November 4 and conditions (46) and (48) and November 25 and the final field equations. For the earlier field equations written exactly in the form of equation (47) appear in Einstein's paper of November 4 as a part of his derivation of conditions (46) and (48). Moreover, we know that Einstein used arguments of exactly this type at that time. In his review article on the theory written early the next year, he generated his field equations by first writing them in their source-free form,

$$\frac{\partial}{\partial x_\alpha}(g^{\sigma\beta}\Gamma_{\mu\beta}^\alpha) = -\kappa\left(t_\mu^\sigma - \frac{1}{2}\delta_\mu^\sigma t\right), \tag{57}$$

and then generalizing to their complete form,

$$\frac{\partial}{\partial x_\alpha}(g^{\sigma\beta}\Gamma_{\mu\beta}^\alpha) = -\kappa\left((t_\mu^\sigma + T_\mu^\sigma) - \frac{1}{2}\delta_\mu^\sigma(t+T)\right), \tag{58}$$

by requiring that $t_\mu^\sigma + T_\mu^\sigma$ replace $t_\mu^\sigma$ everywhere (Einstein 1916a, pp. 806–807). Of course the field equations in form (58) are identical with those to be produced by the modification of equation (47).

It is now known that Hilbert in Göttingen was able to arrive at substantially the same gravitational field equations as Einstein and that these equations were presented to the Göttingen Academy on November 20, 1915, five days prior to Einstein's presentation of his final field equations to the Berlin Academy (Hilbert 1915; see also Mehra 1974). Building on the results of Einstein and Mie, Hilbert used now familiar variational techniques to derive both gravitational and electromagnetic field equations and the associated conservation laws from a combined action density, whose gravitational part was the Riemann curvature scalar. His gravitational field equations took form (53), with the added constraint that the stress-energy tensor on the right-hand side be purely electromagnetic in nature and written in terms of a derivative of the electromagnetic action density with respect to the metric tensor. Apart from the inevitable and fruitless question of priority of discovery, there has been some speculation that Einstein's final formulation of his field equations may have been influenced by a knowledge of Hilbert's field equations. In November 1915, Einstein virtually suspended his usual correspondence, but maintained an active exchange with Hilbert, in which it is possible that Hilbert communicated his field equations to Einstein sometime between November 15 and 18.[57]

We do not know the exact extent of Einstein's knowledge of Hilbert's work in November 1915. Whatever it may have been, however, it seems unlikely that it contributed in any decisive way to Einstein's final formulation of his field equations. I have tried to show here how Einstein's final steps were self-contained. His omission of the familiar trace term in his field equations of November 11 was not the consequence of a simple oversight that could be remedied by a glance at Hilbert's equations. Einstein had good reasons for not admitting any such additional terms. When he realized that these reasons were incorrect, he introduced the new terms by a path that can be fairly readily reconstructed.

There was a brief period of coolness between Einstein and Hilbert immediately after their November correspondence. Einstein's former assistant, E.G. Strauss, attributes this coolness to Einstein's feeling that Hilbert had perhaps unwittingly plagiarized some of Einstein's earlier ideas on the theory from lectures he gave in Göttingen in 1915 (Pais 1982, p. 261). A recently discovered letter of Einstein's to his good friend Heinrich Zangger in late November or early December 1915 supports Strauss's view, although the letter does not mention Hilbert by name (Medicus 1984). Further, Einstein seems to have been unfamiliar with the detailed content of Hilbert's communication of November 20 as late as May 1916, even though he had stayed with the Hilberts some months earlier. Einstein reopened his correspondence with Hilbert that May with a plea for help in understanding Hilbert's paper, which he had to review in a coming colloquium in Berlin.[58] It was only at the end of this exchange that Einstein felt he could state with certainty that his and Hilbert's results agreed.

Einstein's ignorance of the details of Hilbert's work before May 1916 extended to Hilbert's result that was of crucial significance in the context of Einstein's final field equations. Hilbert wrote his gravitational field equations in terms of the variational derivative of the Riemann curvature scalar density with respect to the components of the metric tensor (Hilbert 1915, p. 404). In the line immediately following, he stated without detailed proof that this variational derivative is equal to what we now know as the tensor density corresponding to the Einstein tensor, the tensor on the left-hand side of equation (53). Einstein could not have been aware of this result the following January. Then he wrote to Lorentz that the theory would gain greatly in clarity if a Hamiltonian formulation could be found for its field equations in their general form, that is, in their form prior to the imposition of the constraint $\sqrt{-g} = 1$.[59] This, of course, was what Hilbert had already done. Einstein even described to Lorentz how it appeared to him that the appropriate action density should be the Riemann curvature scalar density, the result Hilbert had already stated, and began to map out the derivation of the field equations from it.[60]

Einstein wrote Lorentz early in 1915:

There are two ways that a theoretician goes astray

1) The devil leads him around by the nose with a false hypothesis (For this he deserves pity)
2) His arguments are erroneous and ridiculous (For this he deserves a beating).[61]

I have tried to show that Einstein went astray in the first way, rather than in the second. The cause of his straying is inseparable from his characteristic methods. On the one hand stood his relentless and uncompromising insistence on certain fundamental physical principles—the requirement of the Newtonian limit, the conservation laws, physical causality. On the other hand was the remarkable flexibility that enabled Einstein to reject even the most cherished of notions if his basic principles seemed to call for it—in this case he was prepared to forfeit general covariance. A lesser physicist might have compromised and faltered. But, eventually and perhaps inevitably, Einstein's same uncompromising relentlessness enabled him to weed out the false hypotheses that had misled him and brought him to his goal, his general theory of relativity.

*Acknowledgments.* I am most grateful for the generous hospitality of the Einstein Project, Princeton, and the Center for the Philosophy and History of Science, Boston University; for the help and advice of John Stachel; for his and Roberto Torretti's comments on earlier drafts of this paper; for many helpful editorial suggestions from Paul Forman; for the sponsorship of the Fulbright Exchange Program during the period of the research and writing of this paper; and for the kind permission to quote from Einstein's unpublished writings given by the Hebrew University of Jerusalem, which holds the copyright.

NOTES

* © 1984 by the Regents of the University of California. Reprinted from *Historical Studies in the Physical Sciences*, Vol. 14, part 2, pp. 253–316, by permission.

[1] Einstein to H.A. Lorentz, August 16, 1913, EA 16-434.

[2] Aspects of Stachel's second thesis had been anticipated in Earman and Glymour 1978a.

[3] EA 23-191.

[4] Einstein to Ehrenfest, May 28, 1913, EA 9-340.

[5] This special case may have been important in the sequence of events in Einstein's transition to the new theory. In an addendum to Einstein 1912b, he noted that the equations of motion of the theory could be written as a variation principle. Formally the equation he gave is identical to the equation for a geodesic in a space-time with the metric (1). See Stachel 1979, pp. 433–434.

[6] I adhere throughout to Einstein's original notation, which is nearly the same as modern notation. However, all indices are written as subscripts, covariant components are indicated by Latin letters, e.g., $g_{\mu\nu}$, and their corresponding contravariant components by Greek letters, e.g., $\gamma_{\mu\nu}$, except for the four contravariant space-time coordinates, written as $x_\mu$. Gothic letters represent tensor densities. The Einstein summation convention is not used. Greek and sometimes Latin indices vary over 1, 2, 3, and 4, with the "4" component representing the time component. I also follow Einstein's use of the term "tensor" by allowing it to describe quantities covariant under limited as well as under arbitrary coordinate transformations. Between 1912 and 1915 Einstein modified his notation gradually until it achieved the now standard form.

[7] In modern notation, these equations are

$$R_{\mu\nu} - \tfrac{1}{2} g_{\mu\nu} R = \kappa T_{\mu\nu},$$

where $R_{\mu\nu}$ is the Ricci tensor with contraction $R$, and $T_{\mu\nu}$ is the stress-energy tensor. The left-hand side of the equation is now called the "Einstein tensor."

[8] See Lanczos 1972, pp. 13–14; Pais 1982, pp. 221–223, 243–244; Vizgin and Smorodinskiĭ 1979, pp. 501–502; Earman and Glymour 1978a, pp. 256–257. Mehra (1974, pp. 11–12) discusses the rejection of the Ricci tensor in terms of Einstein's difficulties in finding a generally covariant conservation law, but these difficulties became important only at a slightly later stage.

[9] Besides the items in note 8, see Hoffmann 1972, p. 161; 1982, pp. 100–102; Zahar 1980, pp. 31–33.

[10] EA 3-006. John Stachel had already noted its mislabeling.

[11] That Einstein sums over $\kappa$ in such terms as $g_{\kappa\kappa}$ indicates that he is considering a weak-field case in a coordinate system with an imaginary $x_4$ ("time") coordinate and in which the metric is Minkowskian and represented by the unit matrix to zeroth order.

[12] Compare this equation (9) with Einstein's equations (42), (43), and (44) later. Note that throughout this period Einstein wrote the Christoffel symbols of the second kind with the upper and lower indices inverted as compared with the modern convention.

[13] Cf. Einstein to Sommerfeld, November 28, 1915, in Hermann 1968, pp. 33–36.

[14] I briefly review the internal evidence for dating the material in the notebook to the period of writing the *Entwurf* paper. Its notation, especially the use of $\gamma_{\mu\nu}$ for the contravariant components of the metric tensor, date it before mid-1914. Its generally elementary content and the similarity of its treatment to that of the *Entwurf* paper place it at the very beginning of Einstein's work on the theory, as does the repeated crediting of the formula for the Riemann curvature tensor to Grossmann. Note also the omission of such concepts as "adapted coordinates" and of the use of variational techniques that became characteristic of Einstein's work on the theory after 1913.

[15] Einstein to M. Besso, ca. March 1914, in Speziali 1972, pp. 52–53.

[16] Einstein to Sommerfeld, November 28, 1915, in Hermann 1968, pp. 33–36. $T$ is the trace of $T_{im}$.

[17] Einstein to Hertz, August 22, 1915?, EA 12-202.

[18] Compare with Einstein's own later writing of the weak-field equations of his final theory in Einstein 1921, p. 86. After reduction with the harmonic coordinate condition, they are

$$\frac{\partial^2 \gamma_{\mu\nu}}{\partial x_\alpha^2} = 2\kappa(T_{\mu\nu} - \tfrac{1}{2}g_{\mu\nu}T),$$

where $\gamma_{\mu\nu}$ are the weak-field deviations of the components of the metric tensor from their Minkowskian values. Here Einstein began with the presumption that all the $\gamma_{\mu\nu}$ were of the same order. He later rejected this presumption. This altered the further development of these weak-field equations and their relationship to the Newtonian case. See Einstein, Infeld, and Hoffmann 1938.

[19] A very similar argument to this appears in Einstein, Infeld, and Hoffmann 1938, pp. 72–73.

[20] Einstein to Hilbert, postmarked November 18, 1915, EA 13-091.

[21] Einstein to Besso, December 1915, in Speziali 1972, p. 61. See also Einstein to Besso, December 10, 1915, in Speziali 1972, pp. 59–60, for a briefer statement of surprise at the variability of $g_{11}$–$g_{33}$ in the weak-field case.

[22] For example, consider the following identity:

$$(2\sqrt{-g}g_{\tau\tau'}g_{,\sigma}^{\tau\alpha}g_{,\beta}^{\tau'\beta} - \delta_\sigma^\alpha \sqrt{-g}g_{\tau\tau'}g_{,\alpha'}^{\tau\alpha'}g_{,\beta}^{\tau'\beta} - 2\sqrt{-g}g_{\tau\beta}g_{,\sigma}^{\tau\rho}g_{,\rho}^{\beta\alpha} + \delta_\sigma^\alpha \sqrt{-g}g_{\alpha'\beta}g_{,\tau}^{\alpha'\rho}g_{,\rho}^{\beta\tau})_{,\alpha}$$

$$\equiv g_{\mu\nu,\sigma}\sqrt{-g}g^{\mu\mu'}g^{\nu\nu'}\left[\frac{1}{\sqrt{-g}}((\sqrt{-g}g_{\alpha\mu'})_{,\sigma'}g_{,\nu'}^{\alpha\sigma'} + (\sqrt{-g}g_{\alpha\nu'})_{,\sigma'}g_{,\mu'}^{\alpha\sigma'}\right.$$

$$-(\sqrt{-g}g_{\tau\mu'})_{,\nu'}g_{,\alpha}^{\tau\alpha} - (\sqrt{-g}g_{\tau\nu'})_{,\mu'}g_{,\alpha}^{\tau\alpha}) + \frac{1}{2}g_{\mu'\nu'}(g_{\alpha\beta}g_{,\tau}^{\alpha\rho}g_{,\rho}^{\beta\tau} - g_{\tau\tau'}g_{,\alpha}^{\tau\alpha}g_{,\beta}^{\tau'\beta})$$

$$\left. + g_{\alpha\mu'}g_{,\tau}^{\alpha\rho}g_{\beta\nu'}g_{,\rho}^{\beta\tau} - g_{\tau\mu'}g_{,\alpha}^{\tau\alpha}g_{\tau'\nu'}g_{,\beta}^{\tau'\beta}\right].$$

This identity can be added to identity (16) to yield a new identity consistent with all the conditions stated so far. Note that the right-hand side of the identity written here vanishes to quantities in the second order of smallness. The discovery of such identities is by no means easy. This one was formed by comparison of different gravitation tensors generated by Lagrangian methods.

[23] Einstein to Lorentz, August 16, 1913, EA 16-434. See also Einstein and Grossmann 1913, pp. 260–261; Einstein 1913b, pp. 1257–1258, and 1914a, p. 178.

[24] Einstein 1914a, p. 178. Einstein did not tell us what this "closer consideration" was. Perhaps he wrote $1/\sqrt{-g}$ times the left-hand side of equation (18) in terms of a covariant divergence in the following way:

$$1/\sqrt{-g}(\sqrt{-g}T_\sigma^v + \sqrt{-g}t_\sigma^v)_{;v} = (T_\sigma^v + t_\sigma^v)_{;v} + \tfrac{1}{2}g_{\mu v,\sigma}(T^{\mu v} + t^{\mu v}),$$

where $t_{v\sigma}$ is assumed to be symmetric. All terms on the right-hand side are generally covariant tensors with the exception of $g_{\mu v,\sigma}$, which is a tensor under linear coordinate transformations only. Thus the left-hand side can only be covariant under linear coordinate transformations as well.

[25] Einstein to Ehrenfest, 1913, EA 9-342.

[26] Ordered by dates of publication, Einstein and Grossmann 1913, pp. 260–261; Einstein 1914a, p. 178; Einstein and Grossmann 1914, pp. 217–218; Einstein 1914b, pp. 1066–1067. The argument does not appear in the body of the published text of Einstein's address to the Congress of German Natural Scientists and Physicians in 1913, although it appears in a footnote to the printed text Einstein 1913b, p. 1257; it occurs in a letter to L. Hopf of November 1913 (EA 13-290), and in the addendum to the journal printing of the *Entwurf* paper and not in the earlier separatum.

[27] See the papers cited in notes 8 and 9.

[28] Einstein later put great stress on this inseparability of metric and manifold. See Einstein 1952, p. 155.

[29] Einstein to Besso, January 3, 1916 in Speziali 1972, pp. 63–64. Cf. Einstein to Ehrenfest, December 26, 1915, EA 9-363.

[30] Einstein treated condition (20) in the same way as coordinate condition (10) in his letter to Paul Hertz of August 22, 1915?, EA 12-202.

[31] Einstein to Besso, March 1914 in Speziali 1972, pp. 52–53.

[32] In this paper, Einstein reintroduced the representation of contravariant components of a tensor by raised indices and covariant components by lowered indices. This convention had been used by Ricci and Levi-Civita. Grossmann described how he and Einstein decided not to use it then because it was too complicated in certain cases (Einstein and Grossmann 1913, p. 246). The term $g_\sigma^{\mu v}$ signifies $\partial g^{\mu v}/\partial x_\sigma$.

[33] Cf. Einstein to Lorentz, January 23, 1915, EA 16-436.

[34] The close connection between the devices employed in the "hole" argument and in the variational treatment of the field equations suggests that the "hole" argument may have occurred to Einstein as a result of early attempts to apply variational techniques to his field equations.

[35] A new derivation of the field equations of Nordström's gravitation theory based on covariance considerations had been presented in Einstein and Fokker 1914, p. 328. The derivation involved postulating a scalar field equation based on the Riemann curvature scalar. They concluded by speculating that a similar derivation of the "Einstein-Grossmann gravitation equations" might also be possible using the Riemann curvature tensor. They observe without further explanation that the reason given in Grossmann's section 4 of the *Entwurf* paper for the nonexistence of such a relationship between the gravitation equations and the Riemann curvature tensor does not hold up under closer consideration. All that Einstein and Grossmann had found in the *Entwurf* paper was that the then obvious methods of forging a link between the Riemann curvature tensor and the gravitation tensor seem to fail on the question of the Newtonian limit. This does not prove that such a connection does not exist and perhaps this was the point of Einstein and Fokker's footnote. Einstein and Fokker do not seem to have doubted the correctness of the *Entwurf* field equations nor did they

repudiate the general arguments against the admissibility of generally covariant field equations.

[36] The most promising of these later comments comes in a letter to Hilbert of March 30, 1916 (EA 13-097), in which Einstein discussed a mistake in his "work of 1914" that Hilbert had pointed out to him. Einstein noted that under the infinitesimal (coordinate?) transformation $\Delta$ the relation

$$\Delta g_\sigma^{\mu\nu} = \frac{\partial}{\partial x_\sigma}(\Delta g^{\mu\nu}) \tag{36}$$

does not hold. "Therefore," he said, "there is no variation in the sense of variational calculations that correspond to the change $\Delta$." This cannot be the confession of a simple blunder in his calculations that might have explained the error in question. For the Einstein of 1914 knew that (36) does not hold for infinitesimal coordinate transformations; he gives the correct relation as his equation (63a) of Einstein 1914b, and uses it correctly throughout the paper. Rather, the problem seems to be associated with the proof of the important result (29). Writing to de Sitter on January 23, 1917 (EA 20-540), Einstein placed a mistake, first found by Hilbert but otherwise unspecified, as somewhere in this proof. There Einstein had divided the variation of the field $\delta$ into two parts, the second of which, $\delta_2$, was a four-parameter variation that could be generated mathematically by an infinitesimal coordinate transformation. Therefore, the variation $\delta_2$ cannot satisfy (36), whereas the variation $\delta$ must, if the gravitation tensor of equation (23) is to be derived from it by the usual methods. Thus, in short, we can say that $\delta_2$ is "no variation in the sense of variational calculations."

[37] Einstein to Sommerfeld, July 15, 1915 in Hermann 1968, p. 30.

[38] EA 16-445.

[39] Einstein to Sommerfeld November 28, 1915, in Hermann 1968, pp. 32–36.

[40] EA 12-445.

[41] Postmarked March 19, 1915, EA 11-208.

[42] I am grateful to John Stachel for drawing this document to my attention.

[43] EA 16-442.

[44] This form of the conservation law follows from a contraction of the field equations with $g_\sigma^{\mu\nu}$ and substitution into the conservation laws in the form of equation (12).

[45] $\begin{Bmatrix} \sigma \\ \mu\nu \end{Bmatrix}$ is the Christoffel symbol of the second kind and summation over repeated indices is implied. This expression results from the Riemann curvature scalar after terms in the second derivatives of the metric tensor are separated out as a total divergence term. See, for example, Dirac 1975, p. 48.

[46] Mehra 1974, pp. 49–50, 78, and Pais 1982, pp. 274–278, discuss the delay in recognition of this connection.

[47] Title from Einstein 1933, pp. 289–290.

[48] Letter of November 28, 1915 in Hermann 1968, pp. 32–36.

[49] Einstein 1915a. As before, I follow Einstein in using $G_{im}$ to refer to the Ricci tensor rather than the modern usage, in which $G_{im}$ would refer to the Einstein tensor. For consistency, I have made Einstein's implicit summation explicit.

[50] Einstein 1915a. Specifically,

$$\kappa t_\sigma^\lambda = \frac{1}{2}\delta_\sigma^\lambda \sum_{\mu\nu\alpha\beta} g^{\mu\nu}\Gamma_{\mu\beta}^\alpha \Gamma_{\nu\alpha}^\beta - \sum_{\mu\nu\alpha} g^{\mu\nu}\Gamma_{\mu\sigma}^\alpha \Gamma_{\nu\alpha}^\lambda,$$

so that

$$\kappa t = \sum_\sigma \kappa t_\sigma^\sigma = \sum_{\mu\nu\alpha\beta} g^{\mu\nu}\Gamma^\alpha_{\mu\beta}\Gamma^\beta_{\nu\alpha}.$$

Note $\Gamma^\tau_{\nu\sigma} = -\begin{Bmatrix} \nu\sigma \\ \tau \end{Bmatrix}$.

[51] Of course this weak-field assumption is itself a coordinate condition of a kind. Strictly speaking, Einstein would have to show that his five coordinate conditions were also consistent with this new constraint. This should not have been a problem, for a coordinate system in which the weak-field assumption holds is still determined only up to a four-parameter infinitesimal coordinate transformation.

[52] Presumably Einstein referred to the general case in which the source of the field is unspecified. Then, from equation (50), $\sqrt{-g}$ cannot have any constant value whose coordinate derivatives all vanish. Einstein's procedure is not "incoherent," contrary to the assertion of Earman and Glymour 1978b, pp. 298–299. For further evidence, note that Einstein chose not to use the familiar covariant divergence of the stress-energy tensor in his conservation laws, but to replace it with a different but closely related quantity, which he demonstrated to be covariant under coordinate transformations with a determinant of one. The distinction between the two divergences drops away when $\sqrt{-g} = 1$, so Einstein's nonstandard choice in Einstein 1915a did not become important in his work over the following weeks.

[53] In Hermann 1968, pp. 32–36. Einstein also mentioned his "prejudice" in a letter to Lorentz, January 1, 1916, EA 16-445.

[54] That this hypothesis resulted from a reexamination of the equations in the earlier theory, rather than from some external source, is suggested by a footnote: "At the writing of the earlier communication, I had not yet become aware of the admissibility in principle of the hypothesis $\sum T^\mu_\mu = 0$" (Einstein 1915b, p. 800).

[55] Even Hilbert was impressed at the speed with which Einstein calculated the perihelion motion, as he told Einstein in his congratulatory postcard of November 19, 1915, EA 13-054.

[56] See the letters cited in note 21.

[57] See especially Earman and Glymour 1978b. They and Pais 1982, pp. 257–261, also outline the extant contents of this exchange, although Pais acquits Einstein of the charge of plagiarism.

[58] Einstein to Hilbert, May 25 and 30 and June 2, 1916, EA 13-099, 13-102, 13-104; Hilbert to Einstein, May 27, 1916, EA 13-056. Einstein complained about the obscurity of Hilbert's work to Hilbert (EA 13-102) and, in stronger terms, to Ehrenfest, where he accused Hilbert of having "pretensions of being a superman by hiding [his] methods" (May 24, 1916, EA 9-378).

[59] Einstein to Lorentz, January 17 and 19, 1916, EA 16-447, 16-449. Einstein's letter to Sommerfeld of December 9, 1915 again suggests a limited knowledge of Hilbert's work (Hermann 1968, p. 36).

[60] See Lorentz 1916; Lorentz described his success in establishing this result in Lorentz to Einstein, June 6, 1916, EA 16-451. Einstein's uncertainty about this result would explain why it did not appear in the review article (Einstein 1916a), where he used methods very similar to those of November 1915 to derive his field equations and to establish their consistency with the conservation laws. These same results can be established much more easily from a variation principle that uses the Riemann curvature scalar as the gravitational field action density and, as Einstein was to show

in Einstein 1916b, this deduction could be achieved with little effort by making use of the mathematical machinery of his investigations of 1914. Presumably Einstein also preferred this latter method: EA 2-077, an early version of Einstein 1916b, seems to have been intended first to replace the derivation in Einstein 1916a and then to be an appendix to that paper.

[61] Einstein to Lorentz, February 2, 1915. A copy of this letter has been recently acquired by the Einstein Project.

REFERENCES

Born, Max (1916). "Einsteins Theorie der Gravitation und der allgemeine Relativität." *Physikalische Zeitschrift* 17: 51–59.

Dirac, Paul A.M. (1975). *General Theory of Relativity*. New York: Wiley.

Earman, John, and Glymour, Clark (1978a). "Lost in the Tensors: Einstein's Struggles with Covariance Principles 1912–1916." *Studies in History and Philosophy of Science* 9: 251–278.

——— (1978b). "Einstein and Hilbert: Two Months in the History of General Relativity." *Archive for History of Exact Sciences* 19: 291–308.

Einstein, Albert (1907). "Über das Relativitätsprinzip und die aus demselben gezogenen Folgerungen." *Jahrbuch der Radioaktivität und Elektronik* 4: 411–462.

——— (1908). "Berichtigungen zu der Arbeit: Über das Relativitätsprinzip und die aus demselben gezogenen Folgerungen." *Jahrbuch der Radioaktivität und Elektronik* 5: 98–99.

——— (1911). "Über den Einfluss der Schwerkraft auf die Ausbreitung des Lichtes." *Annalen der Physik* 35: 898–908.

——— (1912a). "Lichtgeschwindigkeit und Statik des Gravitationsfeldes." *Annalen der Physik* 38: 355–369.

——— (1912b). "Zur Theorie des statischen Gravitationsfeldes." *Annalen der Physik* 38: 443–458.

——— (1913a). "Physikalische Grundlagen einer Gravitationstheorie." *Naturforschende Gesellschaft, Zürich. Vierteljahrsschrift* 58: 284–290 (lecture of September 9, 1913).

——— (1913b). "Zum gegenwärtigen Stande des Gravitationsproblems." *Physikalische Zeitschrift* 14: 1249–1266 (lecture of September 23, 1913).

——— (1914a). "Prinzipielles zur verallgemeinerten Relativitätstheorie." *Physikalische Zeitschrift* 15: 176–180 (received January 24, 1914).

——— (1914b). "Die formale Grundlage der allgemeinen Relativitätstheorie." *Königlich Preussische Akademie der Wissenschaften* (Berlin). *Sitzungsberichte*: 1030–1085 (received October 29, 1914).

——— (1915a). "Zur allgemeinen Relativitätstheorie." *Königlich Preussische Akademie der Wissenschaften* (Berlin). *Sitzungsberichte*: 778–786.

——— (1915b). "Zur allgemeinen Relativitätstheorie (Nachtrag)." *Königlich Preussische Akademie der Wissenschaften* (Berlin). *Sitzungsberichte*: 799–801 (read November 11, 1915).

——— (1915c). "Erklärung der Perihelbewegung des Merkur aus der allgemeinen Relativitätstheorie." *Königlich Preussische Akademie der Wissenschaften* (Berlin). *Sitzungsberichte*: 831–839 (read November 15, 1915).

——— (1915d). "Die Feldgleichungen der Gravitation." *Königlich Preussische Akademie der Wissenschaften* (Berlin). *Sitzungsberichte*: 844–847 (communicated November 25, 1915).

———— (1916a). "Die Grundlage der allgemeinen Relativitätstheorie." *Annalen der Physik* 49: 769–822 (received March 20, 1916).

———— (1916b). "Hamiltonsches Prinzip und allgemeine Relativitätstheorie." *Königlich preussische Akademie der Wissenschaften* (Berlin). *Sitzungsberichte*: 1111–1116 (received October 26, 1916).

———— (1921). *The Meaning of Relativity*. Princeton: Princeton University Press, 1922; page numbers cited from the 5th ed. Princeton: Princeton University Press, 1955.

———— (1933). "Notes on the Origin of the General Theory of Relativity." In *Ideas and Opinions*. Carl Seelig, ed. Sonja Bargmann, trans. New York: Crown, 1954, pp. 285–290.

———— (1952). "Relativity and the Problem of Space." Appendix 5 in *Relativity, The Special and the General Theory: A Popular Exposition*, 15th ed. Robert W. Lawson, trans. London: Methuen, 1954, pp. 135–157.

Einstein, Albert, and Fokker, Adriaan D. (1914). "Die Nordströmsche Gravitationstheorie vom Standpunkt des absoluten Differentialkalküls." *Annalen der Physik* 44: 321–328 (received February 19, 1914).

Einstein, Albert, and Grossmann, Marcel (1913). *Entwurf einer verallgemeinerten Relativitätstheorie und einer Theorie der Gravitation. I. Physikalischer Teil von Albert Einstein. II. Mathematischer Teil von Marcel Grossmann*. Leipzig and Berlin: B.G. Teubner. Reprinted with added "Bemerkungen," in *Zeitschrift für Mathematik und Physik* 62 (1914): 225–261. Page numbers are cited from the latter version.

———— (1914). "Kovarianzeigenschaften der Feldgleichungen der auf die verallgemeinerte Relativitätstheorie gegründeten Gravitationstheorie." *Zeitschrift für Mathematik und Physik* 63: 215–225.

Einstein, Albert, Infeld, Leopold, and Hoffmann, Banesh (1938). "The Gravitational Equations and the Problem of Motion." *Annals of Mathematics* 39: 65–100.

Hermann, Armin, ed. (1968). *Albert Einstein–Arnold Sommerfeld. Briefwechsel*. Basel and Stuttgart: Schwabe.

Hilbert, David (1915). "Die Grundlagen der Physik. (Erste Mitteilung)." *Königliche Gesellschaft der Wissenschaften zu Göttingen. Mathematisch-physikalische Klasse. Nachrichten*: 395–407.

Hoffmann, Banesh (1972). "Einstein and Tensors." *Tensor* 6: 157–162.

———— (1982). "Some Einstein Anomalies." In *Albert Einstein: Historical and Cultural Perspectives. The Centennial Symposium in Jerusalem*. Gerald Holton and Yehuda Elkana, eds. Princeton: Princeton University Press, pp. 91–105.

Lanczos, Cornelius (1972). "Einstein's Path From Special to General Relativity." In *General Relativity: Papers in Honour of J.L. Synge*. L. O'Raifeartaigh, ed. Oxford: Clarendon Press, pp. 5–19.

Lorentz, Hendrik Antoon (1916). "On Einstein's Theory of Gravitation." *Proceedings of the Section of Sciences, Koninklijke Akademie van Wetenschappen te Amsterdam* 19 (1916–1917): 1341–1354; 1354–1369; 20 (1917–1918): 2–19; 20–34 (communication of February 26 and April 1916).

Medicus, Heinrich A. (1984). "A Comment on the Relations between Einstein and Hilbert." *American Journal of Physics* 52: 206–208.

Mehra, Jagdish (1974). *Einstein, Hilbert and the Theory of Gravitation*. Dordrecht: D. Reidel.

Pais, Abraham (1982). *"Subtle is the Lord ..."*: *The Science and the Life of Albert Einstein*. Oxford: Oxford University Press.

Speziali, Pierre, ed. (1972). *Albert Einstein–Michele Besso. Correspondance, 1903–1955.* Paris: Hermann.

Stachel, John (1979). "The Genesis of General Relativity." In *Einstein Symposion Berlin.* H. Nelkowski et al., eds. Berlin, Heidelberg, and New York: Springer-Verlag, pp. 428–442.

——— (1980a). "Einstein and the Rigidly Rotating Disk." In *General Relativity and Gravitation One Hundred Years After the Birth of Albert Einstein.* Vol. 1. A. Held, ed. New York: Plenum, pp. 1–15. See this volume, pp. 48–62.

——— (1980b). "Einstein's Search for General Covariance, 1912–1915." Paper delivered to the Ninth International Conference on General Relativity and Gravitation, Jena, 1980. See this volume, pp. 63–100.

Vizgin, Vladimir P., and Smorodinskiĭ, Ya.A. (1979). "From the Equivalence Principle to the Equations of Gravitation." *Soviet Physics. Uspekhi* 22: 489–513.

Zahar, Elie (1980). "Einstein, Meyerson and the Role of Mathematics in Physical Discovery." *British Journal for the Philosophy of Science* 31: 1–43.

# Max Abraham and the Reception of Relativity in Italy: His 1912 and 1914 Controversies with Einstein

CARLO CATTANI and MICHELANGELO DE MARIA

## 1. Introduction

In the early years of this century, a heated debate on the structure of the electron started in Germany, in which all the unsolved problems of classical electromagnetism were discussed. In this debate, Max Abraham emerged as a prominent figure with the formulation of his theory of the "rigid electron," the mass of which was considered to be totally electromagnetic in nature (Abraham 1903, 1904). On this topic, Abraham came across as the most tenacious supporter of a point of view centered on the reduction of all physical phenomena—including mechanics—to purely electromagnetic effects, and on the defense of the ether. Thus he placed himself in opposition to all those theories that were in conflict with his reductionistic electromagnetic program, either because they introduced physical quantities that were not purely electromagnetic—as in the case of Lorentz and Poincaré's theory of the "deformable electron"—or because they even eliminated the ether, as in the case of Einstein's theory of special relativity.

Abraham had made many adversaries in the German scientific community, and his caustic, critical streak, along with his tendency to issue extremely sharp polemical judgments about his colleagues, certainly did not alleviate the climate of hostility towards him. Also because of this, he had remained at the lowest levels in his academic career, seeing himself passed over several times in the competition for positions as full professor in favor of less-deserving scientists, and had then sought a better situation abroad. After spending a difficult and disappointing semester in the United States at the University of Illinois in 1908, Abraham moved to Italy the following year, thanks to the considerable help of Tullio Levi-Civita. Levi-Civita had invited Abraham to present a paper on "The Theory of the Electron" at the IV Congresso Internazionale dei Matematici, held in Rome in 1908 and attended by the most eminent mathematicians and mathematical physicists in the world, such as: Hilbert, Klein, Poincaré, and Lorentz. On that occasion, Levi-Civita found Abraham "brilliant and likeable," as he wrote in a letter to a colleague in

Turin, in which he recommended Abraham for a position as professor in that university. This letter throws light, among other things, on the reason for Abraham's career difficulties in Germany:

> Kneser from Breslau ... told me that Abraham had made himself disliked by all the experimental physicists because he had repeatedly expressed himself in a manner that was rather unflattering to the "big-shots" and that for this reason (in addition to a widespread anti-Semitism, which we can consider quite probable there) it would have been very difficult for him to become a professor in a German university.[1]

In 1909, Abraham won a post as *Professore Straordinario* (full professor without tenure) of "Rational Mechanics" at Milan Polytechnic (at that time still known as "Istituto Tecnico Superiore"). The chairman-secretary of the judging committee and the compiler of the report in which Abraham was unanimously declared winner of the competition was Levi-Civita himself. His assessment sounded exceptionally flattering:

> We find before us a scientist who deservedly enjoys widespread fame ... an exceptional candidate, an outstanding scientist who is already highly and deservedly acclaimed for his important discoveries.... His research on electromagnetic dynamics has provided a solid and systematic theoretical foundation for the brilliant conceptions of modern physics.[2]

In Italy, Abraham found a scientific milieu that was much more congenial to him than that of Germany. All Italian experimental physicists harbored a deep hostility toward special relativity because they did not understand the conceptual intricacies involved and particularly because they were committed to the concept of the ether, considered by them to be the irreplaceable and necessary substratum for the propagation of electromagnetic waves. The Italian mathematical physicists, although more open to the problem of the "principle of relativity" (especially since they were influenced by the undisputed authority of Poincaré) maintained, by an overwhelming majority, an attitude of cautious expectation with regard to special relativity.[3]

Abraham therefore came to Italy preceded by his reputation as a champion of the ether and a tenacious adversary of special relativity, respected by his mathematician colleagues who saw in him a brilliant theoretical physicist able to move about with competence and technical mastery on their own terrain, and highly esteemed by the experimental physicists who considered him an eminent scientist prepared to give theoretical dignity to their position against "metaphysical prejudices" and in favor of the ether. The scientific work produced by Abraham in his Italian period was not to disappoint their expectations; the field of his research was to shift to the problem of a reformulation of a new theory of gravitation and in this area was to give rise to heated arguments with Einstein regarding the very foundations of general relativity.

## 2. The First Controversy between Abraham and Einstein (1912)

In a series of articles appearing in 1912 (Abraham 1912c–1912g),[4] Abraham had formulated his theory of gravitation, which was based on the hypothesis put forward by Einstein in 1911, according to which the speed of light $c$ depends on a scalar gravitational potential $\varphi$ (Einstein 1911). In the meantime, Einstein had further developed his own theory of gravitation, publishing two articles in 1912 (Einstein 1912a, 1912b). In the first of these papers, he criticized Abraham's theory, describing it as "unsustainable, even from a formal mathematical standpoint" (Einstein 1912a, p. 355). There were two objections raised by Einstein and both were well founded: Abraham's theory did not satisfy the principle of equivalence and it was not invariant under infinitesimal Lorentz transformations (Einstein 1912a, p. 368). These criticisms by Einstein gave rise to a controversy between the two scientists, with an exchange of articles appearing in the *Annalen der Physik*, the tone of which became increasingly heated, especially on the part of Abraham.

In June, incensed by the way in which Einstein had dismissed his theory, Abraham sent a note in which he did not limit himself merely to responding to the criticisms raised by Einstein, but counterattacked strongly, addressing criticisms to Einstein's theory of gravitation and, more generally, to the theory of relativity (Abraham 1912h).

Einstein's second objection was due to the fact that in Abraham's theory the differentials of the space-time coordinates had to be exact differentials under infinitesimal Lorentz transformations, which cannot be true if the speed of light is independent of time, but depends on position. To this criticism Abraham responded, for the most part correctly, that his theory did not require the differentials of the coordinates to be exact and that the question raised by Einstein could be reduced to nothing more than the impossibility of integrating the infinitesimal transformations in question; but, according to Abraham, this was correct, since a substantial and irreducible difference exists between a system of reference in which the gravitational field is static and any other system of reference in uniform motion with respect to the former, in which the gravitational field therefore varies with time. This fact, according to Abraham, rendered "unsustainable any relativistic conception of space and time" and, because of the privileged role of the system of reference in which the field of gravity is static, allowed the motion of bodies relative to it to be called "absolute":

> In the restriction to infinitely small regions the claim is already implicit that in finite regions the invariance under Lorentz transformations should not be valid. In fact, if the gravitational field influences the speed of light, then it is clear that there is an important difference between a system of reference $\Sigma(x, y, z, t)$ in which the gravitational field is static and a system of reference in uniform motion $\Sigma'(x', y', z', t')$ in which the gravitational field and the speed of light change over time.... In fact, as

Einstein observes, the differential equations between $dx'$, $dt'$ and $dx$, $dt$, which contain the Lorentz transformations for infinitesimal regions, cannot be integrated. (Abraham 1912h, p. 1057)

In the same note, Abraham stressed the nonrelativistic character of his theory as if this deserved praise:

Surely any relativistic conception of space-time which is expressed by a relation between the space-time parameters of $\Sigma$ and $\Sigma'$ would be unsustainable. A relativistic conception of space-time of this type is in every way distant from my own.... Since, out of all systems of reference, one distinguishes those in which the gravitational field is static or almost static, it is legitimate to call "absolute" the motion referred to one of these systems. (Abraham 1912h, p. 1057)

In this note, Abraham maintained that the other criticism put forward by Einstein, in which he had claimed that Abraham's theory did not satisfy the principle of equivalence, was not even worth answering. In any case, he observed quite rightly that Einstein had not applied this principle thoroughly and with coherence to his own theory and that therefore Einstein's theory was founded upon "shaky ground" (Abraham 1912h, p. 1058). More generally, Abraham maintained that Einstein's whole concept was incoherent: in fact, if the speed of light depended on position, Lorentz invariance could no longer be valid and therefore the theory of relativity (which, according to Abraham's conception, included Lorentz invariance) was unsustainable. Moreover, a relativistic theory of gravitation, on the basis of which systems of reference in uniform motion are equivalent to each other, could not be correct, in that, as Abraham repeatedly affirmed, the system of reference in which the gravitational field is static constitutes a privileged system. In the end, he put his colleagues on their guard against the fascination deriving from relativity, which threatened to "hinder the healthy development of theoretical physics" (Abraham 1912h, p. 1056).

In his reply, Einstein agreed with Abraham on the issue of invariance under infinitesimal Lorentz transformations; however, he insinuated that Abraham had first raised the question and had defended the noninvariance of his theory only in the reply and not in the original formulation of his theory: "I will not judge whether this [i.e., the hypothesis of noninvariance of the theory under infinitesimal Lorentz transformations] was or was not Abraham's original hypothesis" (Einstein 1912c). Moreover, he pointed out another contradiction present in Abraham's theory: In fact, on the basis of this theory two different expressions for the force exerted on a material point at rest in a gravitational field could be obtained that are inconsistent with each other (Abraham 1912c, p. 1064).[5]

Einstein, in any event, devoted much of his reply not so much to an attack on Abraham's theory as to the defense of his own. In particular, he tried to clarify the distinction between the principle of relativity and that of the constancy of the speed of light, stressing that the principle of relativity is not, *by itself*, sufficient to determine the *type* of the transformation laws between different systems of reference in relative motion, and noting that, for example, the Lorentz transformations had been obtained only by using both principles. Thus, it would be possible and consistent to renounce the principle of the constancy of the speed of light in the presence of nonuniform gravitational fields without having to abandon the principle of relativity. Einstein honestly admitted that he still did not know which general laws of transformation ought to be substituted for those of Lorentz in the case that gravitation were included, but he stressed the importance of the principle of equivalence as "an interesting prospect" in order to define these new laws of transformation; he concluded with a request for help from his fellow physicists "to try to resolve this difficult problem":

> The principle of equivalence offers us an interesting prospect according to which the equations of a theory of relativity which includes gravitation should be invariant in an accelerated and rotational transformation [that is, a transformation between systems of reference in linearly accelerated or rotational relative motion]. Certainly the path leading to this goal seems to be very difficult. It may even be seen in the very particular case dealt with up to now, that of the gravitational field of a stationary mass, that the space-time coordinates lose their simple physical meaning and it is still unclear as to what form the general transformations of space-time can have. I appeal to all colleagues to try to resolve this difficult problem! (Einstein 1912c, pp. 1063–1064)

Abraham's reply to this appeal was, to say the least, contemptuous: "Einstein begs credit for the theory of relativity of tomorrow and appeals to his colleagues so that they may guarantee it" (Abraham 1912i, p. 446).

Einstein, for his part, responded with a very brief note of a few lines in which he informed the readers of his decision to end the controversy, not because he agreed with the theses supported by Abraham, but because each had stated his point of view and any further discussion would have been futile (Einstein 1912d).

Even if the tone of Einstein's replies never reached the provocative tone used by Abraham, Einstein was not sparing of violent criticisms of Abraham's theory in private. Thus, in a letter to Paul Ehrenfest in February 1912, Einstein declared: "Abraham's theory is utterly unsustainable."[6] And, in a letter to Arnold Sommerfeld in October 1912, referring to a slightly modified second version of Abraham's theory published in September of that year (Abraham 1912g), he used even stronger terms: "Abraham's new theory is certainly correct from a logical point of view, but as far as I can see, it's only an embarassing monstrosity."[7]

Certainly, Abraham's theory of gravitation was for Einstein "unsustainable," not so much because of supposed inner logical inconsistencies, as because of the fact that such a theory was not based on the principle of equivalence and, especially, on the principle of relativity, which, as has been seen, Abraham rejected out of hand on the basis of physical argument. That is, while Einstein was seeking equations of the gravitational field that are invariant under transformations between systems of reference in rotational or linearly accelerated relative motion, Abraham, for his part, rejected in principle the very possibility of these transformations, since, as has already been explained, he attributed a priviledged or "absolute" role to that particular system of reference in which the gravitational field is static.

The same ideas had already been defended by Abraham in a lecture he gave in Rome in 1912 at the Società Italiana di Fisica. After having presented his theory of gravitation to the Italian physicists assembled there, Abraham concluded with words that undoubtedly earned him the appreciation of the audience:

> The so-called postulate of the constancy of the speed of light put forward by Einstein in his first theory is fading.... On the other hand we see that among the various systems of reference, the one in which the gravitational field is static or varies little from the static one is distinguished, because in it $c$ [i.e., the speed of light] depends only on place but not on time, while for other systems of reference, in translation or rotation with respect to the former, $c$ will moreover depend on time.... We may therefore call the motion relative to a system in which the gravitational field is static "absolute." (Abraham 1912j, p. 219)

In August 1912, Abraham presented a slightly modified version of his original theory of gravitation (the one that Einstein had described as "an embarrassing monstrosity") to the International Congress of Mathematicians held in Cambridge (see Abraham 1912g), and later to Italian scientists in a lecture given on October 19th of the same year at the Congress of the Società Italiana per il Progresso delle Scienze (SIPS). On this occasion, Abraham also reasserted that his "theory of gravitation based on the variable $c$ hypothesis contradicts the second postulate of the theory of relativity" and that "the very likely hypothesis that the attracting mass be proportional to the energy forces us to abandon the Lorentz group even in the infinitesimal" (Abraham 1912k, pp. 480–481); and he concluded with a series of rhetorical questions in harmony with the mood and expectations of the Italian physicists who, as has already been stressed in the introduction, were all hostile to relativistic theories and were devout supporters of the ether:

> Thus, Einstein's Theory of Relativity (1905) is waning. Will a new, more general principle of relativity arise like a phoenix from the ashes? Or will we return to absolute space? And will we bring back the much disdained ether so that it can take on not only the electromagnetic field, but also the gravitational field? (Abraham 1912j, p. 481)

But in Italy the experimental physicists were not the only ones to receive Abraham's theory of gravitation favorably because of his position against relativity and in favor of the ether. Even Levi-Civita who, at the beginning of 1913, was president of the judging committee involved in promoting Abraham from *Professore Straordinario* to *Professore Ordinario* (full professor with tenure), emphasized in the report written by him how "Abraham's ingenious productivity did not cease after his nomination to the position of 'Professore Straordinario,'" and how in his recent research on the theory of gravitation "he clearly commands the Lagrangian concept, and plausible hypotheses, directly implanted in the usual conceptions of classical mechanics, substitute for the most artificial relativistic suggestions."[8]

## 3. The Second Controversy between Abraham and Einstein (1914)

A second controversy between Abraham and Einstein exploded with violence in the journal *Scientia* in 1914. The polemic was sparked by a review article entitled "La Nouvelle Mécanique" (Abraham 1914a), in which Abraham related the history that had brought about the emergence of a new mechanics (as opposed to the "old mechanics" of Galileo and Newton), a partisan history accepted by the majority of his contemporary physicists.

After looking briefly at both the conceptual and methodological differences between Lorentz's theory of the electromagnetic field and Einstein's first theory of 1905, Abraham presented a series of criticisms of special relativity, explaining its transient success as one similar to that of a trend that, in harmony with the spirit of the times, had essentially won over the young:

> The theory of relativity has attracted the attention of many people to the new mechanics. The distortion of the fundamental conceptions of kinematics and dynamics has surprised those who had not followed the historical evolution of the problems; ... the apparent generality of the solution to the problem of space and time accorded with the desire at that time for the unification and synthesis of science. It is for this reason that the theory of relativity has enthused young people devoted to the study of mathematical physics who, influenced at that time by this theory, flocked to the lecture halls of the universities. Contrary to this, most of the physicists of the preceding generation, whose philosophical conceptions had been formed under the influence of Mach and Kirchhoff, were skeptical of the audacious innovators who, basing themselves on a small number of experiments still being discussed among the specialists, dared to upset the proven foundations of all physical measurements. (Abraham 1914a, p. 18)

Abraham then emphasized how "the impossibility of transferring the notion of a rigid body from classical mechanics to the theory of relativity" (1914a, p. 19) represented a serious limitation of the latter and then he played the old tune of the ether again:

> Many supporters of the theory of relativity have deduced from the first postulate of this theory [i.e., from the principle of relativity] that one could do without a medium which fills space, an 'ether.' And in effect, as a result of this postulate, the 'ether' does not appear in uniform rectilinear motion. On the other hand, as P. Ehrenfest observed, the second postulate, regarding the constancy of the speed of light, could not very well be understood without the help of the theory of (electromagnetic) waves [and therefore, not without the ether, a necessary support for the propagation of these waves, according to the dominant electromagnetic paradigm of the time]. The second postulate testifies that the theory of relativity draws on the theory of the electromagnetic field. Radical relativists, being adversaries of the ether, would gladly deny this origin. (Abraham 1914a, p. 20)

Abraham continued in this vein, emphasizing the superiority of the theory of the electromagnetic field over special relativity, both on a logical-conceptual level and on a predictive one:

> The electromagnetic field theory gives us a logical image of the world and embraces an extremely vast set of facts. Conversely, the theory of relativity is an incomplete theory ... the radical relativists therefore are wrong in claiming that the [electromagnetic] field and its support can be done away with. For the moment, it is impossible that the theory of relativity be freed from the [electromagnetic] field theory to which it is linked through the second postulate. (Abraham 1914a, p. 22)

He concluded his criticism of special relativity by emphasizing how the "inner disagreement" between the two postulates underlying the theory bore within it "a seed of disintegration" and how further attempts at generalizing the theory had brought about the rupture of this "unnatural union":

> In the theory of relativity formulated by Einstein in 1905, a compromise between two heterogeneous sets of conceptions can be recognized. Strictly relativistic thought strives to demonstrate that the only essential thing is the relative motion of matter and that, on the contrary, the [electromagnetic] field and its tenets can be done without. The first postulate draws its origins from this way of thinking while the second is borrowed from the theory of the field. And this inner disagreement contains within it a seed of disintegration. This compromise could last only as long as it was limited to considering solely rectilinear and uniform motion, and the field of phenomena to be treated was chosen with circumspection. Any attempt to extend the theory had necessarily to break this unnatural union. (Abraham 1914a, p. 23)

Then Abraham brought his guns to bear on general relativity, stressing how the very "hypothesis of the dependence of the speed of light on the gravitational potential, which Einstein had formulated (1911), ... had axed one of the roots of the theory of special relativity that he himself had established previously" (Abraham 1914a, p. 24). After having reviewed the new theories of gravitation that had been proposed in those years by several scientists (such as Mie and Nordström) and comparing them with his own, Abraham then examined the latest theory of gravitation put forward by Einstein in 1913:

> Einstein has devoted himself to the theory of gravitation. In his study of the subject, he has commenced from premises which partly correspond to the old mechanics and in part to the theory of relativity of 1905, even though he has renounced the second postulate of the latter theory; he nevertheless leads us to hope for a new theory of relativity that will include gravitation; for the moment we only have an outline of this theory. (Abraham 1914a, p. 27)

Abraham alluded to Einstein's famous article, written in cooperation with his friend, the mathematician Marcel Grossmann, "Outline (*Entwurf*) of a Generalized Theory of Relativity and of a Theory of Gravitation" (Einstein and Grossmann 1913). In this work, Einstein and Grossmann had come within one step of finding the right expression for the equations of the gravitational field with the correct properties of general covariance (i.e., covariance under arbitrary coordinate transformations), equations which Einstein was to succeed in obtaining only two and a half years later, in November 1915; however, they had believed it necessary to give up the requirement of general covariance inasmuch as they were convinced that it is not possible to find the correct Newtonian limit starting from generally covariant equations.[9]

In his critique, Abraham showed once again his mastery and competence in identifying the "mathematical bug" by which the theory of Einstein and Grossmann was affected:

> The authors posit an arbitrary transformation of space and time.... But the very gravitational field, which this "Outline" seems to be trying to incorporate within the theory of relativity, does not fit the fundamental frame of this theory; the differential equations of the gravitational field obtained by the authors are not invariant under general transformations of space and time. (Abraham 1914a, p. 27)

But this fact, according to Abraham, should not have caused any surprise, because the "mathematical bug" of the noncovariance of the equations of the field under general coordinate transformations was the evidence of that very same "physical bug" that he believed he had identified in the previous controversy of 1912: among all the systems of coordinates there is a privileged or "absolute" one, i.e., the one in which the gravitational field is static. Thus Abraham went over his argument on this occasion as well, claiming that "a system of mutually attracting masses animated by a nonuniform or rotational motion is not equivalent in general to a system of masses at rest" (Abraham 1914a, p. 27), and he concluded his criticisms of Einstein's theories with a tone that was more appropriate to a funeral oration than to a scientific article:

> To look for a theory of gravitation that corresponds to the general relativistic pattern would thus be taking a wrong path; even if such a theory were admissable from a mathematical point of view, it would not be true from a physical standpoint. In this way both the special theory of relativity of 1905 and the general theory of 1913 run aground on the reef of gravity. The relativistic ideas are evidently not broad enough to serve as a frame for a complete image of the world.
>
> But there remains a place for the theory of relativity in the history of criticism of the concepts of space and time.... This fact guarantees the theory of relativity an honorable burial. (Abraham 1914a, pp. 28–29)

Einstein's reply was not long in coming. A few weeks after Abraham's article, an article by Einstein, entitled "Sur le Problème de la Relativité," appeared in *Scientia* (Einstein 1914a); in it, Einstein defended his theories, both special and general relativity, in a pacifying and in no way polemical tone, setting them on much different levels, however, as regards their respective corroborations and confirmations:

> The first, which we shall call 'theory of relativity in a strict sense,' rests on a noteworthy set of experiments, and today the majority of theoretical physicists consider it to be the simplest theoretical explanation of these experiments. The second, (which we call 'theory of relativity in a broad sense') has not been confirmed until now by physical experiments. The majority of my colleagues have a skeptical or hostile attitude toward this second theory. (Einstein 1914a, p. 139)

Einstein then went on to examine the two principles that underlie special relativity, emphasizing that "the principle of relativity is as old as mechanics and, from the point of view of experience, nobody could ever have doubted its validity" (Einstein 1914a, p. 140), while "the principle of the constancy of the speed of light, resulting from the electrodynamics of Maxwell and Lorentz, becomes compatible with the principle of relativity if two arbitrary hypotheses are renounced," namely, the hypothesis that the distance between two events is independent of the choice of the reference frame and the hypothesis that the simultaneity of two spatially distinct events is independent of the reference frame. (Einstein 1914a, pp. 141–142). Finally, regarding both the logical coherence and the empirical corroboration of special relativity, Einstein brought to light how "the heuristic value of the theory of [special] relativity lay in the fact that it gives a condition that all the systems of equations that express laws of a general nature have to satisfy" (Einstein 1914a, p. 142), namely, the invariance of the form of these equations (covariance) under Lorentz transformations. He concluded by affirming that "in the application of the theory of [special] relativity neither a logical contradiction nor a conflict with the empirical results has up to now appeared" (Einstein 1914a, p. 143).

Einstein next examined the various theories of gravitation proposed in those years, dispensing with the theories of Abraham and Mie in one single dry sentence: "Abraham's theory is in contradiction with the principle of relativity, Mie's contradicts the condition of equality between inertial and gravitational mass" (Einstein 1914a, p. 145).

He then went on to introduce the conceptual pillars on which general relativity was based, first demonstrating the philosophical limits of both classical mechanics and special relativity:

> Classical mechanics, as is also the case with the theory of relativity in the strict sense, ... has a fundamental defect which is impossible to deny if one is open to philosophical arguments. In his thorough research into the foundations of Newtonian mechanics, E. Mach has already expressed most clearly the insufficiency of our physical image of the world. (Einstein 1914a, p. 146)

"The error of which our physics was guilty against the most elementary postulates of philosophy" consisted, according to Einstein, in the fact that in mechanics one speaks of "the acceleration of a body," while "it will not be possible to define any but relative accelerations, accelerations of bodies in relation to other bodies" (Einstein 1914a, pp. 147–148); and thus he continued:

> Given that not only classical mechanics but also the theory of relativity in the strict sense present the fundamental defect that has just been stated, I have taken as my goal generalizing the theory of relativity so that this imperfection may be avoided. (Einstein 1914a, p. 148)

After having stressed how the principle of equivalence between a gravitational field and an accelerated frame of reference was the main pillar on which his new theory was founded, Einstein showed how general relativity did not at all imply an "abandonment" of special relativity, as Abraham claimed, but represented, on the contrary, an enlargement of it and a "necessary development." In fact, because of the incompatibility of the postulate of the constancy of the speed of light with the principle of equivalence, Einstein had been "led into considering the theory of relativity in a strict sense as valid in those domains within which there were no perceptible differences in the gravitational potential" (Einstein 1914a, p. 149) and thus concluded:

> I have had to substitute for the theory of relativity (in the strict sense) a more general theory which includes it as a limiting case.... The theory of relativity in a broad sense does not at all lead to the abandonment of the previous theory of relativity, but represents a development of it, a development which seems to me to be necessary if seen from the philosophical point of view previously expressed. (Einstein 1914a, pp. 149–150)

Although written in a calm style and completely devoid of polemical tone, Einstein's article aroused the anger of Abraham, who immediately sent *Scientia* a furious note of reply (Abraham 1914b). In it, he characterized Einstein's article as "precious for understanding the psychology of relativity" and accused the author of "not having refuted the serious objections raised by him [Abraham] concerning both special relativity of 1905 and the 'broad' theory of 1913" (Abraham 1914b, p. 101). After emphasizing that experimental measurements of the velocity dependence of the electron's inertial mass had not yet reached the sensitivity necessary for an exact determination of this dependence and that therefore the prediction of special relativity regarding this dependence was still "controversial," Abraham complained about the "scandal" over the parallel established by Einstein between the generality of the principle of relativity and that of the principles of thermodynamics. In fact, Abraham declared, "the latter principles [of thermodynamics] have been verified in all the fields of physics and chemistry while serious difficulties obstruct the application of the principle of relativity to nonelectromagnetic forces, as, for example, gravitation"; and concluded peremptorily: "Because of these difficulties, it is extremely doubtful that mechanics will allow itself to be confined in the straitjacket of the Lorentz group" (Abraham 1914b, p. 101). To Einstein's claim that the majority of theoretical physicists had accepted special relativity, Abraham responded in a steely tone:

> I don't care to find out whether the majority of theorists truly attribute so little importance to these difficulties as to give their support to the definition of space and time that can be found in the 1905 relativity theory, but as for myself, I believe that in an election this theory could not achieve a majority unless more votes were accorded those who shout the loudest. In any case, Einstein himself would be obliged to vote against it, since in his 1913 theory of relativity he abandons the postulate of the constancy of the speed of light and, as a result, the ideas on space and time expressed in the theory of 1905. (Abraham 1914b, p. 102)

Abraham proceeded to express, in similar tones, his criticism of general relativity, and concluded:

> In the beginning, Einstein may have allowed himself to be driven by the hope of reaching a general theory of relativity that embraced rotational motion and accelerated motion, by founding it on his 'hypothesis of equivalence.' But today we must realize that he has not reached his goal at all and that the balance of the theory of relativity here presents a deficit. (Abraham 1914b, p. 103)

With this note from Abraham, the controversy in *Scientia* came to an end because Einstein did not even deign to reply.

## 4. Epilogue

Although in his article in *Scientia*, as in the previous controversy of 1912, Einstein never entered into verbal warfare with Abraham and restrained himself from using tones and epithets similar to those used by Abraham towards him, he continued to express himself in harsher terms in his private correspondence. Thus, in a letter to Ludwig Hopf in 1914, he characterized Abraham's second theory of gravitation as "a stately horse which lacks three legs."[10]

Despite these disagreements, Einstein continued to respect Abraham's scientific mastery. He wrote as much in a letter to his friend Michele Besso in 1914:

> Physicists have such a passive attitude toward [my] work on gravitation. Abraham is still the one who shows the most comprehension. It is true that he complains violently in 'Scientia' against anything to do with relativity, but with understanding [mit Verstand].[11]

Einstein's esteem for Abraham emerges also in the fact that he recommended Abraham to succeed to his chair at the University of Zurich after his departure for Berlin.[12] As proof of this esteem, Einstein continued to write to Abraham, sending him reprints of his articles and asking his opinion of their content.

This is clear from a letter that Abraham addressed to Levi-Civita on February 23, 1915, in which he wrote, among other things:

> Yesterday, I found a letter from Einstein in which he pointed out that in his theory there was relativity with respect to any motion of the origin of the coordinates (??) [Abraham's question marks]. He also asks my opinion on his new work. I really don't understand on what hypotheses his new proof is based. Among the possible invariants ... he chooses rather arbitrarily the one conducive to his field equations.... I have written him to send you his work, so the reprint which I lent you would become free.[13]

In this letter Abraham is no doubt alluding to the long work presented by Einstein to the Preussische Akademie der Wissenschaften in Berlin on October 29, 1914 (Einstein 1914b). In this paper, Einstein had used the variational method in order to rederive the same, nongenerally covariant equations of the gravitational field that he had previously obtained in his work with

Grossmann (Einstein and Grossmann 1913) with a procedure that he wrongly deemed "a purely covariant-theoretical approach" (Einstein 1914b, p. 1030).

Once again, Abraham revealed his mastery by pointing out the main weak point of Einstein's new derivation: of all the possible forms of the "Hamiltonian density" that appeared in the integral of action, Einstein had arbitrarily chosen just that expression that led to the obtaining of his old, nongenerally covariant field equations.

Ironically, it was Abraham himself, the most stubborn opponent of general relativity who, in this letter, urged Levi-Civita to write to Einstein.

Between March and May 1915, Einstein and Levi-Civita discussed in a flurry of correspondence the weak points of Einstein's variational approach to general relativity. Also as a result of this correspondence, Einstein managed to eliminate the "mathematical bug" by which his first formulation of general relativity was affected, and in November 1915, he arrived at the definitive formulation of the equations of the gravitational field with the correct properties of general covariance (Cattani and De Maria 1985).

Forced to leave Italy because of the war, Abraham continued to follow the evolution of the controversy between Levi-Civita and Einstein from a distance and studied the first important works by Levi-Civita on general relativity. In a postcard sent to Levi-Civita from Switzerland in August 1917 (which had been stamped with the phrase "passed through censor," since it was then the middle of the first world war), Abraham wrote with his usual caustic tone:

> Dear friend and colleague, I heartily thank you for your latest works sent to me; the remarks on Einstein's new theory are very interesting. It may be necessary to subject it to preventive mathematical censorship. Hilbert's statement "physics is too difficult for physicists" seems to be true.[14]

NOTES

[1] Draft of a letter from Levi-Civita to a mathematician colleague at the University of Turin (name of recipient unknown), 1909 (?), Levi-Civita Archive, Accademia dei Lincei, Rome.

[2] Draft report of the "Commissione Giudicatrice del Concorso per Professore Straordinario di Meccanica Razionale nel R. Istituto Tecnico di Milano," October 1909, Levi-Civita Archive, Office of Prof. Piervittorio Ceccherini, Rome. (We thank Prof. Ceccherini for his kind permission to consult the material on Levi-Civita in his possession.)

[3] For further details on the early positions of the Italian scientists regarding relativity see De Maria 1983; Battimelli and De Maria 1983.

[4] Abraham's theory first appeared in a series of notes published in the *Rendiconti dell' Accademia dei Lincei* and presented, starting from December 1911, by Levi-Civita: Abraham 1911, 1912a, 1912b.

[5] Abraham's theory of 1912, with its limits and inconsistencies, was analyzed in its technical aspects by Earman and Glymour n.d.; see in particular pp. 3–6 and 20.

[6] Einstein to Ehrenfest, February 12, 1912, EA 9-320.

⁷ Einstein to Sommerfeld, October 29, 1912, in Hermann 1968, p. 26; see also Earman and Glymour n.d., p. 22.

⁸ Draft of the "Relazione della Commissione Giudicatrice della Promozione ad Ordinario del Prof. M. Abraham del R. Ist. Tecnico Superiore di Milano," 1913 (?), Levi-Civita Archive, Office of Prof. Piervittorio Ceccherini, Rome.

⁹ For further details on this point see Earman and Glymour 1978; Stachel 1980; Norton 1984.

¹⁰ Cited in Seelig 1956, p. 142.

¹¹ Einstein to Besso, in Speziali 1972, p. 50.

¹² Einstein refers to this in various letters written between 1914 and 1915 to Heinrich Zangger; parts of these letters are reported in Miller 1976.

¹³ Abraham to Levi-Civita, February 23, 1915, Levi-Civita Archive, Accademia dei Lincei, Rome.

¹⁴ Abraham to Levi-Civita, August 1, 1917, Levi-Civita Archive, Accademia dei Lincei, Rome.

REFERENCES

Abraham, Max (1903). "Prinzipien der Dynamik des Elektrons." *Annalen der Physik* 10: 105–179.

——— (1904). "Die Grundhypothesen der Elektronentheorie." *Physikalische Zeitschrift* 5: 576–579.

——— (1911). "Sulla teoria della gravitazione." *Accademia dei Lincei. Rendiconti* 20: 678–682 (presented by T. Levi-Civita).

——— (1912a). "Sulla teoria della Gravitazione." *Accademia dei Lincei. Rendiconti* 21: 27–29 (presented by T. Levi-Civita).

——— (1912b). "Sulla conservazione dell'energia e della materia nel campo gravitazionale." *Accademia dei Lincei. Rendiconti* 21: 432–437 (presented by T. Levi-Civita).

——— (1912c). "Zur Theorie der Gravitation." *Physikalische Zeitschrift* 13: 1–4.

——— (1912d). "Das Elementargesetz der Gravitation." *Physikalische Zeitschrift* 13: 4–5.

——— (1912e). "Der freie Fall." *Physikalische Zeitschrift* 13: 310–311.

——— (1912f). "Die Erhaltung der Energie und der Materie in Schwerkraftfelde." *Physikalische Zeitschrift* 13: 311–314.

——— (1912g). "Das Gravitationsfeld." *Physikalische Zeitschrift* 13: 793–797.

——— (1912h). "Relativität und Gravitation. Erwiderung auf eine Bemerkung des Hrn. A. Einstein." *Annalen der Physik* 38: 1056–1058.

——— (1912i). "Nochmals Relativität und Gravitation. Bemerkungen zu A. Einsteins Erwiderung." *Annalen der Physik* 39: 444–448.

——— (1912j). "Sulle onde luminose e gravitazionali." *Il Nuovo Cimento* 3: 211–219.

——— (1912k). "Una nuova teoria della gravitazione." *Il Nuovo Cimento* 4: 459–481.

——— (1914a). "La Nouvelle Mécanique." *Scientia* 16: 10–29. ["Die neue Mechanik." *Scientia* 15: 8–27.]

——— (1914b). "Sur le Problème de la Relativité." *Scientia* 16: 101–103.

Battimelli, Giovanni, and De Maria, Michelangelo (1983). "Max Abraham in Italia." In *Atti del III Congresso Nazionale di Storia della Fisica* (Palermo, 1983), pp. 186–192.

Cattani, Carlo, and De Maria, Michelangelo (1985). "The 1915 Epistular Controversy between A. Einstein and T. Levi-Civita." Preprint no. 427 (July 25 1985), Department of Physics, University of Rome "La Sapienza". See this volume pp. 175–200.

De Maria, Michelangelo (1983). "L'impatto della relatività in Italia (echi di una polemica in un paese marginale)." In *Atti del III Congresso Nazionale di Storia della Fisica* (Palermo, 1983), pp. 559–568.

Earman, John, and Glymour, Clark (1978). "Lost in the Tensors: Einstein's Struggles with Covariance Principles 1912–1916." *Studies in History and Philosophy of Science* 9: 251–278.

——— (n.d.). "Abraham and Einstein: Two Theories of Gravitation." Typescript.

Einstein, Albert (1911). "Über den Einfluss der Schwerkraft auf die Ausbreitung des Lichtes." *Annalen der Physik* 35: 898–908.

——— (1912a). "Lichtgeschwindigkeit und Statik des Gravitationsfeldes." *Annalen der Physik* 38: 355–369.

——— (1912b). "Zur Theorie des statischen Gravitationsfeldes." *Annalen der Physik* 38: 443–458.

——— (1912c). "Relativität und Gravitation. Erwiderung auf eine Bemerkung von M. Abraham." *Annalen der Physik* 38: 1059–1064.

——— (1912d). "Bemerkung zu Abrahams vorangehender Auseinandersetzung 'Nochmals Relativität und Gravitation.'" *Annalen der Physik* 39: 704.

——— (1914a). "Sur le Problème de la Relativité." *Scientia* 15 (Suppl.): 139–150. ["Zum Relativitätsproblem." *Scientia* 15: 337–348.]

——— (1914b). "Die formale Grundlage der allgemeinen Relativitätstheorie." *Königlich Preussische Akademie der Wissenschaften* (Berlin). *Sitzungsberichte*: 1030–1085.

Einstein, Albert, and Grossmann, Marcel (1913). *Entwurf einer verallgemeinerten Relativitätstheorie und einer Theorie der Gravitation. I. Physikalischer Teil von Albert Einstein. II. Mathematischer Teil von Marcel Grossman.* Leipzig and Berlin: B.G. Teubner. Reprinted with added "Bemerkungen," *Zeitschrift für Mathematik und Physik* 62 (1914): 225–261.

Hermann, Armin, ed. (1968). *Albert Einstein–Arnold Sommerfeld. Briefwechsel.* Basel and Stuttgart: Schwabe.

Miller, Arthur I. (1976). On Einstein, Light Quanta, Radiation and Relativity in 1905." *American Journal of Physics* 44: 912–923.

Norton, John (1984). "How Einstein Found His Field Equàtions: 1912–1915." *Historical Studies in the Physical Sciences* 14: 253–316. See this volume, pp. 101–159.

Seelig, Carl (1956). *Albert Einstein: A Documentary Bibliography.* London: Staples Press.

Speziali, Pierre, ed. (1972). *Albert Einstein–Michele Besso. Correspondance 1903–1955.* Paris: Hermann.

Stachel, John (1980). "Einstein's Search for General Covariance: 1912–1915." Paper delivered to the Ninth International Conference on General Relativity and Gravitation, Jena, 1980. See this volume, pp. 63–100.

# The 1915 Epistolary Controversy between Einstein and Tullio Levi-Civita

CARLO CATTANI and MICHELANGELO DE MARIA

## 1. Introduction

It is well known that the gravitational field equations that Einstein proposed by the middle of 1913 in the *Entwurf* paper written in collaboration with the mathematician Marcel Grossmann (Einstein and Grossmann 1913) are not generally covariant. More then two and a half years elapsed before Einstein eventually succeeded in obtaining, in November 1915, the correct generally covariant field equations.[1] Various historians of science have tried to reconstruct this dramatic period of Einstein's scientific life.[2]

In particular it has been stressed that in their *Entwurf* paper Einstein and Grossmann had already considered almost generally covariant field equations, correctly based on the choice of the Ricci tensor, which they dismissed since these equations failed to yield the correct Newtonian limit. It has been argued that their failure was due to the fact that they were unaware of the necessity of restricting the choice of the coordinate systems by using a set of additional relations, the so-called coordinate conditions, in order to deduce the Newtonian theory of gravitation (whose equations do not hold in all coordinate systems) from a generally covariant theory (whose equations are valid—by definition—in all coordinate systems). But Norton has recently found convincing evidence, through a detailed analysis of Einstein's "Zurich Notebook,"[3] that

> Einstein was fully aware of his freedom to apply coordinate conditions to generally covariant field equations at the time of the *Entwurf* paper ... and even knew of two different coordinate conditions which could be used to reduce the Ricci tensor to a Newtonian form ... [but] Einstein was not prepared to accept either condition because of a number of related misconceptions.

These misconceptions concerned the problem of the reduction of the ten gravitational potentials to a single potential. Because of his earlier work on gravitation, Einstein

> believed that there was such a reduction in the case of static fields. He chose the simplest and most natural weak field equations and again found that they led to a similar reduction in the number of gravitational potentials.... [But] both his as-

sumptions about static fields and about the weak field equations were inconsistent with his final theory. (Norton 1984, p. 255)

These "misconceptions" induced Einstein to give up the requirement of general covariance. At the same time, Einstein had reached the conclusion that, by imposing the law of conservation of energy and momentum, it is possible to derive unequivocally a set of nongenerally covariant field equations, the *Entwurf* equations.

In the remainder of his article, Norton reconstructs Einstein's efforts to elaborate physical arguments against the requirement of general covariance. In particular, Norton focuses his analysis on the various formulations given by Einstein in 1913 and 1914 of the so-called "hole" argument, by which Einstein strove to demonstrate that generally covariant field equations are in contradiction with the causality principle, since they do not allow a unique determination of the gravitational field in certain particular cases of mass distributions.[4]

In a subsequent section of his article, Norton examines how, in 1914, Einstein rederived his *Entwurf* field equations by means of a variational approach, the equations having, in Einstein's opinion, the maximum degree of covariance compatible with his "hole" argument.[5]

Finally Norton analyzes Einstein's period of "growing dissatisfaction" with his *Entwurf* theory (which Norton dates back to the summer and fall of 1915) and his subsequent "return to the search for generally covariant field equations towards the end of 1915," until his final successful derivation of the correct equations in November 1915.[6]

In this paper, we try to reconstruct a very limited episode of this apparently tortuous return of Einstein to generally covariant field equations through a reconstruction of the epistolary controversy that took place between Einstein and the Italian mathematician, Tullio Levi-Civita, from March to early May 1915.[7] We stress, in particular, how and why Levi-Civita's criticisms contributed to stimulating an early growth of Einstein's "dissatisfaction" with his *Entwurf* theory. As we will show in Section 4, Levi-Civita did not question directly the limited covariance properties of Einstein's *Entwurf* equations; instead, he shot his mathematical darts against the proof of a theorem crucial to Einstein's variational derivation of the *Entwurf* equations and, in particular, contested the covariance properties of the so-called gravitation tensor (see equation (12), Section 2). After many fruitless attempts to rebut Levi-Civita's criticisms and to find a more convincing proof of that theorem, Einstein was obliged for the first time to admit that both the proof of that theorem and its consequences were not correct.

## 2. Einstein's 1914 Variational Approach to Nongenerally Covariant Field Equations

In this section we examine the two articles written in 1914 (Einstein and Grossmann 1914, Einstein 1914b) in which Einstein rediscovered his old *Entwurf* equations by means of variational techniques, emphasizing those points that are relevant to a better understanding of his subsequent correspondence with Levi-Civita.

The first paper was written in collaboration with Grossmann and published early in 1914, when Einstein was still in Zurich. In it, the two authors tried to extend the covariance properties of their *Entwurf* equations. They were perfectly aware that their field equations had to be covariant under general coordinate transformations if their theory was to satisfy the generalized principle of relativity. But their explicit aim was to show that the *Entwurf* equations had the maximum covariance compatible with the "hole" argument, i.e., "that degree of general covariance which is conceivable under the condition that the fundamental tensor $g_{\mu\nu}$ should be completely determined by the gravitation equations" (Einstein and Grossmann 1914, p. 216).

After a brief summary of the "hole" argument, Einstein and Grossmann wrote the *Entwurf* field equations in the form

$$\sum_{\alpha\beta\mu} \frac{\partial}{\partial x^\alpha} \left( \sqrt{-g} \gamma_{\alpha\beta} g_{\sigma\mu} \frac{\partial \gamma_{\mu\nu}}{\partial x_\beta} \right) = \kappa (\mathfrak{T}_{\sigma\nu} + t_{\sigma\nu}), \tag{1}$$

where $g_{\mu\nu}$ and $\gamma_{\mu\nu}$ represent, respectively, the covariant and contravariant components of the metric tensor, and $g$ is its determinant; $\kappa$ is a universal constant; $\mathfrak{T}_{\nu\sigma}$ and $t_{\nu\sigma}$ are the stress-energy tensor densities for matter and for the gravitational field, respectively (Einstein and Grossmann 1914, p. 217). Then, by applying the law of conservation of energy-momentum (Einstein and Grossmann 1914, p. 217),

$$\sum_\nu \frac{\partial}{\partial x^\nu} (\mathfrak{T}_{\nu\sigma} + t_{\nu\sigma}) = 0, \tag{2}$$

they obtained (Einstein and Grossmann 1914, p. 218) the four conditions

$$B_\sigma \equiv \sum_{\alpha\beta\mu\nu} \frac{\partial^2}{\partial x_\nu \partial x_\alpha} \left( \sqrt{-g} \gamma_{\alpha\beta} g_{\sigma\mu} \frac{\partial \gamma_{\mu\nu}}{\partial x_\beta} \right) = 0. \tag{3}$$

These relations are to hold in all those coordinate systems, called "adapted" ("angepaßte") systems (Einstein and Grossmann 1914, p. 221), in which the *Entwurf* field equations are valid. Einstein and Grossmann then showed that $B_\sigma$ is not a generally covariant vector, and they therefore reached the conclusion that "the equations $B_\sigma = 0$ really represent a condition on the choice of the coordinate system" (Einstein and Grossmann 1914, p. 219). The two authors gave a proof by contradiction of the nongenerally covariant character of $B_\sigma$. In doing so they were apparently convinced of the physical necessity of restricting the covariance of the field equations:

> If the $B_\sigma$ were covariant, then all the coordinate systems previously defined as "adapted" would be arbitrary coordinate systems. As a consequence of this circumstance, no part of our proof [i.e., the proof previously given in order to obtain the field equations] would lose its demonstrative power. The final result would be the completely generalized covariance of the field equations. (Einstein and Grossman 1914, p. 224)

As a consequence, the gravitation tensor $\mathfrak{E}_{\mu\nu}/\sqrt{-g}$ would maintain its covariant character for arbitrary coordinate transformations, and its trace would be a scalar invariant under general coordinate transformations. "But as follows from the theory of differential invariants, this quantity [i.e., the trace of the gravitation tensor] does not coincide with the only second order differential invariant" (Einstein and Grossmann 1914, pp. 224–225), namely the Riemann curvature scalar.

From these sentences, it clearly appears that Einstein and Grossmann were perfectly aware of the possibility of making use of the Ricci tensor in order to obtain generally covariant field equations, but that they dismissed it mainly on the grounds of physical considerations, as in the case of the "hole" argument.

In the rest of this article, the authors showed that the relations (3) are also sufficient conditions in order that the *Entwurf* field equations be valid in all "adapted" coordinate systems. More precisely, if equations (3) continue to be satisfied after a transformation of coordinates, then the field equations are still valid in the new coordinate system.

In order to demonstrate this fact, Einstein and Grossmann made use for the first time of a variational principle written in the form (Einstein and Grossmann 1914, eq. V, p. 219):

$$\int_\Sigma \left( \delta H - 2\kappa \sum_{\mu\nu} \sqrt{-g} T_{\mu\nu} \delta\gamma_{\mu\nu} \right) d\tau = 0, \tag{4}$$

where $H$ is defined (Einstein and Grossmann 1914, eq. Va, p. 219) by

$$H = \frac{1}{2}\sqrt{-g} \sum_{\alpha\beta\tau\rho} \gamma_{\alpha\beta} \frac{\partial g_{\tau\rho}}{\partial x_\alpha} \frac{\partial \gamma_{\tau\rho}}{\partial x_\beta}. \tag{5}$$

(It should be emphasized that the two authors did not give any explanation of this particular choice of $H$.)

By requiring that the variations $\delta\gamma_{\mu\nu}$ of $\gamma_{\mu\nu}$ be mutually independent inside the four-dimensional domain of integration $\Sigma$ and vanish on the boundary, they easily recovered their old *Entwurf* equations. In the final sections they studied the limited covariance properties of equations (4) under infinitesimal coordinate transformations, in order to define the "justified" transformations, i.e., those transformations between "adapted" coordinate systems where the conditions $B_\sigma = 0$ hold, and to demonstrate the limited covariance properties of the gravitational equations.

In a letter to Michele Besso, Einstein informed his friend of "the news with regard to the theory of gravitation": from the gravitational equations (1) and the principle of the conservation of energy-momentum, he was able to obtain the four equations $B_\sigma = 0$, "which can be considered as the conditions for the particular choice of the frame of reference."[8] From the conclusion of this letter it appears clear that Einstein had become fully convinced of the validity of his theory in spite of the limited covariance properties of his equations:

I succeeded in showing with a *simple* calculation, *that the gravitational equations are valid in any system that satisfies this condition* [i.e., $B_\sigma = 0$]. Hence, it follows that there are transformations to accelerated frames of many kinds which transform the equations into themselves (for example, rotation too), so that the equivalence hypothesis is preserved in its most original form, that is, even in an unsuspectedly very general way.... Now I am completely satisfied and no longer doubt the correctness of the whole system, whether the observation of the solar eclipse succeeds or not. The logic of the matter is too evident.... The general theory of invariance after all, has functioned as nothing other than an obstacle. The direct approach turned out to be the only workable one. The only thing I do not understand is that I had to grope so long before being able to find what was so near.[9]

Einstein's scientific collaboration with Grossmann ended in April 1914 when he moved from Zurich to the Kaiser Wilhelm Institute in Berlin. In a few months he completed a lengthy new article, which he presented to the Preussische Akademie der Wissenschaften of Berlin, on October 29, 1914 (Einstein 1914b). In that paper he refined and generalized the variational approach, previously developed in collaboration with Grossmann, and found what he (wrongly) believed to be a more satisfactory derivation of his old *Entwurf* field equations; in fact, he thought that he had "succeeded, in particular, in obtaining the gravitational field equations in a purely covariant theoretical approach" (Einstein 1914b, p. 1030).

In this new variational approach, Einstein first discussed the transformation properties of the integral:

$$J = \int_\Sigma H \sqrt{-g}\, d\tau \tag{6}$$

under an infinitesimal transformation $\Delta$ of coordinates $x_\mu \to x'_\mu = x_\mu + \Delta x_\mu$ (where $H$ is a still unspecified "Hamiltonian density," supposed to be a function of the metric tensor $g^{\mu\nu}$ and of its first derivatives $g^{\mu\nu}_\sigma$); the integral is extended to a finite volume $\Sigma$ of the four-dimensional continuum with the boundary conditions $\Delta x_\mu = 0$, $\dfrac{\partial}{\partial x_\sigma} \Delta x_\mu = 0$.

By assuming the invariance of $H$ under linear transformations Einstein was able to specify the forms of $\Delta H$ and $\Delta J$ (Einstein 1914b, pp. 1069–1070):

$$\frac{1}{2}\Delta H = \sum_{\mu\nu\sigma\alpha} g^{\nu\alpha} \frac{\partial H}{\partial g^{\mu\nu}_\sigma} \frac{\partial^2 \Delta x_\mu}{\partial x_\sigma \partial x_\alpha} \tag{7}$$

and

$$\frac{1}{2}\Delta J = \int_\Sigma \sum_\mu (B_\mu \Delta x_\mu)\, d\tau + F, \tag{8}$$

where

$$B_\mu = \sum_{\alpha\sigma\nu} \frac{\partial^2}{\partial x_\sigma \partial x_\alpha} \left( g^{\nu\alpha} \frac{\partial H \sqrt{-g}}{\partial g^{\mu\nu}_\sigma} \right) \tag{9}$$

and $F$ is a surface integral that vanishes when the boundary conditions

$\Delta x_\mu = 0, \frac{\partial}{\partial x_\sigma} \Delta x_\mu = 0$ hold. Then Einstein defined a set of coordinate systems $K, K', K'', \ldots$ infinitesimally near to each other, so that in the passage from one system to the next, the usual boundary conditions $\Delta x_\mu = 0, \frac{\partial}{\partial x_\sigma} \Delta x_\mu = 0$ are valid. This also implies that, for all those systems with coinciding boundary conditions, the relation $F = 0$ holds, so that equation (8) can be written as

$$\frac{1}{2}\Delta J = \int_\Sigma \sum_\mu (B_\mu \Delta x_\mu) \, d\tau. \tag{10}$$

The coordinate systems "adapted" to a given gravitational field are thus defined by Einstein as those systems of the series $K, K', K'', \ldots$ for which $J$ is an extremal, i.e., $\Delta J = 0$.

From the arbitrariness of the variations $\Delta x_\mu$, Einstein easily rederived the relations (3), which he showed to be both necessary and sufficient conditions for the definition of the coordinate systems "adapted" to the gravitational field (that is, those systems that guarantee the maximum covariance properties of the field equations allowed by the "hole" argument).[10]

Next, Einstein considered arbitrary infinitesimal variations of the metric tensor $g^{\mu\nu} \to g^{\mu\nu} + \delta g^{\mu\nu}$ such that $\delta g^{\mu\nu}$ and the derivatives $\delta g_\sigma^{\mu\nu} = \frac{\partial}{\partial x_\sigma} \delta g^{\mu\nu}$ vanish on the boundary of the space-time region of integration $\Sigma$, and he calculated the corresponding increment $\delta J$ of the integral of action $J$

$$\delta J = \int_\Sigma \mathfrak{E}_{\mu\nu} \delta g^{\mu\nu} \, d\tau, \tag{11}$$

where

$$\mathfrak{E}_{\mu\nu} = \frac{\partial H \sqrt{-g}}{\partial g^{\mu\nu}} - \sum_\sigma \frac{\partial}{\partial x_\sigma} \left( \frac{\partial H \sqrt{-g}}{\partial g_\sigma^{\mu\nu}} \right) \tag{12}$$

is the gravitation tensor. His aim was to show the invariance of $\delta J$ under "justified" transformations (i.e., transformations between "adapted" coordinate systems), that is (Einstein 1914b, p. 1073):

$$\Delta \delta J = 0. \tag{13}$$

The demonstration of this theorem is, in Einstein's words, "of fundamental importance for the whole theory" (Einstein 1914b, p. 1071).

In fact, from the invariance of $\delta J$ under "justified" transformations, it follows, according to Einstein, that the expression

$$\frac{1}{\sqrt{-g}} \mathfrak{E}_{\mu\nu} \delta g^{\mu\nu} \tag{14}$$

is also an invariant, since the infinitesimal element of volume $\sqrt{-g} \, d\tau$ is a scalar. Thus, from the tensorial character of $\delta g^{\mu\nu}$, Einstein concluded that $\mathfrak{E}_{\mu\nu}/\sqrt{-g}$ is also a covariant tensor "limited to systems of 'adapted' coordinates and to their substitutions" (Einstein 1914b, p. 1074). In other words, the importance of theorem (13) lies in the demonstration of the tensorial nature of $\mathfrak{E}_{\mu\nu}/\sqrt{-g}$, a crucial result which allowed Einstein to connect the gravitation tensor $\mathfrak{E}_{\mu\nu}$ to the stress-energy tensor of matter (and other physical phenomena) $\mathfrak{T}_{\mu\nu}$ in the familiar relation

$$\mathfrak{E}_{\mu\nu} = \kappa \mathfrak{T}_{\mu\nu} \tag{15}$$

where $\kappa$ is a universal constant (Einstein 1914b, p. 1074).

We want to report in detail Einstein's proof of theorem (13), since, as we shall see in Section 4, most of Levi-Civita's criticisms concerned the validity of that theorem and, consequently, its conclusions regarding the tensorial character of $\mathfrak{E}_{\mu\nu}/\sqrt{-g}$ as well.

Einstein's starting point was the decomposition of the arbitrary variations $\delta g^{\mu\nu}$ into two parts

$$\delta g^{\mu\nu} = \delta_1 g^{\mu\nu} + \delta_2 g^{\mu\nu}. \tag{16}$$

The $\delta_1 g^{\mu\nu}$ represent infinitesimal variations of the metric tensor in the *same system* $K_1$ of "adapted" coordinates. Since $K_1$ is "adapted" to both the original gravitational field $g^{\mu\nu}$ and to the new varied field $g^{\mu\nu} + \delta_1 g^{\mu\nu}$, Einstein drew the conclusion that, in addition to the usual conditions $B_\sigma = 0$, the equations

$$\delta_1 B_\sigma = 0 \tag{17}$$

must also hold. Thus, according to Einstein, the ten $\delta_1 g^{\mu\nu}$ *are not independent*, since they are connected by the four differential equations (17).

The $\delta_2 g^{\mu\nu}$ represent the variation of the *same gravitational field* arising from a change of the coordinate system. As long as, in Einstein's words, the $\delta_2 g^{\mu\nu}$ "are determined by four independent functions (the variations of coordinates)" (Einstein 1914b, p. 1072), the sum of the two variations $\delta_1 g^{\mu\nu} + \delta_2 g^{\mu\nu}$ is determined by $(10 - 4) + 4 = 10$ independent functions, which, according to Einstein, are equivalent to the ten arbitrary variations $\delta g^{\mu\nu}$. As a consequence, Einstein split the proof of equation (13) into two parts:

$$\Delta \delta_1 J = 0 \tag{18}$$

and

$$\Delta \delta_2 J = 0. \tag{19}$$

The proof of equation (18) is straightforward: by applying the variation $\delta_1$ to equation (8), Einstein directly obtained the expression

$$\frac{1}{2}\Delta(\delta_1 J) = \int_\Sigma \sum_\mu (\delta_1 B_\mu \Delta x_\mu) d\tau + \delta_1 F. \tag{20}$$

From the conditions $\delta_1 g^{\mu\nu} = 0$, $\delta_1 g_\sigma^{\mu\nu} = 0$, which hold on the boundary of $\Sigma$, the surface integral $\delta_1 F$ vanishes. Then, from equation (17), it follows

immediately that
$$\Delta\delta_1 J = 0.$$

Einstein's proof of equation (19) is rather obscure and sloppy. Here we present our tentative reconstruction of it.

Einstein first considered the variation $\delta_2 J$, corresponding to an infinitesimal transformation of coordinates $x_\mu \to x_\mu + \delta x_\mu$ that reduces to the identity transformation on the boundary of $\Sigma$. In this case, as a consequence of the definition of $\delta_2 g^{\mu\nu}$, the *same unvaried field* will be expressed as $g^{\mu\nu}$ and $g^{\mu\nu} + \delta_2 g^{\mu\nu}$, respectively, in the "old" and in the "new" infinitesimally varied coordinate system. Since the coordinates of the "old" system are "adapted" to the unvaried field, it follows from the definition of "adapted" systems (where $B_\sigma = 0$), and from the fact that an infinitesimal transformation of coordinates is considered (this is a consequence of equation (10)), that

$$\delta_2(J) = \Delta J = 0, \tag{21}$$

where $\delta_2(J)$ means: $\delta_2(J) = J(g^{\mu\nu} + \delta_2 g^{\mu\nu}) - J(g^{\mu\nu})$; $J(g^{\mu\nu})$ and $J(g^{\mu\nu} + \delta_2 g^{\mu\nu})$ represent the action integrals relative to the *same field*, calculated in the old "adapted" system and in the new infinitesimally varied system, respectively. Then Einstein considered that particular transformation of the field $g^{*\mu\nu} = g^{\mu\nu} + \delta g^{\mu\nu}$, in the same "adapted" coordinate system $K_1$, obtained by choosing those variations $\delta g^{\mu\nu}$ of the field $g^{\mu\nu}$ in $K_1$ that coincide with the $\delta_2 g^{\mu\nu}$ previously obtained as a result of an infinitesimal coordinate transformation; that is to say, the variation of the field in $K_1$ is, in Einstein's words, a "$\delta_2$-variation" (Einstein 1914b, p. 1072). The varied field in $K_1$ will be

$$g^{*\mu\nu} = g^{\mu\nu} + \delta_2 g^{\mu\nu},$$

and from (21) it follows that

$$\delta_2(J_1) = 0, \tag{22}$$

(where $\delta_2(J_1) = J_1(g^{*\mu\nu}) - J_1(g^{\mu\nu}) = J_1(g^{\mu\nu} + \delta_2 g^{\mu\nu}) - J_1(g^{\mu\nu})$; and the two integrals are both calculated in $K_1$ for the unvaried and varied fields, respectively).

Einstein's next step was to show that this variation of the field, considered in another "adapted" coordinate system $K_2$, is still a "$\delta_2$-variation." If this is the case, it follows that in $K_2$ a relation analogous to (22) holds:

$$\delta_2(J_2) = 0. \tag{23}$$

By subtracting (22) from (23), one obtains the desired result (Einstein 1914b, pp. 1072–1073):[11]

$$\delta_2(J_2 - J_1) = \delta_2(\Delta J) = \delta_2 J_2 - \delta_2 J_1 = \Delta(\delta_2 J) = 0.$$

The validity of theorem (13) implies, according to Einstein, the tensorial character of $\mathfrak{E}_{\mu\nu}/\sqrt{-g}$ under "justified" transformations and consequently the "limited" covariance of the field equations $\mathfrak{E}_{\mu\nu} = \kappa \mathfrak{T}_{\mu\nu}$ (see equations (15)).

In the remainder of his article, Einstein tackled the problem of deriving the explicit form of $H$, in order to obtain the gravitational field equations.

As we have already stressed, in his two previous articles written in collaboration with Grossmann (Einstein and Grossmann 1913, 1914), Einstein had chosen the expression for $H$ in the form of equation (5), without giving any explanation of it. But in the paper now under discussion, Einstein believed that he had found a justification of his old choice (5). By substituting the gravitation tensor $\mathfrak{E}_{\mu\nu}$ into the law of conservation of energy-momentum for matter, he was able to derive certain constraints on the Hamiltonian density $H$, which he writes in the form:

$$S_\sigma^\nu = \sum_{\mu\tau} \left( g^{\nu\tau} \frac{\partial H \sqrt{-g}}{\partial g^{\sigma\tau}} + g_\mu^{\nu\tau} \frac{\partial H \sqrt{-g}}{\partial g_\mu^{\sigma\tau}} + \frac{1}{2}\delta_\sigma^\nu H \sqrt{-g} - \frac{1}{2}g_\sigma^{\mu\tau} \frac{\partial H \sqrt{-g}}{\partial g_\nu^{\mu\tau}} \right)$$
$$= 0. \tag{24}$$

Then Einstein made the assumption that $H$ should be a homogeneous second-order function of the derivatives $g_\sigma^{\mu\nu} = \partial g^{\mu\nu}/\partial x_\sigma$. From this hypothesis it follows that $H$ must either coincide with one of the five expressions:

$$\sum g_{\mu\nu} \frac{\partial g^{\mu\nu}}{\partial x_\sigma} \frac{\partial g^{\sigma\tau}}{\partial x_\tau}; \quad \sum g^{\sigma\sigma'} g_{\mu\nu} \frac{\partial g^{\mu\nu}}{\partial x_\sigma} \frac{\partial g^{\mu'\nu'}}{\partial x_{\sigma'}} g_{\mu'\nu'}; \quad \sum g_{\sigma\sigma'} \frac{\partial g^{\sigma\mu}}{\partial x_\mu} \frac{\partial g^{\sigma'\nu}}{\partial x_\nu};$$
$$\sum g_{\mu\mu'} g_{\nu\nu'} g^{\sigma\sigma'} \frac{\partial g^{\mu\nu}}{\partial x_\sigma} \frac{\partial g^{\mu'\nu'}}{\partial x_{\sigma'}}; \quad \sum g_{\alpha\beta} \frac{\partial g^{\alpha\sigma}}{\partial x_\tau} \frac{\partial g^{\beta\tau}}{\partial x_\sigma}, \tag{25}$$

or be a linear combination of them. But according to Einstein, from equation (24) it follows that only the fourth expression of the five expressions (25) can be chosen for $H$. Since this expression is equivalent to his previous choice of $H$ (equation (5)), he could immediately derive his *Entwurf* equations.

Norton stresses that conditions (24) "in no way required that $H$ take on its *Entwurf* form. In effect, all these conditions required was that $H$ be a scalar under linear transformations." Thus Norton concludes that this arbitrary choice of $H$ "remains an outstanding puzzle in the history of Einstein's theory" (Norton 1984, pp. 287–298).[12]

As we shall see in the next section, Max Abraham was compelled to write to Levi-Civita about Einstein's variational derivation of the *Entwurf* equations, since he could not understand this "arbitrary" choice of the form of $H$. This letter of Abraham's was the catalyst initiating the correspondence between Levi-Civita and Einstein.

## 3. The Role of Max Abraham

Max Abraham, a tenacious opponent of Einstein's theories (both special and general relativity), played an important role in directing the attention of Levi-Civita to Einstein's 1914 variational formulation of the *Entwurf* theory.

Abraham's first acquaintance with the Italian mathematician dated back to 1908 when, following Levi-Civita's invitation, Abraham attended the 4th International Congress of Mathematics in Rome. In his letter of invitation, Levi-Civita had expressed a sincere appreciation of Abraham's scientific abilities and high esteem for his "beautiful works on electron theory."[13]

His appreciation and esteem were reiterated in October 1909, when Levi-Civita acted as secretary of the committee that assigned to Abraham the chair of Meccanica Razionale ("Rational Mechanics") at the Politecnico in Milan. In his final report, Levi-Civita described Abraham as "a scientist who deservedly enjoys a wide reputation," whose research on electromagnetic dynamics "provided solid and systematic theoretical foundations of the brilliant concepts of modern physics."[14]

Expressions of the same kind can be found three years later in the report, also written by Levi-Civita, by which the title of Professore Ordinario (full professor) was bestowed on Abraham, by that time well settled in the Italian scientific community. Levi-Civita remarked that "Abraham's genial production did not come to a halt" after his call to Milan and described at length his new theory of gravitation, in which "plausible hypotheses, directly connected to the usual concepts of classical mechanics, replace the more artificial relativistic suggestions."[15]

It is not relevant to discuss in this context the technical details of Abraham's theory, though it is worth recalling the vigourous polemics between Abraham and Einstein in 1912–1914 over their respective gravitation theories. (It is also worth recalling that Abraham was certainly not the kind of person who liked to avoid a heated discussion; it is known that because of his sharp tongue he alienated too many members of his own native academic circles and never found a position in a German university.)[16] The polemics between these two scientists began with a critical exchange in the *Annalen der Physik* regarding the technical limits and deficiencies of their respective theories. Their debate, resumed in 1914 in the Italian journal *Scientia*,[17] soon developed into a clash between the two extremely different personalities, involving both their different conceptions of the physical world and their opposing visions of the construction of a physical theory, with the controversy becoming ever more heated, especially on Abraham's side.

It must be stressed, however, that these rather violent exchanges did not deter Einstein from maintaining his high opinion of Abraham's scientific abilities. Writing to his friend Besso in 1914, Einstein remarked: "Abraham is still the one who shows the most comprehension. It is true that he complains violently in *Scientia* against anything to do with relativity, but with understanding [*mit Verstand*]."[18] Even after these polemics, Einstein kept writing to Abraham, sending Abraham reprints of his papers and asking Abraham's opinion of their content.

On February 23, 1915, Abraham wrote to his friend Levi-Civita:

> I found yesterday a letter from Einstein in which he stressed that his theory of relativity holds with respect to any motion of the origin of the coordinate system (??) [Abraham's underlining and interrogation marks]. He also asks my opinion on his

recent work.... I wrote him to send you his paper. Then you would no longer need the reprint that I lent you.[19]

In Abraham's opinion, both the hypotheses on which Einstein's new variational derivation of the *Entwurf* field equations was based and the reasons for his choice of the Hamiltonian density $H$ were obscure and arbitrary:

> Really I did not understand on which hypotheses his [Einstein's] new demonstration is based. Among all the possible invariants that could be used to construct the function $H$ he chooses very arbitrarily the one that yields his field equations.[20]

## 4. The Controversy between Einstein and Levi-Civita

Abraham's letter certainly attracted the interest of Levi-Civita to Einstein's article. Within a few days, Levi-Civita sent his criticisms from Padua directly to Einstein in Berlin. An intense correspondence started between the two scientists, lasting for two months until early May 1915, a few days before the Italian intervention in World War I. Unfortunately, all of Levi-Civita's letters but one were lost, but Einstein's letters to Levi-Civita are available.[21]

In the remainder of this section we shall try to reconstruct the main objections of Levi-Civita to Einstein's variational approach, in order to stress how and why, in our opinion, this controversy between the two scientists contributed to stimulating Einstein's early dissatisfaction with his *Entwurf* theory and to bringing him back again onto the right path of the requirement of general covariance for his gravitational field equations.

We want to stress here that Levi-Civita never attacked or criticized the "ugly dark spot"[22] common to both the 1913 *Entwurf* paper (Einstein and Grossmann 1913) and the two 1914 variational papers (Einstein and Grossmann 1914; Einstein 1914b), i.e., the limited covariance properties of Einstein's gravitational field equations. Levi-Civita focused his criticisms mainly on the tensorial character of $\mathfrak{E}_{\mu\nu}/\sqrt{-g}$ (see equation (12)), finding fault with Einstein's proof of it. But in doing so, as we shall see later, he brought Einstein to admit the mathematical unsoundness of his variational approach in the framework of a limited-covariance theory.

From Einstein's first answer to Levi-Civita (March 5, 1915) it is easy to reconstruct the essential point in Levi-Civita's letter and to see that his early objections involved the validity of the theorem $\Delta\delta J = 0$ (see equation (13)). In particular, he asserted that the conclusion that Einstein had drawn from equation (11) (i.e., the invariant character of the expression $\delta J = \int_\Sigma \Sigma_{\mu\nu} (\delta g^{\mu\nu} \mathfrak{E}_{\mu\nu}) d\tau$) was wrong. Levi-Civita denied the tensorial character of the variations $\delta g^{\mu\nu}$ and therefore (since $\sqrt{-g}\, d\tau$ is a scalar) the tensorial character of $\mathfrak{E}_{\mu\nu}/\sqrt{-g}$. Moreover, Levi-Civita showed that for a particular choice of the Hamiltonian density, $H = g_{11}$, the expression $\mathfrak{E}_{\mu\nu}/\sqrt{-g}$ is not a tensor.

Einstein was at the same time flattered by Levi-Civita's interest in his work and worried by his objections, as it appears from the beginning of his first letter:

Highly Esteemed Colleague!

It is a great joy for me that you have been so deeply involved in my work. You can imagine how seldom it happens that somebody gets engaged [in my work] in depth, with an autonomous and critical attitude.... But when I saw that you attacked the most important demonstration of my theory [i.e., the theorem $\Delta\delta J = 0$], which I obtained with streams of sweat, I became not a little alarmed, particularly because I know that you have a much better mastery of these mathematical matters than I do. But on thinking it over, I still believe that I am able to justify my demonstration.[23]

Einstein easily dismissed Levi-Civita's last objection by stressing that Levi-Civita's choice of $H = g_{11}$ was incompatible with his basic assumption that $H$ had to be invariant under linear transformations: "Since $g_{11}$ is not invariant with regard to linear transformations, your counterexample does not represent a refutation of my thesis."[24] In regard to the first objection, Einstein's defense appears to be much weaker. His initial claim was that: "In variational calculus people always work with the same methods that I have used."[25] In his answer, Einstein asserted that the following approximation holds:

$$\int_\Sigma \delta g^{\mu\nu} d\tau = \overline{\delta g^{\mu\nu}} \tau$$

where $\tau = \int_\Sigma d\tau$ is the "volume" of the space-time region $\Sigma$, and $\overline{\delta g^{\mu\nu}}$ represent the average values of $\delta g^{\mu\nu}$ when $\Sigma$ is reduced to an "infinitesimal region" coinciding in the limit with a point of the space-time. Thus the integral $\int_\Sigma d\tau \Sigma_{\mu\nu}(\delta g^{\mu\nu} \mathfrak{E}_{\mu\nu})$ can be written as $\tau \Sigma \overline{\delta g^{\mu\nu}} \mathfrak{E}_{\mu\nu}$, since according to Einstein "in the integration the $\mathfrak{E}_{\mu\nu}$ can be considered as constants."[26] Due to the "smallness" of the region $\Sigma$ the transformation properties of $\overline{\delta g^{\mu\nu}}$ coincide with those of $\delta g^{\mu\nu}$ in a given point of the four-dimensional domain, and therefore, with the local tensorial properties of the field $g^{\mu\nu}$. This implies, according to Einstein, the tensorial character of the components of $\mathfrak{E}_{\mu\nu}/\sqrt{-g}$. The weakness of Einstein's defense of his theorem lies, in our opinion, not so much in the "limiting" approximation of $\delta g^{\mu\nu}$ by their average values $\overline{\delta g^{\mu\nu}}$, as in his assumption of the *independence* of the variations $\delta g^{\mu\nu}$.

In reply to Einstein's March 5 letter, Levi-Civita attacked, with a specific example, Einstein's new proof of the tensorial character of $\mathfrak{E}_{\mu\nu}/\sqrt{-g}$ (based as we have seen on the "limiting" replacement of $\delta g^{\mu\nu}$ by their average values $\overline{\delta g^{\mu\nu}}$). His "attack culminates with the statement that when the region $\Sigma$, where the $\delta g^{\mu\nu} \neq 0$, gets smaller and smaller, the average values $\overline{\delta g^{\mu\nu}}$ do not tend, in general, to any definite limit."[27] Einstein did not directly answer this new specific objection of Levi-Civita, considering it "irrelevant,"[28] but gave a new and more accurate proof of the tensorial character of $\mathfrak{E}_{\mu\nu}/\sqrt{-g}$ as a consequence of the invariance of $\delta J$. In his proof Einstein proceeded in the following way:

Letting

$$g^{\rho'\sigma'} = \Sigma \frac{\partial x'_\rho}{\partial x_\mu} \frac{\partial x'_\sigma}{\partial x_\nu} g^{\mu\nu} \qquad (2)$$

then

$$\delta g^{\rho'\sigma'} = \Sigma \frac{\partial x'_\rho}{\partial x_\mu} \frac{\partial x'_\sigma}{\partial x_\nu} \delta g^{\mu\nu}. \tag{2a}$$

I multiply (2a) by $\sqrt{-g}\, d\tau' = \sqrt{-g}\, d\tau$ and integrate it over $\Sigma$.... Since the region of integration $\Sigma$ is to be infinitesimal I can replace the factors $\partial x'_\rho/\partial x_\mu$ and $\partial x'_\sigma/\partial x_\nu$ with those (constant) values assumed by these factors in any point of the region of integration.[29]

In this way it is possible to extract the coefficients of the transformation matrix from the integral $\int \sqrt{-g}\, d\tau \,\Sigma\, (\partial x'_\rho/\partial x_\mu)(\partial x'_\sigma/\partial x_\nu)\delta g^{\mu\nu}$, so it follows that:

$$A^{\rho'\sigma'} = \Sigma \frac{\partial x'_\rho}{\partial x_\mu} \frac{\partial x'_\sigma}{\partial x_\nu} A^{\mu\nu}, \tag{3a}$$

where

$$A^{\mu\nu} = \int_\Sigma \sqrt{-g}\, \delta g^{\mu\nu}\, d\tau. \tag{4}$$

At this point, Einstein drew the conclusion "that $A^{\mu\nu}$ is a contravariant tensor and more precisely a tensor whose components can be independently chosen."[30] Since, by hypothesis, $\delta J = \Sigma(\mathfrak{E}_{\mu\nu}/\sqrt{-g})A^{\mu\nu}$ is invariant (if one neglects infinitesimal terms), Einstein reaches the conclusion that from the contravariant tensorial character of $A^{\mu\nu}$, and because "its components can be arbitrarily chosen, independently of one another, then $\mathfrak{E}_{\mu\nu}/\sqrt{-g}$ is a covariant tensor"[31]

In the meantime, before having received this reply of March 17 from Einstein, with the new demonstration of the tensorial character of $\mathfrak{E}_{\mu\nu}/\sqrt{-g}$, Levi-Civita had worked out another "counter-demonstration against the tensorial character of $\mathfrak{E}_{\mu\nu}/\sqrt{-g}$ based on the case, among all possible 'adapted' coordinate systems, of those systems for which the $g^{\mu\nu}$ are constant."[32]

It is easily shown that the choice $g^{\mu\nu}$ = constant implies the vanishing of $H$ (see equation (5), section 2) and consequently the vanishing of $\mathfrak{E}_{\mu\nu}/\sqrt{-g}$ (see equation (12)). Therefore, if $\mathfrak{E}_{\mu\nu}$ were a tensor, it would follow that if $\mathfrak{E}_{\mu\nu} = 0$ holds in a particular coordinate system, then this result should be valid in all ("adapted") coordinate systems; but this is not the case, as Levi-Civita wrote to Einstein: "This tensor [$\mathfrak{E}_{\mu\nu}$] on the contrary does not vanish identically for *all* the "adapted" coordinate systems; this can be seen, above all, in the Newtonian case."[33] In fact, it can easily be shown that in the Newtonian classical limit, $\mathfrak{E}_{\mu\nu}$ reduces to the Laplacian of $\varphi$, which by Poisson's equation $\Delta\varphi = 4\pi\rho$, does not vanish in all those regions where masses are present. Einstein criticized Levi-Civita's example in the following words:

> I do not believe that this is correct [i.e., Levi-Civita's choice of $g_{\mu\nu}$ = constant]. In fact, one has to consider that in general it is impossible to transform, by means of a change of coordinates, an arbitrarily given field $g_{\mu\nu}$ into a field where the $g_{\mu\nu}$ are constant. This will always be impossible, for instance, for those regions of the Newtonian field that include masses; and even, in general, for massless regions.[34]

Here Einstein hints clearly at the obvious fact that, because of the presence of matter, the space-time is curved and cannot be globally reduced to a flat (Minkowskian) one. After rebutting Levi-Civita's example, Einstein stressed in conclusion his skepticism regarding the possibility of giving a general demonstration or counter-demonstration of the tensorial character of $\mathfrak{E}_{\mu\nu}/\sqrt{-g}$:

> I believe that the demonstration or the counter-demonstration of the thesis cited in your letter are as difficult as the demonstration or the counter-demonstration of the *general* thesis regarding the tensorial character of $\mathfrak{E}_{\mu\nu}/\sqrt{-g}$.[35]

In a later letter, written on March 23, Levi-Civita reconsidered "the special case where the $g_{\mu\nu}$ are constant with an adequate choice of the coordinates," maintaining that "in this case the $\mathfrak{E}_{\mu\nu}$ do not vanish if one refers to another adapted system where the $g_{\mu\nu}$ are *not* constant any longer."[36]

In his reply Einstein denied the validity of Levi-Civita's new argument: "You did not give, however, a proof of your statement and I do not consider it exact as long as you do not give an example or a general demonstration."[37]

In the same letter it appears that Levi-Civita had also raised some objections to the new proof given by Einstein in his previous letter of March 17 regarding the tensorial character of $\mathfrak{E}_{\mu\nu}/\sqrt{-g}$. It seems that Levi-Civita, while accepting both Einstein's result $\Sigma_{\mu\nu}(1/\sqrt{-g})\mathfrak{E}_{\mu\nu}A^{\mu\nu} + \varepsilon =$ invariant (where $\varepsilon$ is an infinitesimal term of higher order) and the proof of the tensorial character of $A^{\mu\nu}$, did question Einstein's choice of $A^{\mu\nu} = \int_{\Sigma} \delta g^{\mu\nu} \sqrt{-g}\, d\tau$ as an infinitesimal tensor. In other words, Levi-Civita did not accept Einstein's use of an *infinitesimal* tensor like $A^{\mu\nu}$ as a "tester" of the tensorial character of a *finite* tensor (in this case $\mathfrak{E}_{\mu\nu}/\sqrt{-g}$) by means of the tensorality criterion. Einstein answered this objection of Levi-Civita's by stating peremptorily:

> Since I have shown that the $A^{\mu\nu}$ quantities transform in a contravariant manner, the fact that the $A^{\mu\nu}$ are infinitesimal is in my opinion totally irrelevant for [the validity of] my proof. This proof, on the contrary, is to be completed by introducing a *finite* tensor $A^{\mu\nu}$, by the construction of a limit [durch Limesbildung]."[38]

In the second part of this letter Einstein repeated his old proof of the tensorial character of $\mathfrak{E}_{\mu\nu}/\sqrt{-g}$, which in his opinion makes "completely superfluous the introduction of any limit whatsoever for $A^{\mu\nu}$," since his deduction of the tensorial properties of $\mathfrak{E}_{\mu\nu}/\sqrt{-g}$ "does not depend on this [limit]," and he concluded that "any consideration where the $\delta g^{\mu\nu}$ are submitted to more conditions than those strictly necessary according to the nature of the problem, is to be rejected as an unnecessary complication."[39]

In a letter of March 28, Levi-Civita reexamined the problem of the existence of "adapted [angepaßte] coordinate systems, in which the tensor $\mathfrak{E}_{\mu\nu}$ does not vanish, while it vanishes when the $g_{\mu\nu}$ are constant."[40]

Levi-Civita agreed with Einstein's criticism (raised in his letter of March 23) that "an arbitrary gravitational field cannot be obtained by means of a coordinate transformation starting from a Euclidean $ds^2$ ($g_{\mu\nu} =$ constant)."[41]

However, Levi-Civita counter-attacked "by showing with a concrete example that by means of some justified transformation [i.e., a transformation between adapted coordinate systems] starting from a Euclidean $ds^2$ one finds some nonvanishing $\mathfrak{E}_{\mu\nu}$ (in contrast with the requirement of the covariance [of $\mathfrak{E}_{\mu\nu}$])."[42]

Levi-Civita's "concrete example" proceeds as follows. He first introduced a coordinate system, where the $ds^2$ takes the canonical Euclidean form

$$ds^2 = dx_1^2 + dx_2^2 + dx_3^2 + dx_4^2, \qquad \text{(i)}$$

the metric tensor being constant: $g_{\mu\nu} = \delta_{\mu\nu}$.[43] Then, by means of an *infinitesimal* coordinate transformation of the kind

$$x'_\mu = x_\mu + y_\mu \qquad \text{(ii)}$$

(where the $y_\mu$ represent arbitrary infinitesimal functions of the $x_\mu$ variables), he obtained the field

$$g_{\mu\nu} = \delta_{\mu\nu} + h_{\mu\nu}, \qquad \text{(iii)}$$

where

$$h_{\mu\nu} = -\left(\frac{\partial y_\mu}{\partial x_\nu} + \frac{\partial y_\nu}{\partial x_\mu}\right). \qquad \text{(iv)}$$

The tensor $g_{\mu\nu}$ can be interpreted, as usual, either as the same *old* field in the *new* coordinates $x'_\mu$ or as a *new* field in the *old* coordinates $x_\mu$. In other words, $g_{\mu\nu} = \delta_{\mu\nu} + h_{\mu\nu}$ can be considered as a *varied* field obtained from the old one ($\delta_{\mu\nu}$) by means of arbitrary infinitesimal variations $\delta g_{\mu\nu} = h_{\mu\nu}$. But if one demands that the old coordinate system $x_\mu$ be adapted also to the new field $g_{\mu\nu}$, it follows that the $\delta g_{\mu\nu}$ are no longer arbitrary, but have to satisfy the "adaptation" conditions $\delta_1 B_\mu = 0$ (see equation (17)), (which means that $\delta g_{\mu\nu} = \delta_1 g_{\mu\nu}$).[44]

Starting from the expression for $H$ introduced by Einstein in his 1914b (eq. (78), p. 1076), and neglecting infinitesimal terms higher than those of the first order, Levi-Civita obtained by means of a straightforward calculation:

$$\frac{\partial(H\sqrt{-g})}{\partial g_\sigma^{\mu\nu}} = \frac{1}{2}\frac{\partial h_{\mu\nu}}{\partial x_\sigma}. \qquad \text{(v)}$$

Then, by inserting this result in the definition of $\mathfrak{E}_{\mu\nu}$ (see equation (12)), he immediately obtained the following expression for the gravitational tensor:

$$\mathfrak{E}_{\mu\nu} = -\tfrac{1}{2}\Delta_2 h_{\mu\nu} \qquad \text{(vi)}$$

(where, according to Levi-Civita's notation, $\Delta_2$ represents the four-dimensional Laplace operator).

Subsequently, by imposing the conditions $B_\mu = 0$ (where $B_\mu$ is given by equation (9)) on the new field $g_{\mu\nu}$, Levi-Civita finally obtained:

$$B_\mu = \frac{1}{2}\frac{\partial}{\partial x_\nu}\Delta_2 h_{\mu\nu} = 0. \qquad \text{(vii)}$$

These additional constraints allowed him to disprove, with a "concrete example," the tensorial character of $\mathfrak{E}_{\mu\nu}$. In fact, since equations (vii) represent, in Levi-Civita's words, "linear, homogeneous equations of the *fourth* order in $y$, ..., therefore [these equations] are satisfied by expressing the $y$ [functions] as arbitrary polynomials of the third order in $x$."[45]

In particular, Levi-Civita chose for the $y_\mu$ the expression:

$$y_\mu = \tfrac{1}{6} c_\mu x_\mu^3, \qquad \text{(viii)}$$

where the $c_\mu$ are (not all vanishing) constants. From equation (viii), it immediately follows:

$$h_{\mu\mu} = -c_\mu x_\mu^2, \qquad \Delta_2 h_{\mu\mu} = -2c_\mu, \qquad \text{(ix)}$$

which implies, as a consequence of equation (vi),

$$\mathfrak{E}_{\mu\mu} = +c_\mu \neq 0. \qquad \text{(x)}$$

It must be stressed that, because of coordinate transformation (ii), $\mathfrak{E}_{\mu\nu}$ can obviously be interpreted as the *old* gravitational tensor in the new coordinates $x'_\mu$. Thus, with his "concrete example," Levi-Civita succeeded in demonstrating the existence of a particular "justified" transformation from a system of "adapted" coordinates $x_\mu$ (where the $ds^2$ is Euclidean, and therefore all the $\mathfrak{E}_{\mu\nu}$ identically vanish) to another "adapted" coordinate system $x'_\mu$, where some components $\mathfrak{E}_{\mu\nu}$ do not vanish, in contradiction with the supposed tensorial character of $\mathfrak{E}_{\mu\nu}$.

This letter of Levi-Civita's "extraordinarily interested" Einstein. As he replied to Levi-Civita:

> For one and a half days I had to reflect unceasingly until I understood how your example could be reconciled with my proof.... Your deduction is completely correct. In the infinitesimal transformation you considered, $(1/\sqrt{-g})\mathfrak{E}_{\mu\nu}$ does *not* have a tensorial character, in spite of the fact that the transformation is between "adapted" coordinate systems.... But strangely, it does not contradict my proof for the following reason: my demonstration does not work only in the special case that you have examined.[46]

Einstein's reply is particularly important since, in order to rebut Levi-Civita's example, he is compelled to admit, for the first time, that his old proof of the crucial theorem $\Delta \delta J = 0$ is incorrect in the case of infinitesimal transformations.

In his answer, Einstein stressed that in order to prove the tensorial character of $\mathfrak{E}_{\mu\nu}/\sqrt{-g}$ from the equation

$$\delta J = \int_\Sigma \sum \delta g^{\mu\nu} \mathfrak{E}_{\mu\nu} \, d\tau = \text{invariant},$$

the ten variations $\delta g^{\mu\nu}$ have to be mutually independent. Or at least, according to Einstein, the mutual independence must be valid for the ten integrals

$$A^{\mu\nu} = \int \sqrt{-g}\, \delta g^{\mu\nu} \, d\tau,$$

which were shown by Einstein in a previous letter to Levi-Civita "to have a tensorial character, if an infinitesimal domain of integration is chosen." On the contrary, according to Einstein, in the special case considered by Levi-Civita in his letter of March 28, "not only cannot the quantities $\int \delta g^{\mu\nu} d\tau$ be freely chosen, but all of them do really vanish."[47]

In order to show this, Einstein rephrased Levi-Civita's example in the old formalism used in his 1914b, starting from the usual splitting (see equation (16))

$$\delta g^{\mu\nu} = \delta_1 g^{\mu\nu} + \delta_2 g^{\mu\nu}.$$

First he showed that $\int_\Sigma \delta_1 g^{\mu\nu} d\tau = 0$. In his own words:

The $\delta_1 g^{\mu\nu}$ must satisfy the condition

$$\delta_1 B_\mu = 0,$$

which, in the special case considered by you, assumes the form:

$$\sum_\nu \frac{\partial}{\partial x_\mu}(\Box \delta_1 g^{\mu\nu}) = 0$$

(see eq. (5) of your letter), where $\Box$ represents the four-dimensional Laplace operator. As it is known, because of the boundary conditions, from that equation it follows:

$$\sum \frac{\partial \delta_1 g^{\mu\nu}}{\partial x_\nu} = 0$$

(this result is characteristic of your special case). Multiplying this equation by $x_\sigma$ and integrating over the whole domain, it follows after a partial integration for any combination of the indices:

$$\int \delta_1 g^{\mu\sigma} d\tau = 0.$$

This result is a consequence of the characteristic degeneration of the equation

$$\delta_1 B_\mu = 0$$

in the special case that you have considered.[48]

As a second step, Einstein argued that, from the definition of $\delta_2 g^{\mu\nu}$, considering an infinitesimal coordinate transformation and integrating $\delta_2 g^{\mu\nu}$ over an infinitesimally small domain, it follows that $\int_\Sigma \delta_2 g^{\mu\nu} d\tau = 0$.[49] Thus, according to Einstein, $\int_\Sigma \delta_2 g^{\mu\nu} d\tau = 0$ for any combination of indices. Since the $A^{\mu\nu}$ all vanish, "the tensorial character of $\mathfrak{E}_{\mu\nu}/\sqrt{-g}$ cannot be proved for an infinitesimal transformation." Although Einstein still believed that his old demonstration mantained its validity for all finite transformations, he was nevertheless obliged to admit that, since $\int_\Sigma \delta_2 g^{\mu\nu} d\tau = 0$ for *every* infinitesimal transformation, the proof of his fundamental theorem had to be revised, at least in the case of infinitesimal transformations, by using only $\delta_1$-variations. In his own words:

My demonstration is still valid for all finite transformations. These considerations suggest a modification of the proof of covariance [of $\mathfrak{E}_{\mu\nu}/\sqrt{-g}$; here Einstein hints evidently at the proof of the theorem $\Delta\delta J = 0$] by using only the $\delta_1$-variations since the $\delta_2$-variations do not give any contribution to the fundamental quantities $\int \sqrt{-g}\,\delta g^{\mu\nu}\, d\tau$.[50]

We want to stress that Einstein's admission of the uselessness of the $\delta_2 g^{\mu\nu}$ variations and his eventual willingness to use only the $\delta_1 g^{\mu\nu}$ variations, means that *the proof of his theorem $\Delta\delta J = 0$ cannot hold within the Entwurf theory*. In fact, if one uses only the $\delta_1 g^{\mu\nu}$, because of the four conditions $\delta_1 B_\sigma = 0$ (see equation (17)), then the *$\delta g^{\mu\nu} \equiv \delta_1 g^{\mu\nu}$ no longer represent ten arbitrary mutually independent quantities*.

The subsequent correspondence between the two scientists continued to touch on the same ideas. In a postcard sent to Einstein on April 2, Levi-Civita continued to criticize Einstein's proof of the tensorial character of $\mathfrak{E}_{\mu\nu}/\sqrt{-g}$ (reported in the letter of March 17), obtained with the help of the $A^{\mu\nu}$, on the grounds that the $A^{\mu\nu}$ are infinitesimal.

Answering this postcard, Einstein strove to give a new, mathematically more accurate version of the same theorem (regarding the proof of the tensorial character of $\mathfrak{E}_{\mu\nu}/\sqrt{-g}$ as a consequence of the tensorial character of $A^{\mu\nu}$).[51] But in the same letter he admitted once again the incompleteness of his old demonstration:

My proof of the invariant character of $\delta J$ does not work for those infinitesimal transformations, starting from an original frame where the $g_{\mu\nu}$ are supposed to be constant, since then the quantities $A^{\mu\nu}$ cannot be freely chosen as they vanish identically. An analogous result, however, is *not* true when the $g_{\mu\nu}$ are variable. Therefore my proof works in general, but it does not work in some special case.[52]

Although Einstein had to admit that his proof of the tensorial character of $\mathfrak{E}_{\mu\nu}/\sqrt{-g}$ is untenable, at least "in some special cases," he nevertheless persisted in his conviction regarding the *correctness* of his result, i.e., that $\mathfrak{E}_{\mu\nu}/\sqrt{-g}$ *must be a tensor*. Thus he concluded in the same letter:

I really must admit that my deeper reasonings provoked by your interesting letters have strengthened my conviction that the proof of the tensorial character of $\mathfrak{E}_{\mu\nu}/\sqrt{-g}$ is correct in principle.[53]

Three days later Einstein, compelled by his inner "conviction" of the tensorial character of $\mathfrak{E}_{\mu\nu}/\sqrt{-g}$, proposed a "nice special case of the statement that $\mathfrak{E}_{\mu\nu}/\sqrt{-g}$ is a tensor." This "nice special case" is based on a particular choice of $H$ that turns out to be *invariant under general coordinate transformations, and consequently it implies generally covariant tensorial properties* for $\mathfrak{E}_{\mu\nu}/\sqrt{-g}$. It is worth reporting this short postcard of Einstein's in full:

Highly Esteemed Colleague,
it is possible to obtain a nice special case of the statement that $\mathfrak{E}_{\mu\nu}/\sqrt{-g}$ is a tensor if one sets:

$$H = \text{const.}$$

The condition $B_\mu = 0$ is then identically satisfied and $H$ is invariant under any transformation, and therefore also for linear transformations. $\mathfrak{E}_{\mu\nu}/\sqrt{-g}$ must have, according to my thesis, a tensorial character with regard to arbitrary transformations. In fact one obtains:

$$\frac{\mathfrak{E}_{\mu\nu}}{\sqrt{-g}} = \text{const.} \cdot g_{\mu\nu},$$

and thus really a covariant tensor.[54]

Although Einstein's choice of $H =$ constant does not have any reasonable physical meaning, we want to stress that in order to prove the tensorial character of $\mathfrak{E}_{\mu\nu}/\sqrt{-g}$ Einstein was obliged to dismiss the requirement of limited covariance implicit in the *Entwurf* theory and to go back to general covariance, since the conditions $B_\sigma = 0$ are automatically satisfied and do not represent additional constraints. We want to emphasize also that this example, $H =$ constant, is incompatible with his previous 1914 choice which, according to Einstein, led uniquely to his old *Entwurf* field equations.

In a subsequent postcard, sent to Levi-Civita on April 14, Einstein once again admitted that his old proof of the tensorial character of $\mathfrak{E}_{\mu\nu}/\sqrt{-g}$ has to be modified, giving implicit credit to the usefulness of Levi-Civita's criticisms: "When there will be an occasion to repeat the proof I will glady introduce those improvements which I have learned in our memorable correspondence."[55] In the same postcard Einstein offered a direct testimony regarding his intellectual isolation:

> At the moment there is a very modest interest in this topic. It is strange how little our colleagues in the field feel the inner necessity of an *authentic* theory of relativity. Unfortunately, the attitude of our fellow human beings who do not work in the field is incomparably more strange. Therefore, it is a double pleasure for me to know better a person like you.[56]

In a subsequent letter, written to Einstein on April 15, (which Einstein characterized as "most interesting"), Levi-Civita insisted on his objections regarding both Einstein's new formulation of the proof of the tensorial character of $A^{\mu\nu}$ (and consequently of $\mathfrak{E}_{\mu\nu}/\sqrt{-g}$) and the old question of the lack of *independence* of the $A^{\mu\nu}$ components.

Eventually, Einstein was obliged to accept the validity of Levi-Civita's last objection and, more generally, he had to admit the "incompleteness" of his proof. Thus on April 20, Einstein wrote to Levi-Civita: "I willingly acknowledge that you have touched the sorest spot of my proof, namely the independence of the $A^{\mu\nu}$. Here my demonstration is lacking in acumen."[57] Finally, in the last letter of their correspondence, Einstein stressed again the same conclusion:

> I also think that we have exhausted our subject up to the limits allowed by our present level of understanding. My demonstration is incomplete, in the sense that the possibility of an arbitrary choice of the $A^{\mu\nu}$ remains unproved.[58]

## 5. Conclusion

Our reconstruction of the elements of this scientific controversy is probably incomplete (especially because all of Levi-Civita's letters but one are missing). Nevertheless, we think that we have been able to single out the main criticisms of Levi-Civita, which eventually obliged Einstein to admit that his old proof of the fundamental theorem $\Delta\delta J = 0$ was untenable, at least in the case of infinitesimal transformations.

In particular, the main bug that Einstein discovered with the help of Levi-Civita (which he called the "sorest spot" of his proof) regards the question of the independence of the $A^{\mu\nu}$. As we have seen at the end of the last section, Einstein was obliged to admit that his demonstration was "incomplete, in the sense that the possibility of an arbitrary choice of $A^{\mu\nu}$ remains unproved." This conclusion, as we have shown, implies the impossibility of proving the tensorial character of $\mathfrak{E}_{\mu\nu}/\sqrt{-g}$ within a limited-covariance theory.

In fact, in his reply to Levi-Civita's "concrete case" of March 28, Einstein became aware, for the first time, that the use of the $\delta_2 g^{\mu\nu}$ variations, (introduced in the splitting $\delta g^{\mu\nu} = \delta_1 g^{\mu\nu} + \delta_2 g^{\mu\nu}$) could no longer be maintained, since in Einstein's words "the $\delta_2$-variations do not give any contribution to the fundamental quantities $\int_\Sigma \sqrt{-g}\,\delta g^{\mu\nu}\,d\tau$." As we have stressed in the last section, Einstein's readiness "to modify [his] proof of covariance [of $\mathfrak{E}_{\mu\nu}/\sqrt{-g}$] by using *only* the $\delta_1$-variations" is equivalent to an admission that his old proof of the theorem $\Delta\delta J = 0$ cannot hold in the framework of his *Entwurf* theory. In fact, in a limited-covariance theory, the conditions $B_\sigma = 0$ for the selection of "adapted" coordinate systems (and consequently $\delta_1 B_\sigma = 0$; see equation (17)), should hold, and this would imply that the $\delta_1 g^{\mu\nu}$ are not independent. If, on the contrary, one requires the $\delta_1 g^{\mu\nu}$ to be *ten independent* variations (as a consequence of the dropping of the $\delta_2$-variations), it follows that one cannot any longer impose the four additional constraints $B_\sigma = 0$, i.e., one has to renounce the requirement of "limited" covariance.[59]

In conclusion, although other new reflections would bring Einstein, during the summer and fall of 1915, to his correct generally covariant formulation of the field equations in November 1915, we can safely state that one of the starting points of his "growing dissatisfaction" with his *Entwurf* theory and of his new interest in general covariance can be traced back to his 1915 correspondence with Levi-Civita.

*Acknowledgments.* We are very grateful to John Norton for his careful reading of a preliminary draft of this article and for many useful suggestions and criticisms. We are also grateful to John Stachel, who helped us to clarify some controversial mathematical points in the Einstein–Levi-Civita correspondence.

An invited paper on the preliminary results of this article (Cattani and De Maria 1986) was presented by the authors at the *4th Marcel Grossmann Meeting on the Recent Developments of General Relativity*, Rome, Italy, June 17–21 1985.

NOTES

[1] Einstein presented four communications to the Prussian Academy of Science, respectively on the 4th (Einstein 1915a), 11th (1915b), 18th (1915c), and 25th (1915d) of November 1915.

[2] See Earman and Glymour 1978a, 1978b; Stachel 1980; Norton 1984.

[3] EA 3-006. This notebook was originally catalogued as containing "notes for Einstein's lectures at the University of Zurich in the period 1909–1911." But Norton has noted its mislabeling: "The major part of the notebook, which extends from pages 5 to 29, clearly belongs to the 1912–1913 period in Zurich, for it contains calculations made by Einstein during his work on the *Entwurf* paper" (Norton 1984, p. 267).

[4] Einstein and Grossmann 1913, Einstein 1914a, Einstein and Grossmann 1914, Einstein 1914b. In his reconstruction, Norton reaches the conclusion, contrary to the usual interpretation, that "the 'hole' argument was not based on the naive misunderstanding that a given gravitational field is somehow physically changed by the transition to a new coordinate system simply because the mathematical functions that represent it have changed.... Rather, the argument amounted to a demonstration that generally covariant field equations cannot uniquely determine the field as long as the point events of the space-time manifold are incorrectly thought of as individuated independently of the field itself" (Norton 1984, p. 256).

[5] Einstein and Grossmann 1914, Einstein 1914b; see Norton 1984, sec. 6, pp. 291–298. Norton stresses in particular how the variational techniques developed by Einstein in 1914 were used "virtually unchanged" in his 1916 analysis of his final generally covariant field equations. In this way Einstein was able to derive a set of identities (the contracted Bianchi identities) from which the conservation laws could be derived.

[6] The main points of dissatisfaction, according to Norton, were:

1. The fact that the *Entwurf* theory did not correctly account for the anomalous precession of Mercury's perihelion (Einstein apparently became aware of it in July 1915);
2. His discovery that the *Entwurf* equations are not covariant under transformations to rotating coordinate systems;
3. His realization that his "approach ... to the determination of the Lagrangian function of the gravitational field was illusory throughout.... so that one needed to apply no limiting condition at all to $H$...." and his related "conviction that the introduction of adapted systems was a false path" (Einstein to H. A. Lorentz, January 1, 1916, EA 16-445; cited in Norton 1984, p. 299).

[7] See Goodstein 1975. Goodstein deserves the credit for having catalogued, in 1973, Levi-Civita's scientific and personal correspondence, presently deposited in the Accademia Nazionale dei Lincei. She was also the first to stress the importance of the 1915 correspondence between Einstein and Levi-Civita. She did not, however, treat the scientific aspects of their controversy.

[8] Einstein to Besso, ca. March 1914, in Speziali 1972, p. 52.

[9] Einstein to Besso, ca. March 1914, in Speziali 1972, p. 53.

[10] As we have already stressed in the introduction, the "adapted" coordinate systems are those compatible with the requirement of the uniqueness of the solution $g^{\mu\nu}$ of the gravitational equations, and thus with the causality principle.

[11] Einstein's proof of equation (23) is straightforward, but can be more easily understood with the help of following diagram:

$$\begin{array}{ccc} K_1 \quad G_1 & \xrightarrow{(\delta_2 J_1 = 0)}{t} & G_1^* \\ (\Delta J = \delta_2 J = 0) \Bigg\downarrow T & & T \Bigg\downarrow (\Delta J = \delta_2 J = 0) \\ K_2 \quad G_2 & \xrightarrow[(\delta_2 J_2 = ?)]{} & G_2^* \end{array}$$

where $K_1$ and $K_2$ are two coordinate systems both "adapted" to the same unvaried field; $G_1$ and $G_2$ are the symbolic expressions of the same *unvaried* gravitational field in $K_1$ and $K_2$, respectively; $G_1^*$ and $G_2^*$ are the expressions of the *varied* field referred to $K_1$ and $K_2$, respectively; $T$ represents the "justified" transformation that maps both $G_1$ into $G_2$ and $G_1^*$ into $G_2^*$; $t$ is the variation that in $K_1$ transforms the *unvaried* field $G_1$ into the *varied* field $G_1^*$; but as a consequence of our comments regarding equation (21), $t$ can be also interpreted as an infinitesimal coordinate transformation which produces the variation $\delta_2 g^{\mu\nu}$ of the *same field*.

Einstein's conclusion is that the transformation from $G_2$ (i.e., the *unvaried* field in $K_2$) to $G_2^*$ (i.e., the *varied* field in $K_2$) can be obtained through the following chain of coordinate transformations:

$$G_2 \xrightarrow{T^{-1}} G_1 \xrightarrow{t} G_1^* \xrightarrow{T} G_2^*.$$

Thus it follows that the transformation $T^{-1}tT$ from $G_2$ to $G_2^*$ is also a "$\delta_2$-variation;" consequently equation (23) $\delta_2(J_2) = 0$ must hold.

[12] For a more detailed derivation of equation (24), see Norton 1984, pp. 295–296.

[13] Levi-Civita to Abraham, February 7, 1907; Levi-Civita Papers, family's collection, Rome. One of us (M.D.M.) is grateful to Professor V. Ceccherini and his wife for their permission to examine Levi-Civita's uncatalogued correspondence still in the family's possession.

[14] Final report of the committee for the "Concorso per Professore Straordinario di Meccanica Razionale," October 1909, Levi-Civita Papers, family's collection, Rome.

[15] Report of the committee for the promotion of M. Abraham to Professore Ordinario, Levi-Civita Papers, family's collection, Rome.

[16] For Abraham's career in Italy and his debate on gravitation with Einstein, see Earman and Glymour n.d., Goodstein 1983, Battimelli and De Maria 1983, De Maria 1986; see also Cattani and De Maria, "Max Abraham and the Reception of Relativity in Italy: His 1912 and 1914 Controversies with Einstein," this volume, pp. 160–174.

[17] See Einstein 1912a, 1912b; Abraham 1912a; Einstein 1912c; Abraham 1912b; Einstein 1912d; Abraham 1914a; Einstein 1914c; Abraham 1914b.

[18] Einstein to Besso, ca. 1914, in Speziali 1972, p. 50. The letter is dated "about the end of 1913," but it must be later than that, since Abraham's first paper in *Scientia* dates from 1914.

[19] Abraham to Levi-Civita, February 23, 1915, Levi-Civita Papers, Accademia dei Lincei, Rome, p. 3 (our translation from the Italian). The term "recent work" clearly

refers to Einstein 1914b. From this quotation it appears that Abraham played a *direct role* in stimulating Levi-Civita's interest in general relativity, since Levi-Civita first received Einstein's paper directly from Abraham.

[20] Abraham to Levi-Civita, February 23, 1915, pp. 3–4.

[21] The originals have been donated by the widow of Levi-Civita, Libera Levi-Civita, to the Einstein Archive. Copies of the originals are in the family's hands (see Goodstein 1975, p. 47).

[22] Einstein to Lorentz, August 16, 1913, cited in Norton 1984, p. 253.

[23] Einstein to Levi-Civita, March 5, 1915 (EA 16-228), p. 1.

[24] Einstein to Levi-Civita, March 5, 1915, p. 2.

[25] Einstein to Levi-Civita, March 5, 1915, p. 2.

[26] Einstein to Levi-Civita, March 5, 1915, p. 3.

[27] Einstein to Levi-Civita, March 17, 1915 (EA 16-231), p. 1.

[28] "I do not want to deal with the matter of the existence or nonexistence of that limit, since this question is, in my opinion, irrelevant." Einstein to Levi-Civita, March 17, 1915, p. 1.

[29] Einstein to Levi-Civita, March 17, 1915, p. 2. We follow the numbering of formulas originally used by Einstein in his letter.

[30] Einstein to Levi-Civita, March 17, 1915, p. 2.

[31] Einstein to Levi-Civita, March 17, 1915, p. 3.

[32] Einstein to Levi-Civita, March 20, 1915 (EA 16-233).

[33] Einstein to Levi-Civita, March 20, 1915.

[34] Einstein to Levi-Civita, March 20, 1915.

[35] Einstein to Levi-Civita, March 20, 1915.

[36] This letter of Levi-Civita is missing, however we have reconstructed its content with the help of Einstein's answer: Einstein to Levi-Civita, March 26, 1915 (EA 16-235).

[37] Einstein to Levi-Civita, March 26, 1915.

[38] Einstein to Levi-Civita, March 26, 1915.

[39] Einstein to Levi-Civita, March 26, 1915.

[40] Levi-Civita to Einstein, March 28, 1915 (EA 16-237), p. 1. This is the only letter from Levi-Civita to Einstein out of their 1915 correspondence that is not missing. Because of the importance of its content, we will examine it in detail.

[41] Levi-Civita to Einstein, March 28, 1915, p. 1.

[42] Levi-Civita to Einstein, March 28, 1915, p. 1.

[43] We will follow *ad litteram* the notation used by Levi-Civita in this letter, although it is slightly different from the modern notation.

[44] It is worth stressing that the old metric tensor $\delta_{\mu\nu}$ trivially satisfies the "adaptation" conditions $B_\mu = 0$.

[45] Levi-Civita to Einstein, March 28, 1915, p. 3.

[46] Einstein to Levi-Civita, April 2, 1915 (EA 16-238), p. 1.

[47] Einstein to Levi-Civita, April 2, 1915, p. 2.

[48] Einstein to Levi-Civita, April 2, 1915, pp. 2–3. Einstein's result, $\int_\Sigma \delta_1 g^{\mu\sigma} d\tau = 0$ is straightforward: by multiplying $\Sigma \delta_1 g^{\mu\nu}/\partial x_\nu$ by $x_\sigma$ and integrating it over the "volume" $\Sigma$, one obtains:

$$\int_\Sigma x_\sigma \frac{\partial}{\partial x_\nu}(\delta_1 g^{\mu\nu}) d\tau = \int_\Sigma \frac{\partial}{\partial x_\nu}(x_\sigma \delta_1 g^{\mu\nu}) d\tau - \int_\Sigma \frac{\partial x_\sigma}{\partial x_\nu} \delta_1 g^{\mu\nu} d\tau = 0. \quad \text{(a)}$$

The first integral on the right-hand side can be transformed into a "surface" integral over the boundary of $\Sigma$, and thus it vanishes because of the boundary conditions $\delta_1 g^{\mu\nu} = 0$; thus from equation (a) it follows that

$$\int_\Sigma \delta_1 g^{\mu\sigma} \, d\tau = 0.$$

[49] The proof of the condition $\int_\Sigma \delta_2 g^{\mu\nu} = 0$ is straightforward. In fact, in the case of infinitesimal coordinate transformations, $\delta_2 g^{\mu\nu}$ takes the explicit form (see Einstein 1914b, eq. (63), p. 1069):

$$\delta_2 g^{\mu\nu} = \Delta g^{\mu\nu} = \sum_\alpha g^{\mu\alpha} \frac{\partial \Delta x_\nu}{\partial x_\alpha} + g^{\nu\alpha} \frac{\partial \Delta x_\mu}{\partial x_\alpha}.$$

Since the region of integration $\Sigma$ is infinitesimal, the following approximation holds in $\Sigma$:

$$g^{\mu\nu} \simeq \delta^{\mu\nu},$$

which implies:

$$\delta_2 g^{\mu\nu} = \frac{\partial \Delta x_\nu}{\partial x_\mu} + \frac{\partial \Delta x_\mu}{\partial x_\nu}.$$

Therefore the integral $\int_\Sigma \delta_2 g^{\mu\nu} \, d\tau$ becomes:

$$\int_\Sigma \delta_2 g^{\mu\nu} \, d\tau \simeq \int_\Sigma \left( \frac{\partial \Delta x_\nu}{\partial x_\mu} + \frac{\partial \Delta x_\mu}{\partial x_\nu} \right) d\tau.$$

Since the last integral can be transformed into an integral over the boundary of $\Sigma$ (where the boundary conditions $\Delta x_\mu = 0$ hold), it vanishes identically.

[50] Einstein to Levi-Civita, April 2, 1915, p. 4. The importance given by Einstein to this correspondence with Levi-Civita appears clearly in a postscript to this letter: "I never had the experience of a correspondence as interesting as this before. You should see how I always look forward to your letters."

[51] Einstein to Levi-Civita, April 8, 1915 (EA 16-240).

[52] Einstein to Levi-Civita, April 8, 1915, pp. 2–3.

[53] Einstein to Levi-Civita, April 8, 1915, p. 3.

[54] Einstein to Levi-Civita, April 11, 1915 (EA 16-242). The proof of Einstein's result $\mathfrak{E}_{\mu\nu}/\sqrt{-g} = \text{constant } g_{\mu\nu}$ is straightforward. Let us assume, in fact, that the Hamiltonian density is constant everywhere inside the integration domain $\Sigma$: $H = \bar{H} = \text{constant}$; from equation (6), it follows that

$$J = \bar{H} \int_\Sigma \sqrt{-g} \, d\tau. \tag{a}$$

For an arbitrary variation $\Delta x_\mu = x'_\mu - x_\mu$ of the coordinate system, because of the invariance of the elementary "volume" of the four-dimensional manifold $\Delta(\sqrt{-g} \, d\tau) = 0$, we obtain from equation (a)

$$\Delta J = \bar{H} \int_\Sigma \Delta(\sqrt{-g} \, d\tau) = 0, \tag{b}$$

that is, *J is invariant under any coordinate transformation.* Therefore, from equation (10), the conditions $B_\sigma = 0$ are identically satisfied in every coordinate system. For arbitrary variations of the potentials $\delta g^{\mu\nu}$, we get from equation (a)

$$\delta J = \int_{\Sigma} \bar{H} \delta(\sqrt{-g})\, d\tau = \int_{\Sigma} \bar{H}(-\tfrac{1}{2} g_{\mu\nu} \sqrt{-g}\, \delta g^{\mu\nu})\, d\tau \qquad \text{(c)}$$

(since with the usual methods of the tensor calculus it is trivial to verify the relation $\delta \sqrt{-g} = -\tfrac{1}{2} g_{\mu\nu} \sqrt{-g}\, \delta g^{\mu\nu}$). If we compare equation (c) with equation (11), we obtain:

$$\frac{\mathfrak{E}_{\mu\nu}}{\sqrt{-g}} = -\frac{1}{2} \bar{H} g_{\mu\nu}, \qquad \text{(d)}$$

that is, $\mathfrak{E}_{\mu\nu}/\sqrt{-g}$ is a *generally covariant tensor*.

[55] Einstein to Levi-Civita, April 14, 1915 (EA 16-244).

[56] Einstein to Levi-Civita, April 14, 1915. He adds: "It will be a pleasure for me to transform our acquaintance by mail into a personal one; [this is] a further reason for me to cross finally the Alps once again. Let us hope that our fatherlands will not rebel against one another."

[57] Einstein to Levi-Civita, April 20, 1915 (EA 16-246).

[58] Einstein to Levi-Civita, May 5, 1915 (EA 16-250).

[59] This would be the approach followed by Einstein in his 1916 variational formulation of his generally covariant final theory, Einstein 1916.

REFERENCES

Abraham, Max (1912a). "Relativität und Gravitation. Erwiderung auf eine Bemerkung des Hrn. A. Einstein." *Annalen der Physik* 38: 1056–1058.

——— (1912b). "Nochmals Relativität und Gravitation. Bemerkungen zu A. Einsteins Erwiderung." *Annalen der Physik* 39: 444–448.

——— (1914a). "Die neue Mechanik." *Scientia* 15: 8–27.

——— (1914b). "Sur le Problème de la Relativité." *Scientia* 16: 101–103.

Battimelli, Giovanni, and De Maria, Michelangelo (1983). "Max Abraham in Italia." In *Atti del III Congresso Nazionale di Storia della Fisica*. (Palermo, 1983), pp. 186–192.

Cattani, Carlo, and De Maria, Michelangelo (1986). "Einstein's Path toward the Generally Covariant Formulation of Gravitational Field Equations: The Contribution of Tullio Levi-Civita." In *Proceedings of the Fourth Marcel Grossmann Meeting on General Relativity*. Remo Ruffini, ed. Amsterdam: Elsevier, pp. 1805–1826.

De Maria, Michelangelo (1986). "Le prime reazioni alla Relatività in Italia: le polemiche fra Max Abraham e Albert Einstein 1912–1914." In *Proceedings of the Conference, La Matematica italiana fra le due querre mondiali, Milano-Gargnano del Garda, October 1986*, pp. 143–159.

Earman, John, and Glymour, Clark (1978a). "Lost in the Tensors: Einstein's Struggles with Covariance Principles 1912–1916." *Studies in History and Philosophy of Science* 9: 251–278.

——— (1978b). "Einstein and Hilbert: Two Months in the History of General Relativity." *Archive for History of Exact Sciences* 19: 291–308.

——— (n.d.). "Abraham and Einstein: Two Theories of Gravitation." Typescript.

Einstein, Albert (1912a). "Lichtgeschwindigkeit und Statik des Gravitationsfeldes." *Annalen der Physik* 38: 355–369.

——— (1912b). "Zur Theorie des statischen Gravitationsfeldes." *Annalen der Physik* 38: 443–458.

——— (1912c). "Relativität und Gravitation. Erwiderung auf eine Bemerkung von M. Abraham." *Annalen der Physik* 38: 1059–1964.

——— (1912d). "Bemerkung zu Abrahams vorangehender Auseinandersetzung 'Nochmals Relativität und Gravitation.'" *Annalen der Physik* 39: 704.

——— (1914a). "Prinzipielles zur verallgemeinerten Relativitätstheorie und Gravitationstheorie." *Physikalische Zeitschrift* 15: 176–180.

——— (1914b). "Die formale Grundlage der allgemeinen Relativitätstheorie." *Königlich Preussische Akademie der Wissenschaften* (Berlin). *Sitzungsberichte*: 1030–1085.

——— (1914c). "Zum Relativitätsproblem." *Scientia* 15: 337–348.

——— (1915a). "Zur allgemeinen Relativitätstheorie." *Königlich Preussische Akademie der Wissenschaften* (Berlin). *Sitzungsberichte*: 778–786.

——— (1915b). "Zur allgemeinen Relativitätstheorie (Nachtrag)." *Königlich Preussische Akademie der Wissenschaften* (Berlin). *Sitzungsberichte*: 799–801.

——— (1915c): "Erklärung der Perihelbewegung des Merkur aus der allgemeinen Relativitätstheorie." *Königlich Preussische Akademie der Wissenschaften* (Berlin). *Sitzungsberichte*: 831–839.

——— (1915d). "Die Feldgleichungen der Gravitation," *Königlich Preussische Akademie der Wissenschaften* (Berlin). *Sitzungsberichte*: 844–847.

——— (1916). "Hamiltonsches Prinzip und allgemeine Relativitätstheorie." *Preussische Akademie der Wissenschaften* (Berlin). *Sitzungsberichte*: 1111–1116.

Einstein, Albert, and Grossmann, Marcel (1913). *Entwurf einer verallgemeinerten Relativitätstheorie und einer Theorie der Gravitation. I. Physikalischer Teil von Albert Einstein. II. Mathematischer Teil von Marcel Grossmann.* Leipzig and Berlin: B.G. Teubner. Reprinted with added "Bemerkungen," *Zeitschrift für Mathematik und Physik* 62 (1914): 225–261.

——— (1914). "Kovarianzeigenschaften der Feldgleichungen der auf die verallgemeinerte Relativitätstheorie gegründeten Gravitationstheorie." *Zeitschrift für Mathematik und Physik* 63: 215–225.

Goodstein, Judith (1975). "Levi-Civita, Albert Einstein and Relativity in Italy." In *Tullio Levi-Civita, Convegno Internazionale Celebrativo del Centenario della Nascita, Roma, 17–19 Dicembre 1973*. Rome: Accademia Nazionale dei Lincei.

——— (1983). "The Italian Mathematicians of Relativity." *Centaurus* 26: 241–261.

Norton, John (1984). "How Einstein Found His Field Equations: 1912–1915." *Historical Studies in the Physical Sciences* 14: 253–316. See this volume, pp. 101–159.

Speziali, Pierre, ed. (1972). *Albert Einstein–Michele Besso. Correspondance, 1903–1955.* Paris: Hermann.

Stachel, John (1980). "Einstein's Search for General Covariance, 1912–1915." Paper delivered to the Ninth International Conference on General Relativity and Gravitation, Jena, 1980. See this volume, pp. 63–100.

# Hendrik Antoon Lorentz, the Ether, and the General Theory of Relativity*

A.J. Kox

## 1. Introduction

From the early days of the development of the general theory of relativity, the Dutch physicist Hendrik Antoon Lorentz showed a lively and active interest in this theory of gravitation. He devoted much time and energy to understanding the theory and made several important contributions himself. In this paper I will discuss Lorentz's work in the field of general relativity; in addition, I will address the question of the apparent discrepancy between Lorentz's enthusiasm for the general theory of relativity and his belief in the existence of an ether. It is well known that until his death in 1928 Lorentz kept insisting on the usefulness of an ether. In spite of his often-expressed admiration for Einstein's special theory of relativity, he preferred his own ether-based "theory of electrons." Lorentz admitted that his theory and the special theory of relativity had the same empirical consequences and that the ether could not be experimentally detected, but he maintained that some kind of ether was needed as carrier of the electromagnetic field. As he said in his *Theory of Electrons*: "I cannot but regard the ether, which can be the seat of an electromagnetic field with its energy and its vibrations, as endowed with a certain degree of substantiality, however different it may be from all ordinary matter."[1] In light of this and many similar statements it seems remarkable that Lorentz occupied himself with the general theory of relativity, in which the ether played no role whatsoever. I will show that, in fact, Lorentz's point of view was not inconsistent, and that he had the same objections against the general theory as against the special theory. He was, on the other hand, so impressed by the beauty and originality of the theory that he almost naturally became involved in its development, and became one of its first ardent propagandists in the Netherlands.[2]

## 2. Lorentz's Early Contributions

Not very long after the publication in 1913 of Albert Einstein and Marcel Grossmann's "Entwurf einer verallgemeinerten Relativitätstheorie und einer Theorie der Gravitation,"[3] Lorentz began to study this first version of the general theory of relativity. He filled many pages of his scientific notebooks with calculations, working hard to understand the mathematical intricacies of the theory and trying to verify its conclusions. He carefully checked the transformation properties of the main formulas and in the process came to the conclusion that the field equations are covariant for arbitrary transformations only if the energy-momentum tensor is symmetric.[4] Lorentz communicated this result to Einstein, as can be inferred from a remark in one of his notebooks and from two letters Einstein wrote to Lorentz, in reply to two letters that are now lost.[5] In his letters, Einstein expressed his pleasure at Lorentz's interest and tried to make plausible that the tensor is indeed symmetric.[6]

For more than a year after this exchange of letters there were no outward signs of Lorentz's actively working on general relativity. But after Einstein published his paper "Die formale Grundlage der allgemeinen Relativitätstheorie" (Einstein 1914) in November 1914, Lorentz set to work again. His activities resulted in an enormous amount of calculation (several hundred pages) and in the publication of a paper in which Einstein's field equations are derived from a variational principle (Lorentz 1915). Although Lorentz's notation is at times somewhat cumbersome, the paper shows a clear insight into the theory and a firm grasp of its formalism. The lack of any criticism of its foundations, moreover, seems to indicate complete agreement with the fundamental assumptions of the theory.

This impression, however, is not quite correct. Just before Lorentz's paper was submitted (at the end of January 1915), an exchange of letters took place between Lorentz and Einstein from which it becomes clear that Lorentz had fundamental objections to Einstein's point of view. Two letters are involved, one from Lorentz, and a reply from Einstein.[7] Only a draft of Lorentz's letter is available, but from Einstein's reply we can conclude that the actual letter was very similar in content to the draft. Both letters merit careful attention, because they very clearly illustrate the fundamentally different attitudes of Lorentz and Einstein toward the foundations of physics.

The first paragraphs of Lorentz's letter contain a rather technical exposition of a mathematical difficulty Lorentz had encountered in Einstein's paper of November 1914. Then the discussion proceeds toward a more fundamental point: the idea of general covariance. This idea plays an important role in the paper of November 1914. In one of the introductory sections, Einstein strongly argues that a theory of gravitation should be generally covariant, in the sense that its laws are invariant under arbitrary coordinate transformations. The theory presented in the paper, however, does not meet this requirement; the field equations allow only a restricted set of coordinate transformations. In order to justify this result, Einstein presents an argument that is known as the "hole argument." The argument, the first version of which dates from 1913 (Einstein and Grossmann 1914, pp. 260–261), runs as follows (Einstein 1914, p. 1067).[8] Consider a finite space-time region $\Sigma$, in which no material processes take place, so that the physical happenings within $\Sigma$ are fully determined by the quantities $g_{\mu\nu}$. In the coordinate system $K$ these quantities are given as functions of $x_\alpha$; symbolically, $g_{\mu\nu} = G(x_\alpha)$. Introduce a new coordinate system $K'$, which coincides with $K$ outside $\Sigma$, but deviates from it inside this region, in such a way that the corresponding field $g'_{\mu\nu}$ and its derivatives are everywhere continuous. It may be written as $g'_{\mu\nu} = G'(x'_\alpha)$. If in $G'$ the argument $x'_\alpha$ is replaced by $x_\alpha$, a new gravitational field relative to $K$ is created that differs from the original one. In the case of generally covariant field equations, both $G(x_\alpha)$ and $G'(x_\alpha)$ are solutions of the field equations with respect to $K$; they describe the same physical situation but are different inside $\Sigma$ (they coincide on its boundary). Thus in the case of generally covariant field equations the source term (the material energy-momentum tensor) does not uniquely determine the gravitational field. Einstein's (incorrect) conclusion (and justification of the failure of the field equations he derived to be covariant) is that covariant field equations are not allowed. One has to restrict oneself to a limited set of coordinate transformations, determined by the demand that the gravitational field be uniquely fixed by the energy-momentum tensor.

In his letter to Einstein, Lorentz brings up the subject of general covariance because he disagreed with Einstein on this point. He claimed that it is always possible to select a coordinate system that is preferable over all others, not only for mathematical reasons (the simplicity of the formulas), but also on physical grounds. As an example he wrote down Newton's second law for a body in the vicinity of the earth:

$$\frac{d^2 x}{dt^2} = -\alpha \frac{x}{r^3}, \quad \text{etc.} \tag{1}$$

In the coordinate system chosen here ("system I") the earth rotates with angular velocity $\omega$. If we transform to a coordinate system that rotates with the earth, the equations become more complicated:

$$\frac{d^2 x'}{dt^2} = -\alpha \frac{x'}{r^3} + 2\omega \frac{dy'}{dt} + \omega^2 x', \quad \text{etc.} \tag{2}$$

Lorentz remarked that the additional terms in equation (2) do not have a clear physical interpretation, for instance, in terms of gravitating bodies. Therefore, system I is to be preferred, not only because the equations are simpler, but also on physical grounds. As Lorentz put it:

> We might imagine that for a long time people were in possession only of equations (2), and had tormented themselves over an "interpretation" of the terms $2\omega \, dy'/dt$, $\omega^2 x'$, etc. If somebody then comes along, and by introduction of coordinate system I reduces equations (2) to those of (1), everyone will hail this as a real solution and would prefer system I.[9]

Lorentz then pointed out that Einstein's theory, since it is not generally covariant, also implies a preference for certain coordinate systems. A little further on, he concluded that Einstein apparently felt more strongly about covariance than he did, and he questioned Einstein's assertion that all coordinate systems should be equivalent with the words: "Are you not going rather too far here, in laying down a personal viewpoint as self-evident?"[10] Not surprisingly, this remark is followed by a defense of the existence of the ether: "You are right in what you say only because you do not wish to hear of an ether at all. This view may eventually be preferable to the earlier one, but it is not the only possible view."[11]

Einstein took Lorentz's objections seriously. The first part of his reply is devoted to Lorentz's mathematical difficulty; it is followed by a lengthy discussion of general covariance. Einstein gives two reasons why general (nonlinear) coordinate transformations should be allowed in physics. The first reason is a physical one: the principle of equivalence demands the admissibility of such transformations. The second reason has, in Einstein's words, an epistemological character. He claims that singling out a particular coordinate system as preferable over all others is arbitrary and therefore undesirable, since one can never give a valid empirical (physical) justification for it. In Einstein's words: "A world-picture that dispenses with such arbitrary choices is in my view preferable."[12] Einstein admits that the restricted covariance of his theory does in fact imply a distinction between various coordinate systems, but the difference is that his "choice of coordinates does not presuppose anything of a physical kind about the world."[13]

## 3. The Final Version of the General Theory of Relativity

At the end of November 1915, Einstein submitted the paper that contained the final, generally covariant form of the general theory of relativity (Einstein 1915d).[14] A month later, both Lorentz and Paul Ehrenfest (Lorentz's successor in Leiden) had already gone deeply into the theory and were exchanging letters on the difficulties they encountered. They also both corresponded with Einstein; although none of the letters they wrote to Einstein during this period has been preserved, the particular "triangular" character of the correspondence allows a partial reconstruction of its contents.[15]

Not surprisingly, the difficulties Lorentz and Ehrenfest struggled with had to do with general covariance. Ehrenfest noticed that the core of the theory lay in two sets of equations: the field equations, and the law of conservation of energy-momentum, which had been postulated separately. He wondered whether one could eliminate the energy-momentum tensor from these two sets and derive an equation that contains the metric tensor only, and he asked whether such an equation would restrict the possible forms of the metric tensor and thus define one or more preferred coordinate systems.[16] Ehrenfest subsequently succeeded in eliminating the energy-momentum tensor, but neither he nor Lorentz could determine the implications of the equation derived in this way.[17] We now know that all metric tensors satisfy Ehrenfest's equation: he had derived the contracted Bianchi identities. Einstein's first reaction was that, since the equation is generally covariant, it cannot impose any restrictions on the choice of possible coordinate systems.[18] Shortly afterwards he reached the conclusion that the relation is in fact an identity.[19] It was not until much later that Einstein and others realized how fundamental the Bianchi identities are: with their help the law of conservation of energy-momentum can be derived from the field equations instead of having to be postulated.[20]

Lorentz's problems with general covariance took the form of a somewhat puzzling objection. In a letter to Ehrenfest he reported on having written to Einstein to ask his opinion on a problem that he had encountered and that he formulated as follows (the equations (A) he refers to are the field equations):

> I confine myself to the "matter-free" field.... From the circumstance that the equation (A)... is covariant with respect to certain substitutions, it follows that from one solution I can deduce others. If, e.g., I have the solution $g_{\mu\nu} = F(x_\alpha)$ (symbolically expressed), and replace $x_\alpha$ by $x'_\alpha$, then by the transformation formulas I can provide the values of $g'_{\mu\nu}$. I can express them in $x'_\alpha$; suppose $g'_{\mu\nu} = F'(x'_\alpha)$. Then $g_{\mu\nu} = F'(x_\alpha)$ will also satisfy equations (A). This is a new solution, differing from the first.[21]

Lorentz then actually constructs such a new solution, starting from a given one, and shows that it is physically different: in the original case, particles move along straight lines, whereas this is not the case for the new solution. His conclusion is:

> From the above it follows, it seems to me, that in the case we are considering of the matter-free field the equations (A) together with continuity and the conditions at infinity are insufficient to determine the field; in contrast with Laplace's equation $\Delta\varphi = 0$, which in view of the subsidiary conditions requires that $\varphi = 0$.[22]

What is puzzling about this objection is that Lorentz here essentially repeats Einstein's "hole argument." Was he aware of this? The phrasing of his letter to Ehrenfest suggests he was not. It also seems unlikely for another reason: If Lorentz had been aware of it, it should have become clear to him that Einstein obviously no longer believed in the "hole argument," since the theory of November 1916 is generally covariant. In any case, one thing is clear: Lorentz still questioned the need for general covariance. Neither Einstein's papers, nor a recent letter in which Einstein explained why he had returned to general covariance had been able to take away his doubts.[23]

The day after Lorentz had written to Einstein, however, he read a letter from Einstein to Ehrenfest that cleared everything up.[24] In his letter, Einstein defended general covariance by pointing out that the only essential elements in physics are coincidences in space-time; coordinates are only of secondary importance. The gravitational field does not have to be uniquely determined, as long as all coincidences, such as the formation of a black spot at a certain point on a photographic plate, are described correctly. This argument quickly convinced Lorentz. As he wrote to Ehrenfest: "I had read only a part of it [i.e., Einstein's letter] when it dawned on me and I saw that he was entirely right. I wrote to him straight away to retract my objections of yesterday."[25] And at the end of the letter he wrote: "I have congratulated Einstein on his brilliant results."[26]

Now that the main obstacle was out of the way, Lorentz wasted no time; during the following months he wrote a series of papers in which he formulated a variational principle for the general theory of relativity and developed the theory on the basis of this principle (Lorentz 1916). Although Einstein himself (Einstein 1915a), and Hilbert (Hilbert 1915), had used variational principles before, Lorentz took a somewhat different approach, which is much more geometrical in nature. One needs much geometrical intuition to follow Lorentz's reasoning; this is perhaps the reason why very few people have used his approach.[27]

The problem of general covariance might have been settled, but Lorentz's ideas about the existence of an ether had not been shaken. For him, admitting general coordinate transformations did not mean that all coordinate systems were fully equivalent. The possibility always remained to choose a preferred coordinate system, which one might then think of as being connected to the "ether." In a letter to Einstein, written in June 1916, Lorentz clearly stated his point of view.[28] He started by describing a "fictional" experiment: in two closed wires that run around the earth along the equator, electromagnetic waves are generated in such a way that the waves in the two wires run in opposite directions.[29] In a coordinate system fixed to the earth the waves propagate with different speeds in the two wires; in a system in which the speeds are equal the earth performs a rotation. A convenient way to describe this phenomenon, Lorentz points out, is to introduce an ether as carrier of the waves. He then went on:

> If we adopt this standpoint, we may say that the experiment has shown us the motion of the earth relative to the ether. If, then, we have thereby acknowledged the possibility of establishing a relative *rotation*, we should not reject in advance the possibility of also obtaining indications of a relative *translation*, i.e., we should not set up the basic principle of relativity theory as a *postulate*. We would need, rather, ... to seek the answer to the question in observation.[30]

According to Lorentz, the relativity principle is a hypothesis, framed on the basis of experimental results, and always open to refutation.

It is worthwhile to analyze Lorentz's argument a little more closely. For Lorentz, the existence of physical effects due to accelerated motion showed that these motions have an "absolute" character, where "absolute" has to be understood as relative to the ether.[31] From this it follows, although it is not explicitly stated, that uniform translations are also "absolute." This argument shows a striking resemblance to the argument Newton gave in the "Scholium" on space and time in Book 1 of the *Principia*.[32] Like Lorentz, Newton first described physical effects due to accelerated motion—in this case the dynamical effects that can be observed in a rotating bucket filled with water. From their occurrence he then infers the existence of absolute space and the absolute character of both accelerated and uniform translatory motion.[33]

Not surprisingly, Einstein was not convinced by Lorentz's reasoning. In his reply he admits that the general theory of relativity is closer to an ether hypothesis than the special theory.[34] But the "ether" he refers to is the metric field, which is something different from the immobile "substantial" ether Lorentz has in mind.[35] As a consequence, one can distinguish between accelerated and nonaccelerated motion: in a part of space where $g_{\mu\nu}$ = constant, a linear coordinate transformation (corresponding to nonaccelerated motion) has no influence on $g_{\mu\nu}$, whereas nonlinear transformations (accelerated motion) change $g_{\mu\nu}$. Thus nonaccelerated motion produces no changes in the gravitational field and cannot be detected.

## 4. The Later Years

In the following years Lorentz inspired several of his students and former students to work in the field of general relativity[36] and made some further contributions himself (Lorentz and Droste 1917; Lorentz 1923). Though he kept insisting on the existence of an ether, he was not dogmatic about it and on many occasions expressed his admiration for Einstein's achievements. His attitude is very clearly illustrated by the statement with which he concluded a series of lectures given at the California Institute of Technology in 1922:

> As to the ether (to return to it once more), though the conception of it has certain advantages, it must be admitted that if Einstein had maintained it he certainly would not have given us his theory, and so we are very grateful to him for not having gone along the old-fashioned roads. (Lorentz 1927, p. 221)

Why Lorentz kept insisting on the existence of an ether is a question that is not so easy to answer. His attitude may show a certain conservatism, perhaps even stubbornness. But it should be kept in mind that from the earliest years of Lorentz's career the concept of an ether had played a fundamental role in his work on electromagnetic theory. Lorentz's idea of a separation between ether and matter, formulated for the first time in the dissertation (Lorentz 1875) and worked out during the following decades, had proven to be immensely fruitful for the development and clarification of electromagnetism.[37] The concept of the ether must have been very dear to Lorentz, and it does not seem to be out of character for a man like him to remain true to it to the very end.[38]

*Acknowledgments.* This work was financially supported by the Netherlands organization for scientific research (N.W.O.). I am grateful to John Stachel, Leendert Suttorp, Leo van den Horn, Henriette Schatz, and especially to Martin Klein for their comments on an earlier version of this paper. Einstein's letters are quoted with the permission of the Hebrew University of Jerusalem.

## Notes

\* Reprinted from *Archive for History of Exact Sciences*, Volume 38, Number 1. © 1988 by Springer-Verlag with permission.

[1] Lorentz 1909, p. 230. There are many more examples of Lorentz expressing himself in this way. For a discussion of Lorentz's electron theory, see Goldberg 1969, Hirosige 1966, 1969, McCormmach 1970, 1973, and Schaffner 1969. In these notes, I use the following abbreviations:

EAL  Ehrenfest Archive. Museum Boerhaave, Leiden, The Netherlands.
ECL  Einstein Collection. Museum Boerhaave, Leiden, The Netherlands.
LAH  Lorentz Archive, Algemeen Rijksarchief, The Hague, The Netherlands.

[2] Lorentz lectured on general relativity from March to June 1916. Among his audience were Paul Ehrenfest and Willem de Sitter; the latter played a crucial role in making the theory known in England through his papers in the *Monthly Notices of the Royal Astronomical Society*.

[3] The paper first appeared as a separatum, Einstein and Grossmann 1913, and then, with some additional remarks, in a journal, Einstein and Grossmann 1914.

[4] LAH 269, pp. 188–201.

[5] LAH 270, p. 65; Einstein to Lorentz, August 14, 1913, and August 16, 1913 (LAH 21).

[6] Einstein used the argument that a symmetric tensor expresses the equivalence of energy and mass in the simplest way. Einstein to Lorentz, August 16, 1913 (LAH 21).

[7] Lorentz to Einstein, January 1915 (draft) (LAH 286); Einstein to Lorentz, January 23, 1915 (LAH 21).

[8] A detailed analysis of the "hole argument" and the role it played in the development of Einstein's thought can be found in Norton 1984.

[9] "Wir können uns vorstellen, man sei eine Zeit lang nur im Besitz der Gleichungen (2) gewesen und habe sich mit einer 'Deutung' der Glieder $2\omega\, dy'/dt, \omega^2 x'$ u.s.w. gequält. Käme dann einer, der durch Einführung des Koordinatensystems I die Gleichungen (2) auf (1) zurückführt, so würde ein jeder das als eine wirkliche Erlösung begrüssen, und jeder würde das System I vorziehen." Lorentz to Einstein, January 1915 (draft) (LAH 286).

[10] "Gehen Sie hier nicht etwas zu weit, indem Sie eine persönliche Auffassung als selbstverständlich hinstellen?" Lorentz to Einstein, January 1915 (draft).

[11] "Sie haben mit Ihrer Bemerkung nur recht, weil Sie von einem Äther überhaupt nicht wissen wollen. Diese Auffassung mag am Ende der Früheren vorzuziehen sein, aber sie ist doch nicht die einzig mögliche." Lorentz to Einstein, January 1915 (draft).

[12] "Ein Weltbild, welches ohne eine derartige Willkür auskommt ist nach meiner Meinung vorzuziehen." Einstein to Lorentz, January 23, 1915 (LAH 21).

[13] "... Koordinatenwahl physikalisch nichts über die Welt voraussetzt." Einstein to Lorentz, January 23, 1915.

[14] This was the last of a series of four papers on general relativity, all published in November 1915 (Einstein 1915a–1915d).

[15] The extant letters exchanged between Lorentz, Ehrenfest, and Einstein during this period are: Lorentz to Ehrenfest December 23, 1915, December 26, 1915, January 9, 1916, January 10–11, 1916, January 12, 1916, January 18, 1916, January 22, 1916, January 28, 1916, February 2, 1916 (all EAL); Ehrenfest to Lorentz December 23, 1915, December 24, 1915, January 9, 1916, January 12–13, 1916, January 25, 1916 (all LAH

20); Einstein to Lorentz January 1, 1916, January 17, 1916, January 19, 1916 (all LAH 21); Einstein to Ehrenfest December 26, 1915, December 29, 1915, January 3, 1916, January 5, 1916, January 17, 1916, undated (winter 1916) (all EAL).

[16] Ehrenfest to Lorentz, December 23, 1915 (LAH 20).

[17] Ehrenfest to Lorentz, December 24, 1915 (LAH 20); Lorentz to Ehrenfest, December 26, 1915 (EAL).

[18] Einstein to Ehrenfest, December 29, 1915 (EAL).

[19] Einstein to Ehrenfest, January 3, 1916 (EAL).

[20] See, e.g., Pais 1982, pp. 274–278.

[21] "Ik bepaal mij tot het 'materievrije' veld.... Uit de omstandigheid dat de vergelijking (A) ... tegenover zekere substituties covariant is, volgt dat ik uit ééne oplossing andere kan afleiden. Heb ik b.v. een oplossing $g_{\mu\nu} = F(x_\alpha)$ (symbolisch voorgesteld), en voer ik in plaats van $x_\alpha$ $x'_\alpha$ in, dan kan ik door de transformatieformules de waarden der $g'_{\mu\nu}$ aangeven. Ik kan die in $x'_\alpha$ uitdrukken; stel $g'_{\mu\nu} = F'(x'_\alpha)$. Dan zal ook $g_{\mu\nu} = F'(x_\alpha)$ aan de vergelijkingen (A) voldoen. Dit is een nieuwe van de eerste verschillende oplossing." Lorentz to Ehrenfest, January 9, 1916 (EAL).

[22] "Mij dunkt dat uit het bovenstaande volgt dat in het beschouwde geval van het materievrije veld de vergelijkingen (A) met de doorlopendheid en de voorwaarden in het oneindige niet voldoende zijn om het veld te bepalen; in tegenstelling met de verg. van Laplace $\Delta\varphi = 0$, die in verband met de bijkomstige voorwaarden eischt dat $\varphi = 0$ is." Lorentz to Ehrenfest, January 9, 1916.

[23] Einstein to Lorentz, January 1, 1916 (LAH 21). In this letter Einstein gave three arguments why he had returned to general covariance: 1. The perihelion motion of Mercury came out too small; 2. The equations were not covariant for transformations corresponding to uniform rotation; 3. The Lagrangian could be chosen entirely arbitrarily. He does not mention the "hole argument."

[24] Einstein to Ehrenfest, January 5, 1916 (EAL). It was enclosed in Ehrenfest to Lorentz, January 9, 1916 (LAH 20).

[25] "Ik had nog maar een gedeelte daarvan gelezen toen mij een licht opging en ik zag dat hij geheel gelijk heeft. Ik heb hem aanstonds geschreven om mijne bedenkingen van gisteren te herroepen." Lorentz to Ehrenfest, January 10–11, 1916 (EAL).

[26] "Ik heb Einstein met zijne schitterende uitkomsten gelukgewenscht." Lorentz to Ehrenfest, January 10–11, 1916.

[27] See, e.g., Fokker 1929.

[28] Lorentz to Einstein, June 6, 1916 (ECL).

[29] Although in his letter Lorentz refers to an experiment performed by Ernst Lecher, his "experiment" resembles one carried out by Georges Sagnac. See Lecher 1890 and Sagnac 1914.

[30] "Stellen wir uns auf diesen Standpunkt, so können wir sagen, der Versuch habe uns die relative Bewegung der Erde gegen den Äther gezeigt. Haben wir dann in dieser Weise die Möglichkeit anerkannt, eine relative *Rotation* zu konstatieren, so dürfen wir nicht von vornherein die Möglichkeit leugnen, auch Andeutungen einer relativen *Translation* zu erhalten, d.h. wir dürfen den Grundsatz der Relativitätstheorie nicht als *Postulat* hinstellen. Wir müssen vielmehr ... die Beantwortung der Frage in den Beobachtungen suchen." Lorentz to Einstein, June 6, 1916 (ECL).

[31] It should be emphasized that Lorentz did not adhere to the idea of absolute space. In Lorentz 1895, sec. 2, for instance, he states that it is meaningless to talk about absolute rest of the ether and that the expression "the ether is at rest" only means that the different parts of the ether do not move with respect to each other.

[32] See Stein 1977 for a discussion of Newton's argument and a critical review of later developments.

[33] There is another instance where Lorentz's reasoning makes one think of Newton. In Lorentz to Einstein, January 1915 (draft) (LAH 286), Lorentz introduces a "world-spirit" ("Weltgeist") that permeates a physical system, without being tied to a particular place. According to Lorentz, such a spirit could directly "feel" all events occurring in the system under consideration, and would therefore be able to single out a preferred coordinate system. In Query 28 of his *Opticks*, Newton formulates the following idea: "... does it not appear from Phaenomena that there is a Being incorporeal, living, intelligent, omnipresent, who in infinite Space, as it were in his Sensory, sees the things themselves intimately, and throughly perceives them...."

[34] Einstein to Lorentz, June 17, 1916 (LAH 21).

[35] This idea was developed in more detail in a lecture Einstein gave in Leiden on October 27, 1920. See Einstein 1920.

[36] E.g., A.D. Fokker and J. Droste.

[37] See, e.g., Hirosige 1966, 1969. Lorentz very consciously separated ether and matter, as becomes clear from an undated unpublished document (LAH 264), in which he states that he made a distinction between ether and matter in the hope that this would facilitate the treatment of light propagation in bodies that move through the ether without dragging it along.

[38] For a very interesting analysis of Lorentz's personality, see McCormmach 1973, p. 490. McCormmach emphasizes that Lorentz was able to make a fair and critical judgement of the work of others, while in his own work he showed strong preferences for certain approaches.

NOTES ADDED IN PROOF:

1. A forthcoming very interesting paper by József Illy ("Einstein Teaches Lorentz, Lorentz Teaches Einstein: Their Collaboration in General Relativity, 1913–1920." *Archive for History of Exact Sciences* 39 (1989): 247–289) treats many of the same topics discussed here.

2. An alternative explanation of Lorentz's unwillingness to dispense with the ether (as well as an overview of the history of the electron theory) is given by Nancy J. Nersessian ("Why Wasn't Lorentz Einstein? An Examination of the Scientific Method of H.A. Lorentz." *Centaurus* 29 (1986): 205–242).

REFERENCES

Einstein, Albert (1914). "Die formale Grundlage der allgemeinen Relativitätstheorie." *Königlich Preussische Akademie der Wissenschaften* (Berlin). *Sitzungsberichte*: 1030–1085 (submitted October 29, 1914).

––––––– (1915a). "Zur allgemeinen Relativitätstheorie." *Königlich Preussische Akademie der Wissenschaften* (Berlin). *Sitzungsberichte*: 778–786 (submitted November 4, 1915).

––––––– (1915b). "Zur allgemeinen Relativitätstheorie (Nachtrag)." *Königlich Preussische Akademie der Wissenschaften* (Berlin). *Sitzungsberichte*: 799–801 (submitted November 11, 1915).

——— (1915c). "Erklärung der Perihelbewegung des Merkur aus der allgemeinen Relativitätstheorie." *Königlich Preussische Akademie der Wissenschaften* (Berlin). *Sitzungsberichte*: 831–839 (submitted November 18, 1915).

——— (1915d). "Die Feldgleichungen der Gravitation." *Königlich Preussische Akademie der Wissenschaften* (Berlin). *Sitzungsberichte*: 844–847 (submitted November 25, 1915).

——— (1920). *Äther und Relativitätstheorie. Rede gehalten am 5. Mai 1920 an der Reichs-Universität zu Leiden.* Berlin: Julius Springer.

Einstein, Albert, and Grossmann, Marcel (1913). *Entwurf einer verallgemeinerten Relativitätstheorie und einer Theorie der Gravitation.* Leipzig: B.G. Teubner.

——— (1914). "Entwurf einer verallgemeinerten Relativitätstheorie und einer Theorie der Gravitation." *Zeitschrift für Mathematik und Physik* 62: 225–261.

Fokker, Adriaan D. (1929). *Relativiteitstheorie.* Groningen: Noordhoff.

Goldberg, Stanley (1969). "The Lorentz Theory of Electrons and Einstein's Theory of Relativity." *American Journal of Physics* 37: 982–994.

Hilbert, David (1915). "Die Grundlagen der Physik. (Erste Mitteilung)." *Königliche Gesellschaft der Wissenschaften zu Göttingen. Mathematisch-physikalische Klasse. Nachrichten*: 395–407 (submitted November 20, 1915).

Hirosige, Tetu (1966). "Electrodynamics Before the Theory of Relativity, 1890–1905." *Japanese Studies in the History of Science* 5: 1–49.

——— (1969). "Origins of Lorentz' Theory of Electrons and the Concept of the Electromagnetic Field." *Historical Studies in the Physical Sciences* 1: 151–209.

Lecher, Ernst (1890). "Eine Studie über electrische Resonanzerscheinungen." *Annalen der Physik und Chemie* 41: 850–870.

Lorentz, Hendrik Antoon (1875). *Over de theorie der terugkaatsing en breking van het licht.* Arnhem: Van der Zande. Reprinted in Lorentz 1934–1939, vol. 1, pp. 1–192

——— (1895). *Versuch einer Theorie der electrischen und optischen Erscheinungen in bewegten Körpen.* Leiden: E.J. Brill. Reprinted in Lorentz 1934–1939, vol. 5, pp. 1–138.

——— (1909). *The Theory of Electrons.* Leipzig: B.G. Teubner.

——— (1915). "Het beginsel van Hamilton in Einstein's theorie der zwaartekracht." *Verslagen van de Gewone Vergaderingen der Wis- en Natuurkundige Afdeeling, Koninklijke Akademie van Wetenschappen te Amsterdam* 23 (1914–1915): 1073–1089 (submitted January 30, 1915). English translation *Proceedings of the Section of Sciences, Koninklijke Akademie van Wetenschappen te Amsterdam* 19 (1916–1917): 751–765. Translation reprinted in Lorentz 1934–1939, vol. 5, pp. 229–245.

——— (1916). "Over Einstein's theorie der zwaartekracht, I–IV." *Verslagen van de Gewone Vergaderingen der Wis- en Natuurkundige Afdeeling, Koninklijke Akademie van Wetenschappen te Amsterdam* 24 (1915–1916): 1389–1402 (submitted February 26, 1916); 1759–1774 (submitted March 25, 1916); 25 (1916–1917): 468–486 (submitted April 28, 1916); 1380–1396 (submitted October 28, 1916). English translation *Proceedings of the Section of Sciences, Koninklijke Akademie van Wetenschappen te Amsterdam* 19 (1916–1917): 1341–1354; 1354–1369; 20 (1917–1918): 2–19; 20–34. Translation reprinted in Lorentz 1934–1939, vol. 5, pp. 246–313.

——— (1923). "De bepaling van het $g$-veld in de algemeene relativiteitstheorie met behulp van de wereldijnen van lichtsignalen en stoffelijke punten, met eenige opmerkingen over de lengte van staven en den duur van tijdsintervallen en over de theorieën van Weyl en Eddington." *Verslagen van de Gewone Vergaderingen der Wis-*

*en Natuurkundige Afdeeling, Koninklijke Akademie van Wetenschappen te Amsterdam* 32 (1923): 383–402 (submitted March 24, 1923). English translation *Proceedings of the Section of Sciences, Koninklijke Akademie van Wetenschappen te Amsterdam* 29 (1926): 383–399. Translation reprinted in Lorentz 1934–1939, vol. 5, pp. 363–382.

——— (1927). *Problems of Modern Physics.* Boston: Ginn. Reprint New York: Dover, 1967.

——— (1934–1939). *Collected Papers*, 9 vols. The Hague: Nijhoff.

Lorentz, Hendrik Antoon, and Droste, J. (1917). "De beweging van een stelsel lichamen onder den invloed van hunne onderlinge aantrekking, behandeld volgens de theorie van Einstein, I–II." *Verslagen van de Gewone Vergaderingen der Wis- en Natuurkundige Afdeeling, Koninklijke Akademie van Wetenschappen te Amsterdam* 26 (1917–1918): 392–403, 649–660. English translation in Lorentz 1934–1939, vol. 5, pp. 330–355.

*Studies in the Physical Sciences* 2: 41–87.

——— (1973). "Lorentz, Hendrik Antoon." *Dictionary of Scientific Biography.* C.C. Gillispie, ed. Vol. 8. New York: Charles Scribner's, pp. 487–500.

Norton, John (1984). "How Einstein Found His Field Equations: 1912–1915." *Historical Studies in the Physical Sciences* 14: 253–316. See this volume, pp. 101–159.

Pais, Abraham (1982). *"Subtle is the Lord ...": The Science and the Life of Albert Einstein.* Oxford: Clarendon Press; New York: Oxford University Press.

Sagnac, Georges (1914). "Effet tourbillonnaire optique. La circulation de l'éther lumineux dans un interférographe tournant." *Journal de Physique* 4: 177–195.

Schaffner, Kenneth F. (1969). "The Lorentz Electron Theory of Relativity." *American Journal of Physics* 37: 498–513.

Stein, Howard (1977). "Some Philosophical Prehistory of General Relativity." In *Foundations of Space-Time Theories.* Minnesota Studies in the Philosophy of Science, vol. 8. John Earman et al., eds. Minneapolis: University of Minnesota Press, pp. 3–49.

# The Early Interpretation of the Schwarzschild Solution

JEAN EISENSTAEDT

On January 9, 1916, Albert Einstein wrote to Karl Schwarzschild:

> I have read your paper with the utmost interest. I had not expected that one could formulate the exact solution of the problem in such a simple way. I liked very much your mathematical treatment of the subject. Next Thursday I shall present the work to the Academy with a few words of explanation. [EA 21-516][1]

This letter concerned of course what has since been called the Schwarzschild solution. The solution represents the gravitational field of a single center in Einstein's theory of gravitation. Schwarzschild was then serving at the front in Russia. There he developed symptoms of a rare and painful skin disease, pemphigus. He died just over seventy years ago, on May 11, 1916.

Schwarzschild's interests were extremely broad. They ranged from celestial mechanics (he wrote his first paper on that topic at the age of sixteen) to observational stellar photometry (one of his most important contributions) and quantum theory (his last paper), and included a great concern for instrumental astronomy. He worked on the theory of stellar structure, on stellar statistics, on Halley's comet in the 1910s, on spectroscopy, and of course on general relativity. "In fact, the many-sidedness of Schwarzschild was astonishing," wrote Ejnar Hertzsprung in the obituary notice he devoted to him, adding, "His high mathematical capacity, so uncommon in the practical astronomer, made him treat the most difficult problem of our science with startling virtuosity from a theoretical point of view" (Hertzsprung 1917).[2]

Schwarzschild published four papers dealing with space curvature or general relativity. The first one, "On the Admissible Curvature of the Universe," was published in 1900 (Schwarzschild 1900). He discussed there the possibility that the geometry of space might be non-Euclidean and suggested two kinds of possible curvatures: elliptic and hyperbolic. He evaluated from available astronomical evidence the value of the curvature of the universe. In 1914 he attempted—unsuccessfully—to observe a gravitational redshift in the spectrum of the sun (Schwarzschild 1914) that had been predicted by Einstein in his 1911 paper in the *Annalen der Physik* (Einstein 1911). During the war Schwarzschild volunteered for military service and was assigned to the headquarters staff of an artillery unit to calculate trajectories for long-range shells. Then, from December 1915 to February 1916, he worked out the two papers in which we are interested.

The Schwarzschild solution was the first and one of the very few exact solutions to be found in the early days of general relativity. Over the next seventy years, thousands of articles, theses, and dissertations were dedicated to it. The gravitational field of a point-mass in empty space is one of the most important models, the main exact solution of Einstein's theory of gravitation. Two of the three classical tests depend heavily on it: to wit, the advance of the perihelion of Mercury and the deflection of light in a gravitational field. But, in fact, the Schwarzschild solution contains much more than a mere generalization of Kepler's law. It exhibits a singular behavior that is odd, strange, and exotic indeed, on which experts would construct, fifty years later, what we now call a "black hole" (so named in Wheeler 1968). But at first, during that long period when relativity was finding its way, this singular behavior of the solution was named the "Schwarzschild singularity," the "magic circle," the "hole in space," or the "Hadamard disaster."[3]

What was the status of such a curious object? How did the experts deal with it? How did the relativists cope with the so-called Schwarzschild singularity? For the most part, I will concentrate here on what I have called the pragmatic view or interpretation of the Schwarzschild solution, which prevailed for fifty years, until the renewal of interest in the 1960s. It represents the dominant interpretation, the standard view of the subject, something like the normal way to treat the problem, although it is not to be thought of as an organized and formalized way to deal with the subject. It has strong points of course, but also some weak points—sometimes erroneous—which leave unanswered questions. At a deeper level, I believe that the pragmatic interpretation of the Schwarzschild solution represents—at least in part—the projection in this particular field of what can be called the "neo-Newtonian" interpretation of general relativity, which prevailed during all these years.[4] But aside from that main topic, I would like to describe and better understand the "everyday life" of general relativity, the way its interpretation was slowly built up.

In the beginning of November 1915, Einstein came back to the principle of general covariance which he had abandoned with "such a heavy heart" in 1913.[5] But, for subtle reasons, he had at his disposal only the restricted field equations:

$$R_v^\mu = \kappa T_v^\mu,$$

where

$$R_{\mu\nu} = -\left\{ \begin{matrix} \rho \\ \mu\nu \end{matrix} \right\}_{,\rho} + \left\{ \begin{matrix} \rho \\ \mu\sigma \end{matrix} \right\} \left\{ \begin{matrix} \sigma \\ \nu\rho \end{matrix} \right\},$$

whose covariance is restricted to systems of coordinates such that $|g| = 1$. Anyway, these field equations are satisfactory outside matter, and Einstein used them in the famous article he published on November 25 to explain the perihelion advance of Mercury (Einstein 1915). To obtain the approximate gravitational field of a spherically symmetrical point-mass in a vacuum, he imposed the following conditions:

—That the determinant $|g|$ be equal to one
—The spatial symmetry of the problem
—The static character of the solution
—The asymptotically Minkowskian character of the line-element

In contrast to Einstein—who used a rectangular system of coordinates—Schwarzschild used quite an unusual system of "pseudo-polar" coordinates, $x_1, x_2, x_3$, and $x_4$, in order to fulfill the condition $|g| = 1$. He then worked out his solution, which assumed the following form:

$$ds^2 = f_4(x_1)\,dx_4^2 - f_1(x_1)\,dx_1^2 - f_2(x_1)\left\{\frac{dx_2^2}{1-x_2^2} + (1-x_2^2)\,dx_3^2\right\},$$

where

$$f_1 = \frac{(3x_1 + \beta)^{-4/3}}{1 - \alpha(3x_1 + \beta)^{-1/3}}, \qquad f_2 = (3x_1 + \beta)^{2/3},$$

$$f_4 = 1 - \alpha(3x_1 + \beta)^{-1/3}, \tag{1}$$

and where both $\alpha$ and $\beta$ were both thought of as "constants of integration" (Schwarzschild 1916a).[6] To Einstein's four conditions, Schwarzschild added a condition of continuity of the gravitational potentials everywhere but at $x_1 = 0$, thus getting $\beta = \alpha^3$ (but not $\beta = 0$!). He then expressed his solution in the standard polar coordinates $R, \theta, \phi$ (with $x_1 = R^3/3$, $x_2 = -\cos\theta$, $x_3 = \phi$, $x_4 = ct$), and obtained

$$ds^2 = \left(1 - \frac{\alpha}{r}\right)c^2\,dt^2 - \frac{dr^2}{1 - \frac{\alpha}{r}} - r^2(d\theta^2 + \sin^2\theta\,d\phi^2), \tag{2}$$

where

$$r = (R^3 + \alpha^3)^{1/3}; \quad \left(\text{with } \alpha = \frac{2Gm}{c^2}\right).$$

He commented on his results in the following terms: "There is no discontinuity at the null point but at $R = (\alpha^3 - \beta)^{1/3}$ so that we must choose $\beta = \alpha^3$ for the discontinuity to be at the null point" (Schwarzschild 1916a, p. 195).[7]

Thus, the "Schwarzschild method," as De Donder called it, provided a first response, something of a primary topological interpretation, to the problem of the discontinuity on the two-sphere $r = \alpha$, $(R = 0)$, the so-called Schwarzschild singularity that appears in the line-element. It was in fact obvious to Schwarzschild that the "true" radial coordinate was not $r$ but $R$, but he used $r$ for easier geodesic calculations only after having noted that $\alpha/r$ is of the order of $10^{-12}$ (for Mercury) so that $r$ and $R$ are almost identical. The Schwarzschild method was used by the experts in the field up to the 1920s (and sometimes, but rarely, later): by the Belgian school (De Donder and his students); by Rice in England; by Mie in Germany; by Straneo in Italy; and even by Schrödinger in Vienna. More recently Abrams has "resurrected" it.[8] Its main virtue lies in its implicit topology: the space-like sections are the usual ones restricted by the condition $r > 2Gm/c^2$. The Schwarzschild singularity, the "horizon" as we now call it, ends the space.

Two months later, Schwarzschild published his interior solution based, of course, on the final field equations, in which the matter is described by a fluid sphere of constant density and radius $r = a$. He noticed that, in his model, the pressure becomes infinite at the center, as soon as what is now called the Schwarzschild limit, $2Gm/c^2 = 8/9$, is reached. During the Paris conference in 1922, at the Collège de France, Jacques Hadamard, the French mathematician, raised the problem of the Schwarzschild singularity. According to Charles Nordmann, a French astronomer, Einstein's reply was as follows:

> If that term could actually vanish somewhere in the universe, it would be a true disaster for the theory; and it is very difficult to say *a priori* what could happen physically because the formula does not apply any more. (Nordmann 1922, p. 156)[9]

This is what Einstein later jokingly named the "Hadamard disaster." But on the following day he came back with the result of a calculation, which certainly refers to the Schwarzschild limit of the interior solution,[10] showing that the Schwarzschild singularity cannot be reached.

In May 1916, a few months after Schwarzschild, Johannes Droste published independently a much simpler derivation of the exterior solution. He was then completing a thesis with H.A. Lorentz in Leyden, the best center of relativity in Einstein's opinion (Pyenson 1975). Two years before, on the basis of Einstein's and Grossmann's "*Entwurf*" theory (Einstein and Grossmann 1913), he had already worked on the same problem (Droste 1915). But he now had at his disposal the final form of the field equations of general relativity. Without using any "determinant condition," and thus keeping the general covariance of the problem, he obtained the well-known form of the Schwarzschild solution:[11]

$$ds^2 = \left(1 - \frac{\alpha}{r}\right)c^2 dt^2 - \frac{dr^2}{1 - \frac{\alpha}{r}} - r^2(d\theta^2 + \sin^2\theta \, d\phi^2), \qquad (3)$$

(with $\alpha = 2GM/c^2$), which should rightly be called the "Schwarzschild solution in Droste's coordinates." In his thesis, Droste expressed the solution in many other systems of coordinates, for example, the well-known isotropic system:

$$ds^2 = \left(\frac{\rho - \frac{mG}{2c^2}}{\rho + \frac{mG}{2c^2}}\right)c^2 dt^2 - \left(1 + \frac{mG}{2\rho c^2}\right)^4 \{d\rho^2 + \rho^2(d\theta^2 + \sin^2\theta \, d\phi^2)\}. \qquad (4)$$

By means of the transformation $r = \tilde{r} + \alpha$ he also wrote the line-element in the following form, in which "the discontinuity goes to zero," as he noted:

$$ds^2 = \frac{c^2 dt^2}{1 + \frac{\alpha}{\tilde{r}}} - \left(1 + \frac{\alpha}{\tilde{r}}\right) d\tilde{r}^2 - (\tilde{r} + \alpha)^2 (d\theta^2 + \sin^2\theta \, d\phi^2). \tag{5}$$

Droste then asked:

> To what formula shall we give the preference, to that of Schwarzschild [2], to [3] or to [5]? It is in fact a matter of personal convenience. But we must remember that the $r$ coordinate doesn't represent the measured interval. We are however free to choose any coordinate (provided that all points may be reached) but some choice of coordinates may appear more appropriate than another. (Droste 1916b, p. 20)[12]

In practice, he would restrict himself—as many other experts would do—to spatial transformations of coordinates. Concerning the problem of the Schwarzschild singularity, Droste's attitude was simply to "discard the region $r < \alpha$" (Droste 1916b, p. 201). Droste's coordinates would play an outstanding role up to the end of the 1950s. They would be the main tool, the fundamental reference of the most current approaches dealing with spherical symmetry, and this in spite of the principle of covariance, the true spirit of general relativity.

In *Raum-Zeit-Materie* (first published in 1918), Hermann Weyl derived the Schwarzschild solution in Droste's coordinates as every expert did. And he added: "The gravitational radius is about 1.47 kilometres for the sun's mass and only 5 millimetres for the earth" (Weyl 1921, p. 256).[13] An argument raised by every author, in every textbook, in every article on the Schwarzschild solution, is that the Schwarzschild singularity cannot exist in nature, that it remains virtual.[14] That is the core of what we will call the "pragmatic interpretation of the Schwarzschild solution."

The representation due to Ludwig Flamm (1916) consists of embedding the spatial part of the Schwarzschild solution in a Euclidean space of four dimensions:

$$ds^2 = \frac{dr^2}{1 - \frac{2Gm}{rc^2}} + r^2 d\theta^2 = dx^2 + dy^2 + dz^2,$$

where:

$$x = r \sin\theta,$$
$$y = r \cos\theta,$$
$$z = \int_0^r \frac{dr}{\sqrt{\frac{rc^2}{2mG} - 1}} = \sqrt{\frac{8mG}{c^2}\left(r - \frac{2mG}{c^2}\right)}, \tag{6}$$

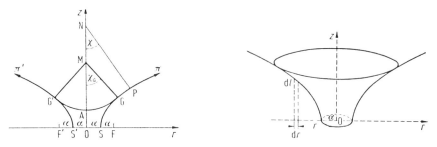

FIGURE 1. The Flamm representation according to L. Flamm and J. Becquerel.

and where $\phi$ has been omitted for simplicity. The representation then is given in Figure 1.

But as is obvious from equation (6), such a representation supposes implicitly that $r > 2m$. Nevertheless, some experts, Max von Laue, for example, used it as an argument showing the singular character of the Schwarzschild sphere. Von Laue wrote that "the singularity of the line-element [3] for $r = 2m$ cannot be eliminated by using any other coordinates; it is essential to the *nature of the thing*" (von Laue 1921, p. 215).[15] And then, to support his nearly metaphysical point of view, he provided the ordinary empirical argument: "Every mass $m$ ... has a radius greater than $2Gm/c^2$." He added that "in fact we do not know as yet any counter-example even in the nucleus of atoms."[16]

Schwarzschild's method would soon be disputed. David Hilbert ended the second part of his well-known "Grundlagen der Physik" (Hilbert 1917) with a detailed account of the Schwarzschild solution. Because of the principle of general covariance there was in the newborn theory of general relativity no intrinsic definition of a singularity available. Hilbert was the first to give a possible definition of regularity: a gravitational field is said to be regular if there exists a system of coordinates in which the potentials $g_{\mu\nu}$ are continuous and differentiable, the determinant being nonzero. That is quite an interesting definition, but Hilbert added a condition of regularity on the transformation leading to such a system: it must be one-to-one and invertible (Hilbert 1917, pp. 70–71). Thus for Hilbert—but for all other experts as well—there were at that time two singularities in the Schwarzschild solution: $r = 0$ and $r = \alpha$. In any case, like Droste, Hilbert did not like Schwarzschild's method. He wrote: "From my point of view, transforming $r = \alpha$ to the origin, as Schwarzschild did is not to be recommended; in any case the Schwarzschild transformation is not the simplest one for reaching that goal" (Hilbert 1917, pp. 70–71).[17]

The first expert to cast doubt on the reality of the singular character of $r = 2m$ was Cornelius Lanczos. But in contrast to George Lemaître—who would definitely prove, ten years later, the nonsingular character of $r = 2m$—Lanczos introduced a new "singularity" at a place where the metric was completely regular before, thus showing the relativity of the then vague concept of a singularity. He started with Droste's solution (3), and via the transformation $\bar{r} = r - \alpha/2$, he got:

$$ds^2 = \left(\frac{\bar{r} - \dfrac{\alpha}{2}}{\bar{r} + \dfrac{\alpha}{2}}\right) c^2\, dt^2 - \left(\frac{\bar{r} + \dfrac{\alpha}{2}}{\bar{r} - \dfrac{\alpha}{2}}\right) d\bar{r}^2 - \left(\bar{r} + \frac{\alpha}{2}\right)^2 (d\theta^2 + \sin^2\theta\, d\phi^2). \quad (7)$$

Then, working out the expression for the determinant of the line-element corresponding to the Euclidean system of coordinates associated with the polar system $\bar{r}$, he obtained: $|g| = (1 + \alpha/2\bar{r})^4$. Consequently, at $\bar{r} = 0$ the determinant is singular, which was not the case at the corresponding point ($r = \alpha/2$) of the Droste representation. Lanczos wrote:

> This example shows how little one can infer an actual singularity of the field from the singular behavior of the functions $g_{\mu\nu}$ since it may be possible to remove the latter by a coordinate transformation. (Lanczos 1922, p. 539)[18]

At that time, however, some problems concerning the Schwarzschild solution were connected to the de Sitter cosmological solution, namely, the question of the singular/regular character of "$r = 2m$," the "Schwarzschild singularity," and of $r = (\pi/2)R$, the "de Sitter horizon." Einstein was then puzzled by the fact that the Schwarzschild solution contradicted his interpretation of the Mach principle because the mere presence of a single body was the only reason for the structure of the whole space. John Stachel has emphasized this point:

> Indeed, from Einstein's point of view it was a scandal that a solution to his field equations should exist which corresponded to the presence of a single body in an otherwise "empty" universe. (Stachel 1979, p. 440)

On January 9, 1916 Einstein wrote to Schwarzschild:

> In the final analysis, according to my theory, inertia is precisely an interaction between masses, not an "action" in which, in addition to the masses contemplated, "space" as such participates. The essential feature of my theory is just that no independent properties are attributed to space as such.
>
> One can express it this way as a joke. If all things were to disappear from the world, then, according to Newton, the Galilean inertial space remains; according to my conception, however, nothing is left. (EA 21-561)[19]

Here is to be found the origin of the so-called Einstein–de Sitter controversy,[20] as de Sitter put it in his first paper on the topic:

> To the question: If all matter is supposed not to exist, with the exception of one material point which is to be used as a test-body, has then this test-body inertia or not? The school of Mach requires the answer NO. Our experience however very decidedly gives the answer YES, if by "all matter" is meant all ordinary physical matter: stars, nebulae, clusters, etc. The followers of Mach are thus compelled to assume the existence of still more matter: the world-matter. (de Sitter 1917b, p. 1222)

In fact, de Sitter's motivation in working out his cosmological solution was to contest Einstein's views concerning that question. He thought at first that his solution was an acceptable empty cosmological solution and that Einstein's point of view was consequently erroneous. Einstein then published an article in which he attempted to prove that de Sitter's solution is not valid everywhere, in that there is some matter present on the "horizon," which therefore is singular. He concluded:

> Truly, the de Sitter system ... is indeed a solution [of the field equations] everywhere, only not on the surface $r = (\pi/2)R$. There—as in the near neighborhood of a gravitating massive point—the $g_{44}$ component of the gravitational potential tends to zero. The de Sitter system cannot by any means describe a world without matter but a world in which the matter is completely concentrated on the surface $r = (\pi/2)R$. (Einstein 1918, p. 272)[21]

To reach such a (wrong) conclusion, Einstein had to define what a regular gravitational field is. His definition was very much the same as Hilbert's, but he demanded as "a requirement of the theory" that the field equations be valid "at every point located at a finite distance." He added, in particular, that "the determinant should never vanish at a finite distance" (Einstein 1918, p. 270). Implicitly, Einstein asked, as Hilbert did, that the allowed transformation of coordinates be one-to-one and invertible.

The controversy then shifted to the problem of defining what a finite distance is. Is it, as Einstein claimed, the proper distance on any time-oriented curve? Or is it to be defined as a "physically accessible point," as de Sitter argued? In any case the controversy ended with de Sitter's acceptance of the singular character of the horizon $r = (\pi/2)R$, but de Sitter emphasized that if the "discontinuity" is located "at a finite distance in space" it is nevertheless "physically inaccessible" (de Sitter 1917c, p. 1309).

Eddington's position was at first much the same as de Sitter's. In his *Report on the Relativity Theory of Gravitation* (1918), Eddington had harsh words for Einstein's universe. He did not like in this model the possibility of observing ghosts, virtual images of real stars: "We regret being unable to recommend this rather picturesque theory of anti-suns and anti-stars" (Eddington 1918, p. 87). This is an observation that the de Sitter horizon prevents:

> There is no anti-sun on de Sitter's hypothesis, because light, like everything else, is reduced to rest at the zone where time stands still, and it can never get round the world. The region beyond the distance $(\pi/2)R$ is altogether shut off from us by this barrier of time. (Eddington 1918, p. 89)

Thus, according to Eddington, the de Sitter horizon is made of light and matter reduced to rest—something like a physical singularity. In *Space, Time and Gravitation* (Eddington 1920a) he gave a very similar picture of a "Schwarzschild singularity" (a term, by the way, he never uses). There he described a thought experiment that consists of approaching the $r = 2m$ surface with a measuring rod:

> We can go on shifting the measuring-rod through its own length time after time, but $dr$ is zero; that is to say, we do not reduce $r$. There is a magic circle which no

measurement can bring us inside. It is not unnatural that we should picture something obstructing our closer approach, and say that a particle of matter is filling up the interior. (Eddington 1920a, p. 98)

In 1924 he obtained—in quite a different context[22]—what we call now the Finkelstein line-element:

$$ds^2 = -dr^2 - r^2(d\theta^2 + \sin^2\theta\, d\phi^2) + c^2 dt^2 - \frac{2mG}{rc^2}(c\,dt - dr)^2, \qquad (8)$$

where the gravitational potentials are everywhere finite (but the $g_{44}$ vanishes at $r = 2m$!). However, Eddington never claimed that the $r = 2m$ surface is regular.

In his *The Mathematical Theory of Relativity* (Eddington 1923) Eddington obtained the well-known generalization of the Schwarzschild solution[23] that includes the cosmological constant, $\Lambda$:

$$ds^2 = \left(1 - \frac{2mG}{rc^2} - \frac{\Lambda r^2}{3}\right)c^2 dt^2 - \left(1 - \frac{2mG}{rc^2} - \frac{\Lambda r^2}{3}\right)^{-1}$$
$$- r^2(d\theta^2 + \sin^2\theta\, d\phi^2), \qquad (9)$$

which contains, as particular cases, the Schwarzschild and de Sitter solutions. Eddington commented as follows:

> At a place where $g_{44}$ vanishes there is an impassable barrier, since any change $dr$ corresponds to an infinite distance $ds$ surveyed by measuring-rods. The two positive roots of the cubic $g_{44}$ are approximately $r = 2mG/c^2$ and $r = \sqrt{3/\Lambda}$. The first root would represent the boundary of the particle—if a genuine particle could exist—and give it the appearance of impenetrability. The second barrier is at a very great distance and may be described as the *horizon* of the world. (Eddington, 1923, pp. 100–101)

But this does not answer the question of what a horizon really is. Further on, Eddington raised the question whether the de Sitter world is empty or not without giving a clear answer. But he pointed out very clearly the difficulties and possibilities concerning the definition of a singularity:

> A singularity of $ds^2$ does not necessarily indicate material particles, for we can introduce or remove such singularities by making transformations of coordinates. It is impossible to know whether to blame the world-structure or the inappropriateness of the coordinate-system. (Eddington 1923, p. 165)[24]

Weyl's opinion was closer to Einstein's than to de Sitter's or Eddington's. He considered the Schwarzschild sphere and the de Sitter horizon as well to be singular. In 1918, in his *Raum-Zeit-Materie*, he performed a calculation aimed at proving that "there must at least be masses at the horizon" of the de Sitter cosmological model (Weyl 1918, p. 225)[25]. But as far as the Schwarzschild solution is concerned, he developed quite an odd set of arguments that should have led to topological implications. He remarked that, in the Flamm representation, the projection on the $z = 0$ plane covers twice the "exterior" of the circle $r = 2m$ (by symmetry $z \to -z$). He then proposed that one of the two mappings be interpreted "by analytical extension" as the "exterior" of the massive point, the other one standing for the "interior": "Both mappings are divided by the sphere $r = 2m$ on which the mass is to be found and where the mass determination is singular" (Weyl 1917, p. 131).[26] He reached the same conclusion using the isotropic form of the Schwarzschild solution (4):

> From the previous interpretation, the domain $\rho > mG/2c^2$ corresponds to the exterior and $\rho < mG/2c^2$ to the interior of the massive point. By analytical extension $\sqrt{g_{44}} = (\rho - mG/2c^2)/(\rho + mG/2c^2)$ will be negative in the interior such that for a point standing at rest, *the cosmic time (t) and the proper time flow in opposite directions.* (Weyl 1917, p. 132; my emphasis)[27]

But, as could be expected, Weyl ended his line of argument by noting that the interior part of his representation will never be relevant in nature. Thus, Weyl's opinion was, at first sight, very much the same as the then prevailing interpretation: the $r = 2m$ sphere, where all the mass is concentrated, is singular in character. Curiously enough, such a singularity did not, however, prevent Weyl from considering the problem of extending the solution beyond it. And this was, anyway, a most important attempt, since analytical extension was at that time quite an uncommon concept, indeed, in the field of general relativity, as was the idea of "analysis situs," which Weyl also used. This was in fact quite a modern way to deal with Einstein's theory. In spite of the differences, an expert on general relativity cannot help seeing some similarity between the naive view of Weyl and some aspects of the modern interpretation. But there does not seem to be more there than the ingenious use of some correct concepts.

The experts, therefore, encountered many difficulties in correctly understanding the general theory. As an example, I just want here to recall briefly the discussions concerning general covariance among Painlevé, Einstein, and Becquerel, just before and during the Paris meeting of Easter 1922. The Schwarzschild solution was Painlevé's primary concern. He wrote it in different systems of coordinates: Droste's system (3); a new system due to A. Gullstrand:

$$ds^2 = \left(1 - \frac{2mG}{rc^2}\right)c^2\,dt^2 + 2\sqrt{\frac{2mG}{rc^2}}\,dr\,c\,dt$$
$$- (dr^2 + r^2(d\theta^2 + \sin^2\theta\,d\phi^2)); \tag{10}$$

and a general class of line-elements.[28] "It seems to me," Painlevé wrote, "that the existence of the formula [10] and the possibility of an infinity of others give a clear indication of the hazardous character of such predictions." He concluded that "it's pure imagination to claim that such consequences can be derived from the $ds^2$" (Painlevé 1921, p. 680).[29]

In a beautiful letter to Painlevé dated December 7, 1921, Einstein apologized for not coming to Paris and gave a clear explanation of the status of coordinates:

When, in the $ds^2$ of the static solution with central symmetry, you introduce any function of $r$ instead of $r$, you do not obtain a *new* solution because the quantity $r$ in itself has no physical meaning.... only conclusions reached after the elimination of coordinates may pretend to an objective significance. Furthermore, the metrical interpretation of the quantity $ds$ is not "pure imagination" but the deep core of the theory itself. (EA 19-004)[30]

Jean Becquerel put forward objections to Painlevé. But for him—and, in fact, for many experts at the time—some systems of coordinates were to be preferred, in particular, those to which one can give a physical significance, and Droste's system qualified as being the "best." During the Paris meeting, Becquerel tried to convince Einstein to share his preference for what he called the Schwarzschild coordinates,[31] in which "the appearance of the universe is the most intuitive to the physicist."[32] According to the report of the Société Philosophique de France, Einstein's reply was that "you can always choose any representation you want if you believe that it is more convenient ... but it has no objective meaning."[33]

Up to this point we have left aside all works concerning the trajectories in the Schwarzschild field. For many authors, the primary analytical interpretation in Droste's coordinates was a sufficient proof of the singular character of the gravitational potential at $r = 2mG/c^2$. And, in any case, the realistic statement according to which "it does not exist in nature" put an end to any possible controversy. When dealing with the trajectories in the Schwarzschild field, however, every expert was aiming at proving the impenetrability, the inaccessibility in a finite "time" of what was called "the Schwarzschild singularity." Let us study now, as one of the most representative examples, the thesis that Droste defended in December 1916. He considered trajectories in the second chapter ("The Motion of a Material Point in the Field of a Center"), and he used, of course, the geodesic equations that Einstein wrote for the first time in 1914:[34]

$$\frac{d^2 x^\rho}{ds^2} + \left\{ \begin{matrix} \rho \\ \mu\nu \end{matrix} \right\} \frac{dx^\mu}{ds} \frac{dx^\nu}{ds} = 0, \tag{11}$$

from which he deduced easily the three well-known equations that govern the trajectory of a test particle, which Schwarzschild had already obtained in his first paper:

$$c^2 \left(1 - \frac{2Gm}{rc^2}\right) \left(\frac{dt}{ds}\right)^2 - \left(1 - \frac{2Gm}{rc^2}\right)^{-1} \left(\frac{dr}{ds}\right)^2 - r^2 \left(\frac{d\phi}{ds}\right)^2 = 1,$$

$$r^2 \left(\frac{d\phi}{ds}\right) = L, \tag{12}$$

$$\left(1 - \frac{2Gm}{rc^2}\right) c \frac{dt}{ds} = E.$$

These were generally named the "energy equation," the "law of areas," and the "equation determining the time." In the nonradial case, Droste used—as did everyone else—the angular parameterization $\phi$. From (12) he derived (13), a generalization of the usual Newtonian equation that he would be the first to integrate exactly with the help of the Weierstrass elliptic function:

$$\left(\frac{du}{d\phi}\right)^2 = \frac{E^2 - 1}{L^2} + \frac{2mG}{c^2 L^2} u - u^2 + \frac{2mG}{c^2} u^3 \quad \text{(where: } u = 1/r\text{)}. \quad (13)$$

But Droste would not base his interpretation on the (too) simple equation (13). In order to determine the "limit of the trajectory" he preferred to write the differential equation for $d\phi/dt$ (a coordinate-velocity), whose interpretation is safer *because* it prevents any orbit from falling into the singular region. He made this clear straight away: "If the time needed to reach *some point* on the trajectory is infinitely long, *such a point* will never be reached nor passed" (Droste 1916b, p. 26; my emphasis).[35] Thus, the $t$-parameterization has really been chosen *in order* to avoid the Schwarzschild singularity: "It turns out that we never have ... $r = \infty$ or $r = \alpha$. We are used to the first possibility; the second one means that the sphere $r = \alpha$ is never reached" (Droste 1916b, p. 26).

Such an analysis—which will in general give more weight to the $t$ time-coordinate—is not at all unique to Droste; it is, in fact, at the bottom of most papers on the subject up to the end of the 1950s.[36] Droste proceeded thus: "If, instead of the $r$ from [3] ... we had used the $\tilde{r}$ that occurs in [5], we would be able to say that the material point does not reach the center."

We must understand Droste's concern that the choice of the coordinates allow "all the points to be reached." In fact, in Droste's mind, space still exists prior to physics—a Newtonian view really, then implicitly shared by many experts. This is precisely the "fundamentally different result" that Droste pointed out and that I emphasized in my first talk:[37] "This result differs fundamentally from what occurs in Newton's theory; it shows us how totally different ... from classical theory the motion in the vicinity of the center becomes" (Droste 1916b, p. 26).[38] This is a most important point to note, really, since for a long time there would not be that many specifically relativistic predictions concerning, implicitly, strong gravitational fields. It is also worth remarking that at that time no expert, indeed, but a student, Droste, was to point out that the approximate correctness of the Newtonian theory, as far as observation is concerned, was in fact the worst threat to general relativity.

In 1923 Carlo De Jans, who was unaware of Droste's work, studied the same problem (1923).[39] His conclusion was the same. But since he was a most conscientious mathematician, his equations—when parameterized by $\phi$—drove him inside of the singularity $r = \alpha$. But since he was—as any expert was then—an adept of the pragmatic interpretation, he did not notice that he had entered the singularity. In fact, as usual, his trajectories were driven, through a special three-velocity $dl/dt$, by the Schwarzschild time-coordinate $t$. Thus, the diagram he published at the end of his contribution (see Figure 2) exhibited an empty hole at the center.

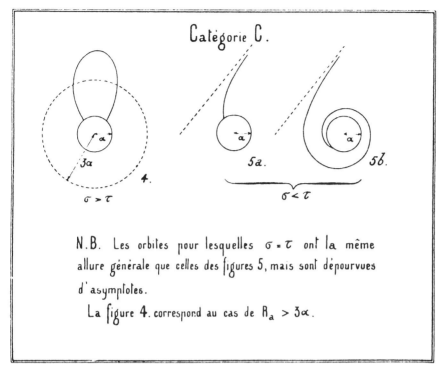

FIGURE 2. Material orbits: De Jans's 1923 negative discriminant ($L/M < 2\sqrt{3}$).

There were, then, not many experts in the field to point out that the trajectories are the right tool for defining physical space as the set of all accessible points.[40] In 1947 E. Rabe (1947), a German astronomer, used the radial trajectories to study the Schwarzschild singularity—an excellent idea indeed. When dealing with the radial case, it was of course necessary to choose between $t$, the Droste time, and $s$, the proper time, as a parameter.[41] However, although the equations parametrized by $s$ are exactly the same as in the Newtonian case; although they are trivially integrable up to $r = 0$ without any singularity at the Schwarzschild sphere $r = 2m$; and although their use was highly recommended by Einstein (and by (11)), nobody integrated them.[42] Maybe this was because the proper time is not an exact differential and has to be defined on every geodesic.[43] Perhaps it was because such an equation drove the test particle into the forbidden region. Almost everyone used the Droste time $t$ as a parameter, either directly or through some coordinate-velocity. Even Rabe wrote down the equation for the acceleration in proper time but did not integrate it, and thus inferred the singular character of what was still called the "Schwarzschild singularity."

More precisely, there is no better example of the difficulties that experts encountered with parametrization—and covariance—than the so-called repulsion. Such a gravitational repulsion was supposed to occur in a region around the $r = 2m$ singularity. Indeed, the acceleration of a test-particle, when computed with the help of the $t$-coordinate ($d^2r/dt^2$), becomes positive, as Droste showed in his thesis. Such an effect is just an artifact of covariance, but Droste was not the last one to misunderstand these tricky problems. In his well-known *Relativitätstheorie*, such an excellent expert as Max von Laue made the same misinterpretation (1921, pp. 223–24). This error was fairly well corrected in 1936 by an unknown relativist, Paul Drumaux, who used "naturally measured quantities." Drumaux's intention, by the way, was to prove that the 1932 Lemaître paper was wrong and that the Schwarzschild sphere is a true singularity (Drumaux 1936, pp. 5–14). The same misunderstanding shows up in Hilbert's 1917 contribution and in a 1920 article by Leigh Page, which has been criticized by Eddington (1920b). And last but not least, as late as 1956 the same misunderstanding was extensively developed in a textbook by George McVittie (1956, p. 85). But such a misunderstanding of covariance is not an isolated case. It shows how difficult it was to construct a truly relativistic paradigm for the treatment of a physical problem in general relativity.

In his *Relativitätstheorie* (von Laue 1921), von Laue classified the trajectories of luminous particles in the Schwarzschild field. The equation of motion in the nonradial case had been obtained by Flamm:

$$\left(\frac{du}{d\phi}\right)^2 = \frac{1}{b^2} - u^2 + \frac{2mG}{c^2}u^3 \quad \text{(where } u = 1/r\text{)}, \tag{14}$$

which depends on two parameters, the central mass $m$ and "the impact parameter" $b$. Von Laue's qualitative analysis is most clearly summarized by his diagram, shown here in Figure 3. The limiting case $A$ represents the last noncaptured orbit: every orbit having an impact parameter $b$ less than that of case $A$ ($b = 3\sqrt{3}m$) is captured. And, as von Laue wrote, the diagram "shows the circle $r = 2m$ on which every incoming ray ends," a conclusion he derived from the observation that "there the light velocity vanishes"—a coordinate-velocity of course (von Laue 1921, p. 226).[44]

De Jans published two detailed memoirs on the topic (1922a, 1922b). In the first one, which is very close to von Laue's analysis, he integrated completely the nonradial equation (but *not* the radial one). Once more some orbits cross the singularity $r = 2m$ (equation (14) is indeed nonsingular at $r = 2m$!) ending in $r = 0$, but de Jans did not notice it. In a second contribution on the subject, he used an "aiming angle" related to the impact parameter.[45] The analysis that Whittaker proposed in 1928 gives a good idea of the representation used at that time of the relation between light rays—and even material particles—and the Schwarzschild singularity:

> If, then, we consider a ray of light coming from infinity and travelling directly towards the point-mass, the velocity of the light will always be $c$; but when it begins

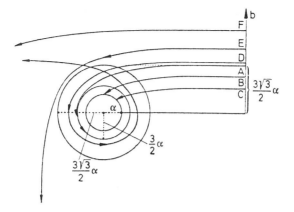

FIGURE 3. The luminous trajectories (von Laue 1921).

to approach the circle of perimeter $2\pi\alpha$, this velocity will only be sufficient to carry it onwards very slowly, if we measure its progress by the rate of diminution of perimeter of the circles it cuts through, and it can never, in any time however great, get nearer to the mass than the circle of perimeter $2\pi\alpha$. (Whittaker 1928, p. 140)

He then adds:

The capture and imprisonment of radiation by the intense gravitational field surrounding a point-mass is a remarkable theoretical possibility, markedly different from anything in prerelativity physics.

Amazingly, as far as I know, the concept of an "intense gravitational field" is here introduced for the first time in this context. This remark must also be connected of course to Droste's anxious comment concerning the place of general relativity in physics.

Thus, the most current model, the main representation of the so-called Schwarzschild singularity, consists of an impenetrable sphere of radius $2mG/c^2$ in curvilinear coordinates: a "singular" sphere on which matter and light aggregate without penetrating it; a "magic circle," as Eddington called it. As we have seen, such a representation is based on many arguments. There are, first, three types of arguments—analytical, topological, and dynamical—which are strongly dependent on the Droste-Schwarzschild coordinates and even more particularly on the Schwarzschild time coordinate. They would gradually be questioned. But there were other types of arguments too. The only physical argument was based on the Schwarzschild interior solution, a too-simple model that was to be discussed in the works of Lemaître and Oppenheimer. As regards the pragmatic argument, which relies on the evidence that such a type of phenomenon "does not exist in nature," it has quite a strange status. It always came as a final argument, as if the other arguments were not strong enough—without forgetting, of course, metaphysical arguments such as Eddington's.

Such was the status of the Schwarzschild singularity before the renewal of interest in the 1960s. We did not consider here the works that were leading the way to the new paradigm;[46] they represented a completely independent line of research, scarcely discussed before the end of the 1950s. But, to conclude, it is necessary to stress, besides the particular status of the Schwarzschild singularity, the importance of the Schwarzschild-Droste system of coordinates, which played de facto the role of absolute Newtonian space, a neo-Newtonian view of the solution that had, of course, many connections with the neo-Newtonian interpretation that was then dominating the theory.

*Acknowledgments.* This article is essentially based on two papers on the early interpretation of the Schwarzschild solution: Eisenstaedt 1982, and Eisenstaedt 1987.

I wish to thank the Hebrew University of Israel for its kind permission to quote from Einstein's unpublished writings. I also want to thank Don Howard and John Stachel for help in translating part of this paper into English, and A.J. Kox for translating the Dutch material.

NOTES

[1] "Ihre Arbeit habe ich mit grösstem Interesse durchgelesen. Ich hätte nicht erwartet, dass man so einfach die strenge Lösung der Aufgabe formulieren könne. Die rechnerische Behandlung des Gegenstandes gefällt mir ausgezeichnet. Nächsten Donnerstag werde ich die Arbeit mit einigen erläuternden Worten der Akademie übergeben."

[2] Concerning Schwarzschild's biography I mainly used Dieke 1975.

[3] The "magic circle" is Edddington's expression; the "hole in space" (*un trou dans l'espace*) was used by J. Becquerel (unpublished, Becquerel's Archives). The "Hadamard disaster" was Einstein's witticism.

[4] Concerning this point, see "The Low Water Mark of General Relativity, 1925–1955," in this volume, pp. 277–292.

[5] For a general view of that period, see Earman and Glymour 1978, Medicus 1984, Norton 1984, Pais 1982, and Stachel 1979.

[6] Schwarzschild did not realize that $\beta$ is due to covariance, $\alpha$ being in fact the only physical constant.

[7] "... die Unstetigkeit nicht im Nullpunkt, sondern an der Stelle $R = (\alpha^3 - \beta)^{1/3}$ eintritt, und dass man gerade $\beta = \alpha^3$ setzen muss, damit die Unstetigkeit in den Nullpunkt rückt."

[8] But de Donder never followed Schwarzschild's topological interpretation. See De Donder 1921, Rice 1923, Mie 1920, Schrödinger 1918, Straneo 1924, and Abrams 1979.

[9] "Si ... ce terme pouvait quelque part dans l'Univers s'annuler, alors ce serait un malheur imaginable pour la théorie; et il est très difficile de dire *a priori* ce qui arriverait physiquement, car alors la formule cesse d'être applicable."

[10] The Schwarzschild limit refers to the interior solution that Schwarzschild found at the beginning of 1916; see Schwarzschild 1916b.

[11] This solution is to be found either in Droste 1916a, or in his thesis Droste 1916b p. 72.

[12] "Aan welke formule men de voorkeur wil geven, aan die van Schwarzschild [2], aan [3] of aan [5], blijft een kwestie vaan smaak. Men moet nl. bedenken, dat de coördinaat $r$ toch den gemeten afstand niet voorstelt. In de keuze van een coördinaat is men echter vrij (mits hare waarden het geheele gebied, waarin de waarneming doordringt, eenmaal omvatten), al zal ook de eene keuze doelmatiger kunnen zijn dan de andere."

[13] Of course the "gravitational radius" is nothing but $2mG/c^2$.

[14] See Eddington 1918, p. 27; and de Sitter 1917a, p. 371.

[15] "... die Singularität der Massbestimmung [3] bei $r = 2m$ nicht durch Einführung anderer Koordinaten beseitigen lässt, sondern durchaus in der Natur der Sache liegt" (my emphasis).

[16] "... jede Masse $m$ ... notwendig einen grösseren Halbmesser hat, als er dem Wert $r = 2Gmc^{-2}$ entspricht. Und in der Tat kennen wir bisher keinen dem widersprechenden Fall, auch nicht bei den Atomkernen." In 1921 E. Kasner worked out the embedding of the complete solution in a six-dimensional flat space. E. Loedel-Palumbo did it again in 1926 and gave a pictorial representation of it. The Schwarzschild two-sphere $r = 2m$ appears as a point in the Kasner-Loedel-Palumbo representation.

[17] "Die Stellen $r = \alpha$ nach dem Nullpunkt zu transformieren, wie es Schwarzschild tut, ist nach meiner Meinung nicht zu empfehlen; die Schwarzschildsche Transformation ist überdies nicht die einfachste, die diesen Zweck erreicht."

[18] "Aber auch dieses Beispiel zeigt, wie wenig man aus dem singulären Verhalten der $g_{\mu\nu}$ Funktionen auf eine tatsächliche Singularität des Feldes schliessen darf, da dieselbe durch eine Transformation der Koordinaten möglicherweise gehoben werden kann."

[19] "Die Trägheit ist eben nach meiner Theorie im letzten Grunde eine Wechselwirkung der Massen, nicht eine Wirkung, bei welcher ausser der ins Auge gefassten Masse der 'Raum' als solcher beteiligt ist. Das Wesentlich meiner Theorie ist gerade, dass dem Raum als solchem keine selbständigen Eigenschaften gegeben werden.

"Man kann es scherzhaft so ausdrücken. Wenn ich alle Dinge aus der Welt verschwinden lasse, so bleibt nach Newton der Galileische Trägheitsraum, nach meiner Auffassung aber <u>nichts</u> übrig."

[20] I just sketch here this important point. A more detailed analysis can be found in North 1965, and Kerszberg 1987.

[21] "In Wahrheit löst das de Sittersche System ... die Gleichungen ... überall, nur nicht in der Fläche $r = (\pi/2)R$. Dort wird—wie in unmittelbarer Nähe eines gravitierenden Massenpunktes—die Komponente $g_{44}$ des Gravitationspotentials zu null. Das de Sittersche System dürfte also keineswegs dem Falle einer materielosen Welt, sondern vielmehr dem Falle einer Welt entsprechen, deren Materie ganz in der Fläche $r = (\pi/2)R$ konzentriert ist."

[22] Eddington 1924. See also Eisenstaedt 1982, note 82.

[23] Which is due to Hermann Weyl.

[24] This is quite an interesting remark, which may point to his 1924 paper. (Cf. Eisenstaedt 1982, note 82, and 1987, p. 324.)

[25] "Zum mindesten am Horizont müssen sich Massen befinden."

[26] "Die beiden Überdeckungen sind durch die Kugel $r = 2m$, auf der sich die Masse

befindet und die Massbestimmung singulär wird, geschieden." See also the first two editions of *Raum-Zeit-Materie*, pp. 205–6; this point is not included, however, in editions following the second.

[27] "Nach der obigen Auffassung würde hier das Gebiet $\rho > mG/2c^2$ dem Äussern, $\rho < mG/2c^2$ dem Innern des Massenpunktes entsprechen. Bei analytische Fortsetzung wird $\sqrt{g_{44}} = (\rho - mG/2c^2)/(\rho + mG/2c^2)$ im Innern negativ, so dass also dort für einen ruhenden Punkt kosmische Zeit ($t$) und Eigenzeit gegenläufig sind."

[28] Note that line-element (10) in fact has the main properties of the Eddington-Finkelstein line-element (8). See Eisenstaedt 1982, p. 174. The main references concerning that problem are to be found there.

[29] "L'existence de la formule [10] et d'une infinité d'autres possibles me paraît suffire à démontrer le caractère plus qu'aventureux de telles prévisions ... c'est pure imagination de prétendre tirer du $ds^2$ des conséquences de cette nature."

[30] "Wenn mann in der zentral-symmetrischen statischen Lösung für $ds^2$ statt $r$ irgend eine Funktion von $r$ einfügt, so erhält man keine *neue* Lösung, da die Grösse $r$ an sich keinerlei physikalische Bedeutung hat,... nur Ergebnisse, die durch Elimination der Koordinaten erlangt sind, können objektive Bedeutung beanspruchen. Die metrische Interpretation der Grösse $ds$ ist ferner keine "pur imagination," sondern der innerste Kern der ganzen Theorie."

[31] But they must be called the Droste-Schwarzschild coordinates.

[32] "Les coordonnées employées par Schwarzschild sont celles avec lesquelles l'aspect de l'univers, pour le physicien, devient le plus intuitif."

[33] "On peut toujours choisir ... telle représentation qu'on veut si l'on croit qu'elle est plus commode qu'une autre pour le travail qu'on se propose: mais cela n'a pas de sens objectif." Comptes-Rendus de la séance du 6 Avril 1922. *Bulletin de la Société Philosophique de France*. 17 (1922). (See Eisenstaedt 1982, p. 177.)

[34] For more details on these points, see Eisenstaedt 1987.

[35] Of course it is the Schwarzschild singularity that is meant here.

[36] For more on this fundamental point, see Eisenstaedt 1987, section 5, "Les relativistes ont-ils peur de la chute?"

[37] "The Low Water Mark of General Relativity, 1925–1955," this volume, pp. 277–292.

[38] "Is nl. de tijd, die voor het bereiken van zecker punt der baan noodig is, oneindig groot, dan wordt zoo'n punt nooit bereikt en evenmin overschreden. ... dat nooit $[r = \infty]$ of $[r = \alpha]$ wordt. ... Het eerste zijn wij gewend; het tweede beteekent, dat bol $r = \alpha$ nooit bereikt wordt. Hadden wij, inplaats van de $r$ uit [3], de $r$ gebezigd, die in [5] optreedt, dan zouden wij kunnen zeggen, dat het stoffelijk punt het centrum niet bereikt. Die resultaat is ten eenenmale verschillend van hetgeen in Newton's theorie optreedt; wij zien er uit, hoe geheel anders de beweging in de nabijheid van het centrum wordt ... dan in de klassieke theorie."

[39] We shall not study here the well-known 1931 Hagihara article, which is much the same (though written independently) as Droste's and De Jans's articles, but, in my opinion, far less interesting. See Eisenstaedt 1987, p. 311.

[40] See, however, Pierre Kerszberg, "The Einstein-de Sitter Controversy of 1916–1917 and the Rise of Relativistic Cosmology," this volume, pp. 325–366.

[41] Note that the angle $\phi$—which cannot be used in the radial case anymore—is there a "proper" parameter, as $s$ (the "proper time") is. Obviously, from (12b), $d\phi/ds$ is everywhere monotonic.

[42] But Robertson did in a work that was published much later. See Eisenstaedt 1987, p. 328.

[43] In some ways, this is the problem that Weyl had in mind in 1917.

[44] "Sie zeigt uns die Kreise $r = 2m$, an welchem jeder herankommende Lichtstrahl endigt (ist doch dort die Lichtgeschwindigkeit 0)."

[45] The models he used and his classification are as modern and as precise as those used in the 1973 textbook by Misner, Thorne, and Wheeler. As far as I know, this excellent analysis has been completely forgotten in the meantime.

[46] A historical article concerning the roots of the new interpretation is planned.

REFERENCES

Abrams, L.S. (1979). "Alternative Space-Time for the Point-Mass." *Physical Review D* 20: 2474–2479.

De Donder, Theophile (1921). *La Gravifique Einsteinienne*. Paris: Gauthier-Villars.

De Jans, Carlo (1922a). "La trajectoire d'un rayon lumineux dans un champ de gravitation à symétrie sphérique." *Académie Royale de Belgique. Classe des Sciences. Mémoires* 6: 1–26.

——— (1922b). "La propagation de la lumière dans un champ de gravitation à symétrie sphérique." *Académie Royale de Belgique. Classe des Sciences. Mémoires* 6: 27–41.

——— (1923). "Sur le mouvement d'une particule matérielle dans un champ de gravitation à symétrie sphérique." *Académie Royale de Belgique. Classe des Sciences. Mémoires* 7: 1–98.

de Sitter, Willem (1917a). "Planetary Motion and the Motion of the Moon According to Einstein's Theory." *Proceedings of the Section of Sciences, Koninklijke Akademie van Wetenschappen te Amsterdam* 19: 367–381.

——— (1917b). "On the Relativity of Inertia. Remarks concerning Einstein's Latest Hypothesis." *Proceedings of the Section of Sciences, Koninklijke Akademie van Wetenschappen te Amsterdam* 19: 1217–1225.

——— (1917c). "Further Remarks on the Solutions of the Field-Equations of Einstein's Theory of Gravitation." *Proceedings of the Section of Sciences, Koninklijke Akademie van Wetenschappen te Amsterdam* 20: 1309–1312.

Dieke, Sally H. (1975). "Schwarzschild, Karl." *Dictionary of Scientific Biography*. C.C. Gillispie, ed. Vol. 12. New York: Charles Scribner's, pp. 247–253.

Droste, Johannes. (1915). "On the Field of a Single Centre in Einstein's Theory of Gravitation." *Proceedings of the Section of Sciences, Koninklijke Akademie van Wetenschappen te Amsterdam* 17: 998–1011.

——— (1916a). "Het veld van een enkel centrum in Einstein's theorie der zwaartekracht en de beweging van een stoffelijk punt in dat veld." *Verslagen van de Gewone Vergaderingen der Wis- en Natuurkundige Afdeeling, Koninklijke Akademie van Wetenschappen te Amsterdam* 25: 163–180; English translation "The Field of a Single Centre in Einstein's Theory of Gravitation, and the Motion of a Particle in that Field." *Proceedings of the Section of Sciences, Koninklijke Akademie van Wetenschappen te Amsterdam* 19: 197–215.

——— (1916b). *Het zwaartekrachtsveld van een of meer Lichamen volgens de theorie van Einstein*. Leiden: E.J. Brill.

Drumaux, Paul (1936). "Sur la force gravifique." *Société Scientifique de Bruxelles. Annales B* 56: 5–14.

Earman, John, and Glymour, Clark (1978). "Einstein and Hilbert: Two Months in the History of General Relativity." *Archive for History of Exact Sciences* 19: 291–308.

Eddington, Arthur Stanley (1918). *Report on the Relativity Theory of Gravitation.* London: Fleetway Press.
—— (1920a). *Space, Time and Gravitation.* Cambridge: Cambridge University Press.
—— (1920b). "Gravitational Deflection of High Speed Particles." *Nature* 105: 37.
—— (1923). *The Mathematical Theory of Gravitation.* Cambridge: Cambridge University Press.
—— (1924). "A Comparison of Whitehead's and Einstein's Formulae." *Nature* 113: 192.
Einstein, Albert (1911). "Über den Einfluss der Schwerkraft auf die Ausbreitung des Lichtes." *Annalen der Physik* 35: 898–908.
—— (1915). "Erklärung der Perihelbewegung des Merkur aus der allgemeinen Relativitätstheorie." *Königlich Preussische Akademie der Wissenschaften* (Berlin). *Sitzungsberichte*: 831–839.
—— (1918). "Kritisches zu einer von Herrn de Sitter gegebenen Lösung der Gravitationsgleichungen." *Königlich Preussische Akademie der Wissenschaften* (Berlin). *Sitzungsberichte*: 270–272.
Einstein, Albert, and Grossmann, Marcel (1913). *Entwurf einer verallgemeinerten Relativitätstheorie und einer Theorie der Gravitation. I. Physikalischer Teil von Albert Einstein. II. Mathematischer Teil von Marcel Grossmann.* Leipzig and Berlin: B.G. Teubner. Reprinted with added "Bemerkungen," *Zeitschrift für Mathematik und Physik* 62 (1914): 225–261.
Eisenstaedt, Jean (1982). "Histoire et singularités de la solution de Schwarzschild (1915–1923)." *Archive for History of Exact Sciences* 27: 157–198.
—— (1987). "Trajectoires et impasses de la solution de Schwarzschild." *Archive for History of Exact Sciences* 37: 275–357.
Flamm, L. (1916). "Beiträge zur Einsteinschen Gravitationstheorie." *Physikalische Zeitschrift* 17: 448–454.
Hertzsprung, Ejnar (1917). "Karl Schwarzschild." *Astrophysical Journal* 45: 285–292.
Hilbert, David (1917). "Die Grundlagen der Physik: zweite Mitteilung." *Königliche Gesellschaft der Wissenschaften zu Göttingen. Mathematisch-physikalische Klasse. Nachrichten*: 55–76.
Kasner, Edward (1921). "Finite Representation of the Solar Gravitational Field in Flat Space of 6 Dimensions." *American Journal of Mathematics* 43: 130–133.
Kersberg, Pierre (1987). "The Relativity of Rotation in the Early Foundations of General Relativity." *Studies in History and Philosophy of Science* 18: 53–79.
Lanczos, Cornelius (1922). "Ein vereinfachendes Koordinatensystem für die Einsteinschen Gravitationsgleichungen." *Physikalische Zeitschrift* 23: 537–539.
Lemaître, Georges (1932). "L'Univers en expansion." *Publication du Laboratoire d'Astronomie et de Géodésie de l'Université de Louvain* 9: 171–205; also in *Société Scientifique de Bruxelles. Annales A* 53: 51–58.
Loedel-Palumbo, Enrique (1926). "Die Form der Raum-Zeit-Oberfläche eines Gravitationsfeldes, das von einer punktförmigen Masse herrührt." *Physikalische Zeitschrift* 27: 645–648.
McVittie, George C. (1956). *General Relativity and Cosmology.* London: Chapman and Hall.
Medicus, Heinrich A. (1984). "A Comment on the Relations Between Einstein and Hilbert." *American Journal of Physics* 52: 206–208.
Mie, Gustav (1920). "Die Einführung eines vernunftgemäßen Koordinatensystems in die Einsteinsche Gravitationstheorie und das Gravitationsfeld einer schweren Kugel." *Annalen der Physik* 62: 46–74.

Misner, Charles W., Thorne, Kip S., and Wheeler, John A. (1973). *Gravitation.* San Francisco: Freeman.

Nordmann, Charles (1922). "Einstein expose et discute sa théorie." *Revue des Deux Mondes* 9: 129–166.

North, John D. (1965). *The Measure of the Universe: A History of Modern Cosmology.* Oxford: Clarendon Press.

Norton, John (1984). "How Einstein found His Field Equations: 1912–1915." *Historical Studies in the Physical Sciences* 14: 253–316. See this volume, pp. 101–159.

Page, Leigh (1920). "Gravitational Deflection of High-Speed Particles." *Nature* 104: 692–693.

Painlevé, Paul (1921). "La mécanique classique et la théorie de la relativité." *Académie des Sciences* (Paris). *Comptes Rendus* 173: 677–680.

Pais, Abraham (1982). *"Subtle is the Lord ...": The Science and the Life of Albert Einstein.* Oxford: Clarendon Press.

Pyenson, Lewis (1975). "La réception de la relativité généralisée: disciplinarité et institutionalisation en physique." *Revue Histoire des Sciences* 28: 61–73.

Rabe, E. (1947). "Zur Singularität der Schwarzschildschen Lösung für $r = 2m$." *Astronomische Nachrichten* 275: 251–255.

Rice, James (1923). *A Systematic Treatment of Einstein's Theory.* London: Longmans, Green and Co.

Schrödinger, Erwin (1918). "Die Energiekomponenten des Gravitationsfeldes." *Physikalische Zeitschrift* 19: 4–7.

Schwarzschild, Karl (1900). "Über das zulässige Krümmungsmass des Raumes." *Vierteljahrsschrift der Astronomischen Gesellschaft* 35: 337–347.

———  (1914). "Über die Verschiebung der Bände bei 3883 Å im Sonnenspektrum." *Königlich Preussische Akademie der Wissenschaften* (Berlin). *Sitzungsberichte*: 1201–1213.

———  (1916a). "Über das Gravitationsfeld eines Massenpunktes nach der Einsteinschen Theorie." *Königlich Preussische Akademie der Wissenschaften* (Berlin). *Sitzungsberichte*: 189–196.

———  (1916b). "Über das Gravitationsfeld einer Kugel aus inkompressibler Flüssigkeit nach der Einsteinschen Theorie." *Königlich Preussische Akademie der Wissenschaften* (Berlin). *Sitzungsberichte*: 424–434.

Stachel, John (1979). "The Genesis of General Relativity." In *Einstein Symposion Berlin.* H. Nelkowski et al., eds. Berlin, Heidelberg, and New York: Springer-Verlag, pp. 428–442.

Straneo, Paolo (1924). *Teoria della Relatività.* Rome: Lib. di Scienze e Lettere.

von Laue, Max (1921). *Die Relativitätstheorie.* Vol. 2, *Die allgemeine Relativitätstheorie und Einsteins Lehre von der Schwerkraft.* Braunschweig: Friedrich Vieweg & Sohn.

Weyl, Hermann (1917). "Zur Gravitationstheorie." *Annalen der Physik* 54: 117–145.

———  (1918). *Raum-Zeit-Materie.* Berlin: Julius Springer.

———  (1921). *Raum-Zeit-Materie,* 4th ed. Berlin: Julius Springer; 6th ed., 1968; English translation *Space-Time-Matter.* H.L. Brose, trans. London: Methuen, 1922; reprint New York: Dover, 1950.

Wheeler, John Archibald (1968). "Our Universe: The Known and the Unknown." *American Scholar* 37: 248–274.

Whittaker, Edmund Taylor (1928). "The Influence of Gravitation on Electromagnetic Phenomena." *Journal of the London Mathematical Society* 3: 137–144.

# The Early History of the "Problem of Motion" in General Relativity*

PETER HAVAS

## 1. Introduction

In Newtonian mechanics, there exists a sharp distinction between laws of motion and force laws. The motions of all bodies have to obey Newton's three laws of motion; in particular, the second law relates the acceleration of any (inertial) mass to the force exerted on it. The latter has to be stipulated by specific force laws, which depend on the particular type of physical characteristics of the bodies under consideration. The most important such laws are Newton's universal law of gravitation, which states that the force between any two mass points (or rigid spherically symmetric bodies) is proportional to the product of their gravitational masses and inversely proportional to the square of the distance (i.e., the difference between their simultaneous positions) between their mass centers, and Coulomb's law for the force between two electrically charged bodies.

In Newton's theory, forces can be superposed, i.e., if more than two bodies are present, the force on any one of them equals the sum of the individual forces exerted by each of the other bodies on it, each given by a particular two-body force law. In the following (using the terminology of Havas and Goldberg 1962), we shall distinguish between "laws of motion" such as Newton's second law, which relate the acceleration of the particle to an unspecified force, and "equations of motion," which relate it to forces given explicitly in terms of particle variables.

In the nineteenth century, it was gradually realized that the interaction between electrically charged bodies could not be described adequately by instantaneous action-at-a-distance forces (see, e.g., Whittaker 1951; Hesse 1961), and finally Coulomb's law was replaced by Maxwell's field equations and Lorentz's force law, which together described the electromagnetic action on a charge; but they were still independent of Newton's (and later Einstein's) second law of motion. Attempts by Lorentz (1909) to unify mechanics and electrodynamics by deriving the law of motion from Maxwell's equations (supplemented by ad hoc assumptions) were not successful. The independence of the laws of motion from the force laws persisted in later special-relativistic field theories developed to describe nuclear and other forces; to remove this independence requires, as a minimum, the additional assumption of the existence of an overall law of conservation of energy-momentum for the system of particles and fields (for reviews, see Havas 1959, 1979b).

Within Newtonian theory, only one force is universal, the gravitational one; but from the point of view of this theory the particular distance dependence of the law of gravitation is accidental, as is the universal proportionality of inertial and gravitational mass. In Einstein's theory of general relativity (Einstein 1913, 1916a), this universal proportionality is a consequence of one of the basic postulates, the principle of equivalence; but, as initially formulated, this theory maintained the Newtonian distinction between laws of motion and force laws. The latter were taken to result from Einstein's field equations. The former appeared to require a separate postulate; in the case of the absence of any nongravitational fields, Einstein postulated (both in the initial 1913 form of the theory and in its final 1916a form) that the motion of a mass point should follow a geodesic in four-dimensional space-time. He was led to this postulate by Newton's first law, the results of special relativity, and the principle of equivalence, which in some sense allowed the motion of a particle to be considered to be force-free.

The geodesic law was taken to be exact for a test particle, i.e., a particle that is subject to the effects of a gravitational field, but whose own effect on the field can be neglected, so that the "force" acting on the particle was determined by solving Einstein's field equations in the absence of the particle. Whether the geodesic law also holds for a particle whose effects on the gravitational field cannot be neglected was not discussed initially. It was clear, however, that for the case of several bodies of comparable masses Einstein's (nonlinear) field equations could not be expected to permit an exact solution and thus no universal force law analogous to Newton's law of gravitation could be hoped for, but only an approximate one, which in lowest order should agree with Newton's.

Subsequently, it was recognized that within the theory of general relativity the laws of motion (and in particular the geodesic law) cannot be stipulated independently of the field equations, but (with certain restrictions not to be discussed here) are a consequence of these equations. This result is generally considered to be one of the major achievements of the theory. Almost universally, this recognition is ascribed to a paper by Einstein and Grommer (1927), which established to a certain approximation that the path of a test particle must be a geodesic; similarly, the derivation of the approximate equations of motion of several, slowly moving, particles of comparable masses is generally ascribed to a paper by Einstein, Infeld, and Hoffmann (1938, in the following referred to as EIH), and the "exact" derivation of the geodesic law to Infeld and Schild (1949). Early reviews of these questions (the "problem of motion") were given by Infeld and Plebański (1960), Bażański (1962a), and Goldberg (1962), all of whom accepted these attributions.

However, these popular attributions of credit are wrong in every instance by at least a decade (Havas 1979a). It will be the aim of this talk to document this through discussion both of published work and of correspondence between Einstein, Infeld, and other scientists working on the problem of motion. Because of the time limitations imposed on my talk, I cannot discuss the mathematical questions involved in any detail or justify many technical assertions; this will be done in a monograph in preparation. Similarly, I will omit any discussion of the question of the equations of motion of fast-moving bodies of comparable masses, of the gravitational radiation-reaction terms, and of the law of motion of a spinning test particle, which will also be discussed elsewhere. I will also limit my talk mostly to work done before 1940, except where later work is relevant for clarification of the questions discussed here.

Any attribution of credits involves a methodological question: implicitly or explicitly, certain criteria are used as to when a result is considered to be "established." The theory of general relativity is still undergoing rapid development, and so are the standards of mathematical rigor; by the standards of the most mathematical school of thought, almost none of even the best-known results of the theory have been established. For the problem of motion, there does exist a consensus that the laws of motion are in some sense a consequence of the field equations, that a nonspinning test particle satisfies the geodesic law, and a somewhat vaguer consensus that slowly moving bodies of comparable masses, to order $c^{-2}$, obey the EIH equations. This last result, being approximate, is on a different mathematical level than the others, which are supposed to be exact, and thus capable of "rigorous" proofs.

In the case of the EIH equations, Einstein himself was dissatisfied with the original derivation and (Einstein and Infeld 1940, 1949) gave two other derivations within the next decade (which were criticized in turn); thus one should not apply higher mathematical standards to earlier contributions to the problem than one applies to the EIH paper. One should also not deride papers of a reasonable mathematical level that happen to contain fairly trivial, easily correctable, mistakes when comparing them to others that avoided trivial mistakes but were mathematically more objectionable.

In the case of the geodesic law, none of the dozens of papers can be considered to be rigorous by the most stringent mathematical criteria, but all of them are reasonably rigorous by the contemporary standards of working physicists. The recognition of the connection between the laws of motion and the field equations hinges on assumptions about the structure of the particles being considered, as well as on an insight into the mathematical structure of the equations. Einstein himself seems to have ascribed the connection entirely to the nonlinearity of the field equations and to have thought that his methods eliminated the need for any energy-momentum tensor of matter, which is incorrect; he also misunderstood the reasons for the differences between his results and those of various other early papers.

However, our interest in this problem should go beyond a simple historical rectification of the facts and some suggested explanations for the reasons why the clearly available evidence of earlier work was ignored by a whole generation of relativists. Multiple discoveries are quite common in the history of physics, although in this instance their extent is unusual. They are surprising, however, in view of the publication of many of the results in easily accessible journals and in standard books on general relativity. Also, a number of authors were acquainted with each other. How could they miss—or misunderstand—the results of their colleagues? And, especially, how could Einstein? That Einstein did so is particularly strange since the available correspondence shows that several colleagues had pointed out their results to him. Thus we will be led to more general problems of objective and subjective difficulties in communication between scientists and of human failings.

## 2. The Geodesic Law and the Connection between Field Equations and Laws of Motion

For the purposes of this presentation, we require only the barest outline of the structure of Einstein's general theory of relativity, without any need to define the various terms involved, since we will not perform any calculations here.[1] In empty space the field equations of this theory are

$$G^{\mu\nu} = 0, \tag{1}$$

where the Einstein tensor $G^{\mu\nu}$ is a geometric quantity constructed from the metric tensor $g_{\mu\nu}$ and its first and second derivatives. The tensor $g_{\mu\nu}$ describes the metric properties of four-dimensional space-time, and represents the four-dimensional generalization of the Newtonian gravitational potential, thus providing a "geometrization" of the gravitational force. In the presence of matter and nongravitational fields, described by an energy-momentum tensor $T^{\mu\nu}$, we have instead of equation (1) (with suitable conventions on units and signs)

$$G^{\mu\nu} = T^{\mu\nu}. \tag{2}$$

All equations here and in the following are tensor equations, i.e., they are of the same form, independent of the choice of coordinates used.

The particular expression for $G^{\mu\nu}$ selected by Einstein was chosen because its "covariant divergence" vanishes identically,

$$G^{\mu\nu}{}_{;\nu} = 0, \tag{3}$$

and thus by equation (2) we have

$$T^{\mu\nu}{}_{;\nu} = 0, \tag{4}$$

which is a generalization of the local conservation law for energy-momentum of the special theory of relativity. This feature of Einstein's theory is patterned after the corresponding feature of Maxwell's electrodynamics, where a similar identity satisfied by the tensor describing the electromagnetic field implies the "equation of continuity" for the charge-current density.

It is of technical importance for the problem of motion that equations (1) and (2) can each be derived from a variational principle (Hilbert 1915; Lorentz 1915a, 1915b; De Donder 1916; Fokker 1917; Weyl 1917; see also Mehra 1974). Equations (3) and (4) can be obtained directly from such a principle without detouring through the field equations.

Newton's theory of gravitation appears to be conceptually incompatible with the theory just sketched. However, it was recognized by Cartan and Friedrichs in the 1920s that it too can be expressed in form (1) in empty space, and this formal analogy was later extended to include equation (2) as well (for an elementary discussion with references see Havas 1967). However, unlike the metric tensor $g_{\mu\nu}$ of general relativity, the four-dimensional tensors needed to describe the metric properties of Newtonian space-time are singular, and equations (3) and (4) do not follow from the field equations, a distinction of fundamental importance for the problem of motion.

We shall first consider Einstein's introduction of the geodesic law of motion as discussed in his survey of the general theory of relativity (Einstein 1916a).[2] We note that the metric tensor $g_{\mu\nu}$ determines the infinitesimal separation $ds$ of two neighboring points with four-coordinates $x^\mu$ and $x^\mu + dx^\mu$ by

$$ds^2 = g_{\mu\nu} dx^\mu dx^\nu, \qquad (5)$$

where summation over repeated indices is understood. A geodesic, i.e., a line whose "length" between two points is an extremum, is described by

$$\frac{d^2 x^\tau}{ds^2} + \Gamma^\tau_{\mu\nu} \frac{dx^\mu}{ds} \frac{dx^\nu}{ds} = 0, \qquad (6)$$

where $\Gamma^\tau_{\mu\nu}$ are the Christoffel symbols (formed from $g_{\mu\nu}$ and its first derivatives). For the Minkowski space-time of special relativity, coordinates can be chosen such that the components of $g_{\mu\nu}$ are constant everywhere; then all components of $\Gamma^\tau_{\mu\nu}$ vanish, and equation (6) becomes the familiar equation for a straight line. Following §13 of Einstein 1916a, we denote such a coordinate system by $K_0$. Einstein then noted that, according to the special theory of relativity,

> a freely movable body not subject to external forces moves ... in a straight line and uniformly. This holds also, according to the general theory of relativity, for a part of the four-dimensional space in which the coordinate system $K_0$ may be, and is, so chosen that the $g_{\mu\nu}$ have ... constant values,

as discussed above (they do not need to have the special values used and referred to by Einstein). He continued:

> If we consider this particular motion from an arbitrarily chosen system of coordinates $K_1$, the body, observed from $K_1$, moves, according to the considerations in

§2, in a gravitational field. The law of motion with respect to $K_1$ results without difficulty from the following consideration. With respect to $K_0$, the law of motion corresponds to a four-dimensional straight line, i.e., to a geodesic line. Now since the geodesic line is defined independently of the system of reference, its equation will also be the equation of motion of the material point with respect to $K_1$ [i.e., our equation (6)][3]. We now make the assumption, which readily suggests itself, that this covariant system of equations also defines the motion of the point in the gravitational field in the case when there is no system of reference $K_0$, with respect to which the special theory of relativity holds good in a finite region.... If the $\Gamma^\tau_{\mu\nu}$ vanish, then the point moves uniformly in a straight line. These quantities therefore determine the deviation of the motion from uniformity. They are the components of the gravitational field.

It is of importance for understanding the problem of motion to note that there are really two assumptions involved in this passage. One was clearly stated by Einstein. The other is implicit—the validity of Newton's first law is taken for granted instead of being recognized as that of a law of motion that does not follow from the space-time structure of the special theory of relativity (or of Newtonian theory). It is not unique; e.g., a "first law" stating that all bodies not subject to external forces move with constant acceleration (or a constant higher derivative of the velocity) would satisfy all formal requirements of these theories, including Poincaré (or Galilei) invariance. It is part of the problem of motion to show that only uniform straight-line motion is compatible with the special relativistic (or Newtonian) limit of general relativity. (We are not concerned here with the physical question of whether or not the usual first law is needed for the definition of an inertial system, but only with problems arising within the formal structure of the general theory considered as given.)

Instead of immediately tracing the historical development of the problem of motion, it will be more instructive first to follow the development of Einstein's thought. It is implicit in his early work on the linear approximation (Einstein 1916b) that he believed that no linear field equations can impose any restrictions on the motion of singularities of these fields, and thus that their motion can be prescribed arbitrarily, which is incorrect; precisely in the case he investigated there, the equations restrict the singularities to motion without acceleration (for a discussion of this point see Havas and Goldberg 1962, section III). This misunderstanding persisted throughout his life; in his scientific autobiography he still wrote:

> If the field law is linear (and homogeneous), the sum of the solutions is still a solution.... In such a theory one cannot deduce from the field law alone an interaction of structures which can be described separately by solutions of the system. Therefore all theories up to now required, in addition to the field laws, separate laws for the motion of the material structures under the influence of the fields. (Einstein 1949, p. 78)

It is quite correct that no *interactions* can arise in linear field theories; but this does not preclude *restrictions* on the motion arising from the linear field equations. In this context, it should be noted that a linearization procedure differing from Einstein's in lowest order still gives no interaction, but allows all bodies to move with constant acceleration; this is one of the possible modified "first laws" alluded to above (Robinson and Robinson 1972).

There is no indication, in spite of all the work done by others (which will be discussed later), that Einstein had investigated the possibility of a connection between the field equations and the laws of motion in any detail until 1926, when he started working on the problem with his assistant Jakob Grommer. It is unlikely that the initiative came from Grommer, who was a mathematician with little background in physics. The "Introduction" of their paper (Einstein and Grommer 1927), however, contains a very clear statement of the problem. First, Einstein and Grommer restated the fact that both in Newton's theory of gravitation, considered as a field theory, and in Maxwell-Lorentz electrodynamics the field equations and the law of motion are independent. Turning to general relativity, they distinguished three ways of approaching the question of whether or not such a dualism between field law and law of motion also exists there.

The first approach, patterned after Newtonian theory, takes the empty-space field equations (1) and the law of geodesics (6) of the material point as independent. The second takes equation (2) as the basic field equation and requires that $T^{\mu\nu}$ be singularity-free and expressible through continuous field quantities that are in turn determined by some partial differential equations; regardless of this, $T^{\mu\nu}$ is subject to restriction (4). Einstein and Grommer then stated (without any calculation):

> If one assumes that matter is arranged along narrow "world tubes," one obtains from this by an elementary consideration the theorem that the axes of those "world tubes" are geodesic lines (in the absence of electromagnetic fields). This means: the law of motion is a consequence of the field law.

This is a rather astonishing statement, considering that Einstein had given no indication in the decade since he had postulated the geodesic law that this postulate might be superfluous. But in any case they immediately rejected this approach since

> All attempts of the last few years to explain the elementary particles of matter by continuous fields have failed. The suspicion that this may not be at all the right way for understanding material particles has become very strong in us.... We are thus led to a third approach, which does not allow any field variables apart from the gravitational and the electromagnetic field (except possibly the "cosmological term"), but assumes singular world lines.... *it has turned out to be probable that the law of motion of the singularities is completely determined by the field equations and the character of the singularities*, without the necessity of additional assumptions.

This is followed by a very revealing paragraph, which shows that Einstein's —still unrecognized—misunderstanding of the implications of the linear approximation of his theory, discussed above, was at the bottom of his difficulties with the problem of motion. Einstein and Grommer wrote:

We had thought much earlier of the possibility that the law of motion of the singularities may be contained in the field equations of gravitation. But the following argument seemed to speak against it and scared us off. The field law of gravitation can be approximated for the actually existing cases very closely by a linear law. The linear field law, however, like the electrodynamic one, allows arbitrarily moving singularities. It seems obvious that one can proceed from such an approximate solution by successive approximations to an exact one differing from it only slightly. If this were the case, then it would be possible to have a field correspond to the exact equations with arbitrarily prescribed motion of the singularities, and thus the law of motion of the singularities would not be contained in the field equations. But that this cannot be so follows from investigations of axisymmetric static gravitational fields for which we can thank Weyl, Levi-Civita, and Bach.

They then referred to these papers (which will be discussed later), and following their methods gave a short proof that a singular point can be in equilibrium only if the external field strength vanishes at this point. Then they proceeded to generalize this result by a method using surface integrals. They summarized their paper as follows:

If one considers masses in the gravitational field as singularities, then the law of motion is fully determined by the field equations [inserted here is a footnote: "However, in this paper this is proved completely only for the case of equilibrium"]. If the total field is approximated by the solutions of linear equations, then the law of motion is that of the geodesic line. In a later work the law of motion of electrons considered as singular points will be derived from the field equations. However, it is well known that electrically neutral atomistic masses do not exist in nature, so that the subject of this paper does not directly correspond to a subject in nature. But the progress made here is that it is shown for the first time that a field theory can include a theory of the mechanical behavior of discontinuities. This can be significant for the theory of matter, or the quantum theory.

The importance of Einstein's basic mathematical misunderstanding of the implications of his linearized field equations (Einstein 1916b) has already been pointed out. But there is another, partly related, puzzling aspect of the considerations of Einstein and Grommer. Their paper appears to be aimed entirely at the problem of the motion of elementary particles. Their "first approach, patterned after Newton's theory," takes the source-free field equations as basic; but clearly, in both Newton's and Einstein's theories, macroscopic bodies must be represented by sources of the gravitational field. Their second approach is abandoned by them because of the absence of a satisfactory theory of elementary particles, and the third considers such particles simply as singularities of the field. The implication of the first as well as the third approach is that they obviate the necessity of introducing any energy-momentum tensor of matter at all—a point that will be encountered again in EIH. But—at least in linear field theories, in particular those leading to Laplace's or d'Alembert's equation—singular solutions correspond not to the absence of sources, but rather to singular source distributions.

Furthermore, in spite of their focusing on elementary structures, it is stated that "for the actually existing cases" the field law can be "approximated ... very closely by a linear law." However, these cases—which were and are considered to furnish the main observational confirmation of the theory—all refer to applications of the theory to the solar system, in particular to the motion of Mercury treated as a geodesic in the solar field. But while it is certainly a legitimate approximation to treat Mercury as a test particle, it most assuredly is not simply a singularity of the field. It is a macroscopic body, for which the disdained second approach is the legitimate one, which by "an elementary conderation" leads to the law of motion, while the considerations of the third approach are irrelevant for it. On the other hand, there was then—and is now—no indication that individual elementary particles correspond to "actually existing cases" of motion to which the geodesic law is applicable. Their behavior is governed by quantum laws—and indeed Einstein hoped until 1938 to be able to obtain quantum conditions from the restrictions on the motion imposed by the field equations.[4]

If one does not insist on basing the theory of general relativity on the vacuum field equations (1), but instead takes equations (2) as fundamental, then the question of whether to obtain the laws of motion via a consideration of the energy-momentum tensor $T^{\mu\nu}$ or via consideration of the Einstein tensor $G^{\mu\nu}$ becomes mainly a technical one. Most early work proceeded via $T^{\mu\nu}$, and indeed it seems to have been recognized by a number of scientists, but apparently not by Einstein, within a very few years after the creation of the general theory that the vanishing of its covariant divergence, equation (4), implied the law of motion. It should be clearly understood, however, that if one uses equation (2) or the corresponding variational principle, then—at least for finite-sized bodies—one has to use specific $T^{\mu\nu}$'s, usually obtained by a generalization of special-relativistic ones; this was also done for mass points by all scientists at that time, and it was only realized later (Mathisson 1931a) that for these the form of $T^{\mu\nu}$ can also be derived. If the special-relativistic $T^{\mu\nu}$'s are used as a crutch, the derivation of the laws of motion is subject to the objection raised above against Einstein's assumption, rather than proof, that the usual first law of motion had to hold in the special-relativistic limit. It was also previously noted that various derivations claiming to use only the vacuum equations (1) really used (2) with singular sources and therefore only appeared to avoid the use of an energy-momentum tensor.

From its inception, the general theory had created enormous interest in the scientific community, and not just physicists, but also mathematicians and astronomers, had started working on it. However, it is difficult to assess questions of priorities for work carried out at that time. Europe had been engulfed by a war that had already claimed millions of victims and touched everyone's life. Unlike the situation in World War II, many scientists had been called into—or had volunteered for—military service (including Schrödinger and Schwarzschild). Nevertheless, even some of these, and many others, continued their research. Again unlike what happened in World War II, communications between scientists of warring countries were not interrupted completely, and scientific journals from the "enemy" side were received, but delays must have been inevitable and are now impossible to trace.

Fortunately for relativity, H.A. Lorentz lived in neutral Holland and could provide a link. He stayed in touch with Einstein in Berlin (who also visited Holland occasionally), and he and his students, colleagues, and collaborators (Droste, de Sitter, and Fokker among them) made many important contributions to relativity during that time. But even communication with neighboring Belgium which, though neutral, had been invaded and occupied by the Germans, was difficult. De Donder, working in isolation in Brussels, seems to have obtained some important results earlier than others, but could not send them to Lorentz immediately because of the occupation. In a notice appended to a review of some of his work (De Donder 1917), he stated that results he had obtained in September 1914 could not be communicated to Lorentz before April 1915, and in a later book (De Donder 1921, p. 22), he noted that De Donder 1916, which contains the general variational principle, was written in September 1915 (i.e., before Hilbert 1915 was submitted), but could not be sent to Lorentz for publication until April 10, 1916, due to the German occupation. (Outlines of his results are given in two follow-up letters to Lorentz on April 17th and 30th [Lorentz Archive, The Hague].) While it is also not clear when the various German publications became accessible to Eddington, an astronomer working in Cambridge and a conscientious objector who barely escaped being jailed (Douglas 1956; Chandrasekhar 1983), he was able to complete a report on the theory of general relativity for the Physical Society of London by June 1918 (Eddington 1918); it is poorly referenced, but mentions a communication by Einstein to the Prussian Academy of February 14 of that year, only four months earlier. Eddington was also in touch with his fellow astronomer de Sitter, whom he thanked for reading the proofs of his report.

In that report, Eddington stated (note that he used the term "law of gravitation" for "field equations"):

> Conversely from the identity [3] we can deduce [4], and hence obtain the equations of hydromechanics and the law of conservation directly from Einstein's law of gravitation. Further, by applying the hydromechanical equations to an isolated particle, we obtain the equations of motion [6]. The mass of a particle has been introduced first as a constant of integration, and afterwards identified with the gravitation-mass by determining the motion of a particle in its field; it now appears that it is also the inertia-mass, because it satisfies the law of conservation of mass and momentum, which gives the recognised definition of inertia.
>
> It is startling to find that the whole of the dynamics of material systems is contained in the law of gravitation; at first sight gravitation seems scarcely relevant in much of our dynamics. But there is a natural explanation. A particle of matter is a singularity in the gravitational field, and its mass is the pole-strength of the singularity; consequently the laws of motion of the singularities must be contained in the field-equations. (Eddington 1918, p. 65)

Thus Eddington clearly recognized the connection between the laws of motion and the field equations. His derivation, preceding the just-quoted passage, is quite adequate; however, the last statement quoted is not, and the rest of the sentence, "just as those of electromagnetic singularities (electrons) are contained in the electromagnetic field-equations," makes no sense at all. No such statement appears in any of Eddington's later discussions of the problem, however.

It appears, though it is not perfectly clear, that Eddington considered this derivation to be an original contribution. No indication is given in the text, but in the "Preface" he states that "Extensive use has been made" of Einstein 1916a and de Sitter 1916, which do not contain such a derivation, and then states modestly, "The principal deviations in the present treatment of the subject will be found in Chapter VI," which contains the derivation and the just-quoted discussion.

We are not concerned here with Eddington's role in the observational confirmation of general relativity, nor with his popularized presentation of the theory in his book *Space, Time and Gravitation*, or elsewhere. What is important is that for the French edition of this book (Eddington 1921), he prepared a 149-page mathematical supplement (apparently completed in October 1920). In its section IV, a much more detailed derivation of the law of motion is given than in Eddington 1918; §42, dealing with a masspoint, concludes:

> The law according to which a free mass point has for trajectory a geodesic and for rest mass a constant, thus appears as a consequence of the law of gravitation [2] and not as an independent hypothesis. One can also remark that while in §29 we had to attribute symmetry properties to the particle, here we made the single hypothesis that the volume occupied is small.

This supplement was the first detailed introduction to general relativity available to French theorists in their own language. Its publication preceded Einstein's March 1922 visit to Paris by almost a year and, unlike Einstein's visit, does not seem to have provoked any chauvinistic reaction either within or outside of French academic circles, possibly because this was an "Allied" contribution. The "Introduction" to Eddington 1921 was written by Paul Langevin, who served as Einstein's mentor on his subsequent visit.

This "mathematical study" was later referred to by Eddington as the first draft of his *The Mathematical Theory of Relativity* (Eddington 1923), which went through many editions and translations and became one of the standard texts on the theory. Its chapter IV is an extended and revised version of section IV of the earlier study; in this later proof of the geodesic law, Eddington retreated from his assertion quoted earlier that no symmetry properties were assumed for the particle, but stated on the contrary that such properties were postulated, as indeed they must be.

Eddington was by no means the only scientist who realized the connection between the field equations and the laws of motion at that time. There are indications that it was recognized in Holland by Lorentz and his school. It was clearly recognized by De Donder in Belgium, whose derivation of the geodesic law from equation (4) (De Donder 1919) is also presented in chapter

III of his exposé of Einstein's theory (De Donder 1921). A derivation from the variational principle was given by a Swiss physicist working in Göttingen (Humm 1918). But the most important contributions are due to Hermann Weyl, a German mathematician who was a professor at the ETH in Zurich from 1913 to 1930 (and was therefore acquainted with Einstein, whom he joined at the Institute for Advanced Study in Princeton in 1933). Initially he was concerned with static axially symmetric exact solutions of Einstein's field equations (Weyl 1917, 1919b, Bach 1922), as was Levi-Civita. The solutions obtained are of no concern for our discussion. However, what is of importance is that in the course of this work he realized that two bodies interacting only gravitationally cannot be in equilibrium (Weyl 1917, 1919b; Addendum to Bach 1922). The nonexistence of equilibrium solutions, unlike all other results obtained for several bodies, is an exact result and thus of crucial importance for the general problem of motion.[5] But it does not imply anything about the form of the laws of motion. This more general problem was first considered in Weyl 1919a in the context of Weyl's own generalization of Einstein's theory. He then discussed it in detail in Weyl 1921a and in the third, more clearly in the fourth, and especially in the fifth edition of his book *Raum-Zeit-Materie* (Weyl 1919c, 1921b, 1923). In the third and fourth editions, this was still done in the context of his own theory, which attempted to geometrize the electromagnetic field in addition to the gravitational field; however, a careful reading of his presentation leaves no doubt that all the mathematical and physical arguments given remain valid in the absence of an electromagnetic field, in which case Weyl's theory reduces to Einstein's. This distinction is important for the subsequent discussion, because Einstein had raised physical objections against Weyl's generalization of his theory (Addendum to Weyl 1918) and indeed this generalization was later abandoned by Weyl and is now mainly of historical interest. In any case, possibly because of these objections by Einstein, in the fifth edition of his book, in §38, Weyl considered the problem of motion only in the context of Einstein's field equations (2), with the electromagnetic field included through $T^{\mu\nu}$ and Maxwell's equations (in the form appropriate for general relativity), and I shall follow the latter presentation here.

Weyl wanted "first of all to derive the basic laws of mechanics without making use of the hypothetical laws in the interior of the elementary particles of matter" on the basis of the Maxwell-Einstein laws, noting: "We know that they hold outside matter, i.e., outside of certain narrow tubes which traverse the world in one-dimensional infinite extension." He then derived the geodesic law (6) and its generalization in the presence of an electromagnetic field, noting that these relations "appear as necessary conditions for the proper field of a mass point to fit into the field prevailing outside the channel," which obeys the Einstein-Maxwell equations, and that the derivation does not make use of any particular assumption about the internal constitution of the particle.

Weyl's and Eddington's books were translated into many languages and became standard texts for scientists studying and working in the theory of general relativity. Another standard text was Pauli's "Encyklopädie" review article (Pauli 1921), which also contained a derivation of the geodesic law (§54) based on equation (4).[6] A similar derivation is contained in a now almost forgotten textbook by a German astronomer (Kopff 1921, based on his lectures given in Heidelberg in 1919–20), which seems to have been popular at the time, because it went through a second edition and was translated into English as well as into Italian; the latter translation contains in addition a number of short contributions evaluating the theory by leading scientists, including Weyl. Similarly, a derivation is contained in the well-known textbook by Laue (1921), with whom Einstein was well acquainted. Also, Gustav Mie, who had been working on a general field theory of matter for many years (which, though now largely forgotten, was frequently mentioned by Einstein and others at that time), derived the geodesic law within this framework in a paper aiming to incorporate the general theory of relativity into his own theory (Mie 1923), and commented: "This well-known law, of which Herr Einstein made use in his first publications on the theory of gravitation, can *of course* be derived much more simply in another way than in that chosen in this paper. Here it serves only as a test for the correctness of my calculations...." (my emphasis). Another derivation, based on Eddington's, is given in the textbook (Becquerel 1922) that grew out of a course given for several years by the French physicist Jean Becquerel, scion of a family that had produced outstanding scientists for several generations. The derivation is found in section 87, unequivocally entitled "The Law of Motion of the Free Material Point is Contained in the Law of Gravitation." It seems hardly possible to avoid the conclusion that by 1923 or 1924 essentially all scientists working in general relativity must have been aware of this fact.

Thus it is somewhat puzzling how Einstein and Grommer could have failed utterly to put their results into this context (quite apart from the question, to be discussed in Section 4, why subsequently an entire generation of physicists would credit their paper with the recognition of the connection between the field equations and the laws of motion). Only in their "second approach" did they show any awareness that the geodesic law can be derived from a consideration of "narrow world tubes," but they implied that this derivation would require a particular field theory of matter, an objection clearly not tenable against Weyl's 1923 derivation. It is not surprising that Weyl was upset when he learned about the presentation of the paper to the Prussian Academy, and responded with a lengthy letter to Einstein on February 3, 1927, less than four weeks after that event:

Dear Colleague,

Herglotz gave me the galleys of your note on the law of motion in general relativity. I thank you very much for this and also for the support you give by it to my old idea about matter. However, I must confess that I did not understand what in it goes beyond my earlier developments.

In my Addendum to the paper by Bach you quoted ... the $\int d\gamma$ ... is interpreted as the force on the body embedded in the gravitational field. Brief review of this in "Raum Zeit Materie" 5th edition p. 267.

Then, apparently well aware of Einstein's reputation for reading very little, Weyl did not simply refer to the relevant section of his book, but gave a detailed outline of it:

> The derivation of the equations of motion, without assumption about what happens "inside" the matter particle, is given by me ibid. §38 (p. 277); the § starts with the passage: [the passage quoted above]. I had explained the matter already earlier, but not that carefully, in Ann. Phys. ("Feld und Materie"), and later also a thick paper of mine on this appeared in Ann. Phys. [This appears to be a reference to Weyl 1919a, which actually preceded "Feld und Materie" (Weyl 1921a).] My method is somewhat different: I use the differential conservation laws in a form in which they are mathematical identities, independent of the field equations, and apply them to the interior of the matter tube, which I take as filled completely by a fictitious field. E.g., to obtain the concept of the charge $e$ with the conservation law $de/dt = 0$, I fill the tube with arbitrary potentials $\varphi_i$ continuously joined to the exterior.... $e$ is independent of the filling.... Similarly one sees that $e$ is independent of the coordinate system. The laws of motion are obtained in fundamentally the same way. This derivation clarifies the reason why it is the "field-producing mass" defined as flux of gravitation which appears as inertial = gravitational mass. I believe that my derivation is as clean as one can make such things, and that it lets the conditions for the validity of the laws of motion emerge clearly. (EA 24-086; in German)

Einstein answered almost three months later, on April 26:

> Because of your friendly letter I have of course carefully studied your proof of the law of motion of the electron and at first was very enthused by its beauty and transparency. But finally I did encounter difficulties which do not appear to me to have a ready solution.
> 1. It cannot be seen readily that terms with $\bar{f}^2$ in the electric field can be neglected....
> This I consider less important. But the following does seem important to me:
> 2. To be able to choose the fictitious field in the interior of the bounding sphere, or channel, using the normal coordinate system, as a *static* one, as you do in your proof, the field must be suitably constituted. But we know nothing about this, especially if the electron is accelerated (is under the influence of an electric field).
>
> I attach so much importance to the entire matter because it would be very important to know whether the field equations as such should be considered as refuted by the fact of quanta or not. One is after all inclined to believe this and most do. But nothing appears to me to have been proved about this until now. (EA 24-088; in German)

Regardless of the validity of Einstein's objections, it should be noted that they are based on a consideration of electrically charged particles. But the paper to which Weyl had addressed his comments in his letter (Einstein and Grommer 1927) was entirely concerned with neutral particles, and (see the footnote to their summary, quoted before) claimed a complete proof only for the case of equilibrium, i.e., no acceleration. Therefore Einstein's objections were entirely irrelevant to Weyl's polite "confession" that he did "not understand what in it goes beyond my earlier development." It seems very unlikely that Weyl would not have responded to Einstein's points and misunderstandings in the letter just quoted and to his failure to acknowledge priority, but I have not been able to locate such an answer or any other exchange between them on this subject.

It appears that it was Einstein's preoccupation with the question whether quantum conditions are compatible with, or even a consequence of, his theory which made him misunderstand the relevance and fail to acknowledge the priority of Weyl's results on the derivation of the geodesic law. He was already at work on a sequel to his paper with Grommer, because barely a week after his letter to Weyl, on May 5, he wrote to Paul Ehrenfest, probably his closest friend in the scientific community:

> I published the paper on relativistic dynamics of the singular point a long time ago. But the dynamic case is still not properly disposed of. I am at the point where I believe that results will emerge here which deviate from the classical laws of motion. The method also has become clear and certain. If I would only calculate better! Grommer is a sick person and is of little help to me in this respect. It would be marvelous if the customary differential equations would lead to quantum mechanics; and I don't consider it to be out of the question at all. (EA 10-162; in German)

No mention is made of Weyl here or in other letters to Ehrenfest on this subject.

Einstein completed his paper without Grommer (who subsequently accepted a position in Minsk, where he died in 1933). He presented the results to the Prussian Academy in November (Einstein 1927). His preoccupation with the quantum problem and misjudgment of Weyl's contributions remain quite evident. After quoting Einstein and Grommer 1927 as investigating "whether the law of motion of singularities is determined by the field equations of the general theory of relativity," he added in a footnote:

> H. Weyl has in the later editions of his books "Raum, Zeit, Materie" already earlier advocated the opinion that the elementary bodies should be considered as singularities of the field. There also he has already attempted to derive the equations of motion from this point of view. Cf also K. Lanczos, Z. f. Phys. 44, 5. 773 [Lanczos 1927a].

Nothing further is said about why he considered this to be only an attempt, which by implication was unsuccessful and unsatisfactory. (The curious use of the plural "books" makes me wonder whether Einstein actually consulted the book.) Then Einstein continued the discussion of his paper with Grommer: "There it was shown that the *until now generally assumed* law of the geodesic line (if necessary complemented by the electromagnetic moving forces) can in the static or stationary case be deduced from the field equations" (my emphasis)—again totally ignoring the vast literature on this subject (discussed above), even Weyl, whom he had just quoted in his footnote; this makes it appear that the footnote had been added as an afterthought, after the paper had been completed. He continued:

> But it also became apparent that this procedure cannot be taken over free of hypotheses for the case that the field surrounding the singularity is not stationary.

Then an additional hypothesis on the character of the singularity present in the external field was required, whose justification could not be proved. This result is of interest from the point of view of the general question whether field theory is in contradiction with the postulates of quantum theory or not. Most physicists today are convinced that the facts of the quanta exclude the validity of a field theory in the usual sense of the word. But this conviction is not based on a sufficient knowledge of the consequences of field theory.

He then proceeded to derive the law of motion of a singularity in an electromagnetic field (which agrees with Weyl's— unmentioned—result and reduces to the geodesic law for an uncharged mass point) under certain assumptions, and concluded:

> Thereby it is shown that the law of motion, until now hypothetically assumed, is a consequence of the field equations, if one bases one's consideration on a point-like singularity of static character. But this does not yet prove that such singularities, subject to the field equations, can carry out *all* motions which satisfy the condition (26b) [the law of motion mentioned earlier]. That is to say, one can imagine that on carrying out this consideration for higher approximations further restrictive conditions might arise.
>
> This investigation does not contribute to the understanding of the quantum facts. However, the result that the law of motion for singularities and the field equations are not independent of each other remains of importance.

Thus, apart from a somewhat grudging and certainly inadequate footnote, the reader is left with the impression that there does not exist any earlier work leading to the conclusion of the last sentence. It is also clear from Einstein's own reservations quoted earlier on hypotheses contained in his calculations, that at least one of his objections against Weyl's method applies to his own calculations. It is also evident that Einstein had still not given up the idea that the "problem of motion" might lead to a resolution of the quantum problem within the theory of general relativity. Two months later, on January 21, 1928, he wrote to Ehrenfest: "I think I told you that the derivation of the law of motion according to the rel. theory has finally succeeded. But it simply comes out classically. I think that Kaluza-Klein have correctly indicated the way to advance further. Long live the 5th dimension" (EA 10-173; in German). And in late summer, on August 23, he again wrote to Ehrenfest: "I have now mastered the problem of motion of singularities so that I can apply the method to any field theory. I believe less than ever in the essentially statistical nature of events" (EA 10-186; in German).

About a month later a new assistant started working with Einstein. Cornel Lanczos was a Hungarian-born physicist who had been working on relativity since 1921. In Lanczos 1927a (the paper quoted without comment in Einstein 1927; for a brief report on the results see Lanczos 1927b), he had treated the problem of motion as a boundary value problem and obtained a new derivation of the geodesic law from the field equations (incidentally, also ignoring much of the earlier literature). However, it was not this work that had induced Einstein to offer the assistantship, but the need for mathematical help in his work on distant parallelism; apparently Lanczos, who had a position at the Goethe University in Frankfurt and had not approached Einstein, was recommended to him by Leo Szilard. This situation and the problems of working with Einstein were discussed much later in a very lengthy and interesting letter to Carl Seelig. Describing his first meeting with Einstein, Lanczos wrote:

> We talked briefly about general things, about the technical side of my employment, and then he brought up his ideas concerning distant parallelism.
>
> I already had a very specific program in mind when I came to Einstein, the so-called "problem of motion" of the general theory of relativity. He had already written a paper with Grommer on this, but the mathematical method used seemed to me insufficient and inconclusive, and I had a plan to illuminate the whole problem from a different side. In this I succeeded, I finally published the results in 1931 in a paper in the *Zeitschrift für Physik* [apparently referring to Lanczos 1930]. But Einstein had the oddity (a peculiarity which showed itself especially in his later years) that he brought almost no interest to thoughts which did not fit exactly into his frame of reference. At that time, I had attacked the dynamical problem using a different method, and there is little doubt that it was at least mathematically superior to Einstein's.... He was looking for somebody who would enter deeply into his train of thought and would not take any detours. But I was working intensely on the "problem of motion," which for the reasons mentioned above did not interest him. On the other hand, I could not become enthused with distant parallelism, since the idea appeared to me as too artificial. So Einstein had gotten the impression that I did not have the necessary "enthusiasm" for the cause, while actually the reasons were more basic, and I really was extremely glad to be able to tackle fundamental problems without interference. (Lanczos to Seelig, September 9, 1955, Seelig Archive, ETH-Bibliothek Zurich)

So they did not hit it off, their collaboration ended after one year, and Walter Mayer took Lanczos's place. But Lanczos never really got over what he took to be Einstein's slighting of his work, as will be discussed in Section 3. Einstein, on the other hand, did not return to the problem of motion for a decade.

Most young scientists starting their careers in the late 1920s preferred to work in the newly created quantum mechanics of Schrödinger and Heisenberg rather than in general relativity. An exception was Myron Mathisson, a young Polish physicist working in complete isolation in Warsaw. By 1927 he had completed, but not published, an important paper providing a new derivation of the geodesic law, and other important investigations were carried out in the next two years, but he had not yet obtained his Ph.D. and was eking out a living mainly by private lessons in Hebrew.

In desperation he turned to Einstein for advice and help in a moving eleven-page letter of December 18, 1929 (EA 18-001; written in beautiful French), detailing his achievements and problems. But it was not the letter of a supplicant. It started out: "*Monsieur*, Your paper [Einstein 1927] concerns a problem for which I have established a more complete solution two and a half years ago." Hardly the deferential approach to the most outstanding scientist of the century one might have expected from a student, but one sure to catch his attention! And indeed, Mathisson's evaluation of his own contribution was quite correct. He first gave a criticism of Einstein's approach, which included the judgment: "one finds again in your work the same defect which is inherent in Weyl's deduction of the equations of dynamics," a recognition both of the mathematical weaknesses of the approaches and of Weyl's priority in the derivation of the law of motion. Then he noted that "my calculations clearly proved that one could push the successive approximations further: they do not yield any additional equations which could make the quantum phenomena conceivable—a possibility you had envisaged near the end of your paper."

After due apologies for having "reproached" Einstein, Mathisson gave a more detailed description of his results and promised to send a first paper on dynamics, ready for the printer, in a week. But he asked for an assured honorarium in case of acceptance because of his precarious financial situation: "How sad life is here for the men of my race!" But due to "unfavorable circumstances," the paper was only sent to Einstein almost two months later (Mathisson to Einstein, February 14, 1930, EA 18-002; in French).

However, Einstein had clearly been deeply impressed by Mathisson's first letter. Unfortunately, his response has been lost, but it is apparent from Mathisson's reply of February 14 that Einstein had invited Mathisson to join him in Berlin to participate in his work. Mathisson was "deeply troubled," because he felt that he was not yet worthy of this. Einstein replied on February 20 that from a superficial reading of the paper he had just received it seemed to him that Mathisson had "an unusual formal talent and that you are made for scientific work" (EA 18-003; in German); on the other hand, it is obvious from his subsequent remarks that he had missed the point of Mathisson's paper and thought that it was only concerned with special relativity. Mathisson answered on February 23 (EA 18-004; in French), giving the requested personal data and responding "with joy" to the advice given by Einstein concerning his doctorate, asking him to write to Professor Czeslaw Białobrzeski in Warsaw. Mathisson also realized that Einstein had skipped over the main point of his paper, and implored him to read its conclusion, quoting from it at great length and elaborating on several points. Einstein replied immediately, stating that now he realized "that you have achieved decisive mathematical progress through the methods used in your paper" and repeated his offer (February 27, EA 18-005; in German). The same day he also wrote a very warm letter to Białobrzeski, asking for acceptance as a thesis of the Polish version of the paper Mathisson had sent him, even offering to pay any necessary expenses himself (EA 18-006). He apparently had also begun inquiries concerning a Rockefeller fellowship for Mathisson, because he asked Białobrzeski whether there was any prospect of an academic career for him, since this was, as he wrote, a necessary condition for obtaining such a fellowship.

Mathisson did obtain his degree, but all efforts to obtain a Rockefeller fellowship for him were unsuccessful. However, in 1933 he obtained an invitation to spend some time with Jacques Hadamard in Paris. Two years later, Einstein tried to contact him via Hadamard, since there appeared to be a possibility to invite him to spend a year in Princeton, where Einstein now was residing permanently (Einstein to Hadamard, November 3, 1935, EA 18-053). In the meantime, Mathisson had returned to Warsaw and then accepted a professorship in Kazan, where Einstein's letter reached him seven months later. He answered that his teaching obligations prevented him from coming to Princeton before the academic year 1937–1938 (Mathisson to Einstein, June 23, 1936, EA 18-054), and Einstein responded expressing his joy "that you have found such a wonderful sphere of action in Soviet Russia. Under these circumstances, it would certainly not be right to invite you here, since there are many capable colleagues who are cut off from any possibility of work" (Einstein to Mathisson, July 7, 1936, EA 18-055; in German). This, after all, was the fourth year of Hitler's Third Reich. But it was also the year in which Stalin's purges were cresting. Some time in 1938, Mathisson had to leave Kazan precipitously, leaving all his "things and books" behind. He returned to Poland and went to Cracow at the invitation of Professor Weyssenhoff, who also helped him to obtain a research grant (Średniawa 1985). He became *Privatdozent*, married, and went to Paris in 1939 on a leave of absence. When the war broke out, he was on a visit in England, where he was in touch with Dirac. He was never able to return to France and, after spending four months in a sanatorium in Cambridge, he died in September 1940 at the age of forty-two.

Mathisson's scientific work, his life, and his correspondence with Einstein will be described in detail elsewhere. Here I shall only discuss those aspects of his work that are relevant for the problem of motion. The paper he had sent to Einstein was published shortly thereafter (Mathisson 1931a). It was written in the spirit of Weyl's approach (although no explicit reference to Weyl is given), and provided an improved derivation of the geodesic law for a test particle. Here Mathisson introduced a new method for dealing with the (singular) $T^{\mu\nu}$ of a point mass $m$ that is a simple pole of the field. He showed that the law of motion could be considered as an integrability condition of the field equations. Furthermore, instead of assuming the form of $T^{\mu\nu}$ as Eddington and many others had done, he started out with an unspecified tensor, split up into components parallel and perpendicular to the four-velocity $v^\mu$ of the particle, and then deduced that only the part $m\, v^\mu\, v^\nu$ survives. This powerful method was used and generalized by him in a number of papers. In the next one (Mathisson 1931b) he dealt with the case of a charged particle, and, apparently for the first time in any paper on the problem of motion, he made use of Dirac's $\delta$-function. In the special-relativistic limit, he obtained the well-known "Lorentz-Dirac" equation, including radiation reaction (first obtained in its relativistic form by Schott 1915), more than seven years before Dirac 1938, the famous paper which rekindled interest in classical special-relativistic equations of motion. The derivation is actually more satisfactory than Dirac's, both because overall conservation of energy-momentum does not have to be assumed, but arises as a consequence of the general-relativistic equation (4), and because the problem of infinities that would have to be "renormalized" is entirely avoided. For arbitrary special-relativistic field theories (not necessarily linear) the method is fully developed in Mathisson 1940 (see also Havas 1962). The charged-particle case is further considered in Mathisson 1931b; there it is also shown that the field equations (2) are not overdetermined, i.e., that only the laws of motion, but no quantum conditions, can be deduced from them. In a further paper (Mathisson 1937), probably written in Kazan, the method of the first paper is generalized to obtain the general-relativistic laws of motion of spinning test particles, i.e., particles that are dipole singularities of the field, and generalization to higher multipoles is indicated.

In this series of papers, Mathisson both obtained new results, and developed much more powerful methods than had ever been used before for the "problem of motion." In spite of his correspondence with Einstein and his personal acquaintance with Dirac, these papers were almost completely overlooked, and had essentially no impact on work on this problem for more than two decades.

In 1936, a new derivation of the law of motion of a test particle, both in the absence and in the presence of nongravitational fields, was given by H.P. Robertson of Princeton University (later of the California Institute of Technology), who started from equation (2), and treated the particle as a singularity (Robertson 1936). He referred to Weyl 1923 and Eddington 1923, but considered his derivation to differ from theirs "mainly in its formulation of the postulate concerning the behaviour of the energy-momentum-stress tensor on the transition to the corpuscular description." Mathisson's work is not mentioned, but—more surprisingly—neither are Einstein and Grommer 1927 or Einstein 1927, although a discussion with Einstein and Mayer on a question relevant for the derivation is acknowledged.

The "Festschrift" on the occasion of Einstein's seventieth birthday contained another derivation of the geodesic law (Infeld and Schild 1949). Infeld, a Polish physicist, had worked with Einstein since 1936, first in Princeton and after 1938 by correspondence (except for brief visits in 1938 and late 1948) from Toronto, where he had accepted a professorship; Schild was one of his students there. Infeld and his contributions will be discussed at length in Section 3. For present purposes it need only be noted that he knew Mathisson from Poland; they were the only two Jewish *Dozenten* in physics in that country, and were both working in general relativity. From a footnote in EIH (to be discussed later) as well as from a book published thirty years later (Infeld 1968) it is also clear that Infeld was aware of Mathisson's publications; however, as he noted in his book, "at that time they were too sophisticated for me. I did not understand them" (p. 204; this passage is not translated in Infeld 1978).

Infeld and Schild's derivation considers matter as consisting of point particles represented as singularities of the field. In spite of this, it considers the vacuum field equations (1) as "alone sufficient to represent neutral matter." Reference is made to Weyl 1923, Eddington 1923, and Robertson 1936, who started from equation (2), and to Einstein and Grommer 1927, and Lanczos 1927a, who started from equation (1), but not to any of Mathisson's papers. There is no discussion of why the derivations in the papers cited are considered to be less satisfactory than that of the authors. There is also no discussion of why point particles should be described by the vacuum field equations rather than by equations with singular sources; this in spite of the fact that such representations of point particles, especially in electrodynamics (which also requires a singular charge-current density) had been used in Mathisson 1931b and had become commonplace since Dirac's 1938 paper, of which Infeld was well aware (Infeld and Wallace 1940). The method used by Infeld and Schild to derive the geodesic law is a limiting process, which considers "a sequence of particles, with masses tending to zero, and a corresponding sequence of gravitational fields." This approach does not differ in principle from the limiting process used in Robertson 1936.

## 3. The Equations of Motion of Interacting Bodies of Comparable Masses

The exact spherically symmetric solution of Einstein's equations, corresponding to the gravitational field of a single body, was found essentially simultaneously by Droste and by Schwarzschild in 1916, very shortly after Einstein had published the final form of his field equations. In Droste 1916a and 1916b, the body was treated as a singularity; in Schwarzschild 1916a and 1916b, the cases of both a singularity and a finite-sized static sphere were treated and the external field was found to be the same. (For detailed technical and historical discussions, see Arzeliès 1963 and Eisenstaedt 1982.) The latter result, analogous to the Newtonian one, was later proved to hold outside any spherically symmetric mass distribution, static or not, and became known as "Birkhoff's theorem" (for a critical review, see Goenner 1970). The geodesic motion of a test particle in this external field was treated approximately and applied to the motion of Mercury by Droste (1916a); the results agreed with those found by Einstein (1915). The geodesic equations were integrated exactly by De Jans (1923).

No such exact solutions can be hoped for in the case of two or more bodies of comparable masses interacting gravitationally. As noted in Section 2, the only exact result to date is a negative one, Weyl's proof of the nonexistence of a static two-body solution. (We are not concerned here with the problem of two electrically charged bodies; then the gravitational attraction and electrostatic repulsion can balance, and exact static solutions have been obtained for this case.) The problem of approximate solutions was considered at great length by the Dutch school in 1916–1917. The first step was taken by Droste (1916a, 1916c). As noted in Section 2, a geodesic is the trajectory for which the length $ds$, defined by equation (5), is an extremum. Droste developed an approximation method to determine $ds$ by calculating the metric $g_{\mu\nu}$ for the case of $n$ slowly moving bodies. Definite assumptions were required for $T^{\mu\nu}$. Droste considered a fluid made up of separate bodies that are spherical at rest. The spherical shape cannot be maintained for bodies in motion, since no rigid bodies are possible in relativity (and thus no "Birkhoff theorem" can exist for a relativistic many-body system); but the effect of the change of shape on $g_{\mu\nu}$ is negligible to the order of approximation considered.

In lowest approximation, Droste obtain the Newtonian results already found by Einstein (1916a). The post-Newtonian corrections to order $v^2/c^2$ showed not only velocity-dependent terms, which were to be expected, but also terms describing three-body interactions (written down by Droste without any comment).

Droste's results were generally accepted and formed the basis for the calculation of relativistic effects in the solar system. Using Droste's metric (obtained by a slightly different method) and the geodesic law, de Sitter (1916)[7] wrote down explicitly the equations of motion for an $n$-body system to order $v^2/c^2$; these differ from the EIH equations only in one term, due to a minor mistake, which went undetected for two decades. No justification was given for accepting the validity of the geodesic law for the $n$-body case; it was provided only several decades later. De Sitter's equations were used by him to calculate various astronomical consequences of general relativity; these consequences were discussed at much greater length by the French astronomer Chazy (1928, 1930) a decade later, using substantially similar methods.

Droste, working with Lorentz, attacked the problem anew from a slightly different angle. Their paper (Lorentz and Droste 1917) was first published in Dutch, like all other communications to the Dutch Academy; however, unlike all other contributions of Lorentz and his collaborators, it was not followed by publication of a translation in the English version of the Academy's transactions, presumably because of the war. Such a translation appeared only in Lorentz's *Collected Papers* in 1939, just before the outbreak of another war, and its relevance escaped the notice even of Fokker, who was one of the editors.[8]

Lorentz and Droste attacked the same problem as Droste had, but stated the assumptions more clearly and carried the calculations further. They assumed

> that the dimensions of the bodies are small in comparison with their mutual distances and that, therefore, the effects of "tidal actions" can be neglected. Further that we can take the motion of each body to be a translation and finally we will assume that the bodies consist of incompressible liquid.... One may expect, however, the final equations to be of the same form in the case of bodies of a different nature ... it proves possible to reduce the equations of motion to the canonical form with a Lagrangian function, depending only on the instantaneous mutual distances and the velocities.

The paper made use of Lorentz's earlier work on variational principles (Lorentz 1915a, 1915b). After rather extended calculations it arrived at a

Lagrangian for interacting particles that is precisely the Lagrangian for the EIH equations (apart from terms that are singular for point particles and can be removed by "mass renormalization"). Starting from these equations, the latter Lagrangian was obtained first by Fichtenholz (1950), thirty-three years after Lorentz and Droste; for finite fluid droplets it was obtained by Plebański and Bażański (1959; see also Bażański 1962b) in a form slightly different from, but equivalent to, that obtained by Lorentz and Droste. The equations of motion, of course, follow trivially from the Lagrangian, and thus the "EIH" equations were actually obtained within two years of the creation of the theory of general relativity, twenty-one years before EIH.

Lorentz, however, communicated their results to Einstein even before their publication (Lorentz to Einstein, March 22, 1917, EA 16-458). Incredibly, Einstein gave away the original letter to an autograph collector without even keeping a copy. After his death, his former secretary, Helen Dukas, wrote the owner for a copy, and unfortunately accepted a transcription from which the formulas had been omitted. However, from an excerpt reproduced in an auction catalog, it is clear that Lorentz both provided the expression for the Lagrangian $L$ that he had obtained, and noted its adequacy for calculating the motion of Mercury. Einstein responded on April 3, 1917:

> Your joint work with Droste is very interesting. It is nice that $L$ contains only first time derivatives. This is an analogue to the Hamiltonian function of the gravitational field. It is too bad that the deviations from Newton are so small, but after all one must be glad that at least the case of Mercury was available. (EA 16-460)

But in spite of these approving comments, Einstein apparently completely forgot Lorentz's letter, because he never again referred to Lorentz and Droste's work. Curiously, neither did de Sitter, who of course had no problems with reading a paper in Dutch, and who should have noticed that the Lorentz-Droste Lagrangian implies slightly different equations of motion than he had used just a few months earlier.

Another approach to the $n$-body problem had been taken by the Italian mathematician Levi-Civita, professor of mechanics in Padua and one of the originators of the absolute differential calculus (Ricci and Levi-Civita 1901), so important for general relativity. Summaries of his results were presented in two lectures at Harvard in 1936 (Levi-Civita 1937a, 1937b). Unfortunately his approximation methods, quite correct in principle, contained some minor mistake, which led him to the conclusion (Levi-Civita 1937b) that there is a secular acceleration of the center of mass of a two-body system. This result, which is also implicit in de Sitter's paper, prompted Eddington and Clark (1938; see also Clark 1941) to reexamine de Sitter's work (de Sitter had died in 1934). They noticed the mistake in his derivation of the equations of motion, and after correcting it, obtained the "EIH" equations (except for some trivial misprints and mistakes in sign [Robertson to Eddington, June 28, 1938, H.P. Robertson Papers, California Institute of Technology Archives]). Their paper, which also contained a calculation of the orbits for the two-body problem, was submitted nine months after Einstein, Infeld, and Hoffmann had submitted theirs.

Leopold Infeld, two months younger than Mathisson, was born in Cracow and obtained his Ph.D. there in 1921. After several years of school teaching he obtained a senior assistantship and in 1931 (with some difficulty due to the prevailing anti-Semitism) the habilitation for docent at the University of Lwów. In 1933 he obtained a Rockefeller fellowship and took a leave of absence to spend two years in Cambridge, working mainly with Max Born. After his return to Lwów, he wrote to Einstein, asking for help; Einstein then obtained a one-year fellowship for him at the Institute for Advanced Study, where he went in 1936. When the fellowship was not renewed, he suggested to Einstein that they write a popular book to obtain the funds necessary for him to stay on; this collaboration resulted in *The Evolution of Physics* (Einstein and Infeld 1938), which was both a popular and a financial success. In 1938 he accepted a professorship in Toronto and left the United States. He visited Princeton again the next year. While driving back to Canada with his wife, news reached them that the Germans had occupied Cracow. Shaken by the occupation and partition of Poland and the growing terror against the Jews, realizing that he "grew up in a world which is being destroyed," he decided that he "must throw one more look upon the image of the path which led me to the present day." He wrote his autobiography, *Quest* (Infeld 1941), almost one third of which is devoted to his three years in Princeton. Thus it should be a prime source of information on the genesis of the EIH paper (Einstein, Infeld, and Hoffmann 1938), the "Note" following it (Robertson 1938), and its sequel (Einstein and Infeld 1940).

Unfortunately, *Quest* is replete with (sometimes obvious) errors of omission. To mention the most glaring ones before proceeding chronologically: Banesh Hoffmann does not appear in it at all; the impression is left that only Einstein and Infeld worked on the problem of motion at that time.[9] Similarly, one is left with the impression that Infeld was Einstein's only collaborator during his Princeton sojourn, although both Valentin Bargmann and Peter Bergmann were working with Einstein in that period. Infeld's earlier acquaintance with Mathisson is also left unmentioned.

In describing (necessarily in laymen's terms) the problem of motion and the beginning of his collaboration with Einstein, Infeld wrote:

> Einstein believed that the equations of motion are contained in the field equations and for years looked for a proof of this statement which would considerably simplify the logical structure of the general relativity theory.... Before we began our thorough investigation of this problem, Einstein already had his idea of a new approach to the field equations by applying a "new approximation method." He hoped to achieve two results: first, to show that the equations of motion are contained in the field equations; second, to find a hidden treasure which would allow us to build a bridge from the classical gravitational theory to quantum theory. Einstein believed that from the laws governing the stars and planets we could deduce laws governing the inside of the atom. (Infeld 1941, p. 280)

Infeld here leaves the impression that only Einstein had been concerned with the problem of motion. EIH was a little more generous. It cited, under "Previous attacks on the problem," Droste 1916b; de Sitter 1916; Mathisson 1931a, 1931b, 1931c, and Levi-Civita 1937a, 1937b, dismissing them all, however, without further comment, as having been "based upon gravitational equations in which some specific energy-momentum tensor for matter has been assumed," i.e., our equation (2). This list—prepared by Infeld according to the story in *Quest*—is very strange indeed. It quotes the wrong paper by Droste (the one giving the "Schwarzschild" solution for a single center), and omits the crucial ones (Droste 1916c, Lorentz and Droste 1917); it quotes Mathisson (the only mention of him in any paper by Infeld), who was essentially only concerned with test particles, but omits all other such papers, including those of Weyl, Eddington, Lanczos, and even Robertson, with whom Infeld was on very friendly terms, and whose paper (Robertson 1938) on the integration of the EIH equations for the two-body problem appeared in the same issue of the *Annals of Mathematics*, immediately following EIH.

EIH reject the use of a specific energy-momentum tensor and "show ... that the gravitational equations for empty space are in fact sufficient to determine the motion of matter represented as point singularities of the field." The approximation method used requires the integration of a sequence of what EIH called Laplace's equations, but which actually are Poisson's equations with singular sources; therefore, use of an energy-momentum tensor is not really avoided, a singular one corresponding to simple poles of the field being implied in the method. Mathisson (1931b) had already used $\delta$-functions to represent such tensors; it is hard to understand why EIH failed to see that they were dealing with such tensors, and to use such a representation.

The claim just quoted is precisely the same as that made in Einstein and Grommer 1927 (which also is not quoted in EIH!). It is of course quite legitimate to restrict oneself to the mathematical problem of whether the vacuum equations (1) alone determine the motion of matter singularities uniquely (although the types of singularities, i.e., whether they are simple poles or also carry dipoles and other multipole moments, would have to be specified). It is quite another question whether the results are also applicable to extended bodies, as noted before in our discussion of Einstein and Grommer 1927. This is taken for granted in EIH, whereas Lorentz and Droste (1917) were fully aware of the appearance of tidal forces in higher approximations in the actual physical problem of describing the motion of a many-body system like the solar one, as was Robertson (1938). Thus the mathematical program of EIH, if carried out successfully to a sufficiently high order of approximation, is bound to come into conflict with the physical program of determining the laws of celestial mechanics following from general relativity.

To carry out their program:

> There remained the problem of special calculations, which were very difficult and required a technique which had to be especially developed for this purpose. Here

> Einstein trusted me completely and took little part in the work. He never made any attempt to check the calculations but was very much interested in discussing all the difficulties which arose. The computations were checked and rechecked many times. (Infeld 1941, p. 283)

There is no reference anywhere to Hoffmann's participation in these calculations. On the other hand, Infeld kept on stressing their difficulty. Even many years later, in the very "Introduction" to Infeld and Plebański 1960, he stated: "The calculations were so troublesome that we had to leave on reference at the Institute for Advanced Study in Princeton a whole manuscript of calculations for others to use" (as was of course also mentioned in EIH), and in his later reminiscences[10] he wrote:

> We had taken only two steps—the first Newtonian and the second, the one that followed it. As a basis of comparison, I might say that if the first step is as difficult as swimming the English Channel then the second is as difficult as swimming the Atlantic Ocean. We took the second step; the third would be as difficult, in comparison, as interplanetary travel.

Surely this is one of the more unusual published evaluations of one's own work!

However, the manuscript deposited at the Institute is actually not very lengthy. It consists of a number of separate, rather short, calculations containing details left out of EIH, and in print would have increased the length of the paper by at most one half; papers of such length are quite common in physics. (Since, according to Hoffmann, the manuscript was completed only after EIH was submitted, possibly it was not included simply to accelerate the publication.[11]) The separate calculations are partly in Infeld's handwriting, partly in Hoffmann's. According to Hoffmann, all calculations were done by both of them. It is not possible now to assess their relative contributions. Hoffmann, almost nine years younger than Infeld, clearly was scientifically the junior partner, but this does not necessarily imply that he did less than his fair share of the mathematical labor, and he contributed some key ideas. In the original manuscript, the listing of the authors had been alphabetical, as is standard, and he was deeply shocked and embittered when he found out that, apparently at Infeld's insistence, the order was changed to "Einstein, Infeld, and Hoffmann" without informing him. His relationship with Infeld, while correct, remained cool for the rest of their lives, although much later Infeld (1978) referred to "my old friend B. Hoffmann"; there he also mentioned him as a coauthor (and also mentioned Bargmann and Bergmann's collaboration with Einstein). Hoffmann left Princeton in 1937 to accept a position at Queens College. He did not collaborate with Einstein again, and in his later extensive research work he rarely touched the problem of motion.

As Infeld had noted, Einstein still had hoped to "build a bridge ... to quantum theory." There is still some hedging in EIH:

> It is of significance that our equations of motion do not restrict the motion of the singularities more strongly than the Newtonian equations, but this may be due to

260     Peter Havas

our simplifying assumption that matter is represented by singularities, and it is possible that it would not be the case if we could represent matter in terms of field theory from which singularities were excluded.

But Infeld claimed credit for converting Einstein:

My contribution concerned one essential aspect only. I furnished the proof that the problem of motion can throw no light on the quantum theory. Here my skepticism won ... the proof held. He said:
"Yes, I am now convinced that we cannot obtain quantum restrictions for motion from the gravitational equations." (Infeld 1941, p. 282)

Neither Einstein nor Infeld remembered that this had already been proved by Mathisson (1931b). But in any case, Einstein, while still pursuing his quantum dream, apparently never again related it to the problem of motion.

The EIH equations were integrated for the two-body problem in Robertson 1938. No secular acceleration of the center of mass was found, in contrast to the previously mentioned results of Levi-Civita and those implied by the equations of motion of de Sitter (1916) and Chazy (1928, 1930). In a footnote Robertson noted that, compared with EIH, "Chazy's equations differ ... only in the absence of the fifth term on the right—an interaction term depending on the product of the two masses which is, as I am informed by Dr. Infeld, the term most difficult to determine," seemingly implying that it could not have been found by Chazy's or de Sitter's method. However, as noted before, Eddington and Clark had corrected de Sitter's mistake and had come to the same conclusion as Robertson (1938). Eddington so informed Robertson, who responded in a polite letter of June 28, 1938 (H.P. Robertson Papers, California Institute of Technology Archives) but wrote to Infeld two days later much less politely:

Dear Sir Arthur and one of his henchmen claim to have discovered and rectified a mistake in de Sitter's $n$-body computations, which when applied to the 2-body problem yields exactly your equations—after making some corrections to the trivial mistakes the dunderheads have made. I have written him the sweetest letter congratulating him on his achievement—having no yen to be taken for either a load of manure or a nigger, I altruistically refrained from anything so vulgar as to claim priority for us. I am enclosing a copy of this masterpiece of gentlemanly correspondence. (Robertson to Infeld, June 30, 1938, H.P. Robertson Papers, California Institute of Technology Archives)

It is not clear what caused the rude tone of this letter and the contemptuous treatment of Eddington (who had contributed far more to astronomy and relativity than Robertson), especially since Eddington had never claimed priority of publication.

In the meantime, Levi-Civita had also seen EIH's and Robertson's papers and received an advance copy of Eddington and Clark's paper. He wrote about this to Robertson, obviously troubled by

these independent warnings [from which] I must unfortunately infer that in my previous calculations ... some numerical coefficient was wrong. I intend to undertake as soon as possible a general revision of my investigation.... I hope to be thus able to establish clearly the conceptual, or, more probably, computational source of the discrepancy. (April 5, 1938, H.P. Robertson Papers, California Institute of Technology Archives)

This was planned as a volume in the Mémorial des Sciences Mathématiques. But before completing this volume, Levi-Civita wrote another paper, which he sent to Robertson (a member of the editorial committee of the *American Journal of Mathematics*), for inclusion in the Hill issue. It arrived too late for that, as Robertson noted in his answer (Robertson to Levi-Civita, October 20, 1938, H.P. Robertson Papers, California Institute of Technology Archives). This paper still claimed a secular acceleration, but Robertson wrote that:

> In view of your suggestion that I might control your computations, with a view of uncovering the discrepancy, I have ventured to go through them with this in mind. I do find one point in which your treatment differs materially from that of the rest of us—and by just enough to account for the discrepancy in our results!

He then gave a three-page discussion proving his point. The paper was never published. Levi-Civita completed the planned volume (Levi-Civita 1950) before his death in 1941, but because of the war, publication was delayed by almost a decade. In that volume he showed that his method too did not yield a secular acceleration. However, his equations of motion do differ from those of EIH. The reason for this is not discussed by Levi-Civita, and there has been no subsequent study comparing the two methods and their respective merits in any detail.[12]

Lanczos's reaction to EIH was, as he wrote to Robertson, that:

> I cannot suppress some bitter remarks ... that I want to express to you, quite confidentially. I made a number of elaborate studies in the motion problem of relativity [listing Lanczos 1927a, 1927b, 1930], which have some bearing on the subject, and still the authors have not found it worth while even to *mention* my name, not even in the bibliography of their paper.

He then noted that in Lanczos 1923 he had developed a perturbation method for the metric and that in Lanczos 1930 he

> was able to show that the "singularity method" is *completely equivalent* with the consideration of the inhomogeneous equations [our equation (2)] avoiding any singularities. I have shown that the divergence-condition for the right side actually permits to deduce a law of motion.... There was not the slightest reason to go back to the "singularity method" again, applying a perturbation method which can hardly be surpassed in clumsiness and inadequateness. Prof. Einstein did not like my method from very beginning—merely by emotional reasons. But it would have been decent at least to *mention* it in this paper. (Lanczos to Robertson. March 24, 1938, H.P. Robertson Papers, California Institute of Technology Archives)

Robertson tried to soothe Lanczos's feelings:

> [I] ... cannot convince myself that things are quite so bad as you seem to feel. In the first place, I am certain that Einstein intended no slight to you by not referring

to your papers, for I know definitely that he has a very high opinion of your work. But as you undoubtedly have noticed, he usually plows along his own row without relying very much on the literature—you will have noticed the paucity of references to other people, such as Weyl, who have also been concerned with the problem. (Robertson to Lanczos, May 3, 1938, H.P. Robertson Papers, California Institute of Technology Archives)

Lanczos seemed pacified:

[I] want to apologize for my foolish remarks in my letter before, with regard to Prof. E. One should not take such personal matters so seriously and at present the whole cause seems to me rather unimportant. But as far as the method of attack is concerned, I do believe that the authors have not chosen the adequate procedure. (Lanczos to Robertson, May 19, 1938, H.P. Robertson Papers, California Institute of Technology Archives)

But it still rankled seventeen years later. In the letter to Seelig quoted before, after stating that his method was superior to Einstein's, he wrote:

Nevertheless, the strange thing happened that Einstein later, when he with Infeld and Hoffmann treated the same problem again, unrolled all the earlier literature, but just my paper he did not consider worthy of mentioning. This I would have considered insulting, if I had not known his peculiarity precisely.

An independent approach to the problem of motion at that period was undertaken by the Russian physicist Fock. Starting from equation (2), he obtained an approximate expression for the metric tensor and derived the Newtonian equations of motion for a system of massive bodies (Fock 1939). In an addendum to the paper he noted that he only became aware of EIH while his article was in press, and briefly commented that while EIH went further in obtaining corrections to Newton's equations,

on the other hand their results are limited by the hypothesis of point masses. It is clear, for example, that this hypothesis excludes the possibility of determining the matter tensor in the interior of the masses ... it seems to us that ... our calculations are much simpler than those of the three authors.

His calculations were extended by one of his students, who obtained precisely the EIH equations in a 1940 dissertation, but publication was delayed due to the war (Petrova 1949); they were also obtained in a slightly different way by Papapetrou (1951). Fock's method is discussed in detail in his book (Fock 1955), which also includes a "critical survey" of the problem of motion; unfortunately, it ignores almost the entire non-Russian literature apart from the papers of Einstein and his collaborators. Papapetrou is charged with an "inconsistency of approach."

Einstein and Infeld clung to their approach of using singularities in their next two papers (1940, 1949). However, a few years later Infeld made his peace with the other approach:[13]

The comparison of Fock's and our method shows one essential similarity. In both methods the same approximation procedure is used. But the two methods also show essential differences. The first consists in the fact that whereas we use the field equations for empty space, regarding masses as singularities, Fock uses the gravitational equations with the energy-momentum tensor. This difference I regard as unimportant ideologically for the following reason. It is obvious that no charge or gravitational body can be properly described by a singularity. We can use the singularity picture only if we describe the field outside the particle, where the energy-momentum tensor vanishes. It is obvious from our derivation that it would have been almost unchanged had we used surfaces covering the regions in which the energy-momentum tensor does not vanish. Indeed Fock's method gives precisely the same result as ours though the mathematical technique is very different. (Infeld 1954)

In the same paper he introduced the use of $\delta$-functions and noted that this "forms a bridge between Fock's method and ours.... *A skilful use of $\delta$-functions in General Relativity Theory simplifies immensely the derivation of the equations of motion*" (his emphasis), without any comment on why it took him seventeen years since starting to work on the problem of motion to arrive at this conclusion, although the tools were available much earlier and had been used for this problem in Mathisson 1931b.

But back to EIH. All three authors had been dissatisfied with the complexity and awkwardness of the calculations in that paper. During his first year in Toronto, Infeld, as he later wrote,

> tried to eliminate special assumptions concerning the coordinate system and to formulate the theory for a general coordinate system.... I sent Einstein the first sketch of my manuscript. He liked the idea but made two comments that immediately changed the appearance of the whole work. (Infeld 1978)

Infeld visited Einstein in May 1939 to discuss this further. They finished the paper by correspondence (Einstein and Infeld 1940). It summarized the new approach thus: "the change introduced here is: nothing is assumed about the coördinate system except that it is galilean at infinity."

In August 1948, Einstein received a letter from the mathematician Horace Levinson (August 25,1948; EA 16-300), whom he had met previously, which contained a detailed criticism of the mathematical method used in EIH and in Einstein and Infeld 1940. Einstein responded immediately (August 31, EA 16-301) and a lengthy correspondence on this ensued. Einstein also wrote to Infeld, describing Levinson's objections, and they decided to accept the criticism and to publish a new derivation of the EIH equations. Some of the way in which this work proceeded by correspondence and during one meeting in New York is described in Infeld 1978. Curiously, the editor of that book misidentified Horace Levinson as the mathematician Norman Levinson, and stated in a note that "Levinson's observations were communicated orally," ignoring the extensive correspondence available in the Einstein Archives; there is also no indication that Einstein and Levinson actually met in 1948.

Levinson had been working in relativity for many years. He obtained a Ph.D. in Mathematical Astronomy in Chicago (Levinson 1922) under F.R. Moulton and a "doctorat d'université" in Paris (Levinson 1923) under M.

Brillouin; the theses were concerned with the gravitational field of $n$ and of two masses, respectively. He had also written a book on *The Law of Gravitation in Relativity* (Levinson and Zeisler 1929). Thus he was no novice in this area, although he had not made any significant contributions and had worked mainly in private business. Einstein, unlike his reaction to previous criticisms, responded by stating that Levinson's "letter was the cause for my reexamining the whole problem and I have found a relatively simple solution of it" (Einstein to Levinson, August 31, 1948, EA 16-301); however, this solution was abandoned and a method suggested by Infeld was adopted. In the "Introduction" to Einstein and Infeld 1949 they acknowledge that

> Mr. Lewison[14] pointed out to us, that from our approximation procedure, it does not follow that field equations can be solved up to arbitrarily high approximation. This is indeed true. We believe that the present work not only removes this difficulty, but that it gives a new and deeper insight into the problem of motion.

Then they described a variation of their previous approximation method, which initially allows for the presence of dipoles. They then interpreted certain constants appearing in the course of the integration as the gravitational masses of the particles and concluded from this: "*Thus we shall have to exclude from our solution negative gravitational masses. But then we must also exclude dipoles and poles of higher order*" (their emphasis).

It had been known from the earliest studies of the spherically symmetric solutions of equation (1) that the sign of the gravitational mass is not determined by that equation, but must be chosen arbitrarily to agree with observation and Newtonian theory; thus the first sentence quoted is quite correct. But the second one is a non sequitur, and reveals a basic misunderstanding of the meaning of multipoles in general relativity. Certain types of dipoles correspond to an intrinsic angular momentum, and similarly, certain quadrupoles to oblateness; both of these properties are of course crucial for the description of real celestial bodies. What is equally disconcerting about their remark is that they had completely forgotten that the significance of multipole singularities had been discussed in great detail in Mathisson 1937. There a clear distinction had been made between nonphysical and physical dipoles as well as higher moments; this paper had been brought to Einstein's attention (Mathisson to Einstein, September 5, 1937; EA 18-061) and had been seen, though admittedly not understood, by Infeld—but the significance of these results must have escaped them, otherwise they would not have forgotten all about them.

Infeld later described that:

> Once, on the way home ... I suddenly had an idea—to introduce gravitational dipoles. Their existence ought to enable a solution at every approximation. Then, since they have no physical meaning, we must eliminate them.... From the conditions at the time when the dipoles vanished we would then obtain the equations of motion. (Infeld 1978, p. 144)

He wrote Einstein about it (November 1, 1948, EA 14-090), but got no reaction. When he visited Einstein to discuss their work and explained his method in detail, "Einstein exclaimed enthusiastically, 'Well, then, our problems are solved. Why didn't you write me about it?'" (Infeld 1978, p. 145). This anecdote agrees with Lanczos's description of the difficulty of getting through to Einstein if his mind was occupied with other matters (Lanczos to Seelig, September 9, 1955). But, alas, Einstein's enthusiasm was misplaced.

The "dipole method" was criticized subsequently by a number of authors (see especially Moffat 1958, 1959, Moffat and Kerr 1958, Kerr 1960), some of whom succeeded in putting the derivation of the EIH equations on a firmer footing, and Infeld too clearly became dissatisfied later and continued trying to improve the derivation (for a summary see Infeld and Plebański 1960); as he noted later: "In 1949, after I had cooperated with Einstein for thirteen years on the problem of motion ... I thought that the problem was completely solved. But even today, a quarter of a century later, I am still pursuing it with my assistants" (Infeld 1978, p. 141).

## 4. Epilogue

World War II had almost totally disrupted the work of the international community of scholars who had been active in relativity. Of the major figures who had been concerned with the problem of motion, only Einstein and Infeld, and to some extent Fock, returned to it after the war. But Einstein's time was more and more taken up with his help for the Zionist cause and his concerns about the consequences of the atomic bomb. Scientifically, he was mainly preoccupied with work on his latest unified field theory, and with the quantum problem. The Einstein and Infeld paper (1949) was a minor diversion from these interests, as is also shown by his initial lack of attention to Infeld's dipole method, mentioned above. This inattention shows also in another example. The "Festschrift" for Einstein's seventieth birthday contains a contribution (Rosen 1949) that proves that for a particle to be at rest in a static gravitational field the force on the particle must vanish. This is precisely the problem treated and the result obtained in Einstein and Grommer 1927, which is not mentioned by Rosen (neither is any other paper on the geodesic law, while EIH is). This omission is of interest here because Rosen had been Einstein's assistant and had collaborated with him on various problems, including the problem of motion (Einstein and Rosen 1936), before the war; clearly the aspect of the problem that had been treated by Einstein and Grommer had not been brought to Rosen's attention by Einstein then or later, because it was no longer of concern to him.

It is not necessary to pay hommage here to Einstein's genius, his independence of thought, and his total concentration on any subject he was concerned with at the moment. This independence, the conviction that he could get to the roots and thereby to the solution of each problem without the need for a literature search or advice from others, had been crucial in leading him to the special and the general theory of relativity. But once these theories had been created, and he was working out their consequences, ignoring the literature had its perils. Some of the best physicists, astronomers, and mathematicians of the time had begun working on these consequences, and some of them had obtained important results, especially on the problem of motion, before Einstein directed his own attention to the same questions. He was well aware of his own lack of interest in the literature. Infeld, in describing in *Quest* the work on EIH, quotes Einstein as saying "I have forgotten how to use books" (Infeld 1941, p. 277), and when Infeld suggested looking up the literature, "Oh yes. Do it by all means. Already I have sinned too often in this respect."

So it was Infeld who put together the references discussed in Section 3. But the rather odd list, containing only authors whose work had been pointed out to Einstein in letters or discussions (and not even all of those) makes it appear that Einstein had suggested a list of names and that Infeld had just looked up some of their papers. Einstein, of course, was the senior author, and thus ultimately responsible; however, most likely he did not attach enough importance to the references to check up on Infeld.

Einstein was equally unconcerned about the historical record of his and other scientists' work. (Unfortunately, this attitude was often reflected in that of his secretary Helen Dukas.) Several examples of this were mentioned in previous sections. And when approached in writing, apparently most of the time he paid sufficient attention only to those arguments that fitted—positively or negatively—into the framework of the problems with which he was concerned at the moment, and even persons of the stature of Weyl could not direct his full attention to other points—which of course he would have understood easily if he had just focused on them.

Einstein's collaboration with Infeld ceased when Infeld returned to Poland in 1950. There he continued work on the problem of motion, and a large number of his students (and, in turn, some of their students) took up various aspects of it, including the motion of rotating bodies and gravitational radiation. He did not attack the third step he had mentioned to be "as difficult ... as interplanetary travel." But, to continue that quotation, "Thus, it did not seem to me possible that anyone would ever want to take the third step. In fact, my student and I later proved that it will never become necessary to take it." What seems to have been on his mind is that it would not be necessary to go beyond the EIH equations in order to describe the planetary motions in the solar system. But even this is questionable because of tidal forces, the effects of rotation, and the possible oblateness of the sun. In addition, one has to go beyond EIH to attack the problem of gravitational radiation. Indeed, there had been several authors who had taken the "third step" by 1964 (with varying success). Furthermore, both in treating astrophysical problems, which involve velocities significantly higher than those in the solar system, and in the general problem of gravitational radiation, approximation methods other than the "slow" one of EIH involving an expansion in $v/c$ might be needed. By 1960, a significant amount of work on such "fast" approximation methods had been undertaken and indeed Plebański, a former student of Infeld, had included a discussion as well as some of his own work on this in the draft of Infeld and Plebański 1960. But then he left for the United States on a Rockefeller fellowship, and Infeld removed all discussion of "fast" approximations from the book without consulting him.[15]

Nevertheless, as the quotation at the end of Section 3 shows, Infeld did not consider all problems to be solved. However, in spite of this, he vetoed the inclusion of any discussion of the problem of motion in the program of the Third International Conference on General Relativity and Gravitation (which took place in Poland in 1962 and of which he was the main organizer), insisting that this problem had been solved. It appears that he did this to avoid any discussion of alternatives to the EIH approach, including that of Fock, who was present at the conference.

Thus, on the one hand, Infeld made an enormous contribution by establishing his own school of relativists in Poland. On the other hand, his domination of it and of GR3 (as the conference became known) had a retarding effect on the development of approaches other than his own, and there was little awareness in Poland of work other than that of Einstein and his collaborators, and of Fock, on the problem of motion. The academic careers of his students were in his hands as much as his own had been in the hands of some professors about whom he had complained so eloquently in *Quest* (where he had also complained about this system). The motivations were different, but the effect may have been worse, because—unlike the experimental physicists with whom Infeld had to deal in the 1930s—he understood, and wanted to influence, the direction in which the research of his ex-students was going. Such a situation is hardly conducive to any investigations into the real history of the research area of one's hoped-for mentor.

The fact that Infeld totally ignored Mathisson's work in all of his own cannot be due to any scientific reasons, because it should have been mentioned even if Infeld had serious objections to it. There must have been some underlying psychological reason due to some conflict prior to 1938. It may possibly have been related to the fact that earlier they both were rival (unsuccessful) candidates for the Chair of Theoretical Physics at the University of Cracow in 1934 (Średniawa 1985), and apparently also later for that in Wilno. Many of Infeld's students were aware of his animosity toward Mathisson, but none of them knew of any particular reason. Neither did Mathisson's widow, who had met him only in 1938.

Infeld, of course, had devoted thirty years of his life to the problem of motion; Mathisson died after thirteen years' work on it. Infeld founded a school; Mathisson's only collaborator on this problem, Lubański, died in 1946. So the work of the two scientists cannot be easily compared. What is quite apparent, however, is that Mathisson's contributions were far more original than Infeld's and introduced far better mathematical methods.

I have discussed briefly various disruptions of the work of the relativists during the first two decades of general relativity due to World War I, the turbulent post-war years, the depression, and the rise of Stalin and Hitler. Obviously, these historical events had a major negative influence, and it is useless to speculate on what might have been if peace and tranquility had reigned. But I cannot resist just noting what effect a minor fluctuation during those upheavals could have had on the development of the problem of motion. What if Einstein had succeeded in getting Mathisson to join him in Berlin or Princeton, instead of in getting the fellowship for Infeld? There would have been no EIH, but presumably an EMH, with much better mathematical techniques, and much more accessible to the relativists working in that area. And then ...?

As discussed before, after World War II it was only Fock and Infeld (apart from one joint paper with Einstein) who remained actively concerned with the problem of motion among all the major earlier contributors. Among the new generations of relativists, there was a major shift of interest, and only a small fraction of their effort (outside of Poland) was devoted to the problem of motion. Einstein's concerns with unified field theory and the quantum problem were considered to be "far out" and were widely ignored. On the other hand, his various contributions to the problem of motion were taken to be in the mainstream of general relativity and generally accepted at face value. Since the statements dismissing the approaches by de Sitter and Levi-Civita as incorrect implied that they could not have succeeded, apparently nobody went to the trouble of taking another look. Eddington and Levi-Civita, who knew better, had died during the war; Clark was still working on the problem, but published in comparatively obscure journals and was ignored. So was Fock, because his various objections against the standard interpretation of general relativity, in particular his stress on a special ("harmonic") coordinate system, had brought him into disrepute and had cast a cloud over all his work in the minds of most relativists.

Thus there existed a true generation gap: the new generation of relativists working on the problem of motion had grown up without any contact with the previous one—apart from Infeld, who was one of the creators and promulgators of the myths surrounding the early work. This, of course, does not absolve the new generation from not having researched the older literature more thoroughly; but there had been no comprehensive survey of the theory of relativity since Pauli's 1921 article, and every single contributor to the problem of motion after that was a "sinner," who either ignored or did not do justice to the contributions of his contemporaries. This had created such a complex story that it is understandable, if not quite excusable, that it was not disentangled by the new generation of contributors to that problem.

*Acknowledgments.* I am deeply indebted to the following institutions and individuals for their help, and for permission to quote from correspondence:

The Hebrew University of Jerusalem, for permission to quote from the unpublished correspondence of Albert Einstein, and Prof. John Stachel, former editor of *The Collected Papers of Albert Einstein*; the California Institute of Technology for permission to quote from the H.P. Robertson Papers, and Dr. Judith Goodstein, Archivist; the Seelig Archive of the ETH Zurich, and Dr. Beat Glaus, Archivist; Mrs. Ellen Weyl-Baer, the widow of Hermann Weyl; Mrs. Alice Lanczos, the daughter-in-law of Cornelius Lanczos; and Mrs. Irena Gill, the widow of Myron Mathisson.

For conversations on various aspects of the subject of this paper, I am also grateful to the late Dr. Cornelius Lanczos, the late Dr. Banesh Hoffmann, Dr. Jerzy Plebański, and Dr. Stanisław Bażański.

Research supported in part by the National Science Foundation.

Notes

* Slightly expanded version of a paper presented at the Conference on History of General Relativity, Osgood Hill, May 1986.

[1] Some calculations sketched in my talk have been omitted.

[2] All translations of German and French texts given here are my own, even if published translations exist, because of various inaccuracies in the published versions.

[3] Einstein, contrary to current practice, used a definition of the Γs with sign opposite that used in equation (6).

[4] It is not clear at what point Einstein arrived at this belief. According to the diary of Rudolf Humm (quoted in Seelig 1954), who as a student had visited Einstein in May 1917, when Humm remarked that Einstein wanted to use the quantum theory to modify the theory of gravitation, while Hilbert wanted to deduce the quantum theory from the theory of gravitation, Einstein "made an impish face: this would hardly work, even though the theory of gravitation was more general. But the idea of relativity could not yield more than gravitation." But as early as March 1920, he wrote to Max Born: "In my free time I always brood over the quantum problem from the standpoint of relativity. I do not believe that the theory can do without the continuum. But I can't succeed in giving tangible form to my favorite idea, to understand the quantum structure through an overdetermination by differential equations" (Einstein to Born, March 3, 1920; in Born 1969). Thus it appears that he had a change of heart some time between 1917 and 1920. His program of obtaining quantum constraints from field theory was quite clearly formulated three years later (Einstein 1923; see also Pais 1979 and Stachel 1986).

[5] This question will be discussed in detail elsewhere; see Havas 1988.

[6] In note 15 to the English translation, prepared many years after Pauli had stopped working on the theory of relativity, he seems to imply that this was not a derivation, by giving credit only to Einstein, Infeld, and their collaborators for proving that the law of motion "must also follow from these fields equations without further assumptions." Quite uncharacteristically, he provided a very incomplete list of references, totally ignoring Eddington, and, though mentioning Weyl, not recognizing the significance of his work.

[7] This paper, incidentally, acknowledges "the great privilege of several conversations with Professor Einstein himself, during a visit of his to Leiden."

[8] Private communication. See also Havas and Stachel 1969; Havas 1979b.

[9] Infeld's version of this collaboration, discussed later in more detail, was accepted at face value in toto in the editor's Introduction to *Why I Left Canada* (Infeld 1978).

[10] These are contain in two books written in Poland (Infeld 1964, 1968), which have been partially translated as *Why I Left Canada* (Infeld 1978) after Infeld's death in 1968. All quotations here are taken from Infeld 1978, but it should be kept in mind that they were written ten to fifteen years earlier.

[11] Interview with P. Havas, 1978.

[12] An as yet unpublished investigation by T. Damour and G. Schäfer criticizing some aspects of Levi-Civita's work became available to me too late for inclusion in this paper and will be discussed elsewhere.

[13] But not with Fock himself. There was a bitter dispute both about the interpretation of the general theory and about priorities, described in Infeld 1978, pp. 77–80.

[14] Note the misspelling of the name and absence of an initial. This error is repeated later in the paper.

[15] It should also be noted that the bibliography of the book was prepared independently by A. Trautman, and that many of the papers included there are not referred to in the text.

REFERENCES

Arzeliès, Henri (1963). *Relativité Generalisée. Gravitation.* Part 2, *Le champ statique à symétrie sphérique.* Paris: Gauthier-Villars.

Bach, Rudolf (1922). "Neue Lösungen der Einsteinschen Gravitationsgleichungen" (with an Addendum by Hermann Weyl). *Mathematische Zeitschrift* 13: 134–145.

Bażański, Stanisław L. (1962a). "The Problem of Motion." In *Recent Developments in General Relativity.* New York: Pergamon; Warsaw: PWN-Polish Scientific Publishers, pp. 13–29.

——— (1962b). "The Equations of Motion and the Action Principle in General Relativity." In *Recent Developments in General Relativity.* New York: Pergamon; Warsaw: PWN-Polish Scientific Publishers, pp. 137–150.

Becquerel, Jean (1922). *Le Principe de relativité et la théorie de la gravitation.* Paris: Gauthier-Villars.

Born, Max, ed. (1969). *Albert Einstein–Hedwig und Max Born. Briefwechsel 1916–1955.* Munich: Nymphenburger. English translation *The Born–Einstein Letters.* I. Born, trans. New York: Walker and Company, 1971.

Chandrasekhar, Subrahmanyan (1983). *Eddington: The Most Distinguished Astrophysicist of his Time.* Cambridge: Cambridge University Press.

Chazy, Jean (1928). *La Théorie de la relativité et la mécanique céleste.* Vol. 1. Paris: Gauthier-Villars.

——— (1930). *La Théorie de la relativité et la mécanique céleste.* Vol. 2. Paris: Gauthier-Villars.

Clark, Gordon L. (1941). "The Derivation of Mechanics from the Law of Gravitation in Relativity Theory." *Royal Society of London. Proceedings A* 177: 227–250.

De Donder, Théophile (1916). "Les équations différentielles du champ gravifique d'Einstein crée par un champ électromagnétique de Maxwell-Lorentz." *Akademie van Wetenschappen te Amsterdam. Zittungsverslag* 25: 153.

——— (1917). "Théorie du champ électromagnétique de Maxwell-Lorentz et du champ gravifique d'Einstein." *Archives du Musée Teyler* 3: 80–179.

——— (1919). "La Gravifique. 2ᵉ communication." *Académie Royale de Belgique. Classe des Sciences. Bulletin* 317–325.

——— (1921). *La Gravifique einsteinienne*. Paris: Gauthier-Villars.

De Jans, Carlo (1923). *Sur le mouvement d'une particule matérielle dans un champ de gravitation à symétrie sphérique*. Brussels: Lamertin.

de Sitter, Willem (1916). "On Einstein's Theory of Gravitation, and its Astronomical Consequences. Second Paper." *Royal Astronomical Society. Monthly Notices* 27: 155–184, 481.

Dirac, Paul A.M. (1938). "Classical Theory of Radiating Electrons." *Royal Society of London. Proceedings A* 167: 148–169.

Douglas, A. Vibert (1956). *The Life of Arthur Stanley Eddington*. London: Thomas Nelson and Sons.

Droste, Johannes (1916a). *Het zwaartekrachtsveld van een of meer lichamen volgens de theorie van Einstein*. Leiden: E.J. Brill.

——— (1916b). "Het veld van een enkel centrum in Einstein's theorie der zwaartekracht en de beweging van een stoffelijk punt in dat veld." *Koninklijke Akademie van Wetenschappen te Amsterdam. Verslagen* 25: 163–180. English translation *Koninklijke Akademie van Wetenschappen te Amsterdam. Proceedings* 19: 197–215.

——— (1916c). "Het veld van $n$ bewegende centra in Einstein's theorie der zwaartekracht." *Koninklijke Akademie van Wetenschappen te Amsterdam. Verslagen* 25: 460–467. English translation *Koninklijke Akademie van Wetenschappen te Amsterdam. Proceedings* 19: 447–455.

Eddington, Arthur S. (1918). *Report on the Relativity Theory of Gravitation*. London: Fleetway Press. Citations are taken from the 2d Ed., 1920.

——— (1921). *Espace, temps et gravitation. La théorie de la relativité généralisée dans ses grandes lignes, exposé rationnel suivi d'une étude mathématique de la théorie*. Paris: J. Hermann. (French translation of *Space, Time and Gravitation: An Outline of the General Relativity Theory*. Cambridge: Cambridge University Press, 1920). Includes a mathematical supplement not contained in the English edition.

——— (1923). *The Mathematical Theory of Relativity*. Cambridge: Cambridge University Press.

Eddington, Arthur S., and Clark, Gordon L. (1938). "The Problem of $n$ Bodies in General Relativity Theory." *Royal Society of London. Proceedings A* 166: 465–475.

Einstein, Albert (1913). *Entwurf einer verallgemeinerten Relativitätstheorie und einer Theorie der Gravitation. I. Physikalischer Teil*. Leipzig and Berlin: B.G. Teubner. Reprinted with added "Bemerkungen," *Zeitschrift für Mathematik und Physik* 62 (1914): 225–261.

——— (1915). "Erklärung der Perihelbewegung des Merkur aus der allgemeinen Relativitätstheorie." *Königlich Preussische Akademie der Wissenschaften* (Berlin). *Sitzungsberichte*: 831–839.

——— (1916a). "Die Grundlage der allgemeinen Relativitätstheorie." *Annalen der Physik* 49: 769–822. Translated as "The Foundation of the General Theory of Relativity." In Lorentz, Hendrik Antoon, et al. *The Principle of Relativity*. W. Perrett and G.B. Jeffery, trans. London: Methuen, 1923; reprint New York: Dover, 1952, pp. 109–164.

——— (1916b). "Näherungsweise Integration der Feldgleichungen der Gravitation." *Königlich Preussische Akademie der Wissenschaften* (Berlin). *Sitzungsberichte*: 688–696.

———— (1923). "Bietet die Feldtheorie Möglichkeiten für die Lösung des Quantenproblems?" *Preussische Akademie der Wissenschaften* (Berlin). *Physikalisch-Mathematische Klasse. Sitzungsberichte*: 359–364.

———— (1927). "Allgemeine Relativitätstheorie und Bewegungsgesetz." *Preussische Akademie der Wissenschaften* (Berlin). *Physikalisch-mathematische Klasse. Sitzungsberichte*: 235–245.

———— (1949). "Autobiographical Notes." In *Albert Einstein: Philosopher-Scientist*. Paul Arthur Schilpp, ed., Evanston, Illinois: The Library of Living Philosophers, pp. 1–95. A corrected text and translation has been issued as *Albert Einstein: Autobiographical Notes*. Paul Arthur Schilpp, ed., La Salle, Illinois and Chicago: Open Court, 1979. Page numbers are cited from the 1949 edition.

Einstein, Albert, and Grommer, Jakob. (1927). "Allgemeine Relativitätstheorie und Bewegungsgesetz." *Preussische Akademie der Wissenschaften* (Berlin). *Physikalisch-mathematische Klasse. Sitzungsberichte*: 2–13.

Einstein, Albert, and Infeld, Leopold (1938). *The Evolution of Physics: The Growth of Ideas from Early Concepts to Relativity and Quanta*. New York: Simon and Schuster.

———— (1940). "Gravitational Equations and the Problem of Motion II." *Annals of Mathematics* 41: 455–464.

———— (1949). "On the Motion of Particles in General Relativity Theory." *Canadian Journal of Mathematics* 3: 209–241.

Einstein, Albert, Infeld, Leopold, and Hoffmann, Banesh (1938). "Gravitational Equations and the Problem of Motion." *Annals of Mathematics* 39: 65–100.

Einstein, Albert, and Rosen, Nathan (1936). "Two-Body Problem in General Relativity Theory." *Physical Review* 49: 404–405.

Eisenstaedt, Jean (1982). "Histoire et Singularités de la Solution de Schwarzschild (1915–1923)." *Archive for History of Exact Sciences* 27: 157–198.

Fichtenholz, I.G. (1950). "Lagranzheva forma uravneniy dvizheniya vo vtorom priblizhenii teorii tyagoteniya Eynshteyna" ["Lagrange Form of the Equations of Motion in the Second Approximation of Einstein's Theory of Gravitation"]. *Zhurnal Eksperimentalnoi i Teoreticheskoi Fiziki* 20: 233–242.

Fock, Vladimir A. (1939). "Sur le mouvement des masses finies d'après la théorie de gravitation einsteinienne." *Journal of Physics* (Moscow) 1: 81.

———— (1955). *Teoriya prostranstva, vremeni i tyagoteniya*. Moscow: Gosudarstvennoye izdatel'stvo fiziko-matematicheskoy literatury. English translation *The Theory of Space, Time, and Gravitation*. N. Kemmer, trans. New York and London: Pergamon Press, 1959.

Fokker, Adriaan D. (1917). "De virtuelle verplaatsingen van het electromagnetische en van het zwaartekrachtsveld bij de toepassing van het variatiebeginsel van Hamilton." *Koninklijke Akademie van Wetenschappen te Amsterdam. Verslagen* 25: 1067–84.

Goenner, Hubert (1970). "Einstein Tensor and Generalizations of Birkhoff's Theorem." *Communications in Mathematical Physics* 16: 34–47.

Goldberg, Joshua N. (1962). "The Equations of Motion." In *Gravitation: An Introduction to Current Research*. Louis Witten, ed. New York and London: John Wiley & Sons, pp. 102–129.

Havas, Peter (1959). "Classical Relativistic Theory of Elementary Particles." In *Argonne National Laboratory Summer Lectures in Theoretical Physics* 1958, ANL-5982, pp. 124–220.

——— (1962). "General Relativity and the Special Relativistic Equations of Motion of Point Particles." In *Recent Developments in General Relativity*. New York: Pergamon Press: Warsaw: PWN-Polish Scientific Publishers, pp. 259–277.

——— (1967). "Foundation Problems in General Relativity." In *Delaware Seminar in the Foundations of Physics*. Mario Bunge, ed. New York: Springer-Verlag, pp. 124–148.

——— (1979a). "The 'Problem of Motion' in General Relativity—Historical Myths and Realities." Talk presented at the Einstein Centennial Symposium, AAAS Annual Meeting, Houston.

——— (1979b). "Equations of Motion and Radiation Reaction in the Special and General Theory of Relativity." In *Isolated Gravitating Systems in General Relativity*. Proceedings of the International School of Physics "Enrico Fermi," Course 67, 1976. Jürgen Ehlers, ed. Amsterdam: North-Holland, pp. 74–155.

——— (1988). "The General Relativistic Two-Body Problem and the Einstein-Silberstein Controversy." Paper presented at the Second Conference on History of General Relativity, Marseilles-Luminy, France, September 1988.

Havas, Peter, and Goldberg, Joshua N. (1962). "Lorentz-Invariant Equations of Motion of Point Masses in the General Theory of Relativity." *Physical Review* 128: 398–414.

Havas, Peter, and Stachel, John (1969). "Invariances of Approximately Relativistic Lagrangians and the Center-of-Mass Theorem. I." *Physical Review* 185: 1636–1647.

Hesse, Mary B. (1961). *Forces and Fields*. London: Thomas Nelson and Sons.

Hilbert, David (1915). "Die Grundlagen der Physik. (Erste Mitteilung)." *Königliche Gesellschaft der Wissenschaften zu Göttingen. Mathematisch-physikalische Klasse. Nachrichten*: 395–407.

Humm, Rudolf J. (1918). "Über die Bewegungsgleichungen der Materie. Ein Beitrag zur Relativitätstheorie." *Annalen der Physik* 57: 68–80.

Infeld, Leopold (1941). *Quest: The Evolution of a Scientist*. New York: Doubleday, Doran & Co.

——— (1954). "On the Motion of Bodies in General Relativity Theory." *Acta Physica Polonica* 13: 187–204.

——— (1964). *Szkice przeszlosci*. Warsaw: Panstwowy Instytut Wydawniczy. German translation: *Leben mit Einstein*. Walter Hacker, trans. Vienna, Frankfurt, and Zurich: Europa Verlag, 1969. Partial English translation in Infeld 1978.

——— (1968). *Kordian, fyzika i ja*. Warsaw: Panstwowy Instytut Wydawniczy. Partial English translation in Infeld 1978.

——— (1978) *Why I Left Canada*. L. Pyenson, ed. Montreal and London: McGill-Queen's University Press.

Infeld, Leopold, and Plebański, Jerzy (1960). *Motion and Relativity*. Oxford: Pergamon; Warsaw: Panstwowe Wydawnictwo Naukowe.

Infeld, Leopold, and Schild, Alfred (1949). "On the Motion of Test Particles in General Relativity." *Reviews of Modern Physics* 21: 408–413.

Infeld, Leopold, and Wallace, P.R. (1940). "The Equations of Motion of Electrodynamics." *Physical Review* 57: 797–806.

Kerr, Roy P. (1960). "On the Quasi-Static Approximation in General Relativity." *Il Nuovo Cimento* 16: 26–60.

Kopff, August (1921). *Grundzüge der Einsteinschen Relativitätstheorie*. Leipzig: S. Hirzel. English translation *The Mathematical Theory of Relativity*. H. Levy, trans.

London: Methuen, 1923. Italian translation *I fondamenti della relatività einsteiniana*. R. Contu and T. Bembo, trans. Milano: U. Hoepli, 1923. Quotations are taken from the English edition.

Lancius [Lanczos], Kornel (1923). "Zur Theorie der Einsteinschen Gravitationsgleichungen." *Zeitschrift für Physik* 13: 7–16.

Lanczos, Kornel (1927a). "Zur Dynamik der allgemeinen Relativitätstheorie." *Zeitschrift für Physik* 44: 773–792.

—— [Cornel] (1927b). "Zum Bewegungsprinzip der allgemeinen Relativitätstheorie." *Physikalische Zeitschrift* 28: 723–726.

—— (1930). "Über eine invariante Formulierung der Erhaltungssätze in der allgemeinen Relativitätstheorie." *Zeitschrift für Physik*: 514–539.

Laue, Max von (1921). *Die Relativitätstheorie*. Vol. 2. Braunschweig: Friedrich Vieweg und Sohn; 2nd ed. 1923.

Levi-Civita, Tullio (1937a). "The Relativistic Problem of Several Bodies." *American Journal of Mathematics* 59: 9–22.

—— (1937b). "Astronomical Consequences of the Relativistic Two-Body Problem." *American Journal of Mathematics* 59: 225–234.

—— (1950). *Le problème des n corps en relativité générale*. Paris: Gauthier-Villars. English translation Dordrecht: D. Reidel, 1964.

Levinson, Horace C. (1922). "The Gravitational Field of Masses Relatively at Rest According to Einstein's Theory of Gravitation." Dissertation. University of Chicago.

—— (1923). *Le champ gravitationnel de deux points matériels fixes dans la théorie d'Einstein*. Thesis. Université de Paris. Paris: Gauthier-Villars.

—— and Zeisler, Ernest B. (1929). *The Law of Gravitation in Relativity*. Chicago: University of Chicago Press.

Lorentz, Hendrik Antoon (1909). *The Theory of Electrons and its Applications to the Phenomena of Light and Radiant Heat*. Leipzig: B.G. Teubner; New York: G.E. Stechert.

—— (1915a). "Het beginsel van Hamilton in Einstein's theorie der zwaartekracht." *Koninklijke Akademie van Wetenschappen te Amsterdam. Verslagen* 23: 1073–1089. English translation: *Koninklijke Akademie van Wetenschappen te Amsterdam. Proceedings* 19: 751–765.

—— (1915b). "Over Einstein's theorie der zwaartekracht." *Koninklijke Akademie van Wetenschappen te Amsterdam. Verslagen* 24: 1389–1402, 1759–1774; 25: 468–486, 1380–1396. English Translation: *Koninklijke Akademie van Wetenschappen te Amsterdam. Proceedings* 19: 1341–1369; 20: 2–34.

Lorentz, Hendrik Antoon, and Droste, Johannes (1917). "De beweging van een stelsel lichamen onder den invloed van hunne onderlinge aantrekking, behandeld volgens de theorie van Einstein." *Koninklijke Akademie van Wetenschappen te Amsterdam. Verslagen* 26: 392–403, 649–660. English translation in H.A. Lorentz. *Collected Papers*. Vol. 5. P. Zeeman and A.D. Fokker, eds. The Hague: Martinus Nijhoff, 1937, pp. 330–355.

Mathisson, Myron (1931a). "Die Beharrungsgesetze in der allgemeinen Relativitätstheorie." *Zeitschrift für Physik* 67: 270–277.

—— (1931b). "Die Mechanik des Materieteilchens in der allgemeinen Relativitätstheorie." *Zeitschrift für Physik* 67: 826–844.

—— (1931c). "Bewegungsproblem der Feldphysik und Elektronenkonstanten." *Zeitschrift für Physik* 69: 389–408.

——— (1937). "Neue Mechanik materieller Systeme." *Acta Physica Polonica* 6: 163–200.

——— (1940). "The Variational Equation of Relativistic Dynamics." *Cambridge Philosophical Society. Proceedings*: 331–350.

Mehra, Jagdish (1974). "Einstein, Hilbert, and the Theory of Gravitation." In *The Physicsts' Conception of Nature*. J. Mehra, ed. Dordrecht and Boston: D. Reidel, pp. 119–178. Also published separately.

Mie, Gustav (1923). "Träge und schwere Masse." *Annalen der Physik* 69: 1–53.

Moffat, John (1958). "On the Motion of Charged Particles in the Complex-Symmetric Unified Field Theory." *Il Nuovo Cimento* 7: 107–109.

——— (1959). "On the Integrability Conditions in the Problem of Motion in General Relativity." *Journal of Mathematics and Mechanics* 8: 771–786.

Moffat, John, and Kerr, Roy P. (1958). "On the Dipole Method in the Problem of Gravitational Motion." Typescript.

Pais, Abraham (1979). "Einstein and the Quantum Theory." *Reviews of Modern Physics* 51: 861–914.

Papapetrou, Achille (1951). "Equations of Motion in General Relativity." *Physical Society of London. Proceedings A* 64: 57–75, 302–310.

Pauli, Wolfgang (1921). "Relativitätstheorie." In *Encyklopädie der mathematischen Wissenschaften, mit Einschluss ihrer Anwendungen*. Vol. 5, *Physik*, part 2. Arnold Sommerfeld, ed. Leipzig: B.G. Teubner, 1904–1922, pp. 539–775. [Issued November 15, 1921.] English translation *Theory of Relativity*. With supplementary notes by the author. G. Field, trans. London: Pergamon, 1958; reprint New York: Dover, 1981.

Petrova, N.M. (1949). "Ob uravnenii dvizheniya i tenzore materii dlya sistemy konechnykh mass v obshchey teorii otnositel'nosti" ["On the Equations of Motion and the Mass Tensor for Systems of Finite Masses in the General Theory of Relativity"]. *Zhurnal Eksperimentalnoi i Teoreticheskoi Fiziki* 19: 989–999.

Plebański, Jerzy and Bażański, Stanisław (1959). "The General Fokker Action Principle and its Application in General Relativity Theory." *Acta Physica Polonica* 13: 307–345.

Ricci, Gregorio, and Levi-Civita, Tullio (1901). "Méthodes de calcul différentiel absolu et leurs applications." *Mathematische Annalen* 54: 125–201. Also published separately Paris: Alb. Blanchard, 1923.

Robertson, Howard P. (1936). "Test Corpuscles in General Relativity." *Edinburgh Mathematical Society. Proceedings. Part II* 5: 63–81.

——— (1938). "Note on the Preceding Paper: The Two Body Problem in General Relativity." *Annals of Mathematics* 39: 101–104.

Robinson, Ivor, and Robinson, Joanna R. (1972). "Equations of Motion in the Linear Approximation." In *General Relativity: Papers in Honor of J.L. Synge*. L. O'Raifeartaigh, ed. Oxford: Clarendon Press, pp. 151–166.

Rosen, Nathan (1949). "A Particle at Rest in a Static Gravitational Field." *Reviews of Modern Physics* 21: 503–505.

Schott, George A. (1915). "On the Motion of the Lorentz Electron." *Philosophical Magazine*: 59–62.

Schwarzschild, Karl (1916a). "Das Gravitationsfeld eines Massenpunktes nach der Einsteinschen Theorie." *Königlich Preussische Akademie der Wissenschaften* (Berlin). *Sitzungsberichte*: 189–196.

——— (1916b). "Über das Gravitationsfeld einer Kugel aus inkompressibler Flüssigkeit nach der Einsteinschen Theorie." *Königlich Preussische Akademie der Wissenschaften* (Berlin). *Sitzungsberichte*: 429–434.

Seelig, Carl (1954). *Albert Einstein. Eine dokumentarische Biographie*. Zurich: Europa Verlag. English Translation London: Staples Press, 1956.

Średniawa, Bronislaw (1985). *History of Theoretical Physics at Jagellonian University in Cracow in XIXth Century and in the First Half of XXth Century*. Warsaw and Cracow: Państwowe Wydawnictwo Naukowe.

Stachel, John (1986). "Einstein and the Quantum: Fifty Years of Struggle." In *From Quarks to Quasars: Philosophical Problems of Modern Physics*. Robert G. Colodny, ed. Pittsburgh: University of Pittsburgh Press, pp. 349–385.

Weyl, Hermann (1917). "Zur Gravitationstheorie." *Annalen der Physik* 54: 117–145.

——— (1918). "Gravitation und Elektrizität. "*Königlich Preussische Akademie der Wissenschaften* (Berlin). *Sitzungsberichte*: 465–480.

——— (1919a). "Eine neue Erweiterung der Relativitätstheorie." *Annalen der Physik* 59: 101–133.

——— (1919b). "Bemerkung über die axialsymmetrischen Lösungen der Einsteinschen Gravitationsgleichungen." *Annalen der Physik* 59: 185–188.

——— (1919c). *Raum-Zeit-Materie*, 3rd ed. Berlin: Julius Springer.

——— (1921a). "Feld und Materie." *Annalen der Physik* 65: 541–563.

——— (1921b). *Raum-Zeit-Materie*, 4th ed. Berlin: Julius Springer. English Translation *Space, Time, Matter*. Henry Brose, trans. London: Methuen, 1922; reprint New York: Dover, 1952.

——— (1923). *Raum-Zeit-Materie*, 5th ed. Berlin: Julius Springer.

Whittaker, Sir Edmund T. (1951). *History of the Theories of Aether and Electricity from the Age of Descartes to the Close of the Nineteenth Century*, 2nd ed. Vol. 1, *The Classical Theories*. London: Thomas Nelson and Sons.

# The Low Water Mark of General Relativity, 1925–1955

JEAN EISENSTAEDT

Starting in the early 1920s, after the period of its reception, the general theory of relativity endured some very lonely years, something like a "desert crossing"; a difficult time, in any case, that was not to end before the late 1950s. From 1907 to 1915, the period of elaboration of the theory, it was above all "one man's work" (Born 1956, p. 252); it was the time of a persevering man, Albert Einstein, who, on the basis of his special theory of relativity, succeeded in unifying many fundamental principles in an unexpected way. But for all that it was unexpected, general relativity was a convincing theory, as far as it is possible for a theory to be, from an epistemological point of view. And it was a theory without concessions from a demanding man, a theory, isolated in competition with other theories, that could play only one empirical trump-card: the 43″ of arc per century advance of Mercury's perihelion, conceded by Newton's theory. During the period of reception, two other experimental cards were played: the remaining two of the three famous tests whose outcomes had been qualitatively predicted by Einstein in the years since 1907 and whose results were then thought to be decisive (the gravitational redshift and the bending of light). This was the period of Einstein's fame, the golden age of the general theory of relativity, the gold of which lay, as has often been said, more in the internal qualities of the theory than in its experimental confirmation.

In fact, the Newtonian theory of gravitation, although falsified, still dominated the field; it continued to cover most of the relevant phenomena—except the three tests—and of course all applications. By its proximity, it threatened to smother Einstein's theory, a most worrisome problem for general relativity, indeed, more especially so, since it added to the many other technical and conceptual difficulties of the theory. Special relativity was turning out to be an everyday tool of physicists; and quantum mechanics, which was undergoing rapid development, became the focus of attention of most theoretical physicists. During these difficult years, the status of general relativity was almost the inverse of that of quantum mechanics. The contrast was evident in almost every respect: concerning their epistemological structures, their relations to experiment and even more to applications, as well as the styles of work they called forth (leaving aside, for the moment, the implicit philosophy each conveyed). And, in some ways, general relativity was considered to be the completion of classical physics, while quantum mechanics was viewed as a dynamic, revolutionary new theory.

Many a theoretical physicist, many an expert on Einstein's theory, deplored the paucity of empirical results in general relativity compared to the elegant structure of its logical and mathematical machinery, which was so difficult to handle. Criticisms were heaped on general relativity. It was accused of being a marginal theory, isolated on a pedestal, poorly understood, slow to develop, fairly unproductive. Einstein's theory was also considered too formal, too mathematical, too speculative. It was a "historical accident," according to one of its most representative students. And those criticisms came not only from physicists who were a priori opposed to Einstein's theory; far from it, they came from experts in the field who were, moreover, often investigating the numerous alternative theories of gravitation available.

In what follows, we will be concerned not so much with a historical review of the subject. We will be interested instead in listening to the points of view of experts of that time, in hearing what were their thoughts and concerns about general relativity.

Roughly, then, two main analyses were put forward. The first was primarily epistemological and insisted on the fact that Einstein's theory is especially convincing, thanks to its logical structure; it emphasized the fact that no other physical theory can better explain the available empirical gravitational data. The second analysis, an "economic" one, stated that general relativity is too sophisticated a theory, exceedingly isolated from other theories, and that it leads to too small a number of empirical predictions in comparison with the effort involved in working on it.

Implicitly or explicitly, the question asked by many authors concerned the profitability of the intellectual investment required by general relativity: the price for three small tests, more or less well verified, was the complex machinery of Einstein's theory. What is the cost of the difficulties one encounters in general relativity compared to the reward of its structural simplicity, of the intellectual satisfaction it brings? Is general covariance, to which Einstein adhered so strongly, really worth so much? The balance between investment and profit was quite difficult to define. In fact, it was in comparison with Newton's theory that general relativity was weighed; and the balance very much favored the Newtonian theory! But, can we really choose the price to be paid? It is certainly not by chance that Einstein stated the main points of the Popperian program many years before the publication of *Logik der Forschung*, as early as 1919: "But the *truth* of a theory will never be proved. Because one can never know if, in the future, some experience will become known that would contradict its conclusions" (Einstein 1919).[1] The example of general relativity shows that satisfying the criterion of falsifiability is not sufficient for a theory to prevail in the relevant scientific and institutional domains: fruitfulness is just as important. Indeed, during these thirty years, general relativity remained in the background, waiting for its time to come.

Some articles have been written on the empirical situation of Einstein's theory of gravitation in the early 1920s.[2] I will just remind the reader briefly that general relativity accounted very precisely and in a convincing way— without any arbitrary constant—for the advance of 43" of arc per century of Mercury's perihelion. The second famous prediction—the bending of light— was verified in 1919 at Principe and Sobral during the eclipse expeditions conducted by Andrew Crommelin and Arthur S. Eddington; this was a crucial moment in the reception of Einstein's theory. Whitehead's description of the meeting of the Royal Society on November 6, 1919, is well known. But Eddington's attitude was even more definite; he wrote then:

> We have tested Einstein's law of gravitation for fast movement (light) and for moderately slow movement (Mercury). For very slow movement it agrees with Newton's law, and the general accordance of the latter with observation can be transferred to Einstein's law. These tests appear to be sufficient to establish the law firmly." (Eddington 1920, p. 126)

There is no doubt that the 1919 eclipse observations represented for many experts an important test, even a decisive one. That was the case, for example, for Max von Laue, who had not been convinced by the first results of Einstein's theory. However, concerning this second test and the third one (the gravitational redshift), the situation was in fact far from being clear. We will come back to this point later.

The reasons for many scientists' adherence to general relativity did not of course rest entirely on empirical results; far from it. It is well known that Einstein's critical analyses and his requirements for an acceptable theory were expressed through principles that have a status of a different kind, based as they are on theoretical necessities: the equivalence principle, the general covariance principle, Mach's principle, and so forth. None of these principles was, strictly speaking, an a priori necessity for building an acceptable theory of gravitation. But the fact that general relativity was constructed on clearly expressed principles, in an especially original and secure way, was for many specialists an additional, essential reason for believing in the theory—at least as long as experiment allowed it, since experiment remained for Einstein, as well as for the other experts on his theory, "the supreme judge." It was a reason, as well, for betting on general relativity, that is, for working on it, or even as in the case of Einstein himself for advancing beyond it. Of course these are structural considerations, but they occupied quite an important place, the formal considerations, but they occupied quite an important place, the place that experiment and observation left unoccupied during these forty years.

In his communication to the Prussian Academy, in November 1915, Einstein wrote:

> Scarcely anyone who has fully understood this theory can escape from its magic; the theory represents a genuine triumph of the method of the absolute differential calculus founded by Gauss, Riemann, Christoffel, Ricci, and Levi-Civita. (Einstein 1915, p. 779)[3]

And Eddington exclaimed, in the preface to his *Report on the Relativity Theory of Gravitation*, the first English book entirely dedicated to general relativity: "Whether the theory ultimately proves to be correct or not, it claims attention as one of the most beautiful examples of the power of general mathematical reasoning" (Eddington 1918); while Hermann Weyl in the preface to the first edition (1918) of his *Raum-Zeit-Materie* wrote emphatically: "It is as if a wall that separated us from truth has suddenly collapsed" (Weyl 1918, p. v).[4] And further on, he wrote:

> The chief support of the theory is to be found less in experience than in its inherent logical consistency, in which it far transcends that of classical mechanics, and also in the fact that it solves the perplexing problem of gravitation and of the relativity of motion at one stroke in a manner highly satisfying to our reason. (Weyl 1918, p. 198)[5]

In the foreword that he wrote for his *Relativitätstheorie*, Max von Laue expressed the hope that general relativity would seduce his readers through its "force of conviction" ("dieser Überzeugungskraft"), and he insisted on the importance of the general covariance principle (von Laue 1921, p. viii). To these few examples, coming from the most eminent authors, it would be easy to add many others: Born, Langevin, Pauli, and so forth. But we must note that a real anxiety concerning the poverty of empirical material was expressed, an anxiety which then allowed, of course, for the hope that the remaining two predictions would be better verified and that new tests would be formulated soon. But only Langevin, speaking to a journalist during Einstein's visit to Paris in 1922, hoped that general relativity could find some application in a practical field like electronics.

In the early 1920s, Einstein's fame was at its height; but he received the Nobel prize in 1922 for his work on the photoelectric effect, while his theories of relativity were not even mentioned. General relativity was still widely misunderstood by many physicists. Allvar Gullstrand, a member of the Nobel prize jury was one of those who misunderstood the theory, as did most of the French physicists—Langevin excepted—not to mention the "anti-relativistic company," as Einstein called the German physicists opposing relativity, led by the anti-Semitic Philipp Lenard. Nevertheless, numerous textbooks were dedicated to general relativity at that time, a fact that gives evidence of the real interest it stirred up not only in scientific circles but also among the public. The number of articles published on general relativity was then almost equal to the number published on Newtonian mechanics, and double the number published on quantum theory. Thus, at first sight, the situation of general relativity in the early 1920s was quite enviable with respect to its epistemological basis as well as with respect to its public standing.

However, general relativity had reached a peak in its history. It controlled most of the available ground: on the theoretical level, most of the easiest results had been obtained; on the empirical level, some results that were to be clarified only much later had already been anticipated. But the retreat of general relativity during the next forty years was betokened first by the domination of the Newtonian theory from many points of view: empirical, theoretical, ideological, and institutional. From an empirical point of view, Newtonian mechanics leaves general relativity only a little room to spread out on the available observational field. From a theoretical point of view, we will see that the approximate empirical validity of Newtonian gravitational theory, but also its ideology, combined with the conceptual and technical difficulties of general relativity to favor what we now call a "neo-Newtonian" practice, that is to say an interpretation of general relativity employing tools borrowed essentially from Newtonian theory. But the retreat was betokened as well by the domination of quantum mechanics, whose roots also grew in empirical ground: the ratio between gravitational and electromagnetic forces at short distances is of the order of $10^{-43}$. On the institutional level the fact of such domination was expressed without the slightest subtlety. Roughly, the question could be formulated in the following way: What is the value of a gravitational theory that the astronomers scarcely need?

The empirical situation of general relativity was in fact more complicated, the results far less clear, than it appeared to the most convinced experts. Such, in any case, is the opinion of Earman and Glymour, who contest Eddington's account of the 1919 Sobral results; but they emphasize also the correlative influence of the eclipse observations on the results of the third test:

> Before 1919 no one claimed to have obtained spectral shifts of the required size; but within a year of the announcement of the eclipse results several researchers reported finding the Einstein effect. The red-shift was confirmed because reputable people agreed to throw out a good part of the observations. They did so in part because they believed the theory; and they believed the theory, again at least in part, because they believed that the British eclipse expeditions had confirmed it. Now the eclipse expeditions confirmed the theory only if part of the observations were thrown out and the discrepancies in the remainder ignored; Dyson and Eddington, who presented the results to the scientific world, threw out a good part of the data and ignored the discrepancies. (Earman and Glymour 1980a, p. 85)

Such an analysis explains why both tests were to be hotly debated long after the 1920s. And we must emphasize the enormous difficulty of the measurements due to the extreme smallness of the predicted effects. Nevertheless, we must also emphasize that the available data were already sufficient to refute the Newtonian theory. Roughly speaking, general relativity remained, beyond all question, the best theory of gravitation, or perhaps just the least of all evils.

Of course the overoptimistic claims for the theory by Eddington and others facilitated its reception. But, through a kind of pendulum effect that is easy to understand, they also explain, at least in part, the ensuing disappointment. Between the 1920s and 1960 the empirical advance of general relativity came to a halt and, in some ways, the advance turned into a retreat. Thus, during a meeting of the Royal Astronomical Society, and in the presence of Erwin Finlay-Freundlich, who was just back from the 1929 eclipse observations, Eddington confessed:

I find it difficult to believe that 1.75″ can be wrong. Light is a strange thing and we must recognize that we do not know as much about it as we thought we did in 1919; but I should be very surprised if it is as strange as all that. (Eddington 1932, p. 5)[6]

In his *Introduction to the Theory of Relativity*, one of the very few textbooks of that time, Peter G. Bergmann, who was Einstein's assistant at the Institute for Advanced Study in Princeton, described the observational situation of the theory. He wrote:

> Each of these effects has been observed; however, two of them are just outside the limits of experimental error, so that the quantitative agreement between observations and theoretical predictions is still doubtful. (Bergmann 1942, p. 211)

This was a cautious comment that Finlay-Freudlich found by far too optimistic. For Finlay-Freundlich, who led the 1929 eclipse expedition testing the general-relativistic light deflection prediction, general relativity was not far from being invalidated. One may wonder whether or not the genuine confidence that he placed in general relativity during its very early years had given way to a measure of bitterness. During the 1950s he was one of the experts most critical of the empirical situation of general relativity. He sharply opposed G. Biesbroeck, who was responsible for the two following expeditions, but also Einstein himself, who replied to an anxious Max Born: "Freundlich, however, does not move me in the slightest" (Einstein to Born, May 12, 1952, in Born 1969). This serenity was justified, however, for the results, if still unclear, were far from refuting general relativity. In any case, it would have been fairer for Finlay-Freundlich to contest the analysis in Tolman's book, published in 1934, an analysis far more optimistic indeed than that of Bergmann. For Tolman, there was no fly in the ointment; each test was said to be satisfactory, and even "exceedingly satisfactory" in the case of the deflection measurements. Tolman did not mention Finlay-Freundlich's results published three years earlier.

In the foreword that he wrote for Bergmann's textbook, Einstein recognized that the theory "has played a rather modest role in the correlation of empirical facts so far" (Einstein 1942). Elsewhere, he explained very clearly the essential question of the surprising near agreement between the predictions given by the Newtonian theory and those derived from general relativity:

> This agreement goes so far, that up to the present we have been able to find only a few deductions from the general theory of relativity that are capable of investigation, and to which the physics of pre-relativity days does not also lead, and this despite the profound difference in the fundamental assumptions of the two theories. (Einstein 1920, appendix 3)

In addition to the three classical tests, quite a few attempts were made before the renaissance of the 1960s to apply general relativity to other problems: to the motion of the perihelion of the earth; to the secular acceleration of the moon; to the displacement of the orbit of Mars; and to the influence of gravitation upon atomic energy levels, and, more realistically, upon rotating disks or gyroscopes. But none of these faint hopes were to be realized in a convincing test, because the predicted effects were far too small. In the memorial note that he wrote for the *Reviews of Modern Physics* at the time of Einstein's death, J.R. Oppenheimer noted: "In the forty years that have elapsed [these three tests] have remained the principal and, with one exception, the only connection between the general theory and experience" (Oppenheimer 1956). The exception was cosmology, a topic we will consider in a moment.

To test general relativity, that was now the first task of the new generation, which, from the 1950s on, was returning to Einstein's theory. No paper has better put into words the wishes of this new generation than that by Alfred Schild (1960), which announced in almost biblical fashion the good news, the promised land:

> These are exciting days: Einstein's theory of gravitation, his general theory of relativity of 1915, is moving from the realm of mathematics to that of physics. After 40 years of sparse meager astronomical checks, new terrestrial experiments are possible and are being planned. (Schild 1960, p. 778)

Indeed, the third prediction had finally been accurately verified through a terrestrial experiment carried out by Pound and Rebka, applying a new effect recently discovered by Mössbauer. For Schild, and for most other relativists, nothing was more urgent than strengthening the empirical basis of the theory. But this desire was accompanied by sharp criticisms of the past development of the theory, criticisms that Robert H. Dicke, an experimentalist coming from quantum mechanics, expressed in 1963 in the following words:

> As an experimentalist I attempted to counteract in some small measure the decided tendency in times past for General Relativity to develop into a formal science divorced from both observations and the rest of physics. A well-known physicist once remarked to me that Einstein's General Relativity was such a beautiful theory that it was a shame that there were so few experiments. An examination of the scientific literature of the past 50 years will testify to the truth of this statement, the number of experimental papers being entirely negligible in comparison with the flood of theoretical publications, mostly formal. (Dicke 1964, p. 7)

That was quite a warranted remark, indeed, but many specialists, such as Andrzej Trautman, did not admit the reproach: "as if this were a fault of the relativists," he protested (Trautman 1966, p. 334).

The general theory of relativity was not only viewed as a difficult theory to test, it was also thought by specialists to be difficult to develop theoretically, especially by comparison with the Newtonian theory of gravitation, which represented the fundamental point of reference. It was of course both together with and in opposition to the Newtonian theory that general relativity had to be developed—conceptually and technically—on the theoretical level. I will give here some examples of the difficulties, interpreted in the same "economic" terms employed by specialists on the theory who were faced with the dazzling fertility of the Newtonian theory.

Thus Bergmann devoted relatively little space to experimental questions in his book (1942), but he explained in detail the very recent and complex method worked out by Einstein, Infeld, and Hoffmann (EIH). This pretty result further improved the logical structure of the theory, since it makes it clear that the equations of motion of mass points are directly determined by the field equations. Concerning this theorem, Infeld wrote:

> Two morals can be drawn from this story. First, it shows how difficult the mathematical deductions are, how complicated are the equations of general relativity theory and how deeply they can hide their secrets. The second moral is of some philosophical significance.... It is logically simpler to assume only field equations and to disregard the equations of motion, but we have to pay for the logical simplicity by increased technical difficulties. (Infeld 1950, pp. 69–70)

This is an important remark, since the EIH method is bound up with the nonlinearity of the field equations.

Quite interesting too is the opinion that Synge expressed in 1950 in the introduction to his important paper, "The Gravitational Field of a Particle." He worried about the lack of theoretical fruitfulness of general relativity compared with the Newtonian theory:

> Einstein's field equations for an empty region, $R_{\mu\nu} = 0$, are the relativistic analogue of Laplace's equation $\nabla^2 V = 0$. Laplace's equation has played a central part in the mathematical sciences for a century and a half; the physicist has used it again and again in various branches of physics, and the pure mathematician has found in the study of harmonic functions the source of many new ideas, so much so that the theory of potential is now regarded as a substantial mathematical subject.
> 
> It is too soon to say whether or not Einstein's equations are destined to play a similar fundamental role. Their study is very difficult for several reasons. Instead of one equation for one unknown, they contain ten equations ... for ten unknowns; they are nonlinear (a formidable difficulty); and the fact that they retain their form under general coordinate transformations is an embarrassment rather than an advantage. Our present knowledge of the properties of the solutions of the field equations is very meager. As particular solutions we have the Schwarzschild solution, corresponding to spherical symmetry, and the solution of Weyl and Levi-Civita, corresponding to rotational symmetry. But even in these solutions there remain some rather fundamental obscurities which it is desirable to remove. (Synge 1950, p. 83)

But Synge was not, by any means, the only expert to worry about the technical difficulties that general relativity entailed. Levi-Civita wrote in 1937:

> Mechanical laws, according to Einstein's theory, are much more complicated in conception than under the assumption of Newton. However, the motion of celestial bodies under ordinary circumstances differs so little from their Newtonian representation that, for astronomical purposes, relativistic effects may be conveniently treated as first-order perturbations. (Levi-Civita 1937, p. 225)

And elsewhere this specialist in Riemanian geometry stated that, apart from the Schwarzschild solution, "it seems that we must give up any hope to approach, in general relativity, in a rigorous way, the questions we consider usually in classical mechanics" (Levi-Civita 1950, p. 2).[7] And further on he adds that "even the two-body problem, formerly worked out by Newton, has no chance of success" (Levi-Civita 1950, p. 2).[8] But since the renaissance, experts on the theory will no longer accede to such an objection. Thus, J.A. Wheeler, on the occasion of the Chapel Hill congress in 1957, was rightly ironical regarding this point:

> An objection one hears raised against the general theory of relativity is that the equations are nonlinear, and hence too difficult to correspond to reality. I would like to apply that argument to hydrodynamics: rivers cannot flow in North America because the hydrodynamical equations are nonlinear and hence much too difficult to correspond to nature. (Wheeler 1957, p. 4)

Thus, the practice of general relativity remained during all these years, to a large extent, neo-Newtonian. This did not give rise to any practical problem, however, since the part of the universe accessible to experiment—problems of cosmology aside—was well described through such a neo-Newtonian interpretation. But, the Newtonian domination was not limited to the empirical level or to theoretical practice. Its effects were felt as well on the interpretation of Einstein's theory. The Newtonian theory did not provide many of the concepts necessary for interpreting general relativity, so the experts were led to construct new ones, whose definitions remained ambiguous. Thus, in 1918 Eddington wrote:

> Mass, time, and distance are all ambiguously defined in Newtonian dynamics, and in defining them for the present theory we have some freedom of choice, provided that our definition agrees with the Newtonian definition in the limiting case of a vanishing field of force. (Eddington 1918, p. 50)

But the problem was far from being solved. In 1956, Felix Pirani stressed the difficulty consisting in "the lack of what might be called a theory of measurement" (Pirani 1956, p. 389), and in 1970, Synge explained:

> In the days when relativity had to win credence in an incredulous world, it was natural to give it respectability by explaining it as far as possible in terms of the old concepts. But this led to getting the concepts mixed up. (Synge 1970, p. 16)

In 1962, at a conference on the "Present State of Relativity," Bondi and Synge discussed the question of the relationship between general relativity and Newtonian theory:

> *H. Bondi*: I should like to take up Professor Synge on one point; I am not entirely in sympathy with his hankering after reality. It seems to me that a great deal of the interest of general relativity lies in asking what the theory would say in conditions which admittedly do not occur in those parts of the universe about which we know much. In conditions which are common, we know that Newtonian theory is a good approximation, except for some minor points of which we had a very clear description this morning. Given that this is so, I feel that Newtonian theory largely satisfies my own hankering after reality.

> *J.L. Synge*: Is Professor Bondi satisfied with Newtonian theory—philosophically —with its absolute time?
> 
> *H. Bondi*: Oh no! I was not looking at Newtonian theory as a satisfying theory; I was looking at Newtonian theory as a particular and well-worked-through approximation to the relativistic equations, and I felt that this particular method of approximation—admittedly invented 250 years before the theory—nevertheless, is highly successful in practically all cases appertaining to reality. (Bondi 1962, p. 325)

Almost fifty years after "its first modification," to quote Whitehead, the Newtonian theory still widely dominated its rival.

At this point we must recall that very few people drew attention to specifically relativistic problems. As early as 1916, Johannes Droste asked, in his thesis, about what is to happen near the center of what is now called the Schwarzschild solution. Hence, he studied the trajectories in empty space in the neighborhood of the Schwarzschild singularity, in the region defined by $r = 2MGc^{-2}$, and he noticed some properties of the motion specific to general relativity. He commented: "This result differs fundamentally from what occurs in Newton's theory; it shows us how totally different ... from classical theory the motion in the vicinity of the center becomes" (Droste 1916, p. 26).[9] His calculations were wrong in many ways,[10] but that is irrelevant to our purposes, because Droste pointed out here a most important fact: between the predictions of the two theories, there was here, for once, something "totally different." And he was the first expert—after Einstein[11]—to understand, or at least to imply, that the Newtonian theory represented the greatest threat to Einstein's theory. Some ten years later, E.T. Whittaker also emphasized what would happen in the vicinity of the Schwarzschild singularity (Whittaker 1928, p. 140), but it was not much before 1965 that the question of a specifically relativistic approach was raised, in opposition to the old neo-Newtonian one, by the leaders of the renaissance of general relativity:

> Except for the prediction of Einstein's theory about the expansion and recontraction of the universe, all the other applications of general relativity (precession of perihelion, redshift, bending of light, gravitational radiation) as normally envisaged have to do with small departures from flatness. Not so here. Collapse produces geometries almost as far as can be from flatness. Any perturbation-theoretic expansion in powers of the departure from flatness is out of place. If one intends to abandon relativity, here is the place to do so. Otherwise he is on the way into a new world of physics, both classical and quantum. Here we go! (Harrison et al. 1965, p. 124)

The large number of articles about the numerous alternative theories of gravitation represent the best evidence of the experts' resistance to accepting general relativity. In a review devoted to these alternative theories in 1965, Whitrow and Morduch wrote:

> The incentive to devise Lorentz-invariant theories of gravitation ... appears to be mainly due to the great difficulty in solving the nonlinear equations that occur in general relativity. (Whitrow and Morduch 1965, p. 15)

But it was also stated that the geometrical character of Einstein's theory was isolating it quite completely from other physical theories, a situation that was later to change completely. Thus, Nathan Rosen, who had been a collaborator of Einstein's, was working in 1940 on a theory "based on flat space"; he wrote then:

> In the theory suggested here ... this geometrization of gravitation has been given up. Perhaps this may be regarded by some as a step backward. It should be noted, however, that the geometrization referred to has never been extended satisfactorily to other branches of physics, so that gravitation is treated differently from other phenomena. (Rosen 1940, p. 153)

However, we must emphasize the fact that, with respect to the empirical evidence, general relativity has—up to now—always triumphed over these alternative theories. Is it not surprising that these three small tests have been sufficient to exclude its rivals, while some alternative theories employed a great number of free parameters? Is this not an indication that the empirical basis for theories of gravitation represented a better sample of the data than one would have thought? In any case, this is evidence of the astonishing resilience, of the staying power of Einstein's theory. But from the point of view of gravitation alone we cannot understand those who are interested in alternative theories; we must, instead, adopt a point of view that includes the other branches of theoretical physics and, above all, of course, quantum mechanics. As John Stachel put it: "Einstein himself was not content with his creation of general relativity. He never believed that the general theory was any more than a way station in his search for a unified field theory" (Stachel 1972, p. 32). And in the foreword to Bergmann's book, Einstein wrote:

> It is quite possible, however, that some of the results of the general theory of relativity, such as the general covariance of the laws of nature and their nonlinearity, may help to overcome the difficulties encountered at present in the theory of atomic and nuclear processes. (Einstein 1942)

Hence for Einstein, the achievements of general relativity did not consist so much in the "correlation of empirical facts," as in the structure of the theory itself, which is a remarkable attitude that is undoubtedly not shared by most physicists, even by many specialists on gravitation. In conclusion, Einstein added: "Apart from this, the theory of relativity has a special appeal because of its inner consistency and the logical simplicity of its axioms" (Einstein 1942). Consistency and simplicity are features of the theory that would remain important for many a relativist.

The status of cosmology is a wholly separate and extremely interesting topic. Since 1917, the time of the first paper on the subject by Einstein, the history of "relativistic cosmology" has undergone steady development. But at first cosmology represented a specialization of general relativity, an application that was often regarded by relativists as, in fact, an independent field of research. Moreover, from the point of view of cosmology, the relationship between general relativity and Newtonian theory appeared to be inverted; here it was the Newtonian theory that borrowed ideas from relativity (through Milne's work, for example). In any case, here more than elsewhere, it was in the empirical domain that the shoe pinched. Cosmology could be considered as an application, but not as a test of the theory. As Hawking and Israel put it in a recent book: "Cosmology was regarded as an area in which wild theoretical speculation was unfettered by any possible observations" (Hawking and Israel 1979, p. 3).

Nevertheless, contrary to most of the other consequences of the general theory of relativity, the cosmological effects, and above all the Hubble effect, are not connected with a Newtonian point of view. In that sense, relativistic cosmology represented for many years the only branch of general relativity that enjoyed some measure of genuine autonomy with respect to the Newtonian theory. But it was still a relative autonomy, because, as McVittie put it, the measurements are so difficult "that numerous different theories may well fit them all within the errors of observation" (McVittie 1956, p. 8).

Strangely, however, the speculative character of cosmology represented quite an advantage, since it made for a field where general relativity was capable of being applied and thoroughly investigated within the framework of a space-time clearly free from the limitations of the Newtonian scheme.

In his important textbook, where cosmology occupies a prominent place, R.C. Tolman wrote:

Since we have based our treatment on acceptable physical theory, we have the right to expect that the theoretical behaviour of our models will at least inform and liberalize our thinking as to conceptual possibilities for the behaviour of the actual universe. (Tolman, 1934, p. 445)

It is in this sense, indeed, that we must understand the significance of cosmology for general relativity. It is not by chance that some of the most important results concerning "the Schwarzschild field" have been achieved by two cosmologists: Georges Lemaître and H.P. Robertson.

During all that period, most astronomers were quite unfamiliar with Einstein's theory of gravitation. Is that a surprise? They probably knew that general relativity explains Mercury's anomaly; maybe they were acquainted with the expression for the deflection of light; but hardly anything more. Some textbooks on astronomy would concede a few pages to Einstein's theory, often less; and many gave it no space at all. But this is not inconsistent with the broad interest that some astronomers displayed in the theory, leading them to be counted, like Eddington, among the foremost experts on it.

Concerning the institutional setting of Einstein's theory during these forty years, all of the evidence from reminiscences and other sources points to the same evaluation. As Infeld recalled:

In any case, the greatest interest in this discipline was evinced by scientists in the 1920s. Then, already in 1936 when I was in contact with Einstein in Princeton, I observed that this interest had almost completely lapsed. The number of physicists working in this field in Princeton could be counted on the fingers of one hand. I remember that very few of us met in the late Professor H.P. Robertson's room and then even those meetings ceased. We, who worked in this field, were looked upon rather askance by other physicists. Einstein himself often remarked to me "In Princeton they regard me as an old fool: Sie glauben ich bin ein alter Trottel." This situation remained almost unchanged up to Einstein's death. Relativity Theory was not very highly estimated in the "West" and frowned upon in the "East." (Infeld 1964, p. xv)

At the time of Einstein's death the relativistic output was only some thirty papers a year, which must be compared to the 10,000 annual references in *Physics Abstracts*. Synge described this situation as an "ivory tower":

> Of all physicists, the general relativist has the least social commitment. He is the great specialist in gravitational theory, and gravitation is socially significant, but he is not consulted in the building of a tower, a bridge, a ship, or an aeroplane, and even the astronauts can do without him until they start wondering which ether their signals travel in.
>
> Splitting hairs in an ivory tower is not to everyone's taste, and no doubt many a relativist looks forward to the day when governments will seek his opinion on important questions. But what does "important" mean? Science has a dual aim, to understand nature and to conquer nature, but in the intellectual life of man surely it is the understanding which is the more important. Then let the relativist rejoice in the ivory tower where he has peace to seek understanding of Einstein's theory as long as the busy world is satisfied to do its jobs without him. (Synge 1960, preface)

And, during the first congress of relativity in Bern, in 1955, just at the time of Einstein's death, Max Born explained the choice he had made with so many other theoretical physicists forty years earlier:

> I remember that on my honeymoon in 1913 I had in my luggage some reprints of Einstein's papers which absorbed my attention for hours, much to the annoyance of my bride. These papers seemed to me fascinating, but difficult and almost frightening. When I met Einstein in Berlin in 1915 the theory was much improved and crowned by the explanation of the anomaly of the perihelion of Mercury, discovered by Leverrier. I learned it not only from the publications but from numerous discussions with Einstein,—which had the effect that I decided never to attempt any work in this field. The foundation of general relativity appeared to me then, and it still does, the greatest feat of human thinking about Nature, the most amazing combination of philosophical penetration, physical intuition and mathematical skill. But its connections with experience were slender. It appealed to me like a great work of art, to be enjoyed and admired from a distance. (Born 1956, p. 253)

This distance would not be overcome before the 1960s, which witnessed the renaissance of the theory.

*Acknowledgments.* I wish to thank the Hebrew University of Israel for its kind permission to quote the material in this paper from Einstein's unpublished writings. I also want to thank John Stachel and Don Howard for comments and help in translating this paper into English.

A more elaborate paper on this topic has already been published in French; see Eisenstaedt 1986, where a detailed bibliography can be found.

NOTES

[1] "Niemals aber kann die *Wahrheit* einer Theorie erwiesen werden. Denn niemals weiss man, dass auch in Zukunft keine Erfahrung bekannt werden wird, die ihren Folgerungen widerspricht." (I owe this reference to John Stachel.)

[2] See for example: Crelinsten 1984, Earman and Glymour 1980a, 1980b, and Roseveare 1982. Concerning the post-reception period very few comments have been made. Some observations can be found in Crelinsten 1984 and Eisenstaedt 1986.

[3] "Dem Zauber dieser Theorie wird sich kaum jemand entziehen können, der sie wirklich erfasst hat; sie bedeutet einen wahren Triumph der durch Gauss, Riemann, Christoffel, Ricci und Levi-Civita begründeten Methode des allgemeinen Differentialkalküls."

[4] "Es ist, als wäre plötzlich eine Wand zusammengebrochen, die uns von der Wahrheit trennte."

[5] "Ihre eigentliche Stütze findet sie aber weniger in der Erfahrung als in ihrer eigenen inneren Folgerichtigkeit, durch welche sie der klassischen Mechanik ganz erheblich überlegen ist, und darin, dass sie in einer die Vernunft aufs höchste befriedigenden Weise das Rätsel der Relativität der Bewegung und der Gravitation auf einen Schlag löst."

[6] $1.75''$ is the general-relativistic prediction.

[7] "Il paraît qu'on doive renoncer à tout espoir d'aborder rigoureusement, d'après la Relativité générale, les questions envisagées ordinairement en Mécanique classique."

[8] "Même le problème des deux corps, épuisé jadis par Newton, n'a quelque chance de succès."

[9] "Dit resultaat is ten eenenmale verschillend van hetgeen in Newton's theorie optreedt; wij zien er uit, hoe geheel anders de beweging in de nabijheid van het centrum wordt ... dan in de klassieke theorie."

[10] Concerning this point, see my "The Early Interpretation of the Schwarzschild Solution" in this volume, pp. 213–233.

[11] Einstein was aware of this problem, but not preoccupied with it. See Eisenstaedt 1986, pp. 148 and 169.

REFERENCES

Bergmann, Peter G. (1942). *Introduction to the Theory of Relativity*. New York: Prentice-Hall.

Bondi, Hermann (1962). "A Discussion on the Present State of Relativity." *Royal Society of London. Proceedings A* 270: 297–356.

Born, Max (1956). "Physics and Relativity." In *Fünfzig Jahre Relativitätstheorie. Bern, 11.–16. Juli 1955. Verhandlungen*. A. Mercier and M. Kervaire, eds. (*Helvetica Physica Acta* 4 (Supplement) (1956).) Basel: Birkhäuser, pp. 244–260.

Born, Max, ed. (1969). *Albert Einstein–Hedwig und Max Born. Briefwechsel 1916–1955*. Munich: Nymphenburger. English translation *The Born–Einstein Letters*. I. Born, trans. New York: Walker and Company, 1971.

Crelinsten, Jeffrey (1984). "William Wallace Campbell and the 'Einstein Problem': An Observational Astronomer Confronts the Theory of Relativity." *Historical Studies in the Physical Sciences* 14: 1–91.

Dicke, Robert H. (1964). *The Theoretical Significance of Experimental Relativity*. New York: Gordon and Breach.

Droste, Johannes (1916). *Het zwaartekrachtsveld van een of meer lichamen volgens de theorie van Einstein*. Leiden: E.J. Brill.

Earman, John, and Glymour, Clark (1980a). "Relativity and Eclipses: The British Eclipse Expeditions of 1919 and their Predecessors." *Historical Studies in the Physical Sciences* 11: 49–85.

——— (1980b). "The Gravitational Red-Shift as a Test of General Relativity: History and Analysis." *Studies in History and Philosophy of Science* 11: 175–214.

Eddington, Arthur Stanley (1918). *Report on the Relativity Theory of Gravitation*. London: Fleetway Press.

——— (1920). *Space, Time and Gravitation*. Cambridge: Cambridge University Press.

——— (1932). "Meeting of the Royal Astronomical Society. Friday, December 11, 1931." *The Observatory* 55: 1–10.

Einstein, Albert (1915). "Zur allgemeinen Relativitätstheorie." *Königlich Preussische Akademie der Wissenschaften* (Berlin). *Sitzungsberichte*: 778–786.

——— (1919). "Induktion und Deduktion in der Physik." *Berliner Tageblatt*. December 25 (Suppl. 4): 1.

——— (1920). *Über die spezielle und die allgemeine Relativitätstheorie. (Gemeinverständlich)*, 5th ed. Braunschweig: Friedrich Vieweg & Sohn.

——— (1942). "Foreword." In Bergmann 1942.

Eisenstaedt, Jean (1986). "La relativité générale à l'étiage: 1925–1955." *Archive for History of Exact Sciences* 35: 115–185.

——— (1988). "The Early Interpretation of the Schwarzschild Solution." See this volume, pp. 213–233.

Harrison, B. Kent, Thorne, Kip S., Wakano, Masami, and Wheeler, John A. (1965). *Gravitation Theory and Gravitational Collapse*. Chicago: University of Chicago Press.

Hawking, Stephen W., and Israel, Werner (1979). *General Relativity: An Einstein Centenary Survey*. Cambridge: Cambridge University Press.

Infeld, Leopold (1950). *Albert Einstein: His Work and Its Influence in Physics*. New York: Charles Scribner's Sons.

——— (1964). In *Conférence internationale sur les théories relativistes de la gravitation*. Warsaw: PWN—Éditions Scientifique de Pologne; Paris: Gauthier-Villars.

Levi-Civita, Tullio. (1937). "Astronomical Consequences of the Relativistic Two-Body Problem." *American Journal of Mathematics* 59: 225–234.

——— (1950). *Le Problème des n corps en relativité générale*. Mémorial des Sciences Mathématiques, no. 41. Paris: Gauthier-Villars.

McVittie, George C. (1956). *General Relativity and Cosmology*. London: Chapman and Hall.

Oppenheimer, J. Robert (1956). "Einstein." *Reviews of Modern Physics* 28: 1–2.

Pirani, Felix. (1956). "On the Physical Significance of the Riemann Tensor." *Acta Physica Polonica* 15: 389–405.

Rosen, Nathan (1940). "General Relativity and Flat Space." *Physical Review* 57: 147–153.

Roseveare, N.T. (1982). *Mercury's Perihelion: From Le Verrier to Einstein*. Oxford: Clarendon Press.

Schild, Alfred (1960). "Equivalence Principle and Red-Shift Measurements." *American Journal of Physics* 28: 778–780.

Stachel, John (1972). "The Rise and Fall of Geometrodynamics." In *PSA 1972: Proceedings of the 1972 Biennial Meeting, Philosophy of Science Association*. Kenneth F. Schaffner and Robert S. Cohen, eds. Boston Studies in the Philosophy of Science, vol. 20. Dordrecht: D. Reidel, 1974, pp. 31–54.

Synge, John L. (1950). "The Gravitational Field of a Particle." *Royal Irish Academy* (Dublin). *Proceedings* 53: 83–114.

────── (1960). *Relativity: The General Theory*. Amsterdam: North-Holland.

────── (1970). *Talking about Relativity*. Amsterdam: North-Holland.

Tolman, Richard C. (1934). *Relativity, Thermodynamics and Cosmology*. Oxford: Clarendon Press.

Trautman, Andrzej. (1966). "The General Theory of Relativity." *Soviet Physics. Uspekhi* 9: 319–335.

von Laue, Max (1921). *Die Relativitätstheorie*. Vol. 2, *Die allgemeine Relativitätstheorie und Einsteins Lehre von der Schwerkraft*. Braunschweig: Friedrich Vieweg & Sohn.

Weyl, Hermann (1918). *Raum-Zeit-Materie*. Berlin: Julius Springer.

Wheeler, John Archibald (1957). "The Present Position of Classical Relativity Theory, and Some of Its Problems." In *Conference on the Role of Gravitation in Physics. Proceedings*. W.A.D.C. Technical Report: 57–216. C. De Witt, ed. Chapel Hill.

Whitrow, G.J., and Morduch, G.E. (1965). "Relativistic Theories of Gravitation." In *Vistas in Astronomy*. Vol. 6. A. Beer, ed. Oxford: Pergamon, pp. 1–67.

Whittaker, Edmund T. (1928). "The Influence of Gravitation on Electromagnetic Phenomena." *Journal of the London Mathematical Society* 3: 137–144.

# The Canonical Formulation of General-Relativistic Theories: The Early Years, 1930–1959

PETER G. BERGMANN

With the birth of quantum field theory in the late 1920s, physicists decided that nature could not be half classical and half quantum, and that the gravitational field ought to be quantized, just as the electromagnetic field had been. This judgment was based not so much on logical as on esthetic grounds. Møller and Rosenfeld pointed out that, from a formal point of view, it is entirely possible to retain a nonquantum geometric background representing gravity, on which all the other forces of nature would rely in playing out their quantum roles. Einstein, who on philosophical grounds never accepted quantum physics as the ultimate truth, remained aloof from all attempts to construct a general-relativistic quantum field. Until the end of the 1950s, quantization remained an endeavor of a few small groups.

Obviously there are different ways for approaching the construction of a quantum theory of gravity. One route is to treat gravity as a small deviation of space-time geometry from Minkowski geometry. The gravitational potentials would then be the components of a symmetric tensor with properties analogous to those of the electromagnetic potentials (which form a vector with respect to Poincaré transformations). This approach leads to a Poincaré-invariant field theory with a gauge group.

Alternatively, one could accept the group of diffeomorphisms as a fundamental characteristic of general relativity (and indeed of all general-relativistic theories), and proceed to construct a quantum field theory that is adapted to that group. Quantization would be attempted by way of a Hamiltonian formulation of the (classical) theory, and quantum commutation relations would be patterned after the Poisson brackets arising in that formulation. This program is usually called the *canonical quantization* program, whereas the weak-field approach is known as *covariant quantization*.

The first step in canonical quantization obviously consists of the identification of the canonical field variables of general relativity, followed by the construction of the Hamiltonian. These first steps, conceived entirely within the framework of the classical theory, turned out to be beset with technical and conceptual difficulties, which today are essentially resolved. In this paper I shall attempt to trace out these initial steps. The period of time to be covered begins with two papers by Rosenfeld of 1930 and 1932 and, except for one later paper by Komar and myself, ends with two papers by Dirac of 1958 and 1959.

## 1. The Constraints

In his very first paper, Rosenfeld discovered that if the canonical momentum densities are defined, as usual, as the partial derivatives of the Lagrangian density with respect to the time derivatives of the configuration variables (the "velocities"), then these momentum densities are not algebraically independent of each other (and of the configuration variables), but by virtue of their defining equations satisfy relationships that do not involve any time derivatives. Such "algebraic" relations between canonical field variables (they may involve derivatives with respect to spatial coordinates) are called *constraints*. Rosenfeld distinguished between algebraic constraints and constraints that involve derivatives. Examples may be found in the canonical formulation of the Maxwell-Lorentz theory. The momentum density conjugate to the scalar potential vanishes (an algebraic constraint); the momentum densities conjugate to the components of the vector potential form a vector (the negative electric field strength), whose divergence equals the negative of the charge density. Even in this early paper, Rosenfeld indicates that constraints that are free of derivatives can be rendered innocuous by appropriate manipulations.

Rosenfeld pointed out that constraints are necessarily associated with the group of diffeomorphisms, in that Cauchy data on a space-like hypersurface cannot determine the values of the field variables on an adjacent hypersurface, but only up to an infinitesimal change of coordinate system. There must be some degree of arbitrariness in the choice of Hamiltonian, enough to make possible this nonuniqueness of propagation. These additions to some basic Hamiltonian are, thus, the generators of infinitesimal coordinate transformations. On account of Noether's theorem, the generators of symmetry transformations are constants of the motion. But as the coordinate transformations involve arbitrary functions, which presumably appear as coefficients in the generators, these generators cannot but vanish.

A somewhat more detailed discussion of the constraints is to be found in papers by Dirac and by our group, beginning approximately in 1949. If constraints are satisfied by the canonical field variables on a given Cauchy hypersurface, they must also hold on adjacent hypersurfaces, that is to say, they must *propagate*. But they may not propagate automatically; the requirement that their time derivatives vanish may lead to additional constraints, to wit, the vanishing of the Poisson brackets of the original constraints with the Hamiltonian. Conceivably, this procedure might have to be iterated more than once, until it leads to Poisson brackets that are algebraically dependent on the preceding constraints, or vanish identically. To distinguish these diverse generations of constraints we coined the expressions "primary constraints," "secondary constraints," etc.

If the constraints are to be added to the Hamiltonian in order to realize the richness of coordinate transformations, it follows that the Poisson brackets between constraints must vanish *modulo* the constraints themselves, that the constraints form a *system of involutions*. To the extent that the constraints generate the symmetry group of a theory, the Lie algebra of the constraints is a realization of the infinitesimal symmetry group, and one might conjecture that all constraints always are in involution with each other. Around 1950, Dirac considered constraints that do not have this property, probably led to them by the constraints encountered when one attempts to subject the Schrödinger equation to second quantization. Constraints that have non-zero Poisson brackets with some other constraints he called "second-class constraints." Constraints that have vanishing Poisson brackets with all other constraints are called "first-class constraints."

The two terminologies, primary versus secondary constraints, and first-class versus second-class constraints, are awkward, to be sure. This confusing state of affairs is due to the fact that, initially, there was almost no contact between Dirac and us, and that by the time, there was contact, these two characterizations had become sufficiently established that we jointly decided to leave well enough alone.

The significance of Dirac's classification of constraints can be visualized in terms of mappings of the phase space of the theory on itself. The set of points of the phase space on which all constraints are satisfied is a subset of the phase space, the *constraint hypersurface*. Viewed as the generator of an infinitesimal mapping of the phase space, first-class constraints map the constraint hypersurface on itself, whereas second-class constraints map it on some other hypersurface. From this it follows immediately that the first-class constraints of the theory form a closed Lie algebra: the Poisson bracket of two first-class constraints is itself a first-class constraint.

As for second-class constraints, Dirac invented an ingenuous modification of the Poisson bracket. By adding to an arbitrary dynamical variable an appropriate linear combination of the second-class constraints, a new variable is obtained, whose value is the same as that of the original variable, but whose Poisson bracket with all second-class constraints vanishes. This procedure, known as "starring," is equivalent to Dirac's original formulation, in which he left the variables unchanged but added to the usual terms of the Poisson bracket some extra terms that are bilinear in the second-class constraints. Because of this equivalence it is easy to prove that the modified Poisson brackets satisfy all requirements of a Lie algebra.

If the infinitesimal group of mappings that is generated by the first-class constraints is expanded into a group, a group of canonical mappings that map the constraint hypersurface on itself, then those orbits of that group of mappings that have at least one point on the constraint hypersurface lie in that hypersurface entirely; we call them *equivalence classes*. These equivalence classes cover the constraint hypersurface, and do so without overlap. Each point on the constraint hypersurface belongs to exactly one equivalence class.

Equivalence classes correspond to sets of values of the canonical field variables that are carried over into each other by members of the symmetry group of the theory. In the language of physics, we should say that the points of one equivalence class represent the variety of ways how one-and-the-same physical situation can be represented. Distinct equivalence classes correspond to inequivalent physical situations. If we construct a new space, a quotient space of the constraint hypersurface, each point of which represents an entire equivalence class, then this new space possesses a symplectic structure, which is a descendant of the symplectic structure of the original phase space; we call that space the *reduced phase space*. By pull-back the symplectic structure of the original phase space induces a symplectic structure on the constraint hypersurface, but this structure is singular: Lagrange brackets are well defined, but Poisson brackets are not. By a second pull-back we obtain the symplectic structure of the reduced phase space, but this symplectic structure turns out to be regular!

Dynamical variables that are defined on the reduced phase space are those that are constant over each equivalence class, and hence, by definition, invariant with respect to the symmetry group of the theory. They are *observables*, in the sense that their values depend only on the physical situation, not on the manner in which we present it. Poisson brackets between observables are well defined, though not between other dynamical variables.

All of these facts were worked out in great detail during the 1950s. Fairly early in that decade it was also established that the role of the primary, secondary, etc., constraints had to do with the form of the (infinitesimal) transformation laws for the field variables. Such transformation laws might contain references to the characteristic functions of a coordinate or gauge transformation (such as the vector field representing an infinitesimal coordinate transformation), to their time derivatives, to their second time derivatives, and so forth. Primary constraints must be multiplied by the highest-order time derivatives, secondary constraints by the next-highest-order time derivatives, and so on, to yield the appropriate generators of the whole transformation.

Most of this discussion applies to the group of diffeomorphisms as well as to any Abelian or noncommuting gauge group that, by itself or in combination with other groups, forms the symmetry group of a physical theory, though the details were worked out primarily in connection with the group of diffeomorphisms. The next section is concerned with a problem that is peculiar to general-relativistic theories.

## 2. Propagation of the Field and the Hamiltonian

The infinitesimal mapping that is supposed to contain the dynamical essence of a physical theory is the displacement in time, that is to say the space-time mapping $\delta t = a$. This mapping certainly is a diffeomorphism; it belongs to the symmetry group of any general-relativistic theory. It follows that in general-relatistic theories the Hamiltonian must belong to the set of first-class constraints.

In the early 1950s, both Dirac and we made an attempt to disconnect the group of diffeomorphisms from the dynamics of the theory by introducing a subsidiary set of coordinates, which we called *parameters*. The idea was to treat the four coordinates $x^\mu$ as functions of the parameters $(t, u^k)$, converting them, in effect, into an additional set of configuration variables, with their own conjugate momentum densities (which are the components of the vector field that generates infinitesimal coordinate transformations). If the action integral is rewritten in terms of the parameters as the independent variables of integration, the Lagrangian becomes a homogeneous function of the first degree of the "velocities," the $t$-derivatives of the augmented set of variables. Accordingly, by Euler's formula, if one attempts to construct the Hamiltonian by the usual formula, $H = \mathbf{p} \cdot \dot{\mathbf{q}} - L$, one obtains identically zero. Nevertheless, a nontrivial Hamiltonian exists. It was constructed first by Pirani and Schild, and independently by our group at Syracuse, in 1950. Subsequently, Dirac found that the introduction of parameters is not necessary for the construction of a Hamiltonian formalism. Without them the whole formalism is much less involved. As it turns out, in Dirac's formulation the primary constraints can be eliminated, and the Hamiltonian density is a linear combination of first-class secondary constraints, the coefficients being arbitrary functions of the canonical field variables and the coordinates.

Before discussing Dirac's 1958–1959 papers, I need to deal with one serious difficulty, which arose in the middle 1950s; this difficulty concerns the representation of the group of diffeomorphisms by canonical transformations.

## 3. The Lie Algebra of Infinitesimal Coordinate Transformations

Consider two infinitesimal coordinate transformations. If they are thought of as vector fields, then their commutator is simply the Lie derivative of one of these fields with respect to the other. Locally, the vector that results depends on the first-order derivatives of the two commuted vector fields with respect to the four coordinates. If we wish to represent this Lie algebra in terms of canonical transformations in some phase space, we need to construct (at least locally) a "space-like" three-dimensional hypersurface, the Cauchy surface, and canonical variables whose transformations in some way reflect infinitesimal coordinate transformations of space-time.

No matter how we choose our phase space (e.g., consisting of the metric tensor and its conjugate momentum density), the transformation law will depend on the components of the vector field associated with the coordinate transformation, and their out-of-the-hypersurface derivatives up to some finite order. Because of the infinitesimal character of the coordinate transformation considered, we may assume that this dependence is linear, that is to say, the components of the vector field, and their "time" derivatives will be multiplied by appropriate generators. Let the highest order that appears in that linear combination be denoted by $k$. Then the expression that generates the commutator vector field will have precisely that same structure, but the components of the vector field will be those of the commutator, and its $k$th-order time derivative will contain $(k + 1)$th-order time derivatives of the commuted fields. As the vector fields themselves are arbitrary in all four dimensions, the $(k + 1)$th order at a point is arbitrary, and cannot depend in any way on the lower-order "time" derivatives. Thus the generator of the commutator must contain elements that cannot possibly be obtained from a canonical commutation of whatever ingredients are available on the Cauchy surface! An adequate description of the symmetry group of general relativity in terms of a canonical formalism appears foreclosed!

This difficulty was overcome by Dirac in his 1958 paper. As long as we deal with the space-time manifold by itself, disregarding its metric structure (which, by assumption, forms part of the dynamical variables), the choice of a three-dimensional Cauchy hypersurface ($t = $ constant) leaves undetermined, and arbitrary, the direction of the local "time axis." Likewise, adoption of an infinitesimal coordinate transformation, or of the equivalent vector field, leaves open the direction and magnitude of the displacement out of the surface. Once we acknowledge the existence of the metric structure, however, we can construct at a given point on the Cauchy hypersurface the unit vector perpendicular to the hypersurface. Instead of describing the four-dimensional infinitesimal coordinate transformation by means of the four components of the displacement vector, we may describe it in terms of the three-dimensional vector tangential to the hypersurface (the "shift" in the terminology coined by Arnowitt, Deser, and Misner), and the magnitude of the displacement perpendicular to the hypersurface (the "lapse"). This description is independent of the choice of coordinate system off the hypersurface. Hence the commutator of two four-vector fields on the three-hypersurface can be obtained in terms of variables defined only on the hypersurface. As Dirac showed, this commutator is generated by the Poisson bracket of the generators associated with the two displacement fields to be commuted with each other.

Dirac's procedure at first appeared like magic. Does the commutator lead to a closed Lie algebra, and does that Lie algebra in turn belong to a group? These questions were answered in a 1972 paper by Komar and myself. There is indeed a group, but it is not the group of four-dimensional diffeomorphisms. Rather, if solutions of Einstein's vacuum equations are described in terms of appropriate data on a Cauchy hypersurface, this can, of course, be done in an infinity of ways, which depend on the choice of Cauchy hypersurface and on the choice of (three-)coordinate system on it. The group in question involves the mappings of all these descriptions of the same solution on each other. There exists neither an isomorphism nor a homomorphism of that group with the group of four-dimensional diffeomorphisms, but in a certain sense the two groups have the same orbits, each orbit representing one physical situation.

## 4. Dirac's Hamiltonian Formulation

Aside from overcoming the seemingly insurmountable obstacle presented by the peculiarities of the group of (four-)diffeomorphisms, Dirac streamlined the identification of the canonical variables. In the earlier formulations there had ben primary and secondary constraints. Almost simultaneously, DeWitt and Anderson had discovered a canonical transformation that simplified the primary constraints, and so had Dirac. By making the primary constraints purely algebraic, Dirac was able to eliminate from the formalism four pairs of canonically conjugate variables, $g_{0\mu}$ and $\pi^{0\mu}$, retaining only the three-metric $g_{mn}$ and their canonically conjugate momentum densities. As for the secondary constraints, these turned out to be four linear combinations of Einstein's field equations, those that are free of second-order $t$-derivatives. The remaining six Einstein equations are reproduced as one-half of the canonical field equations. Because of the contracted Bianchi identities, once the four constraints have been satisfied on a Cauchy hypersurface, then the canonical field equations cause them to be satisfied everywhere else. The Hamiltonian density itself is a linear combination of the constraints, the coefficients being the shift and lapse functions.

Dirac essentially completed the canonical formulation of general relativity. His procedure would presumably also be applicable to modified theories of gravitation, such as the scalar-tensor theories and the Einstein-Cartan theories, provided there is a metric structure. To accomplish the same for theories without a metric would probably require some ingenuity.

To have formulated general relativity in canonical language does not mean that the path to a quantum theory of gravitation is now straightforward. Abhay Ashtekar will speak on the further history of that subject.

EDITORIAL NOTE

Unfortunately, Ashtekar's talk is not published in this volume (see Preface).

# Einstein, Hilbert, and Weyl: The Genesis of the Geometrical Unified Field Theory Program*

VLADIMIR P. VIZGIN

## 1. Introduction

With the recent successes in mastering the problem of unifying the fundamental physical interactions (the Weinberg-Salam theory of electroweak interactions), hope has arisen of our being able to create a unified field theory. At the same time, interest has developed in the history of corresponding earlier attempts. In this connection, it became clear that the concept of a gauge field theory, upon which nearly all current research on this problem is based, was itself built upon the unified field theories of the 1920s and 1930s that were elaborated by Weyl, Eddington, and Einstein, among others.

The geometrical unified field theories of this earlier period also provide abundant material for the study of the interactions between mathematics and physics in the twentieth century. As in the special and general theories of relativity, a new form of this interaction, atypical of classical physics, manifested itself in geometrical unified field theories. Here mathematics was not a mere calculational aid for physics (this view was widespread among physicists in the nineteenth century and even at the beginning of the twentieth century), but far more a structural foundation for the description of physical reality.[1]

In the classical physics of the eighteenth and nineteenth centuries (and at the beginning of the twentieth century), there were two grand strategies for synthesizing physical knowledge, based upon classical mechanics and the theory of the electromagnetic field; that is to say, two global paradigms (if one follows Kuhn) or two global research programs (if one follows Lakatos): the program of classical mechanics and that of the electromagnetic field (Vizgin 1982). In the first two decades of the twentieth century new fundamental programs (in the following the terminology of Lakatos will be employed) took their place, programs that were based upon relativity theory and quantum mechanics. A concrete expression of the relativistic program after the triumph of the general theory of relativity was the program of geometrical unified field theories, which were regarded in the 1920s and 1930s as a real alternative to the quantum-theoretical program for elaborating a unified physical theory of fields and elementary particles (Vizgin 1979).

But for all that it led to a series of mathematically well-founded theories, the program of geometrical unified field theories did not hold its own in competition with the rapidly developing quantum-theoretical research program. The latter led to quantum mechanics, the quantum theory of radiation, relativistic quantum mechanics, and, eventually, quantum electrodynamics; it also showed itself to be more effective in nuclear physics, elementary particle physics, solid state physics, and so forth. Not only were new geometrical structures developed in connection with the geometrical field theory program, but also important physical concepts of modern quantum field theory (charge symmetry, the Klein-Gordon equation, gauge symmetry, and gauge field theory, among others).

In the present article, the first models for unified field theories will be considered, those that constituted the core of the unified field theory program. Above all, Weyl's 1918 theory will be examined. The various factors that conditioned the emergence of this theory will be portrayed: the Göttingen tradition in mathematical physics, the mathematical and physical starting points, Weyl's contacts with Einstein, and his philosophical interests. The article will conclude with discussions of the final stage in the formulation of the geometrical unified field theory program (1921) and its subsequent fate.

## 2. Weyl's Theory

There have been attempts to describe the gravitational and electromagnetic fields in a unified fashion ever since the electromagnetic field theory program came into existence (i.e., since the beginning of the 1890s). We will return to this topic, as we will to Hilbert's 1915 unified field theory, in which the attempt was first made to unite gravitation and electromagnetism on the basis of the general theory of relativity. Both of these approaches were one-sided, in the sense that each resulted in an advantage for only one of the fields, either the electromagnetic or gravitational field. One could consider ponderable matter, i.e., the elementary particles, as either condensations or singularities of the given field, mainly the electromagnetic field. A comparatively constructive procedure for building up particle-like solutions was worked out within the framework of a quite fully developed electromagnetic field theory, the theory of Gustav Mie (1912-1913). This procedure was also employed in Hilbert's theory (a two-fold reduction: the particles led to the electromagnetic field and that, in turn, to the gravitational field) (Vizgin 1981a).

In Weyl's theory (1918) both fields were geometrized in the same way. This became possible through a generalization of Riemannian geometry that allowed one to develop that geometry as a manifestation of the electromagnetic field, just as gravitation was interpreted in terms of the curvature of four-dimensional space. The necessary generalization proceeded perfectly naturally and was connected with a thoroughgoing elimination of elements of "action-at-a-distance" from physical geometry: the parallel displacement of a vector is accompanied not only by a change in the vector's orientation, as in Riemannian geometry, but also by a change in the vector's length. It is not the components of the metric $g_{ik}$, but rather their ratios that have physical significance, which further leads to an extension of the general covariance group, in which scale transformations of the metric $g_{ik}$ are also admissible:

$$g'_{ik} = \lambda(x) g_{ik}. \tag{1}$$

Consequently, local vector spaces (at different points) are connected with one another through transformations belonging to the similarity group. The length of a vector (or the square of its length) varies under an infinitely small parallel displacement according to the law

$$dl = -l\,d\varphi, \tag{2}$$

where $d\varphi$ is a linear differential form

$$d\varphi = \varphi_i\,dx^i, \tag{3}$$

which supplements the quadratic fundamental form (Riemannian line element)

$$ds^2 = g_{ik}\,dx^i\,dx^k$$

in the new geometry. The quantities $\varphi_i$ turn out to be components of a four-vector that transform according to the law

$$\varphi'_i = \varphi_i - \frac{\partial \ln \lambda(x)}{\partial x^i} \tag{4}$$

under the scale transformations (1). In this equation, the vector $\varphi_i$ is analogous to the four-potential of the electromagnetic field. Riemannian geometry is transformed in this fashion into a geometry with an affine connection, whose connection coefficients differ from the Christoffel symbols $\Gamma_{i,rs}$:

$$\Gamma_{i,rs} = \Gamma^*_{i,rs} + \tfrac{1}{2}(g_{ir}\varphi_s + g_{is}\varphi_r - g_{rs}\varphi_i). \tag{5}$$

The electromagnetic field made its own contribution to the affine space-time connection.

In a manner similar to the way in which the metric tensor $g_{ik}$ was identified with the gravitational potential in general relativity, Weyl identified the metric vector $\varphi_i$ with the potential of the electromagnetic field. As a result, the geometrization of the electromagnetic field was achieved within the framework of a unified geometrical description of both fields.

In Weyl's view, the field equations had to be derivable from a variational principle that is invariant with respect to the general covariance group as well as with respect to the similarity transformations (1). The latter demand led to Lagrange functions of the form $R^2$ or $R^i_{jkl} R_i{}^{jkl}$, which leads inevitably to fourth-order differential equations.[2] A series of difficulties follows from this, even though the Schwarzschild solution, as Pauli soon showed, also turns out to be a solution of this equation of Weyl's theory; and thus well-known effects, such as the precession of the perihelion of Mercury and the bending of light rays in the sun's gravitational field, are also predicted by the theory (Pauli 1921).

A grave difficulty, noticed immediately by Einstein (1918), manifested itself in the evident contradiction between the geometrical foundations of the theory and experience, in particular in the fact that chemical elements possess spectral lines of definite frequency. Weyl therefore had to give up the identification of the original geometrical foundations of the theory with actual length and time relationships, which deprived the theory of its "inner power of conviction" (Pauli 1921, p. 763). Nevertheless, Weyl's theory made a considerable impression upon theoreticians. Einstein wrote that its depth and boldness must charm every reader (Einstein 1918, p. 480).

## 3. The Göttingen Tradition of Mathematical Physics

A scholar's belonging to one or another scientific school, to a scientific tradition, determines in significant measure his choice of a research theme and his approach to the solution of scientific problems. Weyl, who concluded his studies at the University of Göttingen in 1907, defended his dissertation in that same year under Hilbert, the leading representative of Göttingen mathematics, and taught there until 1913. Weyl was shaped as a mathematician in the Göttingen atmosphere, where the tone was set in the areas of mathematics and mathematical physics in the first decades of the twentieth century above all by Felix Klein, Hilbert, and Minkowski; and in the first third of the twentieth century this atmosphere contributed essentially to the creation of the relativity and quantum theories. Otherwise, it would be more correct to speak of the continuation or the revival of the famous tradition that is associated with the names of Gauss and Wilhelm Weber, but also with those of Riemann and Dirichlet (Klein 1926). The fundamental core of this tradition consisted in a special linkage of abstract mathematical research with the treatment of physical problems, which proved fruitful for both sciences. This involved not only applied mathematics, but also the employment of the newest branches of mathematics and abstract mathematical structures for the solution of general problems of principle in theoretical physics.

Among the outstanding results of Göttingen mathematical physics were the four-dimensional invariant-theoretical conception of special relativity, created primarily by Minkowski (1907–1908), as well as the work of Hilbert, Klein, and Emmy Noether on general relativity (1915–1918) (Vizgin 1972, 1975, 1981a; Pyenson 1974, 1977). Characteristic of the Göttingen tradition was a great interest in mathematical structures in physics, in the problem of a unified physical theory, and in questions of axiomatics, along with a particular rigor in scientific writings. The experimental and empirical aspects of scientific theories, questions of physical interpretation, and work on special physical problems—all of these were of secondary importance (Pyenson 1977, 1979, 1981). Further references on the topic of this section include: Pyenson 1974, 1975.

## 4. Hilbert's Theory

As already indicated, the first unified field theory based upon the general theory of relativity was Hilbert's; he presented it in a lecture to a session of the Königliche Gesellschaft der Wissenschaften (Royal Academy of Sciences) at Göttingen on November 20, 1915. This theory was at the same time a model of the Göttingen tradition in mathematical physics. Hilbert based himself upon Einstein's papers of 1913–1914 on the relativistic theory of gravitation, which were worked out in part with Marcel Grossmann, and upon Mie's 1912 papers on nonlinear electrodynamics and the electromagnetic theory of matter. Hilbert attempted to master in this way the well-known problem of the axiomatization of physics (Hilbert's sixth problem of 1900).

From the physical standpoint, Hilbert's theory is a classical field theory, constructed in a manner similar to the general theory of relativity. The difference between Hilbert's theory and the general theory of relativity consists in a specialization of the momentum-energy tensor. In this, Hilbert followed Mie's theory, in that he proposed that all matter must be reduced to the electromagnetic field. But in distinction to Mie's theory, the electromagnetic field was not primary for Hilbert. Thus, as in the general theory of relativity, the gravitational field is essentially identical with the Riemannian space-time manifold; and the electromagnetic field turns out to be a peculiar consequence of gravitation, and hence of the geometry. The nature of the connection between gravitation and electromagnetism in Hilbert's theory was plainly characteristic of the Göttingen theoretical procedure: the equations of the electromagnetic field are interpreted as the Euler-Lagrange equations of covariant variational problems (a special case of the Noether theorem on invariant variational problems proved two-and-a-half years later). The deductive-axiomatic standpoint, the employment of abstract mathematical theories (differential geometry, group theory, and the calculus of variations), the goal of constructing a unified physical theory, but also the undervaluation of experimental and empirical issues, of questions of physical interpretation, and of the practical grounding of the initial concepts and initial basis—all of these characteristic traits of Göttingen mathematical physics also appeared clearly in Hilbert's theory.

It should be noted that Hilbert gave his lecture, "Grundlagen der Physik," ("Foundations of Physics") five days prior to Einstein's appearance before a meeting of the Berlin Akademie der Wissenschaften (Academy of Sciences), at which Einstein presented the gravitational field equations:

$$R_{ik} - \tfrac{1}{2} g_{ik} R = -\kappa T_{ik}, \tag{6}$$

that were already contained in Hilbert's lecture. Hilbert had developed them from a generally covariant variational principle, in which he used the corresponding tensor from Mie's theory for $T_{ik}$, thereby incorporating in his rigorously axiomatic-deductive theory Mie's physically inadmissible representation of the structure of matter. On the whole, the Göttingen school felt itself bound to the electromagnetic world picture (or the electromagnetic field program). Remember that researchers like Drude, Abraham, Kaufmann, Wiechert, and other leading representatives of this program worked on it there. A further reference on the topic of this section is Vizgin 1981a.

## 5. The Influence of Göttingen and of Einstein

Weyl spent ten of the most important years of his life in Göttingen (from 1903 to 1913), and he returned there from Zurich in 1930 to take up Hilbert's chair. For Weyl, Hilbert was the ideal of a scholar and a human being, the beloved teacher and tutor. It was under his direction that Weyl defended his doctoral dissertation on singular integral equations in 1907. And it was under the influence of Hilbert and the Göttingen atmosphere that Weyl from the beginning took an interest in mathematical questions in physics (in the first place in the applications of integral equations and the spectral theory of differential equations to physical problems). Klein's influence on Weyl was also substantial: his geometrical conceptions, his efforts on behalf of a close contact between physicists and mathematicians, and his ideas on the role of intuition in mathematics.

In 1913, Weyl came to Zurich, where Einstein was just then creating with Grossmann the foundations of the general theory of relativity. Obviously, the direct contact with Einstein—along with the Göttingen influence—determined Weyl's interest in the general theory of relativity. Beginning in 1916, Weyl gave lectures on this topic at the ETH-Zurich. (Einstein had been in Berlin since early 1914.) The results of Weyl's new investigations concerning general relativity were summarized in his article, "Zur Gravitationstheorie" ("On the Theory of Gravitation") that appeared in August 1917 in the *Annalen der Physik* (Weyl 1917). The Göttingen influence reveals itself in this article in the pronounced interest in mathematical problems of the general theory of relativity: in the formulation of the theory using methods from the calculus of variations, in the derivation of the equations of motion and conservation laws, and in the exact solution of the gravitational equations. Nevertheless, Weyl does not mention Hilbert's unified theory. One can even infer that he considered Hilbert's standpoint on the unification of gravitation and electromagnetism to be false, because Weyl regarded not the equations of electrodynamics but rather the energy-momentum conservation laws as a consequence of general covariance. Carl Seelig published fragments of the correspondence between Einstein and Weyl, from which it emerges that they discussed Hilbert's theory. Einstein expressed a very critical opinion of it, in particular about the employment of Mie's theory and the axiomatic method (Seelig 1960). In all probability, Weyl agreed with Einstein on these points, even though he was obviously well-acquainted with Hilbert's fundamental ideas and considered them very important. The creator of the first geometrical unified field theory, which represented the core of the geometrical unified field theory program, later emphasized that it was precisely Hilbert and his theory that constituted the beginnings of this program.[3]

Weyl's theory falls wholly within the Göttingen tradition of mathematical physics. That becomes evident not only from the fact that Weyl's geometry (the geometry of a semimetrical connection) makes it possible to describe the gravitational field and the electromagnetic field in a unified fashion, but also from the fact that the theory's author devoted insufficient attention to the properly physical, that is, in particular, the experimental or empirical foundations and implications of the theory. In the final outcome, the fundamental difficulties of the theory were connected with this side of it, as was emphasized by Einstein and Pauli. Further references on the topic of this section include: Weyl 1954, Reid 1970, and Newman 1958.

## 6. Differential Geometry and the Preceding Unified Field Theories

Weyl was not only influenced by the Göttingen tradition and by direct contact with Hilbert and Einstein. Mathematical stimuli, above all in the area of differential geometry, were also of great significance for the creation of Weyl's theory, as were works on unified field theory that went beyond the framework of Hilbert's theory, and Weyl's constant and lively interest in philosophy in general, and in the methodology of the exact sciences and mathematics in particular.

After 1916, the general theory of relativity gave new impetus to research in the area of differential geometry. One of the most important results that emerged from this was the concept of infinitesimal parallel displacements of vectors in a Riemann space (Levi-Civita, Hessenberg, Schouten in the years 1916–1918). This concept made it possible for Weyl to generalize Riemannian geometry to a geometry of semimetrical connections on the space-time continuum of his unified field theory (Weyl 1923, Schouten 1926).

The boom in work on the electromagnetic field program, the triumph of the field concept in the special and general theories of relativity, the popularity of Mie's nonlinear electrodynamics, the enthusiasm even of Hilbert for the problem of constructing a unified field theory—all of that advanced the growing prestige and the dissemination of the idea of a "field-theoretic ideal of unity" in physics. Aside from the theories of Mie and Hilbert, a whole series of different unified field theories arose in the years 1912–1918: among others, the theories of Ishiwara, Nordström, and others (in which gravitation was described by means of a scalar potential); the theory of Wiechert (and electromagnetic ether theory); and the theory of Reichenbächer (in which ideas from general relativity and ether-type models were combined eclectically). One should also recall Einstein's unceasing efforts in the years 1908–1909 to find a field-theoretic foundation for the description of the electron and of quanta (McCormmach 1970). It is noteworthy that these investigations of Einstein's were entirely in accord with the spirit of the Göttingen tradition, although it is commonly assumed that, up until the beginning of the 1920s, he stood considerably closer to the tradition in theoretical physics associated with the names Boltzmann, Lorentz, and Planck. Many of the authors of unified field theories named here, who were also cited in Weyl's book, *Raum-Zeit-Materie* (2nd ed., 1919), were otherwise closely connected to Göttingen. If, nevertheless, the differential-geometric investigations mentioned earlier were of fundamental significance for the elaboration of Weyl's theory, then the other works on unified field theory were only a proof of the topicality and difficulty of this problem, and played no essential role.

## 7. Weyl's Philosophical Interests

The significance of this factor is quite problematic. Weyl's work on philosophical problems of physics and mathematics appeared somewhat later, even though his philosophical aspirations are to be recognized already in the first edition of his book, *Raum-Zeit-Materie*. In his lecture, "Erkenntnis und Besinnung" ("Knowledge and Consciousness") (Weyl 1954), which he gave one year before his death, Weyl himself elucidated his development in this connection: from the philosophy of Kant and an enthusiasm for Mach and Poincaré (1905–1913), to the philosophy of Göttingen's Husserl, who also exercised an influence on other Göttingen scholars (Voigt, Wiechert). Weyl considered the relativity theory and, associated with it, the theory of space and time as an example of the phenomenological comprehension of the existent, about which he wrote in the introduction to his book on relativity theory.[4] He conceived of his unified theory as a natural development of this idea, which accorded entirely with the spirit of phenomenological philosophy.

At the beginning of the 1920s, Weyl was impressed by the philosophy of Fichte, on the one hand, while on the other he devoted himself with renewed intensity to the foundations of mathematics. In this connection, Weyl came into conflict with the formal-axiomatic standpoint of Hilbert, after he had subscribed to the intuitionistic ideas of Brouwer. It is to be assumed that Weyl's new areas of interest were also connected with his investigations concerning unified field theory. Fichte's constructivist version of the Kantian critical philosophy, the intuitionistic theory of existence ("to exist means to be constructible"), "the exclusion of superfluous arbitrary elements," and the consideration of the mathematical continuum as "a medium of free becoming" (Weyl 1927, pp. 41–43), all of these come close to some ideas and reflections that Weyl employed in the construction of his unified field theory (e.g., the grounding of the transition from Riemannian geometry to "pure infinitesimal geometry," i.e., Weyl's geometry).

## 8. The Year 1921—The Appearance of the Geometrical Unified Field Theory Program

Weyl's theory astounded his contemporaries "by the grandiosity of its conception that interpreted all appearances in the 'physical world' as law-like regularities of the 'geometrical world'" (Friedmann 1923, p. iii).[5]

In 1919 the general theory of relativity received a splendid experimental confirmation with the observations of the bending of light rays in the gravitational field of the sun during a solar eclipse. These observations were undertaken by two British expeditions under the direction of Crommelin and Eddington. In conjunction with rapid theoretical progress (exact solutions, gravitational waves, cosmology, etc.), the eclipse observations promoted the acceptance and dissemination of the theory of gravitation and its fundamental ideas, which were connected with the generalization of the relativity principle and the identification of the physical field with geometrical structures. The stock of the first unified field theory based upon these ideas rose at the same time.

Although Einstein and Pauli had pointed out the physical dubiousness of Weyl's theory, many theoreticians were impressed by its depth and its thoroughly mathematical character, as well as by its connection with the general theory of relativity, which had achieved great success at the beginning of the 1920s, and these features made the basic idea attractive to them: the creation of a unified field theory incorporating the electromagnetic field on the basis of one or another generalization of four-dimensional Riemannian geometry. That led to the formulation of the already previously contemplated research program of geometrical unified field theories, at the center of which was Weyl's theory.

A clear symptom of such a program's having taken shape was the appearance within a comparatively short time span, during the course of just one year, of a whole series of unified geometrical field theories that were constructed similarly to Weyl's theory. Thus, aside from modifications to Weyl's theory, there appeared in 1921: Eddington's "affine unified field theory," in which nonmetrical quantities—the coefficients of the affine connection—were of fundamental significance (Eddington 1921); Einstein's generalization of Weyl's geometry, which was connected with the abandonment of metrical quantities (Einstein 1921); and Kaluza's five-dimensional theory (Kaluza 1921). Finally, in 1922, Cartan proposed a non-Riemannian geometry that admitted torsion, along with curvature, in order to be able to describe the electromagnetic field (Cartan 1922). Now, for the first time, Einstein began to concentrate more and more on the problem of a geometrical unified field theory and he soon became its uncontested leading advocate.

Until the emergence of quantum mechanics, the great theoretical penetration and mathematical perfection of the geometrical unified field theory program were seen as genuine advantages, notwithstanding these theories' exceedingly weak connection with experience, this by comparison with the theoretical eclecticism of the quantum-theoretical program and the predominance, within it, of the empirical over the theoretical. But after the emergence of quantum mechanics, the theoretical level of the quantum-theoretical program was elevated considerably, and this led to a rapid further development of this program. The shortcomings of the geometrical unified field theory program (its classical character, its abstractness, and its isolation from experiment) became ever clearer. The intensive development of nuclear physics and elementary particle physics, which led to important discoveries at the beginning of the 1930s (relativistic quantum mechanics, quantum electrodynamics, the discovery of neutrons and positrons, the theory of weak interactions and nuclear forces, etc.), thrust the geometrical unified field theory program into the background. Without the authority of Einstein and the mathematical attractiveness of these theories, the geometrical unified field theory program would presumably have fallen into oblivion.

## 9. Concluding Observations

Einstein worked in almost total isolation until the last days of his life on the elaboration of a geometrical unified field theory. But in the 1940s and even more in the 1950s, his course diverged widely from the main line of development in physics, which was connected in the first place with elementary particles and quantum field theory, and, somewhat later, with relativistic astrophysics and cosmology. The majority of the leading theoreticians of this period were of the view that the geometrical unfied field theory program had come to a dead end. It played no further essential role in the development of twentieth-century physics (Vizgin 1981b).

And yet the ways of scientific development are more complicated. The problem is not merely that the actual history of scientific ideas can only be recognized through the study of those lines of investigation that did not lead to success. The problem is that, in spite of their being unsuccessful, these lines of investigation attracted the attention of leading theoreticians—as individuals—and that, all in all, great hopes were invested in them. Careful historical analysis of these "dead ends" or "side roads" can quite unexpectedly reveal their extremely important heuristic function in the elaboration and development of the main lines of research.

As is well known, such a situation exists with regard to the geometrical unified field theory program. Thus, in 1922, Schrödinger attempted to connect the Bohr-Sommerfeld quantum conditions with the geometrical foundations of Weyl's theory. In this way, Schrödinger obtained the resonance formulation of the quantum conditions, in de Broglie's sense. Obviously, it thereby became possible for him to recognize more easily the wave character of elementary particles and to establish wave mechanics (Raman and Forman 1969, Vizgin 1979). In the course of elaborating affine and affine-metrical variants of a unified field theory in the years 1923–1925, Einstein came to the conclusion that one had to demand charge symmetry in any arbitrary generally covariant classical theory that incorporated the electromagnetic field; and this can be regarded as the origin of the idea of the antiparticle, an idea that was only elaborated a few years later by Dirac in the framework of relativistic quantum mechanics (Treder 1975).

In the course of developing five-dimensional field theories in the years 1926–1927, and in the attempts of Klein and Fock to connect them with quantum mechanics, the first relativistic scalar wave equation for particles with a nonzero rest mass was set up (the Klein-Fock-Gordon-Schrödinger equation) (Vizgin 1979). Unified field theories, above all Weyl's theory and the five-dimensional theories, formed important starting points for gauge-field theories, in particular for the gauge transformations (Fock 1926; London 1927).

In 1929 Weyl developed a gauge theory of the electromagnetic field, relying upon his investigations of 1918 and the mentioned works on gauge symmetry. Weyl's papers were also followed by the investigations of Fock and Ivanenko on the formulation of the Dirac equation in Riemann spaces (Fock and Ivanenko 1929a, 1929b). For Weyl, and for the Soviet physicists as well, Einstein's works on unified field theories with "distant parallelism" (Einstein 1929) were of essential significance. Later, at the beginning of the 1950s, the gauge theory of the electromagnetic field was extended by Yang and Mills to the case of isotopic spin symmetry. Thereafter Sakurai and others developed the gauge theory of strong interactions, which played an important role in SU(3) symmetry and the quark hypothesis. The current version of quantum field theory (and gauge theory) was finally formulated in the 1970s. New projects arose for a unified theory of fundamental interactions based upon gauge theory. In this way, Weyl's theory has proved to be a starting point for the theories that are the basis of the enormous changes that have ensued in recent decades in quantum field theory and elementary particle theory (Vizgin 1981b).

Weyl's theory and the geometrical unified field theory program were closely connected with the Göttingen tradition in mathematical physics, and represented a new form of interaction between mathematics and physics that was characteristic of the nonclassical physics of the twentieth century. The program demonstrated not only the great heuristic possibilities of this new form of connection between physics and mathematics, but also certain dangers and difficulties that arose with its realization, these resulting from an overemphasis on the role of mathematical structure and an undervaluation of the experimental and empirical aspects of theories.

NOTES

\* Originally published as: "Einstein, Hilbert, Weyl: Genesis des Programms der einheitlichen geometrischen Feldtheorien." *NTM-Schriftenreihe für Geschichte der Naturwissenschaften, Technik und Medizin*, © Akademische Verlagsgesellschaft Geest & Portig K.-G. (Leipzig). Translated from the Russian by Prof. Dr. V. Wünsch and Dr. R. Tobies. Translation from the German by Don Howard, with the assistance of John Stachel; corrected by the author.

[1] This view of the role of mathematics in physics is advocated by the majority of leading contemporary theoreticians and mathematicians, including Dirac, Wigner, and Dyson, among others. In the works of the well-known Soviet mathematician, Y.I. Manin, this viewpoint is expressed with particular clarity: "The 'mad idea' that will lie at the basis of a future fundamental physical theory will come from a realization that physical meaning has some mathematical form not previously associated with reality" (Manin, 1981, p. 4).

[2] If one chooses as the Lagrange function $R^i{}_{jkl} R_i{}^{jkl}$, which can be represented as the sum of the squares of the components of the Riemann curvature tensor and of the Maxwell-Lagrange function

$$F_{ik}F^{ik} \quad \text{with} \quad F_{ik} = \frac{\partial \varphi_i}{\partial x^h} - \frac{\partial \varphi_h}{\partial x^i},$$

then one obtains both Maxwell's equations and the fourth-order gravitational equations from the corresponding variational problem.

[3] In 1944, Weyl wrote in an obituary for Hilbert:

> In his investigations on general relativity Hilbert combined Einstein's theory of gravitation with G. Mie's program of pure field physics. For the development of the theory of general relativity at that stage, Einstein's more sober procedure, which did not couple the theory with Mie's highly speculative program, proved the more fertile. Hilbert's endeavors must be looked upon as a forerunner of a unified field theory of gravitation and electromagnetism.

He also recalled the enthusiasm that gripped Hilbert's followers on their way to unifying physical knowledge: "Hopes in the Hilbert circle ran high at that time; the dream of a universal law accounting both for the structure of the cosmos as a whole, and of all the atomic nuclei, seemed near fulfillment" (Weyl 1944, pp. 171–172).

[4] "In the essence of a real thing, there lies something the content of which is inexhaustible, something that we can only approach without limit through always new, in part mutually inconsistent experiences and their adjustment. In this sense, the real thing is a limit idea. Upon this rests the empirical character of all knowledge of reality" (Weyl 1923, p. 4). In a footnote to this, Weyl refers to Husserl: "The precise formulation of these ideas follows Husserl as closely as possible, 'Ideen zu einer reinen Phänomenologie und phänomenologischen Philosophie' (Jahrbuch f. Philos. u. phänomenol. Forschung, Bd. I, Halle 1913" (Weyl 1923, p. 325).

[5] Around 1920 there appeared not a few surveys, popular-scientific works, and publications of a philosophical character, under such titles as: "Physics as Geometrical Necessity" (Haas 1920) and "The World as Space and Time" (Friedmann 1923), etc. The Soviet theorists, A.A. Friedmann and V.K. Frederiks, who were charmed by the axiomatic-synthetic ideas of Hilbert, wrote in the preface to their jointly authored book on relativity theory:

> Fortunately, it is not given to us to see the future, and we do not know whether an epoch of axiomatization, an epoch of scepticism, the final hour of knowledge, will appear.... But if it were to appear, then the logical beauty of the end would have compelled us to recognize the appearance of the relativity principle. (Frederiks and Friedmann 1924, p. 27)

REFERENCES

Cartan, Elie (1922). "Sur une généralisation de la notion de courbure de Riemann et les espaces à torsion." *Académie des Sciences* (Paris). *Comptes Rendus* 174: 593–595.
Eddington, Arthur Stanley (1921). "A Generalisation of Weyl's Theory of Electromagnetic and Gravitational Fields." *Royal Society of London. Proceedings A* 99: 104–122.
Einstein, Albert (1918). "'Nachtrag' zu H. Weyl: 'Gravitation und Elektrizität'." *Königlich Preussische Akademie der Wissenschaften* (Berlin). *Sitzungsberichte*: 478–480.
——— (1921). "Über eine naheliegende Ergänzung des Fundaments der allgemeinen Relativitätstheorie." *Preussische Akademie der Wissenschaften* (Berlin). *Sitzungsberichte*: 261–264
——— (1929). "Zur einheitlichen Feldtheorie." *Preussische Akademie der Wissenschaften* (Berlin). *Sitzungsberichte*: 2–7.
Fock, Vladimir (1926). "Über die invariante Form der Wellen- und der Bewegungsgleichungen für einen geladenen Massenpunkt." *Zeitschrift für Physik* 39: 226–233.
Fock, Vladimir, and Ivanenko, Dimitriy (1929a). "Über ein mögliche geometrische Deutung der relativistischen Quantentheorie." *Zeitschrift für Physik* 54: 798–802.
——— (1929b). "Zur Quantengeometrie." *Physikalische Zeitschrift* 30: 648–651.
Frederiks, V.K., and Friedmann, A.A. (1924). *Osnovy teorii otnositel'nosti*. Leningrad.
Friedmann, A.A. (1923). *Mir, kak prostranstvo i vremya*. Petrograd.
Haas, Arthur (1920). "Die Physik als geometrische Notwendigkeit." *Die Naturwissenschaften* 8: 121–127.
Kaluza, Theodor (1921) "Zum Unitätsproblem der Physik." *Preussische Akademie der Wissenshaften* (Berlin). *Sitzungsberichte*: 966–972.
Klein, Felix (1926). *Vorlesungen über die Entwicklung der Mathematik im 19. Jahrhundert*. Part 1. Berlin: Julius Springer.
London, Fritz (1927). "Quantenmechanische Deutung der Theorie von Weyl." *Zeitschrift für Physik* 42: 375–389.
McCormmach, Russell (1970). "Einstein, Lorentz and the Electron Theory." *Historical Studies in the Physical Sciences* 2: 41–98.
Manin, Yuri I. (1981). *Mathematics and Physics*. Boston: Birkhäuser. Translation of *Matematika i Fizika*. Moscow (1979).

Newman, M.H.A. (1958). "Hermann Weyl." *Journal of the London Mathematical Society* 33: 500–511.
Pauli, Wolfgang (1921). "Relativitätstheorie." In *Encyklopädie der mathematischen Wissenschaften, mit Einschluß ihrer Anwendungen*. Vol. 5, *Physik*, part 2. Arnold Sommerfeld, ed. Leipzig: B.G. Teubner, 1904–1922, pp. 539–775. Issued November 15, 1921.
Petrov, Aleksei Z. (1965). "Nekotoriye soobrazheniya o yedinykh teoriyakh polya." In *Gravitatsiya i teoriya otnositel'nosti*. Vol. 2. Kasan, pp. 7–15.
Pyenson, Lewis (1974). "The Göttingen Reception of Einstein's General Theory of Relativity." Dissertation. Johns Hopkins University.
——— (1975). "La réception de la relativité généralisée: disciplinarité et institutionalisation en physique, 1880–1920." *Revue d'Histoire des Sciences* 28: 61–73.
——— (1977). "Hermann Minkowski and Einstein's Special Theory of Relativity." *Archive for History of Exact Sciences* 17: 71–96.
——— (1979). "Physics in the Shadow of Mathematics: The Göttingen Electron-Theory Seminar of 1905." *Archive for History of Exact Sciences* 21: 55–89.
——— (1981). "Preestablished Harmony and Relativity in Late Wilhelmian Germany." *Proceedings of the 16th International Congress for the History of Science. C. Meetings on Specialized Topics. D. Commemorations*. Bucharest, pp. 139–144.
Raman, V.V. and Forman, Paul (1969). "Why Was It Schrödinger Who Developed de Broglie's Ideas?" *Historical Studies in the Physical Sciences* 1: 291–314.
Reid, Constance (1970). *Hilbert*. Berlin and New York: Springer-Verlag.
Schouten, Jan Arnoldus (1926). "Erlanger Programm und Uebertragungslehre. Neue Gesichtspunkte zur Grundlegung der Geometrie." *Circolo Matematico di Palermo. Rendiconti* 50: 142–169.
Seelig, Carl (1960). *Albert Einstein. Leben und Werk eines Genies unserer Zeit*. Zurich: Europa Verlag.
Treder, Hans-Jürgen (1975). "Antimatter and the Particle Problem in Einstein's Cosmology and Field Theory of Elementary Particles." *Astronomische Nachrichten* 296: 149–161.
Vizgin, Vladimir P. (1972). *Razvitiye vzaimosvyazi printsipov simmetrii s zakonami sokhraneniya v klassicheskoy fizike*. Moscow.
——— (1975). *Erlangenskaya programma i fizika*. Moscow.
——— (1979). "Yedinye teorii polya i kvantovaya mekhanika." In *50 let kvantovoy mekhaniki*. L.S. Polak, ed. Moscow, pp. 82–94.
——— (1981a). *Relativistskaya teoriya tyagoteniya (Istoki i formirovaniye, 1900–1915)*. Moscow.
——— (1981b). "Kalibrovochnaya simmetriya ot Weyla do Yanga i Millsa." *Proceedings of the 16th International Congress for the History of Science. A. Scientific Section*. Bucharest, p. 165.
——— (1982). "Nauchno-izsledovatel'skiye programmi i metodologicheskiye printsipi." In *Metodologicheskiye problemy istorikonauchnikh izsledovanyy*, I.S. Timofeyeva, ed. Moscow, pp. 172–179.
Weyl, Hermann (1917). "Zur Gravitationstheorie." *Annalen der Physik* 54: 117–145.
——— (1918). "Gravitation und Elektrizität." *Königlich Preussische Akademie der Wissenschaften* (Berlin). *Sitzungsberichte*: 465–478. Reprinted in Weyl 1968, pp. 29–42.
——— (1923). *Raum-Zeit-Materie*, 5th ed. Berlin: Julius Springer.

——— (1927). *Philosophie der Mathematik und Naturwissenschaft.* Munich and Berlin: R. Oldenbourg.

——— (1944). "David Hilbert and His Mathematical Work." *Bulletin of the American Mathematical Society* 50: 612–654. Reprinted in Weyl 1968, vol. 4, pp. 130–172.

——— (1954). "Erkenntnis und Besinnung (Ein Lebensrückblick)." *Studia Philosophica, Jahrbuch der Schweizerischen Philosophischen Gesellschaft, Annuaire de la Société Suisse de Philosophie.* Reprinted in Weyl 1968, vol. 4, pp. 631–649.

——— (1968). *Gesammelte Abhandlungen*, 4 vols. K. Chandrasekharan, ed. Berlin: Springer-Verlag.

# Inside the Coconut:
# The Einstein-Cartan Discussion on Distant Parallelism

MICHEL BIEZUNSKI

On February 13, 1930, Albert Einstein wrote to Elie Cartan: "For the moment, this theory seems to me to be like a starved ape who, after a long search, has found an amazing coconut, but cannot open it; so he doesn't even know whether there is anything inside."[1] The theory to which Einstein referred was the theory of "Distant Parallelism," an attempt to unify electromagnetism and gravity. This letter was the twenty-first in a series of thirty-nine that Einstein and Cartan exchanged between May 1929 and May 1932. This correspondence illustrates very clearly what Einstein was seeking when he was working towards a unified field theory, and the nature of the difficulties he encountered in discussion with mathematicians. For while Cartan studied the form of the coconut, Einstein wanted to know what was hidden inside. The failure of their joint work would be due precisely to this difference.

Elie Cartan (1869–1951) was a leading French mathematician who made major contributions to Lie group theory and the theory of spinors. He also worked in the field of systems of partial differential equations and the theory of spaces with linear connections.

The exchange between Einstein and Cartan was an example of those numerous and unsuccessful attempts made by Einstein to unify the theory of gravity and electromagnetism. The letters reveal the special features of the Einsteinian approach.

After developing the general theory of relativity, Einstein tried to reach a further level of generalization by unifying the classical physical interactions: electromagnetism and gravitation. From the early 1920s until the end of his life, he would work in this direction. Einstein proceeded according to a scheme that was characteristic of his attempts to derive a unified field theory: the first step was to find a general geometric space in which there exists a symmetric and an antisymmetric tensor field of rank two. The symmetric one would be identified with the gravitational field, the antisymmetric one with the Maxwell field. The second step was to find the field equations, which would represent the dynamics of the physical interactions. In so doing, he repeated his own approach in developing the general theory of relativity, where he had first generalized Minkowski space to a general Riemannian one, and then after much searching had found the correct field equations (see Stachel 1980, 1986; and Norton 1984). His exchange with Cartan was one of the steps in this program. But Einstein would be mostly concerned with a possible physical interpretation, whereas Cartan would be guided by considerations of logical necessity. It should be noted that this joint work was not Einstein's first attempt to find a mathematical expression for his physical dream of the unification of nature, since he had tried originally to follow Weyl's generalization of Riemannian geometry and later to follow the five-dimensional theory formulated by Kaluza and Klein. And it would not be the last. But let us start the story from the beginning.

Together with the mathematician Roland Weitzenböck, who had worked as early as 1913 on a theory of invariants of physical theories (Weitzenböck 1913–1915),[2] Einstein discovered that there existed a mathematical literature on a type of generalized Riemannian space including both curvature and torsion. Torsion is represented by an antisymmetric tensor that generalizes a classical notion in differential geometry. In those spaces that possess a linear connection, there are two limiting cases: if the torsion is zero, but not the curvature, the space is Riemannian; if the curvature equals zero, but not the torsion, it is a space with "distant parallelism" (*Fernparallelismus*) or "absolute parallelism." The name comes from a remarkable property of a space with torsion and no curvature: one can decide whether or not two vectors are parallel even when they are separated by a finite distance. This is due to the fact that the final direction of a vector is independent of the path followed by the vector undergoing parallel transport. In other words, parallel transport is integrable, and the angle between two vectors has an absolute meaning.

In the "old" theory of gravitation (general theory of relativity) Riemannian space-time had curvature but no torsion. Now Einstein suggested a space with torsion and no curvature. Einstein was led in that direction because there is in this case an antisymmetric tensor of rank two, which could be taken to be the electromagnetic tensor, and another symmetric tensor of rank two, which could be used to represent gravitation. Nothing more was needed for him to hope that there might be a basis here for a unified theory of gravitation and electromagnetism.

In reading an article by Weitzenböck (1928) on the new theory, an article containing a supposedly complete bibliography of mathematical works on the subject of distant parallelism, Cartan was surprised and disappointed not to see his own works mentioned, since he had written on this topic a few years before (see Cartan 1923). Distant parallelism was a particular case of the structure he had studied and called a Euclidean connection. He immediately wrote to Einstein on May 8, 1929 asking him whether he remembered their conversation on this very subject seven years earlier at Hadamard's house in Paris: Cartan had then tried to illustrate what a Riemannian space with *Fernparallelismus* looks like.[3] Cartan had explicitly expounded his ideas to Einstein and it is strange to note that the latter, who was already concerned with attempts to find a unified field theory, had paid no attention to Cartan's explanations. In his answer of May 10, Einstein recognized the debt he owed Cartan and proposed several solutions to repair Weitzenböck's omission, for which he felt himself responsible. These included either a postscript to his own forthcoming article, or a longer piece to be written by Cartan and published independently. Einstein added:

> I didn't at all understand the explanations you gave me in Paris; still less was it clear to me how they might be made useful for physical theory. I first remarked last year that it would be quite natural to add the hypothesis of distant parallelism to the Riemann metric. But only in the last few months have I realized that this actually leads to a theory which corresponds to the hitherto existing knowledge of the physical properties of space, i.e. to a useful set of field equations, almost uniquely determined by formal considerations.[4]

This letter marked the beginning of a very rich correspondence and joint work. The most intense period of the correspondence was between December 1929 and February 1930, during which time Einstein and Cartan exchanged twenty-six letters. It often happened that one or the other wrote several letters a week to his colleague. Several times, Einstein, after sending one letter, did not wait for an answer before sending the next. Cartan was no less eager to succeed than Einstein. Such a contribution would have earned him recognition and broken the isolation he felt in the mathematical community. To have contributed significantly to the new Einsteinian unified theory would have helped enlarge his public.[5]

In his answer to Einstein's letter, Cartan accepted Einstein's proposal to add a historical account of distant parallelism (Cartan 1930) to one of Einstein's articles (Einstein 1930).[6] Einstein told Cartan that he was convinced that he had "found the simplest legitimate characterization of a Riemann metric with distant parallelism that can occur in physics."[7] He referred to a system of twenty-two field equations built out of the torsion tensor along with sixteen complementary equations also including this tensor.

But the first difficulty that Einstein and Cartan would encounter concerned the choice of the appropriate set of equations. And the quests of Cartan, a mathematician, and that of Einstein, a physicist, were of different natures. Cartan was concerned with the logical and formal aspects of the theory, Einstein, with its physical interpretation.

Cartan explained in a letter of December 3, 1929 that he had sought but could not find a decisive argument to establish that Einstein's choice of equations was privileged among others, as he had done for the general theory of relativity (Cartan 1922).[8] Furthermore, he was astonished that Einstein had managed to find this particular system.

Cartan inserted with this letter a note in which he defined the concepts he used, concepts that would be the topic of the bulk of his correspondence with Einstein. He introduced, among other things, a measure of the generality of a system of partial differential equations, which he called the "generality index," equal to the number of arbitrary functions on which the general solution of a deterministic system depends. He also explained what he meant by "systems in involution": A system of linear partial differential equations is said to be in involution "if any 0-dimensional solution is part of at least one 1-dimensional solution, if any *arbitrary* 1-dimensional solution is part of at least one 2-dimensional solution, and so on."[9]

Cartan proposed one alternative system of twenty-two field equations, and another one with fifteen equations, which he himself preferred because it gave rise to a richer geometrical scheme. In the same letter he wrote:

> When thinking it over, I believe that the degree of generality of the geometrical scheme corresponding to your system of 22 equations is a bit weak, the old classical theories of gravitation and electromagnetism give to physics a greater degree of generality.

This twenty-two–equation system had a generality index of 12. The fifteen-equation system had a generality index of 18. That is why the second system is richer, Cartan explained.

Einstein answered on December 8 that he disagreed even about the number of equations. This shows how far he was from being convinced by Cartan's reasoning, which was based on a fixed mathematical structure, the number of whose equations could not be modified without a global change of the system. Moreover, Einstein preferred a system with a lower generality index:

> In my opinion, a theory is the more valuable the more strongly it restricts possibilities, without coming into conflict with reality. It is like a wanted poster which is supposed to characterize a criminal; the *more precisely* it points him out the better.[10]

In his next letter, however, Einstein recognized that he had not fully understood Cartan's explanation of the generality index and urged him again to answer the question whether his own system of equations was especially privileged from the point of view of their generality index.[11] Cartan answered that it is impossible to tell whether this system is privileged in any way. But, he added: "*there is no system* MORE *determined than yours.*"

Cartan then raised a question that showed his desire to find a physical concomitant for his generality index:

> Is any solution of Maxwell's equations which is defined only in a small region of space-time, *physically* admissible? To put it another way, let us assume that for [a finite region of space-time] we have functions defining the electromagnetic field in this domain and satisfying Maxwell's equations *in this limited domain*; do you think that such a field can actually exist? The local state is certainly influenced by what happens elsewhere, outside the domain under consideration, but can we deal with the possibilities, infinite in number, that exist outside the domain so as to determine any *local* solution of Maxwell's equations inside the domain?
>
> This question seems important because, if the answer is affirmative, it would give a *physical* meaning and not merely a mathematical one, to the generality index of the system of partial differential equations that determines the field.[12]

Indeed, Einstein was looking for mathematical approval. He had already enjoyed this after the publication of the general theory of relativity. Einstein wanted his mathematical equations to be the only ones. Cartan, on the other hand, was looking for a physical meaning, perhaps to prove to himself that his theory could have some utility to Einstein and was not simply a mathematical exercise. But in reality, it soon became clear that, whereas Cartan's arguments were almost always based on formal considerations, Einstein's sensitivity to problems of physical interpretation led him to stronger restrictions than those imposed by a simple quest for some physical meaning. For example, Einstein objected to Cartan's system that in the latter:

> The $g_{ik}$ alone have an autonomous causality, independent of the parallel structure. This can be expressed as follows: the electromagnetic field has no reaction upon the gravitational field. This completely contradicts physical expectations.[13]

Einstein emphasized that physics requires the field to be free of singularities everywhere. This condition dramatically limits the choice of a set of equations. He raised the question of the existence of nontrivial solutions whose generality index is equal to zero. These systems, like those of classical mechanics, would not be fixed "by an arbitrary choice of given *functions* but by a choice of parameters (numbers). For some time I was convinced that the true laws of nature would have to be of such a kind."[14] But, he added,

> It might also be possible that that high degree of constraint (which I have no doubt is realized in the true laws of nature) is based on something else. The indicated small measure of arbitrariness (in Nature) could also be grounded in the requirement that singularities are to be excluded from all space!

To that question, Cartan answered that for any given torsion there exists a space having that torsion in which the solution of the field equations is singularity free; and these singularity-free solutions correspond to algebraic solutions of the structure equations of the associated Lie group.[15] But Cartan admitted that this is only a particular case, and that he did not know any method allowing one to deal with this problem in general.

Einstein had quite a different point of view on the matter. In a letter of January 7, 1930, he began to distance himself from Cartan's approach, in an elegant way and with humor: "I must further apologize for the thoughtlessness of my questions. There's a lovely old saying: A fool can ask more questions than a wise man can answer."[16]

He added that he was happy that Cartan felt confidence in the fact that there exists no involutive system with a lower degree of freedom. Einstein had tried himself to find one in vain. But he had moved to a concern with the physically acceptable solutions of the field equations. He explained again to Cartan his requirement that the solutions be singularity-free. If they were not, there would exist uncharged point-masses at rest with respect to each other, which "contradicts experience," first because these particles would not be subject to gravitation, and second because they would have no charge. It is interesting to note that, two years before the discovery of the neutron, Einstein had no intuition of the existence of such uncharged particles.

From this point on, the widening of the gap between the two approaches becomes more apparent.[12] The numerous letters in the period between January and February 1930 saw both protagonists go their own way. The details of these letters are less interesting, from a comparative point of view, because they are like two parallel monologues. For example, Einstein closed his letter of January 21, 1930, with this sentence: "Thus you must have *still more* patience and compassion for a poor physicist whose destiny has led him to these troubled pastures."[18]

A month later, Einstein had not gone any further. He was still interested in the existence of singularity-free solutions that could represent electrons and protons, "for without the solution of this difficult problem I feel no judgment can be passed on the usefulness of the theory."[19] It was in this letter that he introduced the metaphor of the coconut quoted at the beginning of this paper. Cartan answered on February 17:

> (The whole problem is to find a singularity-free solution general enough to be physically interpretable. But are we really sure that such solutions exist and that the coconut contains something inside?)
>
> We find ourselves in front of a wall and we mathematicians are quite at a loss as to how make a hole in it. One can only hope for some miracle of divination, but then you already have had several.[20]

But the miracle didn't happen.

Five months passed before Einstein got round to answering this letter. During that period, Cartan continued to pursue his research in this domain. He published a popular article on the new theory (Cartan 1931a) and announced to Einstein that he was just finishing another article on systems in involution and the theory of relativity that he would send on (Cartan 1931b). Einstein wrote that, in the meantime, he had been working with his new collaborator, Walter Mayer, and had decided to abandon the previous field equations:

> The reason is that it appears that, according to those field equations there are no gravitational effects, since static solutions with arbitrarily many point-masses (singularities) exist for which only $h_{44}$ is non-zero....[21]

Einstein was trying, together with Mayer, to find new field equations using another approach. The reason for rejecting the solution was essentially the presence of singularities (Einstein and Meyer 1930). This prompted Cartan to suggest once again his fifteen-equation system and no doubt to feel that his work could serve more than ever as a mathematical basis for Einstein's research. But Einstein waited nearly nine months before replying. On his way to the United States on board the steamer *The San Francisco*, he wrote to thank Cartan for the promised article on involution theory in these terms: "I have read with great enjoyment your work on systems in involution. This seems to me a truly important contribution to the theory of partial differential equations."[22] As the paper was entitled "On the Theory of Systems in Involution *and its Applications to the Theory of Relativity*" (my emphasis), this sentence must have come as a shock to Cartan, because of its implied rejection of the usefulness of his work in the field of relativity and unified field theory. And the rest of the letter left no doubt:

> In any case, I have now completely given up the method of distant parallelism. It seems that this structure has nothing to do with the true character of space. For some years, together with Dr. Mayer, I have pursued another theory, that of the 5-vector and 5-metric in a four dimensional metric continuum. This theory not only yields Maxwell's equations in a natural way but also—as I have recently discovered—an extension of them which admits continuously distributed charged masses. I have hopes that this theory really comes closer to the structure of physical space without its basic laws having to be given a merely statistical interpretation. I absolutely cannot reconcile myself to this dogma of the new generation of physicists, no matter how enticing the arguments are that are made for it....
>
> The main reason for the uselessness of the distant parallelism construction lies, I feel, in that one can attribute absolutely no physical meaning to the "straight lines" of the theory, while the physically meaningful (macroscopic) equations of motion cannot be obtained from it. In other words, the $h_{sv}$ give rise to no useful representation of the electromagnetic field.[23]

The few remaining letters speak only of questions touching purely mathematical matters concerning systems of differential equations.

It is a generally accepted view that during the last thirty years of his life Einstein was constantly changing theories and enthusiasms in the search for a unified field theory. But this work with Cartan reveals there was an underlying unity, that he was using the same methods and the same concepts that he used to derive his general theory of relativity, and he was trying to repeat once again his earlier success. He wanted a theory that is complete, general, aesthetically satisfying, and unified. The means were less important in his eyes. He used a mathematical theory as a tool, and never more. He was concerned with physical problems, and refused to be caught in the trap of formal, mathematical structures.[24]

Even if, as the Einstein-Cartan correspondence shows, these two men followed from the start parallel but distant paths, which never really met, the appreciation that each felt for the other was genuine. As Cartan said in his article popularizing Einstein's distant parallelism theory:

> We see ... the variety of points of view from which the unified field theory can be regarded and also the difficulty of the problems that it raises. But Mr. Einstein is not someone who is frightened by difficulties, and even if his attempt is not ultimately successful, it will have to force us to reflect on the deep questions which lie at the heart of science.[25]

*Acknowledgments.* I would like to thank Jim Ritter, who translated Einstein's correspondence with Cartan into English, for his scientific and linguistic help.

SELECTED BIBLIOGRAPHY OF
EINSTEIN'S WRITINGS ON DISTANT PARALLELISM

Einstein, Albert. "Riemann-Geometrie mit Aufrechterhaltung des Begriffes des Fern-Parallelismus." *Preussische Akademie der Wissenschaften. Physikalisch-mathematische Klasse. Sitzungsberichte*: 217–221 (June 7, 1928).

Einstein, Albert. "Neue Möglichkeit für eine einheitliche Feldtheorie von Gravitation und Elektrizität." *Preussische Akademie der Wissenschaften. Physikalisch-mathematische Klasse. Sitzungsberichte*: 224–227 (June 14, 1928).

Einstein, Albert, and Mayer, Walter. "Systematische Untersuchung über kompatible Feldgleichungen, welche in einem Riemannschen Raume mit Fern-Parallelismus gesetzt werden können." *Preussische Akademie der Wissenschaften. Physikalisch-mathematische Klasse. Sitzungsberichte*: 257–265 (April 23, 1931).

NOTES

[1] Letter 31, Debever 1979, p. 197. All of the letters cited here from the Einstein-Cartan correspondence are published in Debever 1979.

[2] See the discussion in Klein 1927, pp. 57–59.

[3] Letter 1, pp. 4–9.

[4] Letter 2, pp. 11, 13. On Einstein's visit to Paris in 1922 and its impact on French scientists, see Biezunski 1987.

[5] He would write a few months later to Einstein: "I am very happy that my manuscript has interested you and that you think my theory is likely to be of some help. It is relatively little known, probably because I published it in a form that refers to systems of total differentials. Some mathematicians know that I have derived important results from it, for example the theory of the structure of infinite continuous groups. But in the form that I submitted to you in my note it would, of course, reach a wider public, and I shall publish the essentials of it." Cartan to Einstein, Letter 15, January 3, 1930, p. 101.

[6] Letter 3, May 15, 1929, p. 15.

[7] Letter 5, August 25, 1929, p. 19.

[8] Letter 7, p. 27.

[9] Note enclosed with Letter 7, December 3, 1929, p. 37.

[10] Letter 8, pp. 59, 61.

[11] Letter 10, December 18, 1929, p. 73.

[12] Cartan's emphasis. Letter 11, December 22, 1929, pp. 83, 85.

[13] Letter 13, December 27 or 28, 1929, p. 89.

[14] Letter 14, December 29 or 30, 1929, p. 95.

[15] Letter 15, January 3, 1930, p. 103.

[16] Letter 16, p. 109.

[17] Letter 16, January 7, 1930, p. 111.

[18] Letter 23, p. 143.

[19] Letter 31, February 13, 1930, p. 197.

[20] Letter 32, p. 201.

[21] Letter 33, June 13, 1930, p. 203.

[22] Letter 35, March 21, 1932, p. 209.

[23] Letter 35, March 21, 1932, pp. 209, 211.

²⁴ Even in trying to use the ideas of Cartan, Einstein was still a forerunner. For since the 1960s, Cartan's ideas have come into use again, rediscovered by physicists who are looking once more for a unified field theory. For example, at the Einstein Centennial Symposium, held in Princeton in 1979, J.A. Wheeler pointed out the equivalence of the gauge theory of gravitation with Cartan's metric-plus-torsion theory. See Wheeler 1980, note 23, p. 366.

²⁵ "On voit ... la variété des aspects sous lesquels peut être envisagée la théorie unitaire du champ et aussi la difficulté des problèmes qu'elle soulève. Mais M. Einstein n'est pas de ceux à qui les difficultés font peur et, même si sa tentative n'aboutit pas, elle nous aura forcés à réfléchir sur les grandes questions qui sont à la base de la science" (Cartan 1931a).

REFERENCES

Biezunski, Michel (1987). "Einstein's Reception in Paris in 1922." In *The Comparative Reception of Relativity*. Thomas F. Glick, ed. Boston: D. Reidel, pp. 169–188.

Cartan, Elie (1922). "Sur les équations de la gravitation d'Einstein." *Journal des Mathématiques pures et appliquées* 1: 141–203. Reprinted in Cartan 1952–1955, vol. 3.

———— (1923). "Sur les variétés à connexion affine et la théorie de la relativité généralisée." *Annales de l'École Normale* 40: 325–412. Reprinted in Cartan 1952–1955, vol. 3, pp. 616–618.

———— (1930). "Notice historique sur la notion de parallélisme absolu." *Mathematische Annalen* 102: 698–706. Reprinted in Cartan 1952–1955, vol. 3.

———— (1931a). "Le parallélisme absolu et la théorie unitaire du champ." *Revue de Métaphysique et de Morale* 38: 13–28. Reprinted in Cartan 1952–1955, vol. 3.

———— (1931b). "Sur la théorie des systèmes en involution et ses applications à la Relativité." *Bulletin de la Societé Mathématique de France* 59: 88–118. Reprinted in Cartan 1952–1955, vol. 2, pp. 1199–1229.

———— (1952–1955). *Oeuvres Completes*. 3 vols. Paris: Gauthier-Villars.

Debever, Robert, ed. (1979). *Elie Cartan–Albert Einstein: Letters on Absolute Parallelism, 1929–1932*. Jules Leroy and Jim Ritter, trans. Princeton: Princeton University Press; Brussels: Académie Royale de Belgique.

Einstein, Albert (1930). "Auf die Riemann-Metrik und den Fern-Parallelismus gegründete einheitliche Feldtheorie." *Mathematische Annalen* 102: 685–697.

Einstein, Albert, and Mayer, Walter (1930). "Zwei strenge Lösungen der Feldgleichungen der einheitlichen Feldtheorie." *Preussische Akademie der Wissenschaften. Physikalisch-mathematische Klasse. Sitzungsberichte*: 110–120.

Klein, Felix (1927). *Vorlesungen über die Entwicklung der Mathematik im 19. Jahrhundert*. Vol. 2. Berlin: Julius Springer; reprint New York: Chelsea, 1967.

Norton, John (1984). "How Einstein found His Field Equations: 1912–1915." *Historical Studies in the Physical Sciences* 14: 253–316. See this volume, pp. 101–159

Stachel, John (1980). "Einstein's Search for General Covariance, 1912–1915." Paper delivered to the Ninth International Conference on General Relativity and Gravitation, Jena, 1980. See this volume pp. 63–100.

———— (1986). "Solved and Unsolved Problems in the History of General Relativity." Talk delivered at the Conference on the History of General Relativity, Osgood Hill, Massachusetts.

Weitzenböck, Roland (1913–1915). "Über Bewegungsinvarianten." *Kaiserliche Akademie der Wissenschaften* (Vienna). *Mathematisch-naturwissenschaftliche Klasse*.

*Abteilung IIa. Sitzungsberichte* (1913): 1241–1258; 1565–1576; 1577–1594; 1595–1606; (1914): 406–431; 567–581; 679–1697; (1915): 309–331.

——— (1928). "Differentialinvarianten in der Einsteinschen Theorie des Fernparallelismus." *Preussische Akademie der Wissenschaften* (Berlin). *Sitzungsberichte*: 446–474 (October 18, 1928).

Wheeler, John Archibald (1980). "Beyond the Black Hole." In *Some Strangeness in the Proportion: A Centennial Symposium to Celebrate the Achievements of Albert Einstein*. Harry Woolf, ed. Reading, Massachusetts: Addison-Wesley, pp. 341–375.

# The Einstein–de Sitter Controversy of 1916–1917 and the Rise of Relativistic Cosmology

PIERRE KERSZBERG

## 1. Introduction

In textbooks of modern cosmology, and in histories of the subject, the name of Willem de Sitter occurs in relatively few contexts. He is standardly associated with the discovery, in the year 1917, of a very strange and unexpected solution of Einstein's field equations: an *empty* universe need not have the metric of Minkowski flat space-time (see, for instance, Peebles 1980a, p. 4; Whitrow 1980, p. 284). This result is itself a reaction to Einstein's announcement, earlier in the same year, of a metric for the first relativistic model of the whole universe—the so-called cylindrical universe (Einstein 1917). Current literature also mentions an Einstein–de Sitter model of the universe, which has nothing to do with either of these solutions. It refers to a model that de Sitter constructed together with Einstein some fifteen years later, when the rediscovery of nonstatic solutions made it clear that the static solutions of 1917 were particular cases of a much wider class of solutions. In fact, the Einstein–de Sitter model is known for yielding the simplest of all nonstatic cases: its geometry is an expanding Euclidean space, relieved of complicated relations among available variables. This simplicity has made it appropriate to the study of the meaning of these more complicated relations, such as the occurrence of terms representing the pressure.

Historical studies of contemporary cosmology also draw attention to the fact that the role of de Sitter was not restricted to these two contributions. A very decisive role was played by de Sitter's early critique of some of the most fundamental philosophical arguments Einstein had put forward when he constructed the theory of general relativity, prior to the emergence of cosmology. These studies suggest, indeed, that Einstein developed his cosmological ideas as a response to de Sitter, who had expressed objections to Einstein's views on the problem of the relativity of rotation and the origin of inertia. Thus, after having commented on Einstein's first cosmological memoir, Jacques Merleau-Ponty wrote: "It was, in part, to take up de Sitter's challenge that Einstein had sought with so much obstinacy to find solutions to the field equations that would be compatible with the principle of the relativity of inertia." Merleau-Ponty says further that de Sitter was convinced right from the outset that such solutions are an "impossibility," a conclusion which Einstein eventually reached (Merleau-Ponty 1965, p. 52). Similarly, albeit already within an analysis of Einstein's later cosmological memoir, J.D. North hinted that it was from de Sitter that Einstein took the idea "that one must ... refrain from asserting boundary conditions of general validity" (North 1965, p. 72). Neither of the historians, however, dwelt at length upon the alleged influence; they do not demonstrate its existence, nor do they attend to the substance of de Sitter's critique. On the whole, the impression gained from these histories of contemporary cosmology is that de Sitter's ongoing objections were merely repetitions of his earlier criticisms of some of Einstein's philosophical positions; but nothing is really said about the substance of these criticisms in their original form.

In this paper, I wish to reexamine de Sitter's place in the history of early general relativity, looking more closely than has previously been done at de Sitter's criticisms. The validity of the preceding as yet little-supported historical claims can thus be tested, and the crucial nature of de Sitter's influence on Einstein's path to cosmology can be definitively examined. In the first instance, this will help complete the historical record. More importantly, the remarkable aspect of de Sitter's criticisms, those that led to Einstein's cosmology, is that they emphasize an interesting epistemological feature of general relativity.

The historical discussion will focus on a series of published papers by de Sitter that have until now been largely overlooked or simply ignored. These are papers that take on great significance when related to parts of the large body of the as yet unpublished correspondence between Einstein and de Sitter.[1] This correspondence, which deals with both administrative and scientific matters, has only recently been discovered. (The story is narrated in Kahn and Kahn 1975). Though gaps and references in this correspondence suggest that it is seriously incomplete, enough remains to shed significant light on our problem. We will mainly concentrate here on the short period from the fall of 1916 to the publication of Einstein's model of the universe on February 8, 1917, and de Sitter's reaction to it.

Before we examine de Sitter's critical problem as it presented itself in those years, which was so fundamental to what Einstein called his "indirect" and "bumpy" road to cosmology (Einstein 1917), we will start from a broader perspective. The cosmological metric published by de Sitter in 1917 was a static one, but it is now known in a nonstatic form after Friedmann, Lemaître, and Robertson had shown in the years 1922–1929 that it is a limiting case of a class of nonstatic models. From today's standpoint we can look at this metric in the following way. In general terms, Einstein's field equations must be satisfied by both a space-time metric and by an energy-momentum tensor. The form of the metric is simplified in the case of cosmology because of symmetry considerations. Thus the Robertson-Walker metric

$$ds^2 = c^2\,dt^2 - R^2(t)\left(\frac{dr^2}{1 + Kr^2/4} + r^2(d\theta^2 + \sin^2\theta\,d\phi^2)\right)$$

is applicable to all homogeneous and isotropic universes; in this metric the constant $K$ (which specifies the space geometry) and the radius of curvature $R$ (which varies as a function of the time $t$) are a priori independent of one another. The pioneering work of Friedmann in 1922–1924 consisted in investigating the restrictions imposed on the coupling of these variables. He formulated the well-known "equations of motion" for the universe:

$$\dot{R}^2 = \frac{C}{R} + \frac{\Lambda c^2 R^2}{3} - Kc^2,$$

in which $C$ equals $(8/3)\pi G\rho R^3$ and $\Lambda$ is the cosmological constant. Only two models are rigorously static: indeed when $\dot{R} = 0$, we have $K/R^2 - \Lambda/3 = (8/3c^2)\pi G\rho$, and this implies that the density $\rho$ is constant. A positive density yields $K = +1$, which corresponds to Einstein's cylindrical universe; the other possibility is $K = \Lambda = \rho = 0$, with $R$ taken to be any constant, which is the Minkowski metric of special relativity. But this latter metric is not the only empty metric that may be considered. Thus, in the nonstatic case ($\dot{R} \neq 0$), a series of empty models can be described as $\Lambda$ is positive, null, or negative, and $K = 0, +1,$ or $-1$. The de Sitter universe is that for which the cosmological constant is positive and $K = 0$.

However, when de Sitter first set out to write his metric, he began with very different considerations. Consider the Schwarzschild exterior metric (formulated in 1916), which defines the space-time geometry around a single sphere of matter:

$$ds^2 = \left(1 - \frac{2m}{r}\right)dt^2 - \left(1 - \frac{2m}{r}\right)^{-1}dr^2 - r^2(d\theta^2 + \sin^2\theta\,d\phi^2).$$

If we add the cosmological constant we have

$$ds^2 = \left(1 - \frac{2m}{r} - \frac{1}{3}\Lambda r^2\right)dt^2 - \left(1 - \frac{2m}{r} - \frac{1}{3}\Lambda r^2\right)^{-1}dr^2 - r^2(d\theta^2 + \sin^2\theta\,d\phi^2).$$

Supposing that there is no matter,

$$ds^2 = \left(1 - \frac{1}{3}\Lambda r^2\right)dt^2 - \left(1 - \frac{1}{3}\Lambda r^2\right)^{-1}dr^2 - r^2(d\theta^2 + \sin^2\theta\,d\phi^2).$$

With a positive cosmological constant, an unavoidable singularity appears at $r = \sqrt{3/\Lambda}$. Now this value of $r$ is precisely the value of the radius of curvature of the de Sitter universe, which explains why confusion about whether or not this universe is truly empty and/or static long reigned during at least the first two decades of relativistic cosmology. All that can be said is that the de Sitter universe is "static" up to its horizon, the singularity at the horizon being an event horizon. (For an explanation of this concept, see Rindler 1977, pp. 215 and 232-234.) De Sitter's actual route, which is so markedly different from modern presentations of his result, illustrates the simple fact that there was initially no such a thing as a consistent relativistic concept of universe. Einstein's cylindrical universe had created theoretical cosmology virtually out of nothing, the only reference being Mach's critique of Newtonian mechanics in which recourse was had to the "entire universe" in an attempt to get rid of unobservable entities (see Mach 1888, p. 286).

## 2. The Einstein–de Sitter Controversy: The Precosmological Stage

On March 20, 1916, Einstein submitted to the *Annalen der Physik* a memoir in which the aims, methods, and major results of general relativity were synthesized for the first time. The memoir was published in the October issue of this journal (Einstein 1916). There Einstein tackled the whole issue of relative rotation after he had made numerous groping attempts in that direction in the two years preceding this publication. This he did on epistemological grounds in order to justify the very need for a relativity principle more comprehensive than that employed in the special theory. In the introductory part, he came up with a thought experiment that he believed to be a fairly direct illustration of Mach's reflections on rotation. The thought experiment sounds curious and all the more baffling (Einstein 1916, pp. 112–113):

> Two fluid bodies of the same size and nature hover freely in space at so great a distance from each other and from all other masses that only those gravitational forces need be taken into account which arise from the interaction of different parts of the same body.

Einstein posited further that the distance between the two bodies is invariable, adding "in neither of the bodies let there be any relative movements of the parts with respect to one another." Each mass is thus in equilibrium under the action of the gravitation of each of its parts on the other and under the action of other physical forces as well.

> But let either mass, as judged by an observer at rest relatively to the other mass, rotate with constant angular velocity about the line joining the masses.... Now let us imagine that each of the bodies has been surveyed by means of measuring instruments at rest relatively to itself, and let the surface of $S_1$ prove to be a sphere, and that of $S_2$ an ellipsoid of revolution.

Einstein speaks here of a *verifiable* (*konstatierbar*) relative motion, even though the experiment and its results are totally imaginary. Since the difference is observable, a reason for it may be sought.

The defender of Newtonian mechanics would answer that the laws of mechanics apply to the space of reference of the first body $(R_1)$ but not to that of the second body ($R_2$), even though $S_1$ and $S_2$ are both at rest relative to their own spaces of reference. Were this the case, indeed, only the space of reference of the first body would be identified with the Galilean continuum of space and time, that is, with the privileged continuum in which the form of Newtonian laws remains invariant. $S_1$ is at rest in absolute space, whereas $S_2$ is rotating in that space. In other words, the only possible explanation in the framework of Newtonian laws implies that the observer will not regard himself as being at rest *only* with respect to the body he has been surveying by means of measuring instruments. Thus, the observer lies on the line joining $S_1$ and $S_2$, comes to rest successively with respect to $S_1$ and $S_2$, and then records the shapes of the bodies. It cannot be the case that the two bodies will be perfectly spherical, because it has been verified that there is rotational, noninertial motion and, thus, forces. *At least one* of the bodies, therefore, has to become ellipsoidal. On the other hand, *at most one* may be ellipsoidal, since any interaction between the two masses has been excluded by hypothesis. There remains a space both independent of the spaces of reference and capable of exerting physical effects.

In the case considered, nothing enables us to speak of the motion of this or that body, unless the two spaces of reference are distinguished beforehand with respect to their effects on the shape of the bodies. Without such a distinction, $S_2$ could well be *an ellipsoid of revolution in absolute rest*. That is why the appeal to absolute space as a cause is unavoidable: Einstein is entitled to name the space of reference $R_1$ the "merely *factitious* cause" of the difference between $S_1$ and $S_2$ (in the original German: *fingiert*); this cause is not something that can be observed. Yet:

> No answer can be admitted as epistemologically satisfactory, unless the reason given is an *observable fact of experience*. The law of causality has not the significance of a statement as to the world of experience, except when *observable facts* ultimately appear as causes and effects.

The fiction of an absolute space is thus as unsatisfactory as the fiction of a universe deprived of large-scale mechanical effects. The preceding epistemological requirement of observability, which is tantamount to the requirement that we be able to verify that the shapes of the bodies are dependent on their motion, can be satisfied only when the system $S_1 S_2$ is reintroduced in the real universe. The system $S_1 S_2$ offers, by itself, no explanation, and two spheres could equally be "observable" in an otherwise amorphous universe.

Einstein was thus led to see the answer in Mach's theory. The cause must lie *outside* the system $S_1 S_2$ and the spaces of reference:

> We have to take it that the general laws of motion, which in particular determine the shapes of $S_1$ and $S_2$, must be such that the mechanical behavior of $S_1$ and $S_2$ is partly conditioned, in quite essential respects, by distant masses which we have not included in the system under consideration. These distant masses and their motions relative to $S_1$ and $S_2$ must then be regarded as the seat of the causes (which must be

susceptible to observation) of the different behavior of our two bodies $S_1$ and $S_2$. They take over the role of the factitious cause $R_1$.

This answer points to the difficulty with the premises that have been adopted. Two assumptions were involved: (a) there is no interaction between the two bodies under consideration, (b) these two bodies have no interaction with other masses in the universe. Premise (a) can be called *local*, and (b) *global*. This is a set of typically Newtonian assumptions, because inertia is disconnected from gravitational influence. Now, the thought experiment reveals a situation that is inexplicable in terms of this set. In order to remove the difficulty, Einstein maintained (a) and dropped (b). The reason for this is the epistemological law of causality: the system $S_1 S_2$ is the only thing in the experiment that needs to be observed directly, whereas distant matter need not be observed. In this sense, it is assumed that a mistaken conjecture about (a) was absolutely impossible, whereas a mistake about (b) was unwittingly alluring. As a matter of fact, both (a) and (b) taken together do not make *necessary* what is *observable*—the differing behavior within the system $S_1 S_2$ and distant matter itself.

Strictly speaking, Mach's project is an extension of kinematic relativity to the domain of dynamics. Thus, "if we take our stand on the basis of facts, we shall find we have knowledge only of *relative* spaces and motions." Certainly, "when we say that a body $K$ alters its direction and velocity solely through the influence of another body $K'$, we have asserted a conception that it is impossible to come at unless other bodies $A, B, C \ldots$ are present with reference to which the motion of the body $K$ has been estimated" (Mach 1888, pp. 281–283). From this criterion for judgment, Mach concluded that certain intrinsic effects of nature are necessary. In fact, Einstein's experiment is quite the reverse of this procedure. The two masses are not only initially free with respect to distant matter from the observational point of view, but also with respect to one another from the interactive point of view, which is quite different from the bodies $K$ and $K'$ in Mach's reasoning where they are linked mechanically. In other words, Einstein's epistemological requirement is a sort of conjuring trick, intended to implement a true tour de force: distant matter will be absolutely unavoidable if it must be posited even when every means is lacking to form a judgment about the behavior of bodies.

Einstein deduced from this a fundamental principle: "*The laws of physics must be of such a nature that they apply to systems of reference in any kind of motion.* Along this road we arrive at an extension of the postulate of relativity" (Einstein 1916, p. 113). This is the so-called principle of general relativity. It can be asserted that the thought experiment is designed so as to show that this principle rests upon Mach's principle—that the inertia of a single body depends on the totality of other masses in the universe. The thought experiment attempts to provide an epistemological justification for the principle of general relativity on the basis of Mach's principle. But, in fact, yet another principle is needed in order to secure the complete relativity of the phenomena observed in $S_1 S_2$: this is the principle of general covariance ("*The general laws of nature are to be expressed by equations which hold good for all systems of coordinates, that is, are covariant with respect to any substitutions whatsoever,*" Einstein 1916, p. 117), thanks to which the *absolute* difference between $S_1$ and $S_2$ is preserved (that is, the geometry of the ellipsoid is not Euclidean) *without* bestowing any privilege upon either of the two viewpoints.

De Sitter's first paper on the question (1916a) was communicated on September 30, 1916. A second paper of December (de Sitter 1916b) offered a somewhat refined argument, complemented by new insights on the fundamental issues.

De Sitter considered the following coordinate system:

$$x_1 = r, \quad x_2 = \theta, \quad x_3 = z, \quad x_4 = ct,$$

where $z$ is the Earth's axis of rotation, and $r$ and $\theta$ are polar coordinates in the plane perpendicular to that axis. What are the $g_{\mu\nu}$ prevailing on the Earth's surface? Were the Earth not to rotate, the $g_{\mu\nu}$ would take the values:

$$\begin{matrix} -1 & 0 & 0 & 0 \\ 0 & -r^2 & 0 & 0 \\ 0 & 0 & -1 & 0 \\ 0 & 0 & 0 & +1. \end{matrix} \quad (1)$$

This is nothing more than the Minkowski $ds^2$ in terms of polar coordinates:

$$ds^2 = c^2 dt^2 - dr^2 - r^2 d\theta^2 - dz^2. \quad (2)$$

Transforming to rotating axes, the components $g_{\mu\nu}$ for the new system $\theta' = \theta - \omega t$ (where $\omega$ is the angular velocity of rotation) are

$$\begin{matrix} -1 & 0 & 0 & 0 \\ 0 & -r^2 & 0 & -r^2\omega \\ 0 & 0 & -1 & 0 \\ 0 & -r^2\omega & 0 & +1 - r^2\omega^2/c^2. \end{matrix} \quad (3)$$

The interval is now:

$$ds^2 = -dr^2 - r^2 d\theta^2 - dz^2 - 2\omega r^2 d\theta\, dt + (1 - r^2\omega^2)c^2 dt^2. \quad (4)$$

In this expression, if $\omega = 0$, the interval is reduced to (2), with space sections perpendicular to the time axis. If $\omega \neq 0$, orthogonality is no longer preserved and terms appear (such as $g_{24}$) where space and time are interwoven. The aim of the general theory of relativity is to build a world view where the two situations are equivalent.

But de Sitter wrote,

> It is found that the set (1) does not explain the observed phenomena at the surface of the actual earth correctly, and (3) does, if we take the appropriate value for $\omega$. This value of $\omega$ we call the velocity of rotation of the earth. Then relative to the axes

(3) the earth has no rotation, and we should expect the values (1) of $g_{\mu\nu}$. (de Sitter 1916a, p. 528)

Thus, there is a difference between axes "at rest" and "rotating" axes, in the sense that the components of the metrical tensor are not identical. How would it be possible to avoid nonobservable absolute space, granted that this difference be insuperable? Einstein's solution was: "The $g'_{24}$ and the second term of $g'_{44}$ in (3) therefore do not belong to the field of the earth itself, and must be produced by distant masses." Preserving absolute rotation leads to differentiation of the spaces of references (1) and (3). For de Sitter, the introduction of distant masses is as contrary to the spirit of general relativity as the withdrawal of absolute space would be in Newtonian mechanics. These masses, exactly like the absolute space in Newtonian theory, are independent of the reference system.

De Sitter wanted to develop an interpretation more appropriate to the spirit of general relativity. As an example, he considered the component $g_{24}$. The solution of the differential equation that determines that quantity is

$$g_{24} = kr^2, \tag{5}$$

where $k$ is an arbitrary constant of integration. The Einsteinian theory requires $g_{24}$ to be of such a form, but it does not prescribe the value of the constant: "the *differential* equation is the fundamental one, and the choice of the constants of integration remains free" (de Sitter 1916a, p. 529). By contrast the Newtonian theory prescribes the values of the constants, and that is its absolute character. A consistent theory of relativity leaves the constants free; those constants

> must, of course, be so determined that the solution represents the observed relative motions of material bodies and light rays as described in the adopted system of coordinates. The particular solution which does so in one system must, if this system is transformed into another, by the same transformation be reduced to the particular solution which fits the observed phenomena in the new system.

What de Sitter points out is that the requirement of relativity is purely formal: the differential equations that govern all phenomena of rotation keep the same form in every coordinate system. The components of the metric tensor have values that may vary from one system to the other, without jeopardizing dynamic relativity. Thus, general relativity "implies that the constants of integration are also subjected to the transformations, and are therefore as a rule different in different systems of coordinates" (de Sitter 1916b, p. 179). Resorting to distant masses originates in a confusion between the equations and their solutions: "The flaw in the argument used above was that (1) was considered to be *the* solution, instead of (5)" (de Sitter 1916a, p. 529). De Sitter concluded that, if we believe that there is no absolute space, "we must regard the differential equations as the fundamental ones, and be prepared to have different constants of integration in different systems of reference" (de Sitter 1916a, p. 531). The substitution of distant masses for an absolute space comes from the belief that any value of $k$ different from zero in an arbitrary coordinate system should be "explained" by something that is declared to be "observable."

Einstein's differential equations are universally invariant, but this is not true for their solutions, whose constants of integration are different in different coordinate systems. It would be attractive to believe that the system where $k$ takes the value zero has a privileged status analogous to that of inertial systems, since the metric takes a particularly simple diagonal form. The significance of this first contribution by de Sitter lies in its showing that no such thing can happen. The privileged status of inertial systems is based upon the very simple form of the differential equations in these systems. In the case under consideration, the field equations have the same form in the coordinate system associated with the stars and in every other system as well; only the solution is far simpler in the former. This does not confer a privileged status on the system of stars.

In a letter that he sent to de Sitter on October 13, 1916, while awaiting a reprint of de Sitter's paper on relative rotation, Eddington already perceived the limits of Einstein's wish to abolish absolute rotation. The important point, he surmised, is the relation of these limits to the problem of definite boundary conditions for the differential equations of gravitation (which are also, in Einstein's theory, the equations of motion). He gathered that

> when you choose axes which are rotating relatively to Galilean axes, you get a gravitational field which is not due to observable matter; but is of the nature of a complementarity function due to boundary conditions—sources or sinks—at infinity.... That seems to me to contradict the fundamental postulate that observable phenomena are entirely conditioned by other observable phenomena. (Eddington to de Sitter, October 13, 1916, Leiden Observatory.)

These remarks entail what is germane to de Sitter's more sophisticated strategy in the December paper.

That the constants of integration may vary from one coordinate system to the other is instrumental in fostering quite a new picture of the consistency of general relativity. This variability of the values of the constants allows the adjustment of the equations of motion so that a quantity like $\omega$ is always the same for any observer. In particular, no a priori identity governs these values at infinity. As de Sitter said: "The condition that the gravitational field shall be zero at infinity forms part of the conception of an absolute space, and in a theory of relativity it has no foundation." No observation will ever teach us anything about the infinite. Even though the infinite is evidently an "ideality," it is not a priori identical for all observers. This means that the problem of boundary conditions in general relativity cannot be dealt with in classical terms.

Yet, there is something quite attractive in a classical approach to the problem of boundary conditions, particularly in view of Einstein's commitment to Mach's ideas. When, in Poisson's equation (the differential equation of classical mechanics), the boundary conditions at infinity for any finite system are set to zero, it can be asserted that this is done in order for the *whole* of the numerical value of the potential at any point to be derived from the *material sources* only. In the theory of general relativity, such ideal boundary conditions should be replaced by the structural and physical conditions of space-time. In accordance with the usual methodological approach adopted by Einstein, one could well start with the Newtonian approximation. Throughout those portions of space-time that are sufficiently large, the pseudo-Euclidean continuum prevails. Thus, at infinity, we would expect the following values for the potentials:

$$\begin{matrix} -1 & 0 & 0 & 0 \\ 0 & -1 & 0 & 0 \\ 0 & 0 & -1 & 0 \\ 0 & 0 & 0 & +1. \end{matrix} \quad (6)$$

Of course, all "real" potentials differ from the values at infinity. But, as in Poisson's equation, they are all determined by differential equations that represent matter. In this sense, set (6) seems to meet the requirements of Mach's principle.

Once again, de Sitter raised the question of a genuinely relativistic consistency:

> Thus matter here also appears as the source of the $g_{\mu\nu}$, i.e., of inertia. But can we say that the *whole* of the $g_{\mu\nu}$ is derived from the sources? The differential equations determine the $g_{\mu\nu}$ apart from constants of integration, or rather arbitrary functions, or boundary conditions, which can be mathematically defined by stating the values of $g_{\mu\nu}$ at infinity. Evidently we could only say that the whole of the $g_{\mu\nu}$ is of material origin if these values at infinity were *the same for all systems of coordinates*.

Were the values of $g_{\mu\nu}$ at infinity the same for all coordinate systems, i.e., were they both *arbitrary* and *generally covariant* constants of integration, then, and only then, would any observer distinguish local values from values at infinity in one and the same manner. But set (6) is undoubtedly not generally covariant; these values only satisfy the principle of special relativity.

Because the search for a physical interpretation of boundary conditions is now seen as vital to the survival of Mach's principle, Einstein was obstinate and made a last attempt to find boundary conditions that are generally convariant. De Sitter related a conversation he had with Einstein, shortly after de Sitter's first paper on the relativity of rotation became known to Einstein (and prior to its publication). Einstein had suggested that the $g_{\mu\nu}$ converge at infinity toward the following values:[2]

$$\begin{matrix} 0 & 0 & 0 & \infty \\ 0 & 0 & 0 & \infty \\ 0 & 0 & 0 & \infty \\ \infty & \infty & \infty & \infty^2. \end{matrix} \quad (7)$$

This is a sort of complete separation of space and time. At infinity, it is still possible to introduce arbitrary functions $t'$ of $t$, but no longer of $x_1, x_2, x_3$. At a finite distance, relativity is total—all four dimensions are affected by it—but a kind of "absolute time" reappears at infinity. The components of the metric that entail time become infinite at infinity: the longer the time interval, the larger the distance from a point-mass taken as center. Space is, so to speak, "engulfed" in time. Because the relative measurements of time at infinity are infinite, time is a sort of "bottleneck" of space. As a result, boundary conditions (7) form a "nothing" that is not merely a mathematical device, but a "nothing" that should be physically representable as being different from the empty Minkowski space-time.

In Einstein's eyes, set (7) was the condition for an ultimate version of Mach's principle. In his cosmological memoir, he recalled it as being a most decisive step (Einstein 1917, pp. 179–183). First comes the mathematical formulation of Mach's requirement that a test body, sufficiently distant from all other masses in the universe, have zero inertia. Spatial isotropy, that is, the invariability of the solution of the field equations under an orthogonal transformation of $x_1, x_2, x_3$, allows the infinitesimal space-time interval to be written:

$$ds^2 = -A(dx_1^2 + dx_2^2 + dx_3^2) + B(dx_4^2).$$

Moreover, the condition for local Euclideanicity can be written as

$$\sqrt{-g} = 1 = \sqrt{A^3 B},$$

while the assumption of small stellar velocities leads to

$$ds^2 \approx \sqrt{g_{44}}\, dx_4.$$

These restrictive conditions have the following consequences for the form of the components of the energy-momentum tensor for a point-mass. The components of momentum are given by the three first components of the covariant tensor multiplied by $\sqrt{-g}$, i.e., $m\sqrt{-g} \cdot g_{\mu\nu}(dx_\mu/ds)$:

$$m\frac{A}{\sqrt{B}}\frac{dx_1}{dx_4}, \quad m\frac{A}{\sqrt{B}}\frac{dx_2}{dx_4}, \quad m\frac{A}{\sqrt{B}}\frac{dx_3}{dx_4}.$$

The component of energy, which is the fourth component of the covariant tensor, is (in the static case), $m\sqrt{B}$. These are thus the components of the energy-momentum tensor in their geometrical expression ($A$ and $B$ are metric elements). In order to have a full physical meaning, they are not to be contradicted by the conditions pertaining to the material tensor. Mach's principle acts at that level. The quantity $A/\sqrt{B}$ in the expression for the momentum plays the role of the coefficient of inertia, while $mA/\sqrt{B}$ represents mass at rest in the static metric. Einstein reasoned:

> As $m$ is a constant peculiar to the point of mass, independently of its position, this expression, if we retain the condition $\sqrt{-g} = 1$ at spatial infinity, can vanish only when $A$ diminishes to zero, while $B$ increases to infinity. It seems, therefore, that such a degeneration of the coefficients $g_{\mu\nu}$ is required by the postulate of relativity of all inertia.

This justifies the set of values (7), which Einstein had suggested in his conversation with de Sitter. These values would enable us to speak of our measurements anywhere in the universe with uniformity. Such uniformity implies that the $ds^2$ should retain the same numerical value throughout space and in particular, at infinity. Thus, if $\sqrt{-g} = 1$ is retained at infinity, and if $A$ tends to zero for a test body very remote from all other matter, $B$ must simultaneously tend to infinity; only this allows the $ds^2$ to remain invariant.

If the hypothesis (7) is to be adopted, de Sitter said,

> then in any system of coordinates the $g_{\mu\nu}$ would have exactly determined values, there being no constants of integration left by which they can be made to fit the observed phenomena. (de Sitter 1916b, p. 182)

In other words, this ultimate version of Mach's principle gets around de Sitter's earlier objections, by pointing out that the constants of integration are fixed by the total material universe.

Now observation shows that the values of $g_{\mu\nu}$ are very different from (7) at the remotest distances. This fact forms the basis for de Sitter's new counterargument, which, as will be seen, is not at all restricted to the rather short distances then explored by astronomers. Concerning the observed values, de Sitter commented:

> On Einstein's hypothesis these are special values which, since they differ from (7), must be produced by some material bodies. Consequently there must exist, at still larger distances, certain unknown masses which are the source of values (6), i.e., of all inertia.

If the infinite is a "nothing" different from an empty, Minkowski space-time and if, furthermore, this hypothesis

> has arisen from the wish to explain not only a small portion of the $g_{\mu\nu}$ (i.e., of inertia) by the influence of material bodies, but to ascribe *the whole of the $g_{\mu\nu}$* (or rather the whole of the difference of the actual $g_{\mu\nu}$ from the standard values (7)) to this influence (de Sitter 1916a, p. 531, note 2),

it then follows that the $g_{\mu\nu}$ pertaining to the Minkowski metric must be produced by material sources. The values at infinity are postulated in order to have distant masses, but they only succeed in moving these masses beyond *any* field of effective observation. Indeed, all measured values must be different from (7), since the infinite is not measurable. Distant masses will always fulfill the same task, however large our picture of the measured universe may be, that is, the task of explaining the values (6) (or any other local values that could well be measured in the future) for the $g_{\mu\nu}$ at a finite distance. Rather than being real objects, these masses, therefore, are nothing more than objects always beyond the field of effective observation. And yet, in principle, because the infinite is "physically" nothing, the whole hypothesis

> implies the finiteness of the physical world, it assigns to it a priori a limit, however large, beyond which there is *nothing* but the field of the $g_{\mu\nu}$ which at infinity degenerate into the values (7).

The limit is that formed by the hypothetical masses. But, even though a priori fixed, the limit remains as large as we like, as large as any possible experience. The theory provides no means to fix it.

On the whole, by seeking to make Einstein's theory truly consistent, de Sitter had pointed to the consequences that would result if one were to accept and maintain both premises (a) and (b) in the thought experiment of the two rotating fluid bodies. Rather than dropping one of the premises on the grounds of the problematic status of distant matter, as Einstein did, de Sitter showed that the proper way of regarding the possible existence of distant matter would be *not* as the system of directly visible stars, but as some other matter that is even farther away. Because he was concerned with the whole problem of the origin of inertia, and precisely because he made his criticism a matter of principle, de Sitter was inclined to adopt an epistemological position rather similar to that which requires the dynamic equivalence of coordinate systems in the problem of rotation. In the latter problem, the belief in a true value of $k$ is conducive to granting a privileged status to the system in which $k$ takes that value; as a result, an explanation of the other values of $k$ in other systems is sought. But the dynamical equivalence of coordinate systems blocks any explanation of the divergence from a strict diagonal metric by means of some causal influence. And similarly, with regard to an entirely material origin of inertia: "practically it makes no difference whether we explain a thing by an uncontrollable hypothesis invented for the purpose, or not explain it at all" (de Sitter 1916a, pp. 531–532).

## 3. Einstein's Very Last Doubts

In the note that he added to his paper on relative rotation on September 29, 1916, after a conversation with Einstein, de Sitter mentioned set (7) as implying the finiteness of "the physical world," with the degenerate field of $g_{\mu\nu}$ making "space and time absolute at infinity." Accordingly, "if we wish to have complete four-dimensional relativity for the actual world, this world must of necessity be finite." A shift in the use of words is quite noticeable. De Sitter's account of Einstein's hypothesis (7) is in terms of the *physical* world, while his reference to four-dimensional relativity is to the *actual* world, thus stressing the fulfillment of relativity as accommodating both physics and geometry.

What is the meaning here of "finite"? In fact, Einstein wrote as early as May 14, 1916, to Michele Besso, well before he discussed the matter with de Sitter, that he was trying in the quiet period following the establishment of the field equations, "to determine the boundary conditions at infinity." He added: "It is interesting to ask how a *finite* universe can exist, that is, a universe whose finite extent has been fixed by nature and in which all inertia is truly relative" (Speziali 1972, p. 69). But by the end of October, he wrote again about the problem, only to confess that the distinguishing characteristic of general relativity as local physics was its independence from the specification of boundary conditions (p. 98). Clearly the problem was now how to articulate the relation of geometry to physics in a novel way. As Einstein himself recalled in his Princeton lectures of 1921: "The question as to whether the universe as a whole is non-Euclidean was much discussed from the geometrical point of view before the development of the theory of relativity" (Einstein 1921b, p. 109). Einstein thus explicitly acknowledged his debt to the mathematicians of the second half of the nineteenth century, without mentioning their names. He went on to say that "with the theory of relativity, this problem has entered upon a new stage." True, Riemann himself had best depicted the nineteenth century divorce of both field physics and differential geometry from large-scale considerations, when he proclaimed that the "questions concerning the incommensurably large are, for the explanation of Nature, useless questions.... Knowledge of the causal connection of phenomena is based essentially upon the precision with which we follow them down into the infinitely small" (Riemann 1854, pp. 423–424).

In March 1917, de Sitter acknowledged in a revealing note that the idea of conceiving of a spherical four-dimensional world had been suggested by Paul Ehrenfest in a conversation a few months earlier (de Sitter 1917a, p. 1220). It would appear, therefore, that Einstein was very probably aware of the idea. At any rate, it is obvious that the word "finite" could carry, as far back as these early discussions, a decidedly non-Euclidean meaning. In fact, in a letter to de Sitter dated April 18, 1917, Ehrenfest confessed that the whole idea of a curved four-dimensional universe had crossed his mind as early as summer 1912, when he was still in Russia. He subsequently abandoned the problem rather quickly, however, because he could see the overwhelming difficulty this kind of curvature would present. It is true that Ehrenfest did not publish a single line on the subject, neither at that time nor later when the difficulty had apparently been overcome, partly, it seems, because he thought the very concept of time would become quite unintelligible. Yet he still believed the curvature of time together with that of space to be the straightforward consequence of what he called Lorentz-Einstein-Minkowski relativity. It was the only possible answer to the following problem, which Ehrenfest put like this:

> Newton-Galileo relativity (with uniform translation) will be completely annihilated, if we pass from Euclidean to non-Euclidean space (for instance, the *spherical* space). Now I asked myself: what will become of Lorentz-Einstein-Minkowski relativity if we transform to curved space? I saw then, that the whole four-dimensional world had to be curved. (Ehrenfest to de Sitter, April 18, 1917, Leiden Observatory).

That Einstein had a strong *theoretical* motivation for assigning a special role to time is shown by his reaction in a letter to de Sitter, dated November 4, 1916 (EA 20-539). The letter presents a partly corrective response to de Sitter's views in the note of September 29. A few days before Einstein's letter arrived, on the 1st of November, de Sitter had worked out a lot of interesting further details on how he understood the proposed boundary conditions (7).

If the $g_{\mu\nu}$ degenerate to these values when both the space *and the time* variables become infinite, then "the hypothesis would make the world finite not only in space but also in time." De Sitter added that nothing can be known from the infinite past or from the infinite future. Does this make the hypothesis comply with the requirement of observability? No, because the limits "always remain hypothetical, and can never be observed." De Sitter called these limits the "shell," and said that this "is *not* a physical reality," since we will never have the right to assume that the knowledge of larger portions of the universe (of what lies outside the Milky Way) will confirm the already computed relations within the known part.

In fact, de Sitter was ready to accept the limitedness of the universe as a general principle (*"die prinzipielle Beschränktheit"*), but he could not see how this stops the endless chain of questions: *Where* are the distant masses? *What* is their constitution? *How* do they influence the inertia here? He ended up suggesting that a true explanation of inertia might be found in the infinitely *small* rather than in the infinitely *large*, but shrugged it off: "I am not a physicist, and this is probably just hallucination." De Sitter was perhaps echoing what *should* have happened to field physics after the revolution brought about by differential geometry. Weyl described this in terms of a theoretical requirement: "The transition from Euclidean geometry to that of Riemann is founded in principle on the same idea as that which led from physics based on action-at-a-distance to physics based on infinitely near action" (Weyl 1918, p. 91). Obviously, the proposed explanation of the origin of inertia falls short of this principle, and de Sitter even seems to insinuate that Mach's principle is no better than action-at-a-distance.

Einstein's reply of November 4 shows how vexing the whole question had become. Toward the end of the letter, he almost apologized for his Machian commitment, saying that it is no more than one of his innocent fads, and adding that in no way "do I demand that you share my curiosity." The question of boundary conditions, Einstein then said, is a question of taste, which apparently cannot have a scientific meaning in any case. Yet, he insisted that he would never have thought of the world as being of finite *temporal* extent. Nor is there a finite extent with regard to *space*. Einstein generalized de Sitter's conception of the "shell." Take a purely spatial shell (without matter); in four dimensions, this becomes a "tube" (*Schlauch*) since time is *not* curved. The shell can be taken to be so large as to include most of the inertia in the world; at any rate, the inertia outside it will be as small as we want. It can now be said that, *inside* the shell, only the existing masses will determine the inertia. Einstein was well aware that, if his argument is to have any validity in regard to observable facts, it must be assumed that the observed part of the world is *extremely small* relative to its actual size. Although this is all he said, we can easily infer that what he had in mind here was, actually, one of the strongest arguments that has ever been put forward in favor of Mach's principle: the isotropy of inertia is observed to a very high degree of accuracy in our part of the universe, but the matter in our *immediate* vicinity (planets, sun, stars) is patently not isotropic; for this reason, if inertia is produced by matter, an overwhelming part of the effect must come from *distant* matter which must, of necessity, be isotropically distributed. About the constitution of these masses Einstein was not worried, because he readily assumed that all the sources are stars. The shell, as he says, has nothing distinctive (*besondere*) about it. We find in these statements the first hints of what was later conceptualized as *the cosmological principle*: either one begins with the assumption that the basic constituents of the universe are the same, in which case there is no reason to deny a uniform distribution, or one begins with the uniform distribution and finds the constitutive uniformity quite plausible.

Einstein went on to distinguish the practical from the theoretical implications of the problem of boundary conditions. On practical grounds, he was reluctant to concede any validity to the problem of the totality. In any designated part of the world that we wish to single out, only the totality of masses that happen to exist in it may be said to determine the inertia of a given body. In the delineated section, the $g_{\mu\nu}$ are determined by these masses and by the $g_{\mu\nu}$ at the periphery. The inertia of the masses and the boundary conditions thus depend either on the system that happens to be selected, or more simply on the part of the world that we can describe. But Einstein clearly glimpsed the only way meaning can be bestowed on the theoretical aspect of the problem of the totality, when he showed that it is tantamount to the following question: Can I possibly think of the world in terms that would make the inertia depend on the masses and not on the boundary conditions? Nonetheless, because the observed part of the universe is so small, Einstein was led to undermine the strength of his own desire to articulate a full relativity of inertia. He explained that this smallness was the *psychological* motivation that dictated the drive to secure generally covariant laws.[3] General covariance having been gained, "there is no longer any reason to attach such a big weight to the full relativity of inertia." Behind his apparent carefulness, one can see that Einstein is now putting into practice the remark he made to Besso in October. Indeed he now raises the question of whether the same kind of procedure (independence from boundary conditions) could be applied to the entire universe. As he says to de Sitter, "You must not scold me if I am still quite interested in that kind of question."

In the cosmological memoir Einstein recapitulated how the problem of boundary conditions was then perceived by de Sitter and himself (Einstein 1917, pp. 182–183). Two possibilities are available. Firstly, the universe as a whole cannot be a problem because only systems within the universe are properly speaking the subject matter of physical science. Einstein viewed this as a mere clinging to the hitherto assumed validity of general relativity. His words are virtually a repetition of his November letter to de Sitter: "at the spatial limit of the domain under consideration we have to give the $g_{\mu\nu}$ separately in each individual case." In other words, because the proposed boundary conditions (7) do not make sense, a metric for the whole universe cannot make sense either. The second way of evading the difficulties with the boundary conditions is to assume that the pseudo-Euclidean values (6), which can be taken as the boundary conditions of the system of planets around the sun, are also applicable to the entire universe. Of course, as de Sitter had already argued, this presupposes a determinate choice of the reference system. But Einstein now realized that it is equally in contradiction with his own Machian conception of the relativity of inertia:

> For the inertia of a material point of mass $m$ (in natural measure) depends upon the $g_{\mu\nu}$ but these differ but little from their postulated values, as given above, for spatial infinity. Thus inertia would indeed be *influenced*, but would not be *conditioned* by matter (present in finite space).

Any single body would possess an amount of inertia that is virtually independent of the actual quantity of matter in the universe as a whole. This would amount to a restricted version of Mach's principle, the inertia of a single point-mass being a limited variation around some otherwise determined value.

Einstein's obstinate wish to incorporate a complete relativity of rotation in the theory of general relativity thus results in his being left with a fairly unsatisfactory alternative. This was the situation he was in by the end of 1916; then on February 2, 1917, Einstein announced in a letter to de Sitter that he was doing fresh work on the problem of boundary conditions, in which he claimed to have overcome finally the awkward degenerating field of $g_{\mu\nu}$: "I am curious to know what you will have to say about this somewhat fanciful conception" (EA 20-541). Two days later, he wrote to Ehrenfest: "I have ... again perpetrated something about gravitation theory which somewhat exposes me to the danger of being confined to a madhouse" (quoted in Pais 1982, p. 287).

## 4. "... A Method Which Does Not Itself Claim to be Taken Seriously"

The essence of Einstein's new argument is quite the opposite of the point of view he expressed to Besso at the end of October: there *is* a way of bringing together the absence of boundary conditions with a consistent discourse about the universe envisaged as a totality. Einstein achieved this by taking a fresh look at his recent investigations of boundary conditions and their physical nature.

It had gradually become apparent that no boundary condition could be harmonized with a "sensible" conception of the universe. Einstein's breakthrough was to think of a third possibility, quite distinct from the unsatisfactory alternative he was left with: "For if it were possible to regard the universe as a continuum which is *finite (closed) with respect to its spatial dimensions*, we should have no need at all for any such boundary conditions" (Einstein 1917, p. 183). This was the spark, the new way of looking at the same thing. Einstein's discovery was to see that there is an immense conceptual difference between disregarding the boundary conditions or treating them with indifference and explicitly rejecting them; this was the needed leap that carried Einstein from general relativity to the idea of a relativistic physics of the universe.

In the third edition of his famous popular exposition of relativity, Einstein explained that the infinite had long been acknowledged as a nonentity, both mathematically and physically (1918a, p. 108). He sent his reader back to the much earlier discussions of Riemann, Helmholtz, and Poincaré, in accordance with which the noninfinity of space does not necessarily stand in contradiction with the laws of thought or even less with the facts of experience. Applying these geometrical ideas to his new cosmology, Einstein seems to make implicit use of an idea due to Riemann in his well-known 1854 lecture. The idea is that the homogeneity and isotropy of geometrical figures are, in Riemannian manifolds, *global* properties of these manifolds that can be obtained only when the space under consideration is of constant curvature (Riemann 1854, p. 419; see the discussions in Eisenhart 1926, p. 86 and Torretti 1978, p. 100—constant curvature is the only case in which homogeneity and isotropy mutually imply one another). Setting about the task of making physical sense of the finiteness of space, Einstein's problem was now to find out how the distribution of matter could be made responsible for space being "re-entrant," as Eddington so vividly described it (Eddington 1928, p. 83).

Two requirements had to be set down (Einstein 1917, pp. 183–184). The first is a sort of principle, stating the possibility of disregarding the local lack of uniformity in the matter distribution: over enormous spaces, it should be possible to assume a uniform distribution. The second requirement is the fact of experience: "the relative velocities of the stars are very small as compared with the velocity of light"; this, in fact, becomes something like another principle, since it enables Einstein to state that "there is a system of reference relative to which matter may be looked upon as being permanently at rest." From these two requirements, Einstein set out to calculate the corresponding components of the field equations.

For the energy-momentum tensor, rest implies that only $T_{44}$ is nonvanishing: it is equal to the density of matter. The density, Einstein says, is not a function of the space coordinates *because* space is taken as being finite. In a sense, he actually seeks to derive the uniform distribution from the finiteness of space. The determination of the purely spatial components of the metric tensor $(g_{11}, g_{12}, \ldots g_{33})$ is, in turn, based on this uniformity, for, from this assumption, "it follows that the curvature of the required space must be constant" (p. 185), that is, the space must be spherical. The chain of reasoning is as follows: finiteness of space → uniform distribution of matter → sphericity (constant curvature) of space. And the calculation of the temporal components amalgamates all of these assumptions at one stroke. Matter at rest implies no acceleration for a test body, that is, the global contribution of the gravitational field is null. Like $T_{44}$, $g_{44}$ is thus independent of position. We can write $g_{44} = 1$

and, because the static field involves time being orthogonal to space, $g_{14} = g_{24} = g_{34} = 0$ for all spatial relations. The constancy of $g_{44}$ means that there is a *cosmic time* superimposed on all proper times. Cosmic time is a new concept that facilitates Einstein's getting over his earlier problem with the boundary conditions. The direction of time is the same from all points: the paths of material particles or light rays are geodesics along the time axis of space-time; cosmic time means that, the "length" of geodesics being measured along this direction, there can be neither convergence nor divergence of world-lines. When the space and time components are taken together, the figure of the universe becomes the *cylindrical universe*, a representation first introduced by the mathematican Felix Klein (1918, p. 408): the universe is a cylinder whose axis of symmetry is time; each circular slice represents a contemporaneous space. In fact, in trying to bring about the connection between the geometrical and the physical, Einstein also seems to make an implicit use of one of the theorems demonstrated by Schur (1886) stating that a Riemannian manifold that is isotropic about each point is also homogeneous; as Einstein had shown in the first part of his cosmological memoir, the flaw in Newtonian cosmology is that, if the matter distribution around one point is isotropic, this point is unique, i.e., there is no homogeneity. It is this mutual implication of large-scale homogeneity and large-scale isotropy that has become known as the assumption formulated as the cosmological principle. That this idea remained, for the most part, implicit at this stage is seen in the fact that only in the mid-1930s did Milne first introduce the idea by name; and Milne repeatedly accorded the honor to Einstein 1917 (see Milne 1935, pp. 24 and 60ff).

Schur's theorem, it is worth noting, says nothing about constant curvature. So the actual progression of Einstein's demonstration can be represented in the following way:

1. The very idea of the finiteness of space is, for Einstein, tantamount to isotropy about each point. In a finite space, the existence of a center has no raison d'être. Here is the crucial difference from Newton, since the Newtonian prediction of a finite distribution of matter in space implies isotropy about one point only.
2. The implicit use of Schur's theorem implies that homogeneity has to be added to isotropy for the distribution of matter.
3. Riemann's idea of global homogeneity and isotropy makes it possible to speak of constant curvature for a finite space.

The role of linear, uncurved time is justified in the same way: $g_{44}$ is constant *because* it is independent of position. In consequence, only a spherical space could make the mutual implication of rest and isotropy valid for the whole of space, in accordance with the general assumption of uniformity. This accord embodied an interaction between three features that Einstein wanted to combine: Mach's principle, the possibility of one reference system covering the whole of space, and the fact of the small stellar velocities. As a result, rest itself is now more than just a fact: it shares in the theoretical framework because it cannot be removed without annihilating the entire interaction.

The mathematical representation of the three-dimensional spherical space was not essentially different from that by which Schwarzschild arrived at his interior metric in 1916. The similarity was, of course, limited to the spatial geometry, since both the time and the material tensor were not the same. In fact, the proposed eradication of boundary conditions in the case of cosmology was the major difference, but it also proved the most extraordinary stumbling block to Einstein's innovation. In cosmology, there would be an enormous price to pay for getting rid of boundary conditions, and this price was connected with an apparently inoffensive question—that of the pressure. Indeed, the correct understanding of the status of boundary conditions required a careful distinction between the solutions of the field equations and the equations themselves, a fact that de Sitter had stressed with considerable brilliance. But what about the consequences of spatial finiteness for the equations themselves? This was the price to pay:

> If it were certain that the field-equations ... which I have hitherto employed were the only ones compatible with the postulate of general relativity, we should probably have to conclude that the theory of general relativity does not admit the hypothesis of a spatially finite universe. (Einstein 1917, p. 186)

A priori, the metric in both the Schwarzschild interior solution and the cosmological solution is of the form

$$ds^2 = -e^\nu\, dr^2 - r^2\, d\theta^2 - r^2 \sin^2\theta\, d\phi^2 + e^\mu\, dt^2,$$

where $\nu, \mu$ are functions of $r$ only. This is reducible to the Minkowski metric when both $\nu$ and $\mu$ are zero, so that the problem is to determine $\nu$ and $\mu$ in relation to the actual distribution of matter. Both the interior and the cosmological metrics yield the following system of equations:

(a) $8\pi p = e^{-\nu}(\mu'/r + 1/r^2) - 1/r^2$

(b) $8\pi\rho = e^{-\nu}(\nu'/r - 1/r^2) + 1/r^2$

(c) $dp/dr = [(p + \rho)/2]\mu',$

in which $p$ denotes the pressure; (c) is a fundamental equation for perfect fluids. The two metrics differ in regard to the nature of the boundary conditions. Firstly, the cosmological metric demands that $\nu$ and $\mu$ tend to zero when $r$ itself tends to zero, while the interior metric demands the same when $r$ tends to infinity. Secondly, there is a definite option for the boundary condition on the pressure in the interior case (a decreasing pressure from center to periphery), while the cosmological metric requires a constant pressure throughout space. A null pressure at the periphery leads to

$$p + \rho = Ae^{\mu/2}$$

(where $A$ is a constant of integration), the solution of which provides the well-known interior metric. An overall constant pressure, on the other hand, leads to

$$(p + \rho)\mu' = 0.$$

As a consequence of $\mu' = 0$, we have $\mu =$ constant. The function $\mu$ even vanishes, since $e^\mu = 1$ for $r = 0$. From equation (a),

$$(8\pi p)r^2 = e^{-\nu} - 1,$$

i.e., $e^{-\nu} = 1 - r^2(-8\pi p)$. With this relation, the metric of the cylindrical universe can be written as

$$ds^2 = -(-dr^2)/[1 - r^2(-8\pi p)] - r^2\, d\theta^2 - r^2 \sin^2\theta\, d\phi^2 + dt^2.$$

The coefficients of the metric will have the appropriate Minkowskian signature only for $r^2$ less than $-1/8\pi p$. In other words, the radius of curvature $R$ of the whole world is

$$R^2 = 1/-8\pi p.$$

Indeed, that is the only value that satisfies the general condition for a spatially finite universe. This is also the fullest expression of the problem, since the right-hand side of this equation should be positive like $R^2$. The cosmological constant $\Lambda$ enables us to make the right-hand side positive:

$$R^2 = 1/(\Lambda - 8\pi p).$$

The fact that Einstein took the pressure to be zero is not really an accident, because the radius of curvature now becomes

$$R^2 = 1/\Lambda.$$

With zero pressure, the value of the cosmological constant is not confined a priori to a particular set of values. As Einstein put it to Besso as early as December 1916, the closed universe "is also suggested by the fact that *the curvature has the same sign everywhere, since experience teaches us that the energy-density does not become negative*" (Speziali 1972, p. 97).

That the value of the new constant should not be constrained in any way tallies with Einstein's early doubts about its physical nature. For how is it possible to justify its existence? At first sight, it seems that the problem of a negative pressure has been simply transferred to $\Lambda$; so much so that, less than two years after the cosmological memoir, Einstein tried to show that $\Lambda$ is comparable to a mere constant of integration and that it has nothing to do with a constant of nature (Einstein 1919, p. 354). As he made clear in a letter of March 24, 1918 to Felix Klein, the point of having zero pressure is that it makes cosmological considerations theoretically distinct from local ones: in the case of the Schwarzschild sphere, "$g_{44}$ is variable and equilibrium is possible only with spatially variable pressure," while in his solution, as Einstein went on to say, the cosmological constant makes two things possible at one stroke, namely, "the constancy of $g_{44}$ and the vanishing of the pressure" (EA 14-432). Yet, of course, the relation of $\Lambda$ to $R$ remains very puzzling indeed. It is really unclear that any serious reason can be invoked why $\Lambda$ should not be fixed, on account of its direct relation to $R$. It seems difficult to speak of a definite size for the universe if $\Lambda$ is to remain something like an arbitrary constant of integration.

The introduction of $\Lambda$ into the field equations transforms the original equations

$$R_{\mu\nu} = -k(T_{\mu\nu} - \tfrac{1}{2}g_{\mu\nu}T)$$

into

$$R_{\mu\nu} - \Lambda g_{\mu\nu} = -k(T_{\mu\nu} - \tfrac{1}{2}g_{\mu\nu}T).$$

Amazingly, Einstein closed his memoir by emphasizing a completely different function:

> It is to be emphasized ... that a positive curvature of space is given by our results, even if the supplementary term is not introduced. That term is necessary only for the purpose of making possible a quasi-static distribution of matter, as required by the fact of the small velocities of the stars. (Einstein 1917, p. 188)

This testifies to Einstein's quite fantastic wavering over the true significance of his own labors. This statement is in explicit contradiction to his earlier view of the introduction of $\Lambda$ as tantamount to the hypothesis of a finite space. On the one hand, it highlights the fact that the finiteness of space follows from the purely geometrical part of the new cosmological considerations: this is achieved by the decision, independent of Newton's theory, to incorporate non-Euclidean geometry in order to describe large-scale portions of spacetime. On the other hand, it lends support to an interpretation of the constant that derives clearly enough from Newtonian theory: quite apart from its mathematical role as a constant, $\Lambda$ is the *physical* force of repulsion that counteracts the effects of gravitation; and contrary to all known forces, the repulsion increases with distance so as to cancel the additive (and devastating) effects of gravitation.

Einstein was well aware that the introduction of $\Lambda$ in Poisson's equation (Newton's law of gravitation expressed in differential form) is already a covert way of removing any necessity for the rigorous statement of boundary conditions in the classical, nonrelativistic theory, since then no cosmic periphery need be considered: "A [Newtonian] universe so constituted would have, with respect to its gravitational field, no centre" (p. 179). So a theory incorporating the new constant undoubtedly parallels the Newtonian theory insofar as the *form* of the laws is concerned, but the *interpretation* of it differs sharply from a mere formal analogy, since, in the relativistic case, the purpose of $\Lambda$ is to fix a definite periphery.

Warning against too serious a comparison between Newton's theory and his own theory, Einstein therefore with some justification finally referred to his model as "logically consistent" (p. 188). The smallness of stellar velocities and the vanishing of the pressure are just good approximations to the facts. But the invention of cosmology has made their insertion in the corresponding equations *structural*, in the sense that without them the whole theoretical edifice would have fallen away. The new laws of gravitation had already succeeded in securing the interdependence of previously unrelated terms, or even entire laws of nature, like the identity of inertial and gravitational mass or Poisson's equation and the conservation laws. It was certainly no little satisfaction for Einstein to realize that his cosmological model achieved an interdependence between variables of an entirely different kind. However, the logical consistency of the new cosmology rested, in its turn, upon $\Lambda$. Because Einstein was loath to spell out a wholly convincing and unequivocal interpretation of it, he made the entire logical consistency of his system rest on the decision to incorporate it: "In order to arrive at this consistent view, we admittedly had to introduce an extension of the field equations of gravitation which is not justified by our actual knowledge of gravitation." What exactly is the ground for making $\Lambda$ essential to the logical consistency of the new theory? This was the task that fell to the astronomer, de Sitter.

## 5. The Almost Full and the Almost Empty

Einstein sent a letter to de Sitter on March 12, 1917, probably along with his cosmological memoir, which revealed how burning the question had become for him either way: whether the construction of a model for the whole universe "offers an extension of the relativistic way of thinking, or whether it leads to contradiction." It was a relief, he concluded, that all obstacles had been overcome. From the standpoint of astronomy, he surmised that what he had built was of course "a spacious castle in the air" (EA 20-542). But the more important thing was the theoretical background, for to Besso he spoke of his model as "a proof that general relativity can lead to a noncontradictory system" (Speziali 1972, p. 102). Also in his letter to de Sitter, Einstein went on to insist that the model entails no violation of the principle of relativity, inasmuch as only the statical conception is possible.

De Sitter wrote a brief postcard on March 15, 1917, thanking Einstein for his explanations. He agreed with Einstein's opinion to the extent of thinking that the model should not be seen as an attempt to "force" reality. He had no objection on the grounds that it is an "uncontradicted train of thought." But he stressed that he had strictly astronomical views on the problem of the total mass of the stars: he claimed that "the simple fact that we can identify spectral lines proves the potentials among all stars and nebulae ... to be of the same order as here. This proof is stronger than that of the small stellar velocities" (EA 20-543).

This point was to be the prelude to a most baffling intellectual journey on de Sitter's part. He was inclined to believe that this knowledge of the total mass of the stars confined within the limits of observation is the *only* knowledge we have of what lies outside the solar system. So, in Einstein's new model, the theoretical determination is exposed to the dangers of extrapolation. And because of the peculiar status of time in this model, de Sitter was also eager to emphasize the dangers of extrapolation in both space and time from the "photographic image" we have of the universe. The image is a static one; but in a letter to Einstein (April 1, 1917), de Sitter says that

> all extrapolations beyond the region of observation are insecure.... Your premise that the world is mechanically quasi-static I contest with the utmost energy. We have only a snapshot of the world, and we must not conclude from the fact that the picture shows no great transformations that everything will always remain as it was when the photo was made. (EA 20-551)

From this critique, de Sitter was led to formulate his own model of the universe. It is almost as if he became a cosmologist against his own will, finding no better way of criticizing Einstein's new hopes than constructing a counterexample; this became, unexpectedly, an alternative model of the universe.

De Sitter's approach to the problems of cosmology was strikingly reminiscent of his earlier criticisms, in which the concept of distant matter was opposed to the real premises of general relativity. He conceived of the consistency of general relativity in terms of a theory that dispenses with the need to explain inertia, and he now used this general epistemological standpoint to attack Einstein's cosmology. In a letter to Einstein dated March 20 (EA 20-545), de Sitter said of his new conjectures that he simply did not know whether they could help explain inertia; and a few days later (April 1) he was almost apologetic for having dared to build a theoretical construction that Einstein ventured to call a "world." All extrapolations remained dangerous, and de Sitter found it difficult to take Einstein at his word when he spoke of "your" model and "my" model.

De Sitter began by objecting to the alleged noncontradiction with the relativity postulate: "From the point of view of the theory of relativity, it appears at first sight incorrect to say: the world *is* spherical" (de Sitter 1917a, p. 1218). He does not appear to balk at the idea of finiteness, provided it is used for purely geometrical purposes. Faced with Einstein's extension of it so that it acquires physical significance, however, de Sitter pointed to the residual distinction between the strictly formal requirement of relativity and any assertion as to the actual existence of postulated entities. His misgivings concerned the relation of generalized mathematical transformations to the existence of some unique, worldwide space-time. In order to "test" the claim that a physics of the whole universe could not be different from the proposed one, he used a stereographic projection that maps the Einstein universe onto the Euclidean plane; this projection leaves the relevant invariants unaltered but throws some interesting light on the behavior of the old "boundary conditions." Indeed, in the new coordinates, the values for the $g_{\mu\nu}$ at infinity for this *finite* universe can be written as

$$\begin{matrix} 0 & 0 & 0 & 0 \\ 0 & 0 & 0 & 0 \\ 0 & 0 & 0 & 0 \\ 0 & 0 & 0 & 1. \end{matrix} \quad (8)$$

de Sitter immediately pointed to the ambiguity of this set:

> Einstein only assumes *three*-dimensional space to be finite. It is in consequence of this assumption that in [8] $g_{44}$ remains 1, instead of becoming zero with the other $g_{\mu\nu}$. This has suggested the idea to extend Einstein's hypothesis to *four*-dimensional space-time. (p. 1219)

The potentials at infinity would then degenerate to the values:

$$\begin{matrix} 0 & 0 & 0 & 0 \\ 0 & 0 & 0 & 0 \\ 0 & 0 & 0 & 0 \\ 0 & 0 & 0 & 0. \end{matrix} \quad (9)$$

Having gone so far in his attempt to discover the essence of Einstein's strategy, de Sitter could no longer hesitate in the face of the formidable idea of spelling out the actual meaning of a fully relative time. In his letter of March 20 to Einstein, he wrote that only these $g_{\mu\nu}$ make it possible to consider the infinite as "either spatial, or temporal, or both," in accordance with the original notion of a relativistic continuum that had been defined by Minkowski.

Inserting the new $g_{\mu\nu}$ into the field equations with the cosmological constant, de Sitter went on to discuss the *physical* difference between Einstein's continuum (labeled $A$) and his own continuum ($B$). In view of the smallness of $\Lambda$, as indicated by our knowledge of the perturbations in the solar system, the predicted quantity of matter in the Einstein universe is truly enormous; the rough estimate amounted to something like 10,000 Milky Ways, which did not really trouble Einstein himself, as he found that "the exact figure is a minor question" (Moszkowski 1921, p. 127). An astronomer like de Sitter would have found this quite unacceptable, given that at the time the very question of whether even one nebula is actually extra-galactic had not been settled. (For a comprehensive discussion of the observational context, see Smith 1984.) De Sitter stated this problem of the quantity of matter in $A$ in the following terms:

> It is found necessary to suppose the whole three-dimensional space to be filled with matter, of which the total mass is so enormously great that, compared with it, all matter known to us is utterly negligible. This hypothetical matter I will call the "world-matter." (1917a, pp. 1218–1219)

This world-matter is of course quite reminiscent of the earlier distant masses, which are always found to lie beyond any given field of observation. For the sake of comparison, de Sitter selected for his continuum $B$ those very hypotheses that Einstein had already chosen, namely, the universe is a smoothed-out system, quite similar to a perfect fluid at rest. He found that "for $B$ we have $\rho = 0$: the hypothetical world-matter does not exist" (p. 1221). This argument indicates that there is no necessary relation (in the logical sense outlined by Einstein) between the curvature of space and the particular form of the material tensor chosen by Einstein. Remove the physical, actually observed masses from your universe, de Sitter said to Einstein, and the result is the same: a test body still has inertia. Only the reason for this assertion now appears in a new light: it is the purpose of model $B$ to show that in $A$, the argument for a causal connection between the spatial curvature and the uniform distribution of matter rests upon a strange form of reasoning, namely, that the neglect of distant masses is as impossible as a physical proof that would begin with the statement (these are de Sitter's words): "If the world was not there." De Sitter then made a tentative suggestion that metric $B$ could also be interpreted as a spatially finite world, with a radius of curvature $R = \sqrt{3/\Lambda}$. But at this early stage he was evidently uneasy about this. He simply claimed that, without "supernatural" (*übernatürliche*) masses and only by keeping $\Lambda$, he has reached a conclusion as to the inertia of a test body analogous to Einstein's.

A few days after de Sitter's letter, Einstein did not show as much caution as de Sitter regarding the physical meaning of the concepts, for, taking de Sitter at his word, he investigated his result as if it were a new form of the spatially closed universe. The difficulty lay, of course, in the theory of time. In a new letter, Einstein pointed out a troublesome singularity occurring at a finite distance in model $B$.

Comparison between the two models implied some symmetry, and de Sitter had used imaginary time in $B$ in order to compare the results directly. But because Einstein tried to make more physical sense of what de Sitter had found, he used real time so that the imaginary hypersphere becomes a real hyperboloid. Intuitively, this real form of the de Sitter universe is just Einstein's original cylinder twisted about its axis because of the curvature of time. Viewed thus, the de Sitter universe is now a one-sheet hyperboloid, with its two ends open in the time direction. But Einstein believed that the stereographic projection of the hyperboloid reveals that it is affected by a singularity. As he put it in a letter to de Sitter of March 24: "The surface with singular properties is ... in finite physical space. Therefore it seems to me that your solution corresponds to no physical possibility" (EA 20-547).

It is this argument that prompted de Sitter to reconsider the implications of model $B$. Einstein's discrediting of its physical significance would simply have tended to reinforce de Sitter's conviction that this significance was not settled in any case, were it not for the fact that de Sitter discovered a fallacy in the argument. For Einstein was wrong in his view of the nature of the singularity, so it is ironic that de Sitter's reply was concerned with an attempt to promote the physical interpretation of $B$. And from then on, de Sitter began to suspect that his model could sustain such an interpretation. Surprisingly, in the following months a whole series of different events had the effect of confirming this suspicion.

The fallacy in Einstein's argument is that it confuses two types of singularity: one is produced by the coordinates, the other by the intrinsic features of space-time. De Sitter claimed that the singularity pointed out by Einstein is of the coordinate kind. Einstein was not to be won over: "I have not yet understood your remark about the singularity at finite distance as being only apparent, due to the choice of the coordinates... I am waiting for clarification" (Einstein to de Sitter, April 14, 1917, EA 20-553). To which de Sitter replied: "The question of the discontinuity... is properly speaking not very interesting." (April 18, EA 20-554). Einstein did not give up so quickly, and came back with new arguments some two months later: "Your four-dimensional continuum lacks the property of having all its points similar." (June 14, EA 20-556).

The first stage of the dispute is reprised in two footnotes written by de Sitter, in his 1917a (p. 1220) and 1917b (p. 229). Through projection, the points at infinity of the hyperboloid are brought back at finite distance, but this has nothing to do with a place where space would be so distorted that its geometrical properties break down. As de Sitter said in a letter to Einstein of April 18, "the natural distance of those points from the origin is... *not* finite, contrary to what you thought at first" (EA 20-554). But then he made a curious mistake in asserting that his hyperboloid has two sheets instead of one, and that he was not really bothered by these two sheets since "the formulae embrace both sheets, but only one of them represents the actual universe" (de Sitter 1917a, p. 1220). This mistake determines the second stage of the dispute, because Einstein was very puzzled by this strange and unjustified argument. It embodies yet another way of seeing that something is wrong with the geometry of model *B*: by looking not at the periphery, but rather at the center. Einstein now believed that his argument was capable of transcending the distinction between coordinates and structure: there is a paradox with time because "the spatial extent of your world depends on $t$ in a peculiar manner. For sufficiently early times one can put a rigid circular hoop into your world which has no place in it at time $t = 0$" (June 14, EA 20-556). This is due to the fact that, if the world is only one of the two sheets, it is difficult to find a justification for the "privilege" of either one; the two sheets must be divided by a puzzling line $t = 0$ if they are to be symmetrical, and nothing seems to exist at $t = 0$ since the line divides two equally possible worlds.

At this stage, it becomes extremely difficult to reconstruct the actual significance of the arguments, since a certain number of gaps appear in the correspondence available to us. In another letter (June 22, 1917), Einstein discussed the problem anew. Perhaps we should understand this discussion as a reaction to de Sitter's later correction—model *B* is not in fact a two-sheet, but a one-sheet hyperboloid. (De Sitter himself, it should be noted, realized the mistake only when Ehrenfest pointed it out to him.) At any rate, Einstein was now eager to make it clear that the privileged status of the central point is certainly independent of all coordinates. Consider the light-cone originating from any point on the hyperboloid, and let $H$ stand for "the infinitely remote (in natural measure)." All light-cones form "light-surfaces" (*Lichtfläche*), that is, they follow the hyperbolic shape of the four-dimensional world. The privilege of the central, null point comes from the fact that the light-surface originating from that point does *not* intersect $H$; it approaches it only asymptotically. As a result, Einstein says, "this point is de facto preferred.... This is of course no disproof, but the circumstance irritates me" (EA 20-559). In today's terminology, Einstein was irritated by the fact that the particle horizon (which divides all actually observable particles for a given observer into two classes, one containing the already observed particles and the other all those yet to come within reach) is also an event horizon (that is, the particles of the latter class will *never* come within reach).

In spite of its clumsy terminology, this debate between Einstein and de Sitter certainly represents the first explicit recognition of the problem of horizons in modern cosmology. What bothered Einstein in this early discussion is that the real form of the de Sitter universe revives the old problem of boundary conditions. Indeed, whatever the justification for the change from set (8) to (9), the curvature of time appears to go hand in hand with the positing of definite boundary conditions for time, namely, a natural origin of time. Thus, Einstein may well have seen that, far from representing the last remnant of a classical conception of the world, linear time in model $A$ is the only property of time that does away *completely* with boundary conditions. As he wrote to Ehrenfest as early as February 14, 1917, it was quite ironic to realize that finally all the requirements of relativity would be satisfied with "a new quasi-absolute time and a preferred coordinate system" (EA 9-398).

## 6. Mathematical and Physical Postulates of Cosmology

In fact, Einstein's discussion of singularities was but one aspect of a broader problem. Already in his first reaction to model $B$ (the letter to de Sitter of March 24), he was raising the issue of the implications of Mach's principle. This part of the letter is so crucial that de Sitter asked Einstein's permission to quote an excerpt as a postscript to his article on the relativity of inertia (de Sitter 1917a, p. 1225). Einstein had written: "In my opinion it would be unsatisfactory to think of the possibility of a world without matter. The field of $g_{\mu\nu}$ ought to be conditioned by matter, otherwise it would not exist at all." This line of reasoning was called by de Sitter the "*material postulate* of the relativity of inertia." But model $B$ shows that the relation of $\Lambda$ to the world-matter is not a necessary one (postcard of April 18). This led him to conclude that any connection on a large scale between physics and mathematics has to be relaxed: "We can also abandon the postulate of Mach, and replace it by the postulate that at infinity the $g_{\mu\nu}$, or only the $g_{ij}$ of the three-dimensional space, shall be zero, or at least invariant for all transformations" (de Sitter 1917a, p. 1222). De Sitter called this condition, in contrast to the material postulate, the "*mathematical postulate* of the relativity of inertia."

Einstein seemed to think that de Sitter's skepticism blocked the way to any understanding of the origin of inertia. Replying to de Sitter on April 14, he asked that all belief be suspended: "Conviction is a good mainspring, but a bad regulator." Reflection should be confined to the actually available possibilities, whereas de Sitter was indulging in extrascientific speculations when he wrote: "The question thus really is: how are we to extrapolate outside our neighborhood? The choice can thus not be decided by physical arguments, but must depend on metaphysical or philosophical considerations, in which of course also personal judgment or predilections will have some influence" (de Sitter 1917a, p. 1222). By contrast, in the earlier discussions on relative rotation, de Sitter had strongly opposed philosophical concerns getting mixed up with particular solutions to the generally covariant field equations. The true issue, however, was all a matter of what changes need to be made to our basic conception of physics when we pass from local questions to the question of cosmology.

The opposition between a local perspective and the cosmic viewpoint was first discussed in a note that Einstein added to his letter of March 24. Here he claimed that the argument based upon small stellar velocities is still to be preferred to that based upon the absence of a violet shift in the spectral lines if the static nature of the universe is to be established. As he put it, "if we come up with the hypothesis of a mechanical, quasi-stationary behavior for the matter," the small velocities prove that large differences of potential cannot occur at all. As a result, the observation of spectral lines may play some role in determining the value of the cosmological constant. For if $\Lambda$ were zero, the mean density of matter would become such that considerable violet shifts would occur. De Sitter was totally dissatisfied with these explanations. In the margin of Einstein's letter, he wrote: "This train of reasoning is completely erroneous ... it supposes that the average density here remains the same until infinity. This is certainly not true." In his reply to Einstein (April 1), he said:

> Of the world we have only a snapshot [*Momentphotographie*]. Because we do not see many changes in the photographs, it would be too easy to conclude that it will always remain the same. Even the Milky Way does not seem to be a stable system, for the stars [the visible ones, not the world-matter] are obviously *not* distributed homogeneously.

The next exchange of letters (April 14 and 18) brought to light what de Sitter called "the difference in belief" between Einstein and himself. Einstein repeated that the constant $\Lambda$ is offered only as a tentative extension of general relativity, observation having the last word on whether it should disappear or not. And among these observations, the fact that the spectral lines are a function of the distance from us has a decisive significance. De Sitter, on the other hand, spoke of the actual *value* of $\Lambda$ as being beyond the possibilities of observation; he was not concerned with its *existence*, which he seems to take as part and parcel of Einstein's cosmological considerations. The divergence between the two authors was now complete, since de Sitter was willing to extrapolate in time but not in space, whereas Einstein extrapolated in space but not in time; de Sitter assumed $\Lambda$ to be part of the theory of the universe, whereas Einstein expected observation to settle the question of its existence. In fact, de Sitter reserved the terms "existence" and "nonexistence" for the world-matter as opposed to what he called "ordinary matter."

This, indeed, is how he understood the opposition between the local and the cosmic. His distinction between physical and mathematical postulates was now translated into this opposition:

> To the question: If all matter is supposed not to exist, with the exception of one material point which is to be used as a test-body, has this test-body inertia or not? The school of Mach requires the answer *No*. Our experience however very decidedly gives the answer *Yes*, if by "all matter" is meant all ordinary physical matter: stars, nebulae, clusters, etc. The followers of Mach are thus compelled to assume the existence of still more matter: the world-matter. (de Sitter 1917a, p. 1222)

Were the world-matter some kind of ideal arrangement of ordinary matter, it is clear that the de Sitter universe would be *completely empty* by virtue of its equations. What de Sitter wanted to prove is that, if the world-matter is to exist at all, it cannot be such an arrangement of ordinary matter, i.e., the former is necessarily added to the latter. Suppose that the only existing matter is the ordinary kind. The circumstances under which this ordinary matter is supposed not to exist are identical with those found in the exterior metric, where the matter surrounding a unique body such as the sun is neglected. The inertial field created by the sun exists, even though the sun is artificially isolated from the planets. A Machian conception is thus forced to include other, remoter matter: were this distant matter done away with, the inertial field around the sun could not exist at all. De Sitter stressed, not without irony, that "this world-matter ... serves no other purpose than to enable us to suppose it not to exist" (de Sitter 1917a, p. 1222), and in 1917c he added: "and to assert that in that case there would be no inertia" (p. 5). In short, according to Mach's theory, the fiction by which we suppress all ordinary matter *is not the same* as the fiction by which the world-matter is supposed not to exist. However, there is nothing that allows us to believe that the two fictions differ, save the belief that this world-matter is more than a mere idealization.

Einstein did not like this reasoning. He threw de Sitter's objection back at him, by showing that there is no world-matter outside the stars (June 14, 1917). He claimed that $\rho$ is, as far as he could see, nothing more than a uniform distribution of the existing stars. De Sitter was of course unconvinced. First in a letter of June 20 (EA 20-557), and then with additional detail in de Sitter 1917b, he offered an in-depth analysis of the true difference between $A$ and $B$, based on the implications arising from their physical properties rather than from "philosophical predilections." Einstein was so impressed that, in a reply dated June 22, he wrote how delighted he was that de Sitter and he now shared an awareness of common problems in cosmology. He praised de Sitter for his ability to scrutinize what was "intellectually possible" in the construction of the field in the large.

De Sitter's starting point was the necessity of overcoming the contradiction that, in $A$, we have either $\rho = 0$ or $g_{44}$ = constant. How is it possible to have, at the same time, a stationary equilibrium and ordinary matter, since ordinary matter is primarily responsible for deviations from $g_{44}$ = constant? This is possible, de Sitter argued, only if the material tensor is modified so as to allow for the existence of internal pressure and stress. The pressure and stress would compensate for the deviations from $g_{44}$ = constant. But this is only valid if the world-matter is identified with a continuous fluid; were the world-matter to be compared to separated material points, that is, were it to consist of discrete entities, the internal forces would not have the desired effect and would prevent this world-matter from remaining at rest. In either case, the metric $A$ is inappropriate, since it is supposed to describe a pressureless, stationary world-matter.

Of course, ordinary matter consists of discrete entities. In his synthesis prepared for the end of 1917, de Sitter summed the situation up in the following terms: "In $A$ there is a world-matter, with which the whole world is filled, and this can be in a state of equilibrium without any internal stresses or pressures, if it is entirely homogeneous and at rest" (de Sitter 1917c, p. 20). In this case, where ordinary matter is dispensed with, it is impossible to go back to it. In contrast, the occurrence of ordinary matter in system $B$ makes it impossible to go back to the world-matter:

> In $B$ there may, or may not, be matter, but if there is more than one material particle these cannot be at rest, and if the whole world were filled homogeneously with matter this could not be at rest without any internal pressure or stress; for if it were, we would have the system $A$, with $g_{44} = 1$ for all values of the four coordinates. (de Sitter 1917c, p. 20)

Thus, system $B$ is *never* equivalent to system $A$, whatever the quantity or nature of matter (whether continuous or discrete) that is introduced in system $B$. De Sitter's discovery is that Einstein's universe can in no way be compared to an extended ordinary universe. As he had claimed earlier about the reality of distant masses, de Sitter now claimed that the new world-matter "takes the place of the absolute space in Newton's theory, or of the 'inertial system.' It is nothing more than this inertial system materialized" (de Sitter 1917c, p. 9). De Sitter's skepticism acted at that level: he was willing to reject both $A$ and $B$, to return to the original field equations without a cosmological constant and without generally invariant $g_{\mu\nu}$ at infinity, and so to leave unexplained the mystery of inertia. True, the whole difficulty of conceiving of system $B$ as somehow derivable from system $A$ comes from the status of the cosmological constant. An overall curvature of space is possible without matter, provided

Λ is not set to zero. Thus, the fulfilment of what de Sitter referred to as the mathematical postulate "is brought about by the introduction of the term with Λ, and not by the world-matter which, from this point of view, is not essential" (p. 6). It is only in system $A$ that the field equations are not satisfied when the density is zero. The consequence is that "supposing it [the world-matter] not to exist thus appears to be a logical impossibility; in the system $A$, the world-matter *is* the three-dimensional space, or at least is inseparable from it" (de Sitter 1917a, p. 1222). This is the curious logic of Einstein's universe: the only basis for asserting that the model is logically consistent (as Einstein dearly wished) is that, if we give credence to the supposition that its matter does not exist (system $B$), there is no bridge allowing us to go back to the model in its original form.

## 7. Einstein's New Reply

In the current state of the records detailing the correspondence between Einstein and de Sitter, we have to wait until April 10 of the following year, 1918, before there is the trace of a new letter from de Sitter (EA 20-565). This gap of almost a year makes it very difficult to evaluate how he gradually came to accept the existence of a higher form of connection between the physics and the geometry of his own model. Two letters and three postcards from Einstein during this period are the only guides here. Fortunately, in one of them, de Sitter made notes in the margin. On June 28, 1917 (EA 20-560), Einstein reacted to the conclusions de Sitter had drawn in his 1917b. The reaction was as short as it was lucid: Einstein realized that his earlier speculations about the occurrence of a singularity in model $B$ were by no means completely wrong after all. There was now a new way of seeing things, which, Einstein believed, made his intuition an inescapable certainty.

The discussion reached a pitch of excitement when Einstein wrote to de Sitter on July 31: "Your system is not a physical possibility" (EA 20-561). Take the expression for the energy of a material particle, $m\sqrt{g_{44}}$. In the metric $B$, $dt^2$ is affected by a coefficient $\cos^2(r/R)$. Therefore, at the distance $r = (1/2)\pi R$, this particle has simply lost its energy and it can no longer exist. This argument refers to an interpretation of the horizon of visibility that de Sitter had worked out in his model $B$ (see 1917c, pp. 16–17); in this model, there is a temporal periphery in the sense that, at the distance $r = (1/2)\pi R$, all physical phenomena cease to have duration. De Sitter explained that results of that kind "sound very strange and paradoxical," but the reason is only that:

> They are, of course, all due to the fact that $g_{44}$ becomes zero at $r = (1/2)\pi R$. We can say that on the polar line the four-dimensional time-space is reduced to the three-dimensional space: *there is no time*, and consequently no motion. (p. 17)

But Einstein now argued that it is certainly hard to maintain that the periphery itself exists. A week after this, on August 8, he managed to find a bridge between models *A* and *B*. The tendency of $g_{44}$ to decrease to zero is, as he now said, analogous to the phenomenon of clocks slowing down near large masses. In other words, all the matter of the de Sitter universe has been concentrated on the equator of the spherical space. The nonphysicality of the model is therefore due to the following fact: "The heterogeneity of the different points of space is not, in this conception, an 'autonomous' property of space" (EA 20-562). Einstein preferred an energy that is everywhere finite.

It is here that de Sitter added two comments in the margin of Einstein's letter. The first is a response to the supposition of matter concentrated at the equator: "That would be distant masses yet again." His position is quite understandable, and a general comment by de Sitter is pertinent: "If $g_{44}$ must become zero for $r = (1/2)\pi R$ by 'matter' present there, how large must the 'mass' of that matter then be? I suspect $\infty$! We then adopt a matter [which is] *not* ordinary matter." And he concluded with a pugnacious pun: "It is a *materia ex machina* to save the dogma of Mach." Just as de Sitter had given a version of the Einsteinian model in *his* own image, it seems that Einstein was now taking over de Sitter's model as part of his own imperial theme.

Alas, de Sitter's reply is not available, but a new postcard from Einstein on August 22nd (EA 20-564) seemed to widen the gap between them. Einstein reiterated his view that all points of the metric *B* are spatially identical, but that the frequency of a clock changes with the place, so that it reaches the inadmissible value of zero at the equator. As a result, the total energy of a particle would indeed disappear there. Einstein gave a clear-cut presentation of his objections in a paper that was published somewhat later by the Berlin Academy (Einstein 1918b). There he said that a fundamental requirement of the equations of general relativity, even when they are supplemented by the cosmological constant, is an absence of discontinuity at any finite distance. The $g_{\mu\nu}$ and their derivatives should be everywhere continuous and differentiable, but this is not the case with the de Sitter metric. The determinant of this metric is

$$g = R^4 \sin^4 \frac{r}{R} \sin^2 \psi \cos^2 \frac{r}{R},$$

and continuity requires that this determinant be everywhere different from zero. In fact, $g$ already vanishes for $r = 0$ (as well as for $\psi = 0$). However, this is only an apparent violation of continuity, since the choice of a new reference system (a new center) transforms it away. But the determinant also vanishes for $r = (1/2)\pi R$ and, as de Sitter had demonstrated himself, this distance from the origin represents a finite distance in coordinate measure. What Einstein argued is that no change of coordinates can prevent this periphery from being a singularity—it is a singularity of the field itself. If no inertial field is possible without matter, then all the matter must be concentrated at the singularity, since this can account in physical terms for why the clocks slow down.

It was bound to take some time for de Sitter to draw up a consistent, i.e., non-Machian interpretation of the singularity. In his reply to Einstein's paper

(April 10, 1918), de Sitter wrote that he did not know whether the matter of his model is concentrated at the equator, that this was something on which he could not commit himself. But the actual constitution of the periphery and the relation of the periphery to the center (that is, to observability) were two different things. He referred to a remark that he had included in his 1917c, which he thought of as a decisive reply. There is no possible connection between the center and the periphery because:

> A particle which has not always been on the polar line can ... only reach it after an infinite time, i.e., it can never reach it at all. We can thus say that all paradoxical phenomena (or rather negations of phenomena) which have been enumerated above can only happen after the end or before the beginning of eternity. (de Sitter 1917c, pp. 17–18)

De Sitter himself was never completely satisfied with this counterargument, and he formulated a more detailed version of it in a new article. It would appear that Einstein wanted to show that the singularity in the metric $B$ did not really differ from the better-known, supposedly intrinsic singularity at $r = 2m$, which occurs in the Schwarzschild exterior metric. But, de Sitter argued, the postulate of continuity enunciated by Einstein

> is a *philosophical*, or metaphysical postulate. To make it a *physical* one, the words "*all points at finite distance*" must be replaced by "*all physically accessible* points." And if the postulate is thus formulated, my solution $B$ does fulfill it. (de Sitter 1917d, p. 1309)

## 8. The Outcome of the Einstein–de Sitter Controversies

From the sequence of controversies between Einstein and de Sitter (ranging from the first epistemological debate in 1916 up to these cosmological speculations of 1918), it is clear that there is a seemingly irreconcilable conflict between what is regarded as physics and what is regarded as philosophy. Although this conflict does something to explain de Sitter's caution, his skepticism, and his humility, it has by no means hindered the practical development of ideas; what originated in the two protagonists' minds from then on gained a quasi-autonomous status when it was taken up by other scientists, who called for an increasingly detailed interpretation of the available models as well as for a radical solution to the problem. It took some fifteen years for the problem of singularities to be properly understood thanks to such people as Klein (1918), Lanczos (1922, 1923), Weyl (1918, 1923, 1930), Eddington (1920, 1923, 1930), Friedmann (1922, 1924), Lemaître (1925, 1927), and Robertson (1928, 1929). The key to its resolution was the recognition that changes of space-time coordinates (rather than of space alone) transformed what had been taken for intrinsic singularities into apparent ones. Space-time coordinates create a truly dynamic picture of the universe, one that overcomes most of de Sitter's obsessions with such oppositions as motion and rest, discrete and continuous matter distributions, and the general tension between physics and geometry. Where de Sitter's feeling for dichotomies allowed him to see contradictions in Einstein's model as well as those between models *A* and *B*, it was Einstein's uncovering of a possible contradiction within model *B* that was decisive in determining a new strategy. Indeed, as de Sitter was quick to recognize, the fact that $g_{44}$ diminishes with increasing distance from the origin implies that "in *B* the lines in the spectra of very distant objects must appear displaced towards the red" (de Sitter 1917b, p. 235). He implicitly assumed that there is a kind of equivalence between this metric shift and a Doppler shift: he took for granted, somewhat uncritically, that if the displacements are systematic, the velocities are therefore only apparent and need be interpreted only as properties of the inertial field. It was left for Eddington and Weyl to see the implications of this assumption in explicit terms, and this, in turn, paved the way for the formulation of a truly dynamic space-time (see Kerszberg 1986). It is well known that, in the Friedmann-Lemaître generalization in terms of a nonstatic metric, the Einstein and de Sitter universes appear as two limiting cases of an evolution that leads from the former to the latter. In this last section I would like to examine how our two authors developed their ideas before the advent of such a generalization.

On May 5, 1920, Einstein returned to Leiden and gave a lecture in which he argued that departures from Euclidean space must necessarily take place even in the case of the *smallest* density of matter (Einstein 1920, p. 20). In actual fact, the main purpose of the paper was to show the implications of a new meaning for the ether in relativistic physics: this is the notion of a field, a physical medium "which is itself devoid of *all* mechanical and kinematical qualities, but helps to determine mechanical (and electromagnetic) events" (p. 19). The point was that this ether conditions the inert masses but is also conditioned by their behavior, which leads to the recognition of an interaction between metric and matter rather than a complete subordination of the former to the latter. Clearly, at the time of the "Cosmological Considerations" (1917), Einstein would have said that the metric quantities are *totally* conditioned by matter.

De Sitter should obviously have been satisfied with Einstein's new ideas, but he chose to dispute Einstein's brief statement about the density of matter. He argued that, in order to arrive at a consistent conception of the relation of the curvature of space to the smallest quantity of matter, there is another hypothesis that Einstein had not mentioned and that had to be introduced— the notion of a statistical equilibrium for matter. He now compared the world-matter at rest in the cylindrical model with a state of statistical equilibrium. He did not explicitly develop what he meant by this, but instead immediately applied himself to criticizing it:

> Now the possibility of statistical equilibrium of large portions of the universe is, to my mind at least, by no means self-evident, or even probable. The idea of evolution in a determined sense appears to me to be rather opposed to the actual existence, if not to the possibility, of equilibrium. (de Sitter 1921, pp. 866–867)

The notion of statistical equilibrium is enmeshed with evolutionary processes in a determinate direction of time. Such a laconic comment makes it difficult to understand what is here meant by evolution. But in a letter to Einstein of November 4, 1920 (EA 20-571), he explained that one of the least attractive features of Einstein's model was the possibility for an observer to perceive several images of one and the same star. Those images depict the star at epochs separated by intervals of time during which light travels once entirely around the world. Thus, the spherical space would be filled with a multitude of "ghost images." Of course, this kind of difficulty with spherical space could be overcome with the aid of purely topological considerations. There is no doubt that both de Sitter and Einstein knew very well of these considerations, and in fact they had already subscribed to Felix Klein's views (see Klein 1918), according to which Einstein's model is best represented by means of an elliptical space. But the topology only modifies the magnitude of the relevant quantities, and what was afoot here was the attempt to account for physical predictions in general. De Sitter calculated that the universe should present a greater number of young stars than old ones, but observation suggests the contrary. He considered two possible explanations for this: either the process of star creation is practically over, or the first phases of star formation develop more quickly than subsequent ones. Favoring the latter, that is, the idea that the universe is endlessly involved in the evolutionary process of its fundamental constituents, he tended to identify the existence and activity of these processes with the reality of movements that tend to disturb statistical equilibrium. In Einstein's metric there is no provision for taking into account the pressures and internal forces at work any more than there is any allowance for large-scale motions; as a result, all mechanisms of evolution that initiate motion have to be thought of as neutralized, and statistical equilibrium is precisely a state that borders on a final one.

A response from Einstein (dated November 24, 1920), followed by de Sitter's reply five days later, on the 29th (EA 20-573), represents the final episode in the controversy between the two scientists. Einstein rejected the arguments concerning the possibility of the sun's ghost images in model $A$: the matter distribution is not quite homogeneous, so such images could never take shape. To this de Sitter replied, basing himself on the (at the time) recently proved existence of extragalactic nebulas: "The world is incredibly empty." Because of this relative emptiness, the absorption of light by matter is certainly not sufficient to prevent the formation of images. But Einstein's criticism of the statistical argument was more devastating. It proceeded quite simply. Consider an isolated system: according to the virial theorem of Newtonian mechanics, the average kinetic energy of each particle is equal to half of the total potential energy; consequently, if the system is to be stationary, it cannot be so in terms of the kinetic theory of gases, which is statistical. Only gravitational forces are required to hold the system together and to prevent it from dispersing itself. If it could be demonstrated by observation that stellar velocities are compatible with an equilibrium of this type, Einstein wrote, then the value of the cosmological constant could be estimated. Dealing then with model $B$, Einstein for the first time stated quite unambiguously that it predicts a repulsion that increases with distance: this is the effect of retaining $\Lambda$ in a space-time whose structure is conceived of as independent of the presence of matter. In these conditions, Einstein suggested, "would it not be more satisfactory to take $\Lambda = 0$," since no value for the mean density was available? Einstein thought that, in $B$, setting $\Lambda = 0$ would be the only way to avoid cosmic repulsion, and thus also the only way *to preserve inertia as an interaction between bodies*. It other words, cosmic repulsion would contradict the concept of inertia as interaction, and here may lie one of the motivations for Einstein's resistance to early suggestions of an expanding universe (see, in particular, his ambiguous replies to Friedmann's pioneering work in 1922a and 1923).

In his reply of November 29, 1920, de Sitter immediately stressed the fact that the observations of star velocities in the Milky Way were certainly in Einstein's favor: it seemed that the gravitational pull alone was enough to account for the stability. But the basic problem remained unresolved: "In order to know how the Milky Way holds together, $\Lambda$ is of no use," for if $\Lambda$ is derived from the mass of the Milky Way, space becomes too small to contain it, and if $\Lambda$ is derived from the density of matter around the sun, the cosmos becomes so large that the Milky Way could no more hold together than if $\Lambda$ were left out. In view of the useleness of $\Lambda$, de Sitter once more underlined his preference for the theory without cosmological constant. "But ... the existence of an apparent force of repulsion seems to be truly confirmed!" he exclaimed immediately. He specified very carefully, however, that the apparent force of repulsion is nothing but an effect of the metric. He also added that the extragalactic nebulas, whose velocities are radial, seem to be scattered randomly throughout the universe. In other words, for de Sitter, the existence of extragalactic nebulas meant that it is ordinary matter that shows its true distribution, not the world-matter that would start making itself visible.

The disagreement between Einstein and de Sitter was thus never more extreme than at this time. The subsequent evolution of observational cosmology has repeatedly proved, in a virtually systematic way, that with regard to the distribution of matter at large Einstein was right and de Sitter was wrong. But if Einstein was right, it was clearly not for the reasons he gave on this occasion, since he confined his argument specifically to an analysis of the Milky Way, where the configuration of stars presents a conspicuous lack of uniformity. In fact, he did not even bother to extrapolate from observations, leaving this up to de Sitter, who had always been reluctant to do so. True, Einstein could see nothing decisive in de Sitter's reports on the reality of extragalactic systems. Even the possibility of systematic movements of recession did not bother him, because, as he argued in his famous lecture of January 27, 1921, "Geometry and Experience," the only experimental technique suitable for proving the spatial finiteness of the universe depends not on the idea of homogeneity, but on the static nature of the universe (Einstein 1921a, pp. 43–45). Certainly, de Sitter himself did not interpret the new discovery as indicating movements, and in any case, Einstein believed that cosmic repulsion as a consequence of $\Lambda$ occurred *only* in model *B*. The upshot was that Einstein never again mentioned the cosmological constant as a "real" feature of *his* cylindrical model, neither in this address nor in the lectures delivered at Princeton in May 1921. Instead, he talked about theoretical proofs lending credence to spatial finiteness, using such terms as "mass-density of negative sign" or "hypothetical pressure" that cannot vanish (Einstein 1921a, p. 44, 1921b, pp. 117–118). Furthermore, he explicitly inferred a relation between pressure and negative density, thereby reverting to the first premises of the problem he tackled in 1917, when he postulated a cosmological constant to prevent the effect of a negative pressure.

These 1921 addresses point to Einstein's new awareness of the problems, an awareness that defined his attitude toward cosmology for the next ten years. What is noticeable is that he gradually came to seek a unification of various areas of physics, and in this project the achievements so far in cosmology formed the background that Einstein had no wish to modify significantly. Thus, he hoped to wrest from a better theoretical knowledge of the electromagnetic field the complete vindication of the pressure term, the physical nature of which remained unclear (Einstein 1921b, p. 117). As to the application of the principles of Riemannian geometry "outside the domain of their physical definitions," he argued that "success alone can decide as to the justification of such an attempt" with regard to the submolecular domain, but he found that things are "less problematical" when one is dealing with the other extreme, cosmology (Einstein 1921a, p. 40).

On the other hand, if Einstein, the theoretician, was right about the observational knowledge that could be derived from the universe, de Sitter, the astronomer, was right when he placed the cosmological debate at the level of an examination of the nature of the concepts implicated in the new physics. Already in his letter of November 4, 1920, de Sitter had argued that "the ether carries the inertia.... The field itself is the real." As he wrote these lines, he was only foreshadowing Einstein's later attitude towards a unified field theory; it became obvious that the remaining duality of metric and matter had to be overcome by deriving the properties of matter from those of space-time, rather than the other way around. But a concept of the ether in general relativity is extremely hard to understand, for all that can be said about it is what it is not: "This ether may not be thought of as endowed with the quality characteristic of ponderable mass, as consisting of parts which may be tracked through time" (Einstein 1920, pp. 23–24). And in a remarkable letter to Arnold Sommerfeld, dated November 28, 1926, Einstein criticized Eddington's conception, which presented the theory of relativity in such a way that it might lead scientists to believe that it is a logical necessity, saying that "God could very well have chosen an ether completely at rest, rather than a relativistic ether" (Hermann 1968, pp. 109–111). An ether at rest—something like the one Lorentz had imagined before the advent of the theory of relativity—would have been inevitable, Einstein went on to argue, if God "had created an ether according to de Sitter's model, which is essentially independent of matter."

We can see how, with the passage of the years, Einstein lost much of his early confidence. Once everything was reduced to divine choice, the logical consistency of general relativity that Einstein had dreamed of when he created cosmology in 1917 seemed forever out of reach. And we may reflect on the fact that de Sitter, Einstein's early sparring partner, chose to remain silent throughout the rest of the 1920s. He made no public statement about cosmology until 1930, when other scientists had found the supposed solution to his problem. The silence has its own significance because it testifies to the growing relevance of purely conceptual considerations that have no immediate place within the workings of the formal apparatus.

NOTES

[1] My sources are the Einstein postcards and letters that have been kept at the Leiden Observatory. Copies of most of these items are in the Einstein Archive at the Hebrew University in Jerusalem. I wish to thank Prof. John Stachel (Boston University, Center for Einstein Studies) for his assistance, as well as Dr. E. Dekker (Museum Boerhaave, Leiden) and Prof. H. van der Laan (Leiden Observatory) for permission to consult the de Sitter archives.

[2] This is a conversation of September 29, reported in an addendum to the paper on the relativity of rotation (de Sitter 1916a, p. 531). A short commentary on these debates can be found in Peebles 1980b.

[3] A year later, Erich Kretschmann argued that general covariance has no physical meaning by itself: any equation can be expressed in a generally covariant form. In his reply to Kretschmann, Einstein stated that general covariance had played a powerful *heuristic* role in the establishment of the theory of general relativity.

REFERENCES

de Sitter, Willem (1916a). "On the Relativity of Rotation in Einstein's Theory." *Proceedings of the Section of Sciences. Koninklijke Akademie van Wetenschappen te Amsterdam* 19: 527–532.

———— (1916b). "On Einstein's Theory of Gravitation and Its Astronomical Consequences. Second Paper." *Monthly Notices of the Royal Astronomical Society* 77: 155–183.

———— (1917a). "On the Relativity of Inertia." *Proceedings of the Section of Sciences. Koninklijke Akademie van Wetenschappen te Amsterdam* 19: 1217–1225.

———— (1917b). "On the Curvature of Space." *Proceedings of the Section of Sciences. Koninklijke Akademie van Wetenschappen te Amsterdam* 20: 229–242.

———— (1917c). "On Einstein's Theory of Gravitation and Its Astronomical Consequences. Third paper." *Monthly Notices of the Royal Astronomical Society* 78: 3–28.

———— (1917d). "Further Remarks on the Solutions of the Field Equations of Einstein's Theory of Gravitation." *Proceedings of the Section of Sciences. Koninklijke Akademie van Wetenschappen te Amsterdam* 20: 1309–1312.

——— (1921). "On the Possibility of a Statistical Equilibrium of the Universe." *Proceedings of the Section of Sciences. Koninklijke Akademie van Wetenschappen te Amsterdam* 23: 866–868.

Eddington, Arthur Stanley (1920). *Space, Time and Gravitation*. Cambridge: Cambridge University Press.

——— (1923). *The Mathematical Theory of Gravitation*. Cambridge: Cambridge University Press.

——— (1928). *The Nature of the Physical World*. Cambridge: Cambridge University Press.

——— (1930). "On the Instability of Einstein's Spherical World." *Monthly Notices of the Royal Astronomical Society* 90: 668–678.

Einstein, Albert (1916). "Die Grundlage der allgemeinen Relivätstheorie." *Annalen der Physik* 49: 769–822. Translated as "The Foundation of the General Theory of Relativity." In Lorentz, et al. 1923, pp. 109–164.

——— (1917). "Kosmologische Betrachtungen zur allgemeinen Relativitätstheorie." *Königlich Preussische Akademie der Wissenschaften* (Berlin). *Sitzungsberichte*: 142–152. Translated as "Cosmological Considerations on the General Theory of Relativity." In Lorentz, et al. 1923, pp. 175–188.

——— (1918a). *Über die spezielle und die allgemeine Relativitätstheorie (Gemeinverständlich)*, 3rd ed. Brauschweig: Friedrich Vieweg & Sohn. English translation *Relativity: The Special and the General Theory*. R.W. Lawson, trans. London: Methuen, 1920; New York: Holt, 1921. [The 1918 3rd German edition was the first in which the cosmological chapters were incorporated.]

——— (1918b). "Kritisches zu einer von Herrn de Sitter gegebenen Lösung der Gravitationsgleichungen." *Königlich Preussische Akademie der Wissenschaften* (Berlin). *Sitzungsberichte*: 270–272.

——— (1919). "Spielen Gravitationsfelder im Aufbau der materiellen Elementarteilchen eine wesentliche Rolle?" *Preussische Akademie der Wissenschaften* (Berlin). *Sitzungsberichte*: 349–356.

——— (1920). *Äther und Relativitätstheorie. Rede gehalten am 5. Mai 1920 an der Reichs-Universität zu Leiden*. Berlin: Springer. Translated as "Ether and Relativity." In Einstein 1922b, pp. 3–24.

——— (1921a). "Geometrie und Erfahrung." *Preussische Akademie der Wissenschaften* (Berlin). *Sitzungsberichte*: 123–130. Translated as "Geometry and Experience." In Einstein 1922b, pp. 27–56.

——— (1921b). *The Meaning of Relativity: Four Lectures Delivered at Princeton University*. E.P. Adams, trans. London: Methuen, 1922.

——— (1922a). "Bemerkung zu der Arbeit von A. Friedmann: Über die Krümmung des Raumes." *Zeitschrift für Physik* 11: 326.

——— (1922b). *Sidelights on Relativity*. W. Perrett and G.B. Jeffery, trans. London: Methuen.

——— (1923). "Notiz zu der Arbeit von A. Friedmann 'Über die Krümmung des Raumes.'" *Zeitschrift für Physik* 16: 228.

Eisenhart, Luther P. (1926). *Riemannian Geometry*. Princeton, New Jersey: Princeton University Press.

Friedmann, Aleksander Aleksandrovich. (1922). "Über die Krümmung des Raumes." *Zeitschrift für Physik* 10: 377–386.

——— (1924). "Über die Möglichkeit einer Welt mit konstanter negativer Krümmung des Raumes." *Zeitschrift für Physik* 21: 326–332.

Hermann, Armin (1968). *Albert Einstein–Arnold Sommerfeld. Briefwechsel.* Basel and Stuttgart: Schwabe.
Kahn, Carla, and Kahn, Franz (1975). "Letters from Einstein to de Sitter on the Nature of the Universe." *Nature* 257: 451–454.
Kerszberg, Pierre (1986). "Le Principe de Weyl et l'Invention d'une Cosmologie Non-Statique." *Archive for History of Exact Sciences* 35: 1–89.
Klein, Felix (1918). Über die Integralform der Erhaltungssätze und die Theorie der räumlich-geschlossenen Welt." *Königliche Gesellschaft der Wissenschaften zu Göttingen. Nachrichten*: 394–423.
Lanczos, Cornelius (1922). "Bemerkung zur De Sitterschen Welt." *Physikalische Zeitschrift* 23: 539–543.
——— (1923). "Über die Rotverschiebung in der De Sitterschen Welt." *Zeitschrift für Physik* 17: 168–189.
Lemaître, Georges (1925). "Note on de Sitter's Universe." *Journal of Mathematics and Physics* 4: 37–41.
——— (1927). "Un Univers homogène de masse constante et de rayon croissant, rendant compte de la vitesse radiale des nebuleuses extragalactiques." *Société Scientifique de Bruxelles. Annales A* 47: 49–59.
Lorentz, Hendrik Antoon, et al. (1923). *The Principle of Relativity: A Collection of Original Memoirs on the Special and General Theory of Relativity.* W. Perrett and G.B. Jeffery, trans. London: Methuen; reprint New York: Dover, 1952.
Mach, Ernst (1888). *Die Mechanik in ihrer Entwickelung, historisch-kritisch dargestellt,* 2nd ed. Leipzig: Brockhaus. Page numbers cited from the English translation *The Science of Mechanics: A Critical and Historical Account of Its Development,* 6th ed. T.J. McCormack, trans. LaSalle, Illinois: Open Court, 1960.
Merleau-Ponty, Jacques (1965). *Cosmologie du XXème Siecle.* Paris: Gallimard.
Milne, Edward A. (1935). *Relativity, Gravitation and World-Structure.* Oxford: Clarendon Press.
Moszkowski, Alexander (1921). *Einstein the Searcher: His Work Explained from Dialogues.* H.L. Brose, trans. London: Methuen.
North, John D. (1965). *The Measure of the Universe. A History of Modern Cosmology.* Oxford: Clarendon Press.
Pais, Abraham (1982). *"Subtle is the Lord ...": The Science and the Life of Albert Einstein.* Oxford: Oxford University Press.
Peebles, Phillip J.E. (1980a). *The Large-Scale Structure of the Universe.* Princeton, New Jersey: Princeton University Press.
——— (1980b). "Comment on 'The Size and Shape of the Universe' by M. Rees." In *Some Strangeness in the Proportion: A Centennial Symposium to Celebrate the Achievements of Albert Einstein.* Harry Woolf, ed. Reading, Massachusetts: Addison-Wesley, pp. 302–305.
Riemann, Bernhard (1854). "On the Hypotheses Which Lie at the Foundations of Geometry." H.S. White, trans. In *A Source Book in Mathematics.* Vol. 2. D.E. Smith, ed. New York and London: McGraw-Hill, 1929, pp. 418–428.
Rindler, Wolfgang (1977). *Essential Relativity.* New York: Springer-Verlag.
Robertson, Howard P. (1928). "On Relativistic Cosmology." *Philosophical Magazine* 5: 835–848.
——— (1929). "On the Foundations of Relativistic Cosmology." *National Academy of Sciences. Proceedings* 15: 822–829.
Schur, Friedrich H. (1886). "Ueber die Zusammenhang der Räume constanten

Riemann'schen Krümmungsmaasses mit den projectiven Räumen." *Mathematische Annalen* 27: 537–567.

Smith, R.W. (1984). *The Expanding Universe: Astronomy's Great Debate 1900–1931*. Cambridge: Cambridge University Press.

Speziali, Pierre, ed. (1972). *Albert Einstein–Michele Besso. Correspondance 1903–1955*. Paris: Hermann.

Torretti, Roberto (1978). *Philosophy of Geometry from Riemann to Poincaré*. Dordrecht: D. Reidel.

Weyl, Hermann (1918). *Raum-Zeit-Materie*. Berlin: Springer. English translation from the 4th German ed. (1921) *Space-Time-Matter*. H.L. Brose, trans. London: Methuen, 1922; reprint New York: Dover, 1950.

—— (1923). "Zur allgemeinen Relativitätstheorie." *Physikalische Zeitschrift* 24: 230–232.

—— (1930). "Redshift and Relativistic Cosmology." *Philosophical Magazine* 9: 936–943.

Whitrow, G.J. (1980). *The Natural Philosophy of Time*. 2nd ed. Oxford: Clarendon Press.

# The Expanding Universe: A History of Cosmology from 1917 to 1960

GEORGE F.R. ELLIS

The history of cosmology since 1917 shows many great discoveries and an increasing understanding of the universe, particularly in recent decades; but there have also been many missed opportunities, and a reluctance or inability to seize new ideas. There is considerable difficulty in looking back at this history without (with the benefit of hindsight) reading what was not there into what happened. This problem is exacerbated because contemporary summaries do not necessarily give a true historical perspective on what was happening at the time; for example, the justly celebrated review by Robertson (1933) is quite misleading about historical events by giving the impression that the "expanding universe" idea was accepted before it really was.

On various occasions in the history of cosmology the subject has been dominated by the bandwagon effect, that is, strongly held beliefs have been widely held because they were unquestioned or fashionable, rather than because they were supported by evidence. As a result, particular theories have sometimes dominated the discussion while more convincing explanations were missed or neglected for a substantial time, even though the basis for their understanding was already present. This review of the history of cosmology, as well as tracing the history of some of the ideas that dominate our view of the universe today, will attempt to bring out the story of some of these missed opportunities and misleading avenues; this may have important lessons for us even today.

The body of this review is divided into consideration of three periods. The *initial period* (1917–1930) was when quantitative cosmological models were being developed but their significance, and particularly the idea of the expanding universe, was not understood. At the end of this period the expanding universe idea suddenly became part of the scientific consciousness. The *second period* (1930–1945) was a time of consolidation, when the geometrical and dynamical aspects of the idea of the expanding universe were extensively explored, and observational relations in these models were developed. The *third period* (1945–1960)[1] was a time when mathematical developments, observational improvements, and investigations of the astrophysical aspects of the expanding universe laid the foundations for the dramatic developments of the following years (particularly from 1965 on), when the detection of the microwave background radiation, together with the agreement between theory and observations of primordial nucleosynthesis, vindicated the idea of the expanding universe and the hot big bang. While the period after 1960 is not explicitly dealt with as a historical period, of necessity many themes discussed in the previous periods continue into that one, and various topics only attain their true significance in the light of later developments, so some mention of later developments (with associated references) has been included.

The review and references attempt to be fairly complete in the early years and are progressively less complete at later times, the period after 1960 being only briefly touched on (a complete bibliography of the later and current periods would be an enormous task). Even in the main period covered, from 1917 to 1960, selection has had to be made because a complete bibliography would be very unwieldy through inclusion of material that is basically uninteresting, either because it is repetitive of other work, or because it has turned out in the end both to be misleading and not to have been particularly influential in the historical development of the subject. Thus the choice of papers has of necessity been made on the basis of the author's judgement of their significance. Other writers would doubtless make different choices.

The bibliography refers to various papers that contain historical reviews of aspects of cosmology and have served as vital sources in drawing up this review. Particularly worth mentioning as historical sources are Barrow and Tipler 1985; Berger 1984; Berendzen, Hart, and Seeley 1976; Harrison 1981; Harwit 1981; Pais 1982; Peebles 1971; Reines 1972; Robertson 1933; Ryan and Shepley 1975, 1976; Smith 1979, 1982; Tipler, Clarke, and Ellis 1980; and Weinberg 1972. The general relativity bibliography in Synge 1960 is also invaluable.

## Preliminaries

Once Einstein had established the field equations of his new theory of gravity, the general theory of relativity (Einstein 1916), the scene was set for its first applications: to stars and the solar system, through examining static spherically symmetric solutions (Droste 1915, 1917; Schwarzschild 1916a, 1916b; de Sitter 1916b), and to the universe itself (Einstein 1917; de Sitter 1917b, 1917c). The latter is the concern of this article. Before looking at that history, we note that the most widely accepted universe models at the current time are the Friedmann-Lemaître-Robertson-Walker (FLRW) universe models which are spatially homogeneous and isotropic (see, e.g., Weinberg 1972; Ellis 1987 for reviews). The metric tensor can be written in the form

$$ds^2 = -dt^2 + R^2(t)\{dr^2 + f^2(r)(d\theta^2 + \sin^2\theta\, d\phi^2)\}, \qquad (1)$$

where $t$ is the cosmic time and $R(t)$ the "radius" or "scale" function determining the time evolution of the universe. The spatial sections $\{t = \text{constant}\}$ are positively curved (locally spherical), negatively curved, or flat according to whether a parameter $k$ takes the values $\{+1, -1, \text{or } 0\}$ respectively; then $f(r) = \{\sin r, \sinh r, \text{or } r\}$, respectively. The spatial sections are necessarily closed (compact) if $k = +1$. The average motion of matter in the universe model is represented by a four-velocity vector $u^a$ given in these normalized comoving coordinates by

$$u^a = (1, 0, 0, 0). \tag{2}$$

In the standard FLRW model it is assumed that the only contribution to the redshift is the cosmological expansion; then the observed redshift in light received at time $t_0$ directly measures the expansion of the universe since the light was emitted at time $t_g$:

$$1 + z = R(t_0)/R(t_g). \tag{3}$$

Given equations of state for the matter in the universe, relating the pressure $p$ and energy density $\rho$, the application of Einstein's field equations to these universe models determines the time evolution of the function $R(t)$ and so the evolution of the universe. In particular they imply that provided matter obeys certain "energy conditions" (whose form depends on the value of the cosmological constant $\Lambda$, if nonzero), the universe had its origin at a singularity preceding its hot initial expansion phase. The present-day state of such a universe is characterized by three parameters: the Hubble constant $H_0 = (R^{\cdot}/R)_0$, the deceleration parameter $q_0 = -(R^{\cdot\cdot}/R)_0(H_0)^{-2}$, and the present total density parameter $\rho_0$ (which may represent contributions from various matter components); these quantities being related to $\Lambda$ by

$$q_0 = (\kappa \rho_0 c^2/2 - \Lambda)/3H_0^2,$$

where $\kappa$ is the gravitational constant and $c$ the speed of light.

## 1. Initial Period (1917–1930)

Main Themes of the Period:

1. Quantitative self-consistent GR cosmological models possible.
2. Observational discovery of distance and distribution of nebulas, and redshift-distance relation.
3. Is model $A$ (Einstein static universe) or $B$ (de Sitter universe) a better model?
4. Discovery of expanding universe model, and of the fact that the instability of the Einstein static universe implies either expansion or contraction.

The attempt to construct infinite Newtonian cosmological models ran into problems because the gravitational potential cannot be assigned a definite value in such universes (Neumann 1874, 1896; von Seeliger 1895, 1896; see North 1965 for a discussion). However Newton's theory of gravity was in any case soon to be replaced by Einstein's general theory of relativity, which became the accepted theory of gravitation and therefore the correct theory to use in discussing the large-scale structure of the universe.

By 1916 the main foundations needed had been laid: the idea of a curved space was introduced by Riemann (1854), the mathematics of the tensor calculus developed (Christoffel 1869; Ricci and Levi-Civita 1901). Special relativity (Einstein 1905) and the idea of a space-time (Minkowski 1909) led to a reevaluation of the nature of space and time measurements, causality, and simultaneity. Einstein's major insight in 1907, the equivalence principle, led to the concepts of light bending and gravitational redshift (Einstein 1907, 1911, 1915a; de Sitter 1916a), and eventually to the general relativity field equations (Hilbert 1915; Einstein 1915b, 1916). Shortly afterwards, confirmation of the predicted light bending (RAS 1919) led to widespread acceptance of the general theory (see Pais 1982 for a superb review of all these developments).

### 1.1 Einstein Static Universe (1917: Solution $A$)

Given what we now know, it is hard to look back in retrospect and recall that at the time Einstein proposed his static universe model (Einstein 1917), not only was there no evidence available that the universe might be spatially homogeneous, but even the nature and distances of galaxies ("nebulas") were unknown. Indeed it was plausibly thought by many that they might all be subsystems of the Milky Way—a manifestly anisotropic and inhomogeneous structure. The foundations of distance measurement (particularly, the calibration of Cepheid variables as distance indicators), which would clarify this issue, were being laid; for example, Shapley (1918a) was using Cepheids in detailed determinations of the distances of globular clusters and showed that they are subordinate to the general galactic system (Shapley 1918b). Einstein's universe model implied a completely uniform matter distribution despite the observational evidence then available:

> If we are concerned with the structure only on a large scale, we may represent matter to ourselves as being uniformly distributed over enormous spaces.... The curvature of space is variable in time and space, according to the distribution of matter, but we may roughly approximate it by a spherical space. At any rate, this view is logically consistent, and from the standpoint of the general theory of relativity lies nearest at hand; whether from the standpoint of present astronomical knowledge, it is tenable, will not be discussed here. (Einstein 1917)

As this universe model is spatially homogeneous and isotropic, it is a particular FLRW universe model. Einstein's paper (1917), resulting from his reflections on the origin of inertia ("no doubt motivated by Machian ideas," Pais 1982), is concerned with the problem of boundary conditions at spatial infinity for both Newtonian theory and general relativity, which he had been debating with de Sitter. With the mathematician Grommer he tried solutions that are degenerate at infinity, and found them unsatisfactory; so he then solved this problem by positing a spatially closed universe model:

> If it were possible to regard the universe as a continuum which is finite (closed) with respect to its spatial dimensions, we should have no need of any such boundary conditions.

(This is a viewpoint discussed by Weyl (1918, 1922, see section 34), and later adopted to much effect by Wheeler and Hawking.) He proceeded to show that this idea is compatible with the postulates of general relativity, by deriving the Einstein static solution. This is an exact solution of the field equations for pressure-free matter ($p = 0$) that has a spherical geometry ($k = +1$) and is static ($R(t)$ = constant). There was no systematic evidence to contradict the last assumption, which was deeply ingrained in the foremost scientific minds of the age. Indeed, in order to achieve a static solution, Einstein introduced into the field equations the cosmological constant $\Lambda$, without which such solutions were impossible:

> [To obtain the solution] we admittedly had to introduce an extension of the field equations of gravitation which is not justified by our actual knowledge of gravitation.... The term is only necessary for the purpose of making a quasi-static distribution of matter as required by the fact of the small velocities of the stars.

The introduction of this term allowed this first quantitative cosmological model to be uniform in space and time, with no evolution taking place.

This model introduced the "fluid" description of the matter in the universe, which has been used ever since in virtually all quantitative universe models; the fluid four-velocity at each space-time point represents the average motion of matter in the universe there. As the "fundamental observers" in the Einstein static model (those moving with precisely the fluid four-velocity) move geodesically and without expansion, no redshifts are predicted in this universe model. The model predicts a unique relation between the mass density $\rho$ in the universe and its radius of curvature $R$:

$$\kappa \rho c^2 = 2/R^2 = 2\Lambda. \tag{4}$$

Einstein wrote to de Sitter giving the value of $10^7$ light years for the radius, determined from an estimate of $10^{-22}$ gm/cc for the density (Kahn and Kahn 1975), but did not publish it, probably because of unease over the value obtained. He wrote in 1921:

> The exact figure is a minor question. What is important is to recognize that the universe may be regarded as a closed continuum as far as distance measurements are concerned. (Smith 1979)

1.2 DE SITTER UNIVERSE (1917: SOLUTION $B$)

As soon as this model was published, and in response to a request from Eddington, the Dutch astronomer Willem de Sitter published in the *Monthly*

*Notices of the Royal Astronomical Society* a series of papers on the astronomical implications of general relativity. The third of these papers (de Sitter 1917c) described the Einstein static universe ("system $A$") including a pressure term, and gave the field equations for this case (Einstein only considered the case $p = 0$), quoting a paper by Levi-Civita (1917) as independently deriving this result. He used estimates of the distances up to which galaxies had been observed to obtain minimum estimates for the radius of that universe, and he estimated the radius from density estimates by King, concluding that the density ($10^{-12}$ gm/cc) was too high and the consequent radius ($10^9$ light years) too small. He also stated a preference for the elliptical rather than spherical topology for the Einstein static universe.

De Sitter went on to give a summary of his work on a second cosmological solution of Einstein's field equations with cosmological constant ("system $B$," now called the de Sitter universe), apparently first suggested by Ehrenfest (Robertson 1933). As this is a space-time of constant curvature, it is uniform in space and time, and additionally the field equations imply that the fluid energy density has to be zero.[2] Consequently in this case there is no relation between the density of matter and the radius of the universe; also the motions of the "fundamental particles" in the universe are not uniquely determined by the space-time geometry, and could be chosen in different ways (simply related to various possible natural coordinate systems in the de Sitter space-time). See Schrödinger 1950 for a clear exposition.

*1.2.1 Static Frame.* De Sitter (1917c) used a static (Schwarzschild-like) frame,[3] in which the distribution of world-lines is inhomogeneous and the distances between the fundamental particles remain constant. Nevertheless, as he himself pointed out, there would be observed redshifts in this universe:

> The frequency of light vibrations diminishes with increasing distance from the origin of coordinates. The lines of very distant stars or nebulae must therefore be systematically displaced towards the red, giving rise to a spurious radial velocity.

These are gravitational redshifts, consequent on the fundamental observers' moving on nongeodesic orbits in space-time (just as in the case of static observers in the Schwarzschild space-time or uniformly accelerated observers in Minkowski space-time; see Ellis, Maartens, and Nel 1978 for a modern revival of the idea of cosmic gravitational redshifts). De Sitter gave a relation between apparent diameter and linear diameter and used this, plus a parallax-distance formula due to Schwarzschild (1900), to estimate the radius of his model (system $B$). He gave data for three galaxies whose redshifts had been measured at that time (by Slipher), and commented that continued observation of systematic positive radial velocities would indicate adoption of model $B$ rather than $A$. Thus his were the first published attempts to relate the geometry of the universe (models $A$ and $B$) to observational evidence.

De Sitter's papers were very influential, for it was through them that scientists in England and the United States first learned the details of general relativity (the first papers on which had been published in Germany during the First World War[4]). Einstein himself reacted against this universe model, regarding it as anti-Machian because it had no matter content (Einstein letter in Smith 1982, p. 172; Kahn and Kahn 1975). He then went on to question the cosmological constant because it allowed this solution (Einstein 1918b). Soon Eddington (1924a) pointed out that the "scattering of particles" moving on geodesics (due to the nongeodesic motion of the fundamental particles in this world model) would also lead to Doppler shifts in the radiation received from freely moving particles (but this also did not carry with it a connotation of the expansion of space-time).

*1.2.2 Nonstatic Frames.* Alternative frames were soon discovered for the de Sitter universe.

1. First Lanczos (1922) found a frame expressing it as a FLRW geometry with $k = +1$ and a "cosh" expansion; this is the global coordinate system eliminating the "mass horizon" (where the mass apparently diverges, due to the nature of the coordinates) in the static coordinates. He discussed the redshift that would occur in this form (Lanczos 1923).
2. Then Lemaître (1925a, 1925b) and Robertson (1928) expressed it in the "steady-state" form: a FLRW geometry with $k = 0$, in which the fundamental observers are expanding exponentially away from each other—but in a static space-time, the whole universe model being stationary (Tolman and Robertson in 1929 showed that this is the only stationary form of the de Sitter universe). Weyl (1923) effectively found this form by presenting the universe as the imbedding of a hyperboloid in a five-dimensional flat space-time; he derived it from causal considerations, and showed (without using explicit coordinates) that there would be a linear velocity-distance relation in this case. When Lemaître (1925a) obtained the metric explicitly, he deduced from the metric that the velocity-distance relation would be linear (Lemaître 1925b). After Robertson's paper, Weyl (1930) confirmed the result in a different way, this time using the steady-state coordinates. Thus, in this frame there were expected redshifts representing a relative velocity of matter, but not an expansion of the space-time itself (which is locally unchanging in time).

It was already clear to Weyl that the metric of the "stationary form" ($k = 0$) covers only half of the whole de Sitter hyperboloid. He characterized this frame by the fact that its geodesic world-lines "form a system that has been causally interconnected since its origin" (1930), i.e., have a mean motion such that all particles have a common origin with causal futures all the same. It is implied by this work that this frame is geodesically incomplete in the past (Penrose 1968). Weyl (1924) also related this universe model to Olber's problem. This stationary model later resurfaced as the steady-state model of Bondi, Gold, and Hoyle (Bondi and Gold 1948, 1954; Hoyle 1948), and even later as the inflationary universe of Guth (1981).

*1.2.3 Redshift in the de Sitter Universe.* As mentioned earlier, it was clear that free particles would not remain stationary in the static frame but rather would be seen from that frame to start moving apart (because they move on geodesics, and that frame is nongeodesic), resulting in Doppler shifts. Thus the situation was rather confused: three different causes of redshift could occur in the de Sitter universe: (1) Static fundamental observers perceive a gravitational redshift; and in this frame, (2) freely falling particles would additionally produce Doppler shifts associated with the local motion of this matter being different from that of the static fundamental observers. (3) In the expanding frame there would be redshifts representing the expansion of the fundamental observers, but not of the space-time itself. A particular source of confusion was the representation of the gravitational redshifts in terms of equivalent velocities. The situation in the static frame is described by Eddington thus:

> De Sitter's theory gives a double explanation of this motion of recession; first there is a general tendency to scatter ... second there is a general displacement of spectral lines to the red in distant objects owing to the slowing down of atomic vibrations which ... would erroneously be interpreted as a motion of recession. (Eddington 1924b)

To clarify the situation, it is perhaps useful to note that in a general universe model, the predicted redshifts are given by[5]

$$(1 + z) = (1 + z_{D_S})(1 + z_{G_S})(1 + z_{D_C} + z_{G_C})(1 + z_{D_O})(1 + z_{G_O}). \quad (5)$$

The first two terms ($z_{D_S}$ and $z_{G_S}$) represent the Doppler (or expansion) and gravitational redshift contributions due to relative motion and the inhomogeneity of matter at the source; the last two terms ($z_{D_O}$ and $z_{G_O}$) the Doppler and gravitational redshift contributions due to the peculiar velocity of the observer (as indicated by the microwave background radiation anisotropy) and inhomogeneities in the gravitational field near the observer. The remaining two terms ($z_{D_C}$ and $z_{G_C}$) represent cosmological contributions to the observed redshift, the first due to the expansion of the universe (as in the FLRW universe models) and the second due to large-scale gravitational inhomogeneities in the universe resulting in a cosmological gravitational redshift (as in Ellis, Maartens, and Nel 1978). In the de Sitter universe, (1) the static frame sets $z_{D_C}$ to zero, all the redshift $z_{G_C}$ being due to the acceleration of the observers (as in the Schwarzschild solution). This alone gives a quadratic redshift-distance relation near the origin. (2) The scattering of test particles relative to the static frame refers to objects not moving with the fundamental four-velocity and so in effect attributes the redshift to $z_{D_S}$. This adds a linear term to the redshift-distance relation near the origin. (3) The FLRW ($k = 0$) frame sets $z_{G_C}$ to zero, all the redshift being attributed to the expansion of the universe. This gives a linear redshift-distance relation near the origin. One can choose between these interpretations because the motion of matter is not uniquely determined by the space-time geometry (for the de Sitter universe is a space-time of constant curvature).

Weyl's postulate (1923) was intended to supply a way of uniquely choosing the family of stationary expanding geodesic world-lines. However its statement in causal terms was not well understood by Weyl's contemporaries. Robertson (1933) paraphrased it as simply postulating "the existence of a coherent flow of matter converging towards the past" and then stated that it proves the existence of a cosmic time. However this only follows because the flow is vorticity-free, which happens to be true for the family of geodesics picked out in de Sitter space-time by Weyl's causal postulate, but does not follow in general from Robertson's statement (the notion of vorticity and its relation to the existence of a cosmic time were only developed much later (Gödel 1949))

On comparing the Einstein static solution ("solution $A$") and de Sitter solution ("solution $B$"), Eddington wrote in 1918 to Shapley:

> De Sitter's hypothesis does not attract me very much, but he predicted this (spurious) systematic recession before it was discovered definitely; and if, as I gather, the more distant spirals show a greater recession that is a further point in its favour. (Smith 1979)

However he really preferred the Einstein static universe, although it could not explain the observed redshifts, because it offered the chance of relating the ratio of the electron's radius to the number of particles in the universe (Eddington 1924b).

1.3 EXPANDING SPACE-TIME (FRIEDMAN)

In 1922, Alexander Friedman, a Russian meteorologist, first found the expanding FLRW universe models with positive spatial curvature ($k = +1$) and nonzero cosmological constant (Friedman 1922). Thus he was the first person to propose a mathematical model of an evolving universe. He commented on the density and age of the universe, but did not determine if there were any expected redshifts in these universe models; according to his student Gamow, he derived the models as a mathematical exercise rather than a model of the real universe (Gamow 1970). Einstein published a brief note claiming that there was an error in Friedman's paper invalidating the concept of an evolving universe (Einstein 1922), but a year later, after comments from a Russian scientist ("Herr Krutkoff"), he withdrew this claim, stating that Friedman's results showed that

> as well as the static solutions, the field equations admit dynamic (that is, varying with the time coordinate) centrally symmetric solutions for the space-time structure. (Einstein 1923)

Thus Einstein was certainly then aware of the concept of an expanding universe, but did not see it as a useful model of the real universe ("examination of Einstein's retraction of his criticism of the Friedman paper shows that it originally ended with a line to the effect that of course the solution, while mathematically correct, was of no physical significance! Fortunately it was crossed out in the manuscript"; Stachel, 1986, also quoted in Smith 1982, p. 199, note 97). Friedman pressed ahead and found the expanding FLRW universe models with hyperbolic space sections ($k = -1$) (Friedman 1924), but again without any attempt to relate these models to observations (the $k = 0$ expanding FLRW universes were only found in 1929 by Robertson). The calculation of aberration and parallaxes in these universes was examined in an unremarked paper by Freedrichsz and Schechter (1928).

1.4 EARLY OBSERVATIONAL EVIDENCE

At this time observations were proceeding apace. The critical problem on the observational side was determination of reliable distance indicators for distant objects; in particular the nature of spiral and elliptical galaxies was unclear until the mid-1920s.

*1.4.1 Spiral Nebulas: "Island Universes" or Gas in our Galaxy?* The debate on the nature of spiral nebulas has been well documented (Berendzen, Hart, and Seeley 1976; Smith 1982). Harlow Shapley believed they were mere nebulous objects; H.D. Curtis that they were "inconceivably distant galaxies of stars or separate stellar universes so remote that an entire galaxy becomes but an unresolved haze" (Berendzen, Hart, and Seeley 1976, p. 28). Early measures concentrated on angular size. The 1885 supernova in Andromeda and the discovery of novas in spirals (Curtis 1917; Ritchey 1917a, 1917b) implied that they are very distant. By 1920 Shapley was convinced he had been wrong, and then he (Shapley 1923) estimated the distance to NGC 6822 as 300 kpc from morphology and apparent size. Van Maanen made contrary measurements; later it was suggested that he was "seeing what he expected to see" (Smith 1982, p. 129). However it was Hubble who made the decisive discovery. The period-luminosity relation for Cepheid variables had been recognized by Henrietta Leavitt in 1908 and calibrations for them were established by 1912. Hubble discovered a Cepheid in Andromeda in 1923 and then set out to use them systematically as distance indicators of "nebulas." Using Cepheids as primary distance indicators, supported by others such as novas, he obtained the first definitive proof of the extragalactic nature of astronomical objects: NGC 6822 (Hubble 1925a), M33 with fifteen cepheids (Hubble 1926a) and the Andromeda nebula (with ten cepheids) (Hubble 1929c), were each shown to be separate stellar systems. The initial results were presented at the American Astronomical Society meeting in 1925.

*1.4.2 Uniform Distribution of Galaxies.* Hubble also provided the first evidence for the uniform distribution of galaxies (Hubble 1926b), including an estimate of the density of matter in the universe that was much lower than that given by de Sitter in 1917. Hubble then applied the density-radius relation (4) of the Einstein static universe to determine the radius of the universe. He found an average density of $\rho = 1.5 \times 10^{-31}$ gm/cc, a total mass in the universe of

$M = 9 \times 10^{22}$ solar masses, and a radius of $R = 2.7 \times 10^{10}$ pc. The first two figures are remarkably modern but the third is not (de Sitter's figure for the density was $10^{-26}$ gm/cc).

*1.4.3 Systematic Redshifts.* The spectrum of a typical spiral nebula, the Andromeda nebula, was obtained in 1898 by Julius Scheiner. Already on January 2, 1913, V.M. Slipher measured its redshift (Smith 1979). Interpreting the redshift as a radial Doppler shift, he calculated that the Andromeda nebula is rushing towards the sun at 300 km per second (Slipher 1914). By 1914 he had collected radial velocities for fifteen spiral nebulas (Slipher 1915) and by 1922, he had measured the redshifts of forty galaxies ("nebulas"), as quoted by Eddington in his book on general relativity (Eddington 1923), where he stated "the great preponderance of positive (receding) velocities is very striking" (thirty-six were positive). Slipher's work, using the Lowell 24-inch refractor, came to an end about 1926.

Paddock commented in 1916 "The average velocity (of spirals) is decisively positive" and introduced a constant redshift "$K$-term" to describe this effect. Wirtz (1918, 1922) also suggested a general recession. A velocity-distance relation for spiral nebulas was obtained by Wirtz (1922, 1924) (note also van Maanen's (1922) table). Silberstein (1924) tried fitting the redshifts of globular clusters, which exhibited equal numbers of positive and negative redshifts, generating much controversy. Lundmark (1920, 1924) failed to obtain definite relations but later determined both a linear and quadratic term in the relation (Lundmark 1925). Stromberg (1925) stated that there was no clear relation. The major problem at this time was that the distance measures used were unreliable (Berendzen et al. 1976).

As mentioned earlier the de Sitter universe predicted an increase of redshift with distance (de Sitter 1917c), quadratic in the static frame. Weyl, in the fifth edition of *Raum-Zeit-Materie* (Weyl 1923), Appendix 3, used spectral shifts to derive a linear relation with a value for the Hubble constant of 1,000 km/sec/Mpc and commented that neither cosmology ($A$ or $B$) leads to such a redshift law. However the expanding steady-state frame implicitly introduced by him (Weyl 1923) predicted a linear relation at small distances. Robertson (1928) also derived a linear velocity-distance relation for the expanding frame in a de Sitter universe and compared the values of distances for galaxies given by Hubble in 1926 and Slipher's radial velocity figures. He decided that "we have a rough verification of the linear law and a value of $R = 2 \times 10^{27}$ cm," hence the redshifts implied by the de Sitter universe were "in accord with known facts concerning the radial velocities of spiral nebulae."

All this work was supplanted by Hubble's work based on observations with the 100-inch telescope at Mount Wilson from 1925 on, using Cepheids as distance indicators. His famous paper (Hubble 1929b), establishing a linear velocity-distance relation for "extragalactic nebulas" for the range $220 < v < 1{,}100$ km/s with a value of 500 km/sec/Mpc for the constant of proportionality (the "Hubble constant"), was based on forty-six redshifts and twenty-four accurate distances to $2 \times 10^6$ pc. It should be noted here that in all these cases, although the redshifts are expressed in terms of equivalent velocities, there is no implied commitment to the notion that these represent real velocities. According to Humason (1929), the program was carried out

> to determine if possible whether the absorbtion lines in these objects show large displacements towards longer wave-lengths as might be expected on de Sitter's theory of curved space-time.

Hubble's contribution was not so much to propose the relation (which had already been done), as rather by using accurate distance estimates (in particular, Cepheids) to persuade astronomers that there was solid evidence for it (Smith 1979). There are two particular things one might note here. The first is that while Hubble fitted a linear relation to the data, a quadratic relation would have been at least as plausible for the data he had available at that time. The second is that he did not at that time interpret the data as implying an expansion of the universe. Indeed the last paragraph of his paper shows that insofar as he believed that what he had discovered related to the curvature of space-time, it related to the curvature of de Sitter space-time (in the static frame) and the associated (gravitational) redshift:

> The outstanding feature however is the possibility that the velocity-distance relation may represent the de Sitter effect and hence that numerical data may be introduced into discussions of the general curvature of space. In the de Sitter cosmology, displacements of the spectra arise from two sources, an apparent slowing down of atomic vibrations and a general tendency of material particles to scatter. The latter involves a separation and hence introduces the element of time. The relative importance of these two effects should determine the form of the relation between distances and observed velocities; and in this connection it may be emphasized that the linear relation found in the present discussion is a first approximation representing a restricted range in distance.

It is uncertain if Hubble was aware of the concept of an expanding and evolving universe as proposed by Friedman and Lemaître (Smith 1979). Curiously he did not mention that among others, both de Sitter's 1917 paper and his own written in 1924 already tried to relate observational evidence (the density of matter) to the curvature of space-time, and additionally de Sitter had used redshifts (of $B$ stars, the lesser Magellanic cloud and three spiral nebulas) to limit the radius of curvature. In any case (although not stated in the famous 1929 paper, Hubble 1929b) he believed he was conducting a critical test that would enable him to discard either solution $A$ or solution $B$, and stated: "The necessary investigations are now under way with the odds, for the moment, favouring de Sitter" (Hubble 1929a). Hubble's observational work was queried by Shapley (1929) but supported by de Sitter (1930c).

### 1.5 Expanding Universe Concept—Lemaître

Meanwhile Georges Lemaître, apparently unaware of Friedman's work, rediscovered the $k = +1$ expanding and evolving universe models (with nonzero cosmological constant) (Lemaître 1927). He worked out the consequences of energy conservation in these models, solved the field equations, and determined the expected redshift. Thus he became the first person seriously to propose an expanding universe as a model of the real universe. He even gave a value for the rate of expansion (630 km/sec/Mpc),[6] close to the value of 500 km/sec/Mpc announced by Hubble in 1929.

The model he considered is one with a nonsingular origin; it starts off like an Einstein-static universe and asymptotically approaches a de Sitter universe in the future; that is, it lies "between" the Einstein and de Sitter universes. Thus, although it evolves, it does not have a singular origin. Lemaître showed there are effective horizons in this universe ("Note that the largest part of the universe is forever outside our reach"). This outstandingly original paper (which discussed the Olbers problem for the de Sitter universe, and gave a value for the mean density of matter) was unread or forgotten by the astronomers of the time, and did not influence events for a while. The accepted wisdom was that the universe must be unchanging in time.

1.6 QUESTION AT ISSUE: IS IT UNIVERSE $A$ OR $B$?

From 1917 until 1930, with Friedman's (1922, 1924) and Lemaître's (1927) papers remaining unrecognized, the issue that was in the forefront of observational cosmology was whether system $A$ (the Einstein static universe) or $B$ (the de Sitter universe) is a better model of reality.

Eddington (letter to de Sitter, see Smith 1982) and Jeans (1923) favored $B$ because of the observed redshifts, but soon Eddington (1924b) stated that despite the redshifts he really preferred $A$ because of its closed space sections. de Sitter (1922) was uncertain. Hubble (1926b, 1929a) favored $B$. Weyl (1923) stated that the closed space sections in $A$ separate the past and future (which are equal projectively) but allow one to see "ghosts," i.e., multiple images of each galaxy. Since he regarded this as undesirable, he preferred $B$. Robertson (1928) preferred solution $B$, and gave a radius value for this universe.

Robertson (1929) wrote down at this time the general line element of an expanding FLRW model with arbitrary spatial curvature, and the Einstein field equations for that model with a perfect fluid matter source, without realizing its significance;[7] for he was looking only for the static universes among these models (although he was aware of Friedman's papers, for he refers to them in a footnote). The paper is remarkable in that he realized that all these FLRW models are both conformally flat and imbeddable in a flat five-dimensional space-time. He and Tolman (1929a) proved there are only three static FLRW solutions: solution $A$, solution $B$, and the flat space-time of special relativity. It was thus still at this stage taken for granted that the universe must be static—despite data being available that would shortly be taken to prove the contrary, with at least three published papers proposing the idea of an expanding universe, and incipient hints that one should look at expanding models, e.g., Tolman's comment: "It should be noted that our assumption of a static line element makes no explicit recognition of any universal evolutionary process which may be going on. The investigation of nonstatic line elements would be interesting" (1929a). After Hubble's data was publicized, Tolman published a major paper analyzing how one could fit the data to the de Sitter universe (Tolman 1929b), and concluded:

> The conclusion is drawn that the de Sitter line element does not afford a simple and unmistakably evident explanation of our present knowledge of the distribution, distances and Doppler effects for the extragalactic nebulae.

### 1.7 DISCOVERY OF EXPANDING UNIVERSE

The situation changed at a critical meeting of the Royal Astronomical Society in January 1930, where de Sitter informally communicated some of his results. As described by Eddington,

> making (as I think) a rather excessive estimate of the masses of the nebulae, de Sitter propounded the dilemma that the actual universe apparently contained enough matter to make it an Einstein world and enough motion to make it a de Sitter world. This naturally called attention to the need of intermediate solutions for handling such a question. (Eddington 1931a)

Following this line of thought, Eddington remarked at this meeting: "One puzzling question is why there should be only two solutions. I suppose the trouble is that people look for static solutions" (Eddington 1930a). He then began investigating the stability of the Einstein solution with G.C. McVittie, using Robertson's paper as a basis. Meanwhile Lemaître, a former post-doctoral student of Eddington's, read this remark in a report of the meeting published in *Observatory*, and wrote to Eddington pointing out his 1927 paper. McVittie, who in 1929 was a research student of Eddington's, wrote of

> the day when Eddington, rather shamefacedly, showed me a letter from Lemaître which reminded Eddington of the solution to the problem which Lemaître had already given. Eddington confessed that although he had seen Lemaître's paper in 1927 he had forgotten completely about it until that moment. The oversight was quickly remedied by Eddington's letter to *Nature* of 1930 June 7, in which he drew attention to Lemaître's brilliant work of three years before. (McVittie 1967)

Eddington was greatly impressed; he wrote to *Nature* commending Lemaître's work (Eddington 1930b) and arranged for a translation to be printed in the *Monthly Notices of the Royal Astronomical Society* (Lemaître 1931a). It was only because of this, and through Eddington's proof of the instability of the Einstein static universe (Eddington 1930c), that the concept of an evolving universe began to receive wide acceptance. The central equation of Eddington's proof (equation (4)) is the restriction to the FLRW case of Raychaudhuri's equation (Raychaudhuri 1955b; Ehlers 1961); on the basis of this equation Eddington made the fundamental point that in a FLRW universe, pressure cannot *cause* the expansion of the universe but rather retards it (contrary to the view of Tolman (1930a) and a suggestion by Lemaître (1927)).

De Sitter corroborated Hubble's observational conclusions (de Sitter 1930c) and accepted Lemaître's (1930) solution, writing:

> Lemaître's theory not only gives a complete solution of the difficulties it was intended to solve, a solution of such simplicity as to make it appear self-evident ... there cannot be the slightest doubt that Lemaître's theory is essentially true, and must be accepted as a very real and important step towards a better understanding of Nature. (de Sitter 1931c)

CONTEMPORARY SUMMARIES

Eddington, 1924b, 1933, Robertson 1933, and Lemaître 1933a are all excellent contemporary summaries, the first written before the discovery of the expanding universe (expounding only the Einstein static and de Sitter models) and the others going somewhat beyond this period and into the next. They do not give one an accurate view of the history of the first period, however; each of the later papers focuses on the new idea of the expanding universe, and does not make clear that for over a decade astronomers had been trying to choose between the two static possibilities (model $A$ or model $B$) as the better universe model, ignoring or overlooking the idea of an expanding universe. Particularly interesting is what is omitted in these reviews. For example, Eddington's (1924) penetrating book did not mention the Friedman papers, and the same is apparently true for all relativity treatises published in the 1920s (Stachel 1986). On the other hand Robertson's review (1933), written after the discovery of the expanding universe, omits from its bibliography Einstein 1922, 1923, Tolman 1929b, and Hubble 1929b, thus hiding the strand of history that resisted this new idea. He also omits Lemaître 1931c and Kasner 1925, thus missing papers containing ideas that would be of importance much later.

## 2. Consolidation (1930–1945)

Main Themes of the Period:

1. Expanding universe idea vindicated for many, but origin unclear and age is a problem.
2. Observational consolidation of redshift and distance observations.
3. Theoretically, various alternatives, particularly kinematical relativity (Milne).

After Lemaître's work was called to the attention of the astronomical community by Eddington (1930b), the expanding and evolving universe concept became widely accepted (de Sitter 1930b, 1931c; Eddington 1931a; Einstein 1931; Robertson 1932; Tolman 1930a–1930d), and an explosion of papers then explored this concept. However, initially they concentrated on nonsingular universe models "between the Einstein static universe and the de Sitter universe," as summarized by Eddington:

> The radius of the universe has expanded by one part in 2000 in the last million years. The result is impressive. It indicates that the radius of space his doubled within ordinary geological time.... We conclude that the radius of space was originally about $1,2.10^9$ l.y. ... that it has since expanded considerably, but to an amount practically undeterminable, and that the present rate of expansion is 1 per cent in about 20 million years—a rate which will continue indefinitely.... In $10^{10}$ years the spiral nebulae will be 10 magnitudes fainter than they are now. With a time scale of billions of years astronomers must count themselves extraordinarily fortunate that they are just in time to observe this interesting but evanescent feature of the sky. (Eddington 1930c)

Thus while the concept of an evolving universe gained widespread currency, the idea of a singular origin of the universe had still to be accepted. Lemaître was again at the forefront (Lemaître 1931c), suggesting the possibility of expansion from a singularly dense state, the primeval atom (later termed the big bang by George Gamow). A period of consolidation now began, as various properties of these universes and their relation to other models were studied.

2.1 THE FAMILY OF EVOLVING WORLD MODELS

The basic geometry and simple dynamic properties of the FLRW universe models were reasonably well understood within a short while; they are very well summarized in Robertson's (1933) review article characterizing the dynamic behavior of all models with $p \geq 0$. The underlying symmetry structure, based on the work of Bianchi (1897, 1918) on groups of isometries (which had been applied by Eiesland (1925) to the Schwarzschild geometry), was applied by Robertson to the FLRW geometry. He also emphasized the conformal invariance of these spaces and that they could all be embedded in a flat five-dimensional space.

The possible dynamic behaviors of all the FLRW models were systematically investigated (e.g., Tolman 1930d; Heckmann 1932; Robertson 1933); in particular Tolman (1931a, 1931b) investigated the effect of pressure and discussed universes containing black-body radiation (proving in particular that a black-body spectrum remained a black-body spectrum as the universe expanded, see section 171 of Tolman 1934a). Tolman 1931c looked at conditions for the recurrently popular idea of a "bouncing" or oscillatory ($k = +1$) universe, showing that for entropy reasons strictly periodic reversible models are not possible. A major issue was how the universe began.

2.2 BEGINNING OF THE UNIVERSE

Particularly due to the influence of Lemaître and Eddington, many workers (e.g., Robertson 1932; Eddington 1933) believed that the universe began as an

Einstein static universe that started expanding due to its instability and would end as a de Sitter universe, and so was "between" solutions A and B. Thus these universes had no beginning: they had existed forever. Various attempts were made to see whether inhomogeneities in an Einstein static universe would be more likely to initiate a phase of expansion or contraction. McCrea and McVittie (1931a) concluded that a contraction process is as probable as an expansion one:

> It seems therefore that a condensation starting in an Einstein universe would cause this universe to contract.... We must conclude that if the actual universe started as an Einstein universe and is now expanding, as the work of Lemaître and Eddington suggests, this expansion cannot have been caused by the redistribution of matter into massive nuclei. Thus as yet no mechanism for setting up the expansion has been found.

Einstein, however, now proposed that since nonstatic solutions are possible, the cosmological constant ("which is theoretically unsatisfactory anyway") should be set to zero (Einstein 1931); this implies acceptance of a singular origin of the universe. Eddington was horrified by the suggestion that the cosmological constant should be dropped (see, e.g., Chandrasekhar 1983; Stachel 1986); for Einstein the loss was rather Mach's principle (Pais 1982, p. 288). His later view on the cosmological constant is summarized as follows:

> If Hubble's expansion had been discovered at the time of the creation of the general theory of relativity, the cosmologic member would never have been introduced. It seems now so much less justified to introduce such a member into the field equations, since its introduction loses its sole original justification—that of leading to a natural solution of the cosmologic problem. (Einstein 1945, p. 127)

It has been suggested by R.W. Smith that this event was important in Einstein's thinking about scientific methodology, for

> the cosmological constant had the status of an *ad hoc* hypothesis, not required by the kinds of simplicity considerations that were otherwise such an important driving force in Einstein's work. Realizing that the constant was not needed could well have renewed Einstein's faith in simplicity as a criterion of theory choice, and in this light it is perhaps no accident that this theme emerged much more prominently than before in Einstein's writings in the early to middle 1930s. (Don Howard, private communication)

Einstein and de Sitter emphasized this new position with their publication of the Einstein–de Sitter universe (Einstein and de Sitter 1932),[8] the simplest expanding FLRW universe model, with flat spatial sections, $k = 0$; vanishing pressure, $p = 0$; and zero cosmological constant, $\Lambda = 0$. This model also has the virtue of predicting a simple relation between the Hubble constant $H_0$ and the current density $\rho$ of matter, namely, that the density should take the critical value $\rho = 3H_0^2/8\pi$ (or in modern notation, $\Omega = 1$). Because it has a beginning the universe in this model has existed for a finite time; in fact the age of such a universe is $T_0 = (2/3)H_0^{-1}$, so one of the major problems continually to plague the simple expanding universe models, the timescale problem, surfaced (de Sitter 1933): some stars appeared to have lived longer than the age of the universe.

From then on, those who believed in the expanding universe were engaged in the continuing debate on the Hubble constant and the age of the universe, Lemaître (1933), for example, preferring a universe with finite spatial sections ($k = +1$), but finding important constraints from the radioactive decay of terrestrial minerals, which (with current estimates of the Hubble constant) exceeded the maximum age $(2/3)H_0^{-1}$ of a high-density universe with vanishing $\Lambda$ term. He commented:

> From a purely aesthetic point of view that perhaps is regrettable. The solutions where the universe alternatively expands and contracts to an atomic state with the dimensions of the solar system have an incontestably poetic charm, bringing to mind the legendary phoenix. (Lemaître 1934, as translated in Peebles 1984)

He avoided the timescale problem by adopting a closed universe with a positive cosmological constant.

Eddington rejected the singular origin, retaining the cosmological constant for theoretical reasons. Tolman and Ward (1932) however published the first "singularity theorem" indicating how some "energy conditions" (in this case, $p \geq 0, \Lambda \leq 0$) make an initial singularity inevitable in a ($k = +1$) FLRW universe. This result should have been obvious to Eddington when he demonstrated the instability of the Einstein static universe (Eddington 1930c), if his view had not at that time been strongly set against the idea of the creation of the universe at an initial singularity. The energy conditions are not fulfilled in the cases he considered, because of a large positive cosmological constant.

### 2.3 Thermodynamics and Entropy

Perhaps the area where the most was "missed out' during this era was in understanding the thermodynamics and physics of the expanding universe. Tolman, who was an expert in statistical mechanics, was the lone pioneer investigating these topics, e.g., the entropy of the universe and the work done in expansion (Tolman 1931a, 1931b, 1934a). He pointed out the possibility of radiation-dominated universes (1931a) and worked out the thermal expansion of black-body radiation in an expanding universe; thus by implication he predicted the hot early phase of the universe. He also examined the relativistic thermodynamics of a monatomic gas in equilibrium with black-body radiation (1931b). He apparently did not, however, make the jump to the idea of a decoupling of matter and radiation.

While there were some conjectures during this period on element formation (Tolman 1922; Sterne 1933; von Weizsäcker 1938), it was the equilibrium or "$e$-process" that was considered; the importance of the thermodynamic effect of the expanding universe on primordial nucleosynthesis was not yet realized.

However one key concept was already being developed, albeit in the wrong context. The idea of background radiation had already surfaced in the context of "Olbers's paradox" (Halley 1720; de Cheseaux 1744; Olbers 1823; Thomson 1901; Charlier 1908; Weyl 1924; Lemaître 1927; see also Jaki 1969; Harrison 1984, 1986), but this had not been related to the origin of the universe. Lemaître characteristically was ahead of the field in understanding the broad idea of a hot big bang and its aging remnants:

> The evolution of the universe can be compared to a display of fireworks that has just ended: some few wisps, ashes and smoke. Standing on a well-chilled cinder, we see the slow fading of the suns, and we try to recall the vanished brilliance of the origin of the worlds. (Lemaître 1931c; as translated in Peebles 1984)

He then started a search for remnant radiation from the big bang (Lemaître 1931d). This led him to his extended study of cosmic rays, which he hoped might be such remnants, not knowing that they are probably galactic in origin. Perhaps if he had not been misled in this way, he would have predicted the remnant radiation we now measure as microwave background radiation.

2.4 ORIGIN OF STARS AND GALAXIES; OTHER GEOMETRIES

Galaxy formation was also under discussion. Jeans (1902, 1928a) had developed the theory of gravitational instability in a uniform, nonexpanding medium, and defined the Jeans length (but this theory was not self-consistent). McVittie (1932) and Lemaître (1933a–1933c) extended this work to the context of an expanding universe. Lemaître presented the exact spherically symmetric solution of Einstein's equations for pressure-free matter, which later became known as the Tolman-Bondi model, found the $t^{2/3}$ growth law for linear perturbations of an Einstein–de Sitter model (Lemaître 1933c) and used it as the basis of a theory of galaxy formation (see Peebles 1984). Tolman (1934b) investigated these inhomogeneous, spherically symmetric, pressure-free universes (whose equations had already been written down by de Sitter (1917c), see equation (29) in that paper); these models have been studied again and again for the light they throw on inhomogeneities in FLRW universe models in general and on galaxy formation in particular (Bondi 1948a gave a very readable and clear account of them, but without application to particular astrophysical problems). Dingle (1934) also looked at these inhomogeneous models, presenting "the general nearly homogeneous universe." He dissented from Milne's (1933a) view that homogeneity, expressed in a "cosmological principle," is included in the definition of the universe:

> We take it to be perfectly conceivable that an increase of telescopic power may reveal a variation of material density with distance, and the denial of this possibility connoted by Milne's statement seems to us to be inconsistent with the fundamental principles of science.... Whatever philosophical objections may be advanced against the inhomogeneity of space tell with equal force against the inhomogeneity of time, and that time is not homogeneous (i.e., that space is not static) is already acknowledged in the acceptance of the Friedmann-Lemaître model of the universe.... The assumption of a nonhomogeneous universe would give us a no more improbable

position in space than the Friedmann-Lemaître universe gives us in time.... We have no grounds for assuming that the part of the universe which we observe is typical of the whole ... the phenomena we embody in our models may be purely local characteristics.

It is remarkable that many years before, Kasner (1925) had already examined the geometry and dynamics of spatially homogenous anisotropic empty universes, which we now know as the Bianchi I universes. The fact that these are significant models in cosmology was only to be fully understood much later, but already in 1933, Lemaître, at Einstein's instigation (Lemaître 1933a), had shown that models with this geometry do not avoid the initial singularity (Einstein had suggested that the singularity might be a result of the high symmetry of the FLRW models). Together with Tolman and Ward's (1932) result, this work foreshadowed the famous singularity theorems of later years.

Finally, the general properties (in any space-time) of the fluid description that underlies standard cosmological models were investigated in two clear papers on relativistic hydrodynamics by Eisenhart (1924) and Synge (1937), examining the energy and momentum conservation equations in depth, and laying the ground for interesting investigations much later on.

2.5 THEORY: ALTERNATIVES

During this period various alternatives to the standard theory were proposed.

*2.5.1 Hierarchical Universe.* Charlier (1908, 1922) proposed the idea of a hierachical model even before the expanding universe idea was known. This was adapted to the observed redshifts by Moessard (Curtis 1933).

*2.5.2 Tired Light.* Zwicky (1929a, 1929b) introduced the "tired light" hypothesis, a similar proposal being put by MacMillan (1932).

*2.5.3 Milne's Kinematical Relativity.* Milne mounted a sustained challenge to the standard theory through proposing "kinematic relativity," based on a thorough rethinking of the geometry of cosmology from the ground up (Milne 1933b, 1935). His kinematical theory did not assume the Einstein field equations, but rather emphasized the importance of symmetry assumptions (homogeneity and isotropy), which he introduced as a fundamental principle of cosmology (the "cosmological principle"). He demonstrated that the existence of a space-time metric follows from the properties of null rays in this case, and that the Hubble law follows from this symmetry. He also discussed the observational problem arising from evolution of observed sources and background radiation (Olbers) limits.

However Robertson (1936a, 1936b) and Walker (1936a, 1936b) showed that his universe models were just special cases of the general FLRW family, which embraced all spatially homogeneous, isotropic universe models. These papers also introduced the kinetic theory description of matter in these curved space-times. Walker (1944a) later proved the important general result that if a universe model is everywhere isotropic then it is necessarily spatially homogeneous. This, together with observational evidence of the isotropy of the universe about us, was later to be taken as one of the best justifications for the "cosmological principle"—the assumption that the universe is indeed spatially homogeneous.

*2.5.4 Newtonian Cosmology.* Milne (1934a) and McCrea and Milne (1934) then showed that one could adapt Newtonian theory to obtain quantitative models of an infinite expanding universe. It is curious that it took so long for these dynamic models to be discovered after the (more complex) general relativity models were known, and since Cartan in 1922–1924 had pointed out the structural similarities of Newton's and Einstein's theories. However these models could not explain all of the expected observational relations, since they could not incorporate a satisfactory theory of light propagation.

*2.5.5 Dirac's Theory.* Eddington had long thought about the "large numbers" of nature and their possible causal relations. Now Dirac (1937, 1938) published two influential papers on the foundations of cosmology, discussing his "large numbers" hypothesis and proposing the possibility of a varying gravitational constant $G$. This idea was taken up by Jordan (1937, 1938) and others.

## 2.6 OBSERVATIONAL CONSOLIDATION

On the observation side, great progress was made, first, through the investigations many made of observational relations in relativistic cosmology and the related issue of different definitions of distance in a FLRW space-time, and, second, through further observational evidence.

*2.6.1 Distance and Luminosity.* The nature of the redshift had been clarified particularly through the work of Weyl (1923), who showed how it has its origin in the ratio of observed proper times between the source and observer; Lemaître (1927) gave the basic formula for redshifts in a FLRW universe. These contributions were based on the geometric-optics approximation; von Laue (1931) introduced an examination of the problem from the view of the wave equation, deriving the redshift from Maxwell's equations and their conformal invariance.

De Sitter (1917c) had already discussed aberration and angular size in his universe model; Freedericsz and Schechter (1928) discussed aberration and parallax in the Einstein static, de Sitter, and Friedman universes, also commenting on the problems raised for observations by absorption. Whittaker (1931) gave a definition of area distance and applied it to the de Sitter universe; the expected angular diameter and luminosity of a source in a FLRW space-time was derived by Tolman (1930b). The first person to obtain an explicit magnitude-redshift relation was Kohler (1933), in a far-seeing paper that also touched on entropy and the energy of radiation in FLRW universes and horizons.

Hubble was aware of the problem of selection effects occurring as images are lost due to their low surface brightness (Hubble 1932). With Tolman, he gave a careful theoretical analysis of the nature of observational relations based on source apparent size or on number counts (Hubble and Tolman 1935), without committing themselves to the FLRW geometry or the idea of an expanding universe. They explored the practicalities of observations of surface brightness and photographic magnitudes of nebulas, also introducing the "$K$-correction" term allowing for spectral effects and formulas for number counts to a given limiting magnitude.

Hubble was also apparently the first person to use number counts systematically in a cosmological test (Hubble 1936b) and to realize that the Hubble diagram could be used in principle to determine the current value of the deceleration parameter $q_0$ in a FLRW universe and so to determine the density of matter and the eventual fate of the universe (Tammann 1984). The theoretical relations were extended and corrected by many, e.g., McVittie (1938a); Eddington (1937a); and Robertson (1938). McVittie (1938b) discussed power-series expansions of the observational relations; Heckmann (1942) determined a series expansion for the magnitude-redshift relation that could be compared with observations to determine the second derivative of $R(t)$, i.e., the quantity $q_0$: the apparent magnitude $m$ is given by

$$m = 5\log z + 1.086(1 - q_0)z + O(z^2) + \text{constant}. \tag{6}$$

This relation (later extended to an exact form by Mattig (1958)) was to be the basis of enormous observational effort aimed at determining $q_0$.[9]

Some more fundamental bases of understanding observational relations in general space-times were laid by Kermack, McCrea, and Whittaker (1933) and Etherington (1933), who derived the "reciprocity relation" underlying the relation between area distance and luminosity distance, and so also underlying the brightness relations (see Ellis 1971 for a summary). Temple (1938) introduced optical coordinates for cosmology. McCrea (1935, 1939) in a series of far-sighted papers examined observational relations in general cosmological models, giving power series expansions for possible observations (cf. Kristian and Sachs 1966).

*2.6.2 Redshift-Distance Relation.* Meanwhile observations to consolidate the data on the Hubble diagram, confirm the linear velocity-distance relation, and determine the slope of this relation ("the Hubble constant") proceeded apace. The linear relation was criticized by Shapley (1929) but confirmed by de Sitter (1931b). The data were extended by Hubble and Humason (1931), using a chain of distance indicators and observations extending the redshift range by a factor of 5 to $32 \times 10^6$ pc. They determined a value of $H = 558$ km/sec/Mpc for the Hubble constant. However Oort (1931) suggested revision of this value by a factor of $1/2$.

By 1936, Hubble had accumulated considerable data, summarized in his superb book *The Realm of the Nebulae*, determining a value of 526 km/sec/Mpc for the Hubble constant. However in all this work the interpretation of the observed relation was still uncertain; the velocities presented were regarded as *apparent velocities* (see Section 2.7). Humason (1936) advanced the idea of "standard candles" by publishing a Hubble diagram for the fifth brightest cluster members, with 100 new redshifts including the Ursa Major 2 cluster; with available detector technology, this rather exhausted the observational possibilities of the Mount Wilson 100-inch reflector and further progress stagnated until the 200-inch Hale reflector went into operation in 1949 (Tammann 1984).

*2.6.3 Age Evidence.* As mentioned earlier, on the basis of a "big bang" beginning to the universe, the different time scales in the universe had to be consistent; whether or not they were was already investigated (de Sitter 1933; McVittie 1937). The ages of the stars were estimated by Eddington (1924a), of the earth by Holmes and Lawson (1927), Jeffreys (1928), Aston (1929), and Rutherford (1929), using measurements of the uranium/lead ratio. Consistency with the Hubble constant values was a problem if the cosmological constant was assumed zero (de Sitter 1933), which was one reason for preferring the "Eddington-Lemaître" universes with a postive cosmological constant and long "coasting period" near the Einstein static radius.

*2.6.4 Number Counts, Matter Density.* A direct density estimate by Shapley (1933) suggested the value of $10^{-30}$ gm/cc on average. Hubble extended the observations of galaxies to a thorough investigation of their distances and a demonstration of the near-uniformity of their distribution (Hubble 1934). He suggested on the basis of number counts that the universe must be a high-density universe with a density of $10^{-30}$ gm/cc, but he did not relate this to a particular universe model. Later density estimates taking into account redshift effects in expanding universes led him to even higher density estimates ($10^{-26}$ gm/cc); if anything, he doubted the expanding universe models because of the high densities they implied (Hubble 1936b). All of this was complicated by the absorption and reddening in our galaxy demonstrated by Trumpler (1930).

*2.6.5 Virial Theorem.* The virial theorem method of estimating density was introduced by Zwicky (1933). This was the start of estimates leading to the deduction that there is present hidden (i.e., nonluminous) matter, a possibility emphasized, e.g., by Einstein (1945).

*2.6.6 Extragalactic Gravitational Lenses.* Zwicky (1937a, 1937b) also discussed the idea of gravitational lensing in the context of cosmology, again foreshadowing a method of investigation of mass density that would be significant later.

*2.6.7 Horizons.* Finally, within the standard models, one of the most perplexing issues was that of horizons. It was known that observational horizons existed in the de Sitter universe, but their nature was not well understood. However Weyl (1930) was very aware of the nature of the causal limits implied by the null-cone structure of space-time, and Kohler (1933) looked at the concept of horizons in FLRW universes.

2.7 REPRISE

Thus, during this period, the theoreticians were coming to terms with the new concept of an evolutionary universe. But this was by no means overwhelmingly believed. Thus when Hubble and Humason (1931) published their analysis of the redshift-luminosity data they cautiously proclaimed that they did not wish to stand by any particular explanation of the redshift-distance relation they had observed. They were not yet ready to back the idea of an expanding universe. The idea was clearly still viewed skeptically by many astronomers, and some alternative theories were available.

With hindsight, it is easy to read into the situation what was not there, and to assume that everyone believed in the expanding universe idea during this period; so it is salutary to note that Heber D. Curtis (1933) in his review article, "The Nebulae," in the *Handbuch der Astrophysik* wrote:

> Whether one prefers to assume that light vibrations are slowed up by their passage through space (velocity-distance correlation; spiral velocities apparent) or to postulate a universe that is actually expanding (Lemaître's and other relativity universes; spiral velocities real) or to accept a universe of the Charlier type with the Moessard modification (spiral velocities in part real, in part apparent) is doubtless entirely a matter of individual choice and belief.

Again, the published version of Hubble's superb Silliman memorial lectures *The Realm of the Nebulae* (Hubble 1936c) relegated cosmological theory to the last four pages of a 200-page book, and stated (pp. 120–122):

> The velocity distance relation is ... a general characteristic of our sample of the universe.... If it could be fully interpreted the relation would probably contribute an essential clue to the problem of the structure of the universe.

This same kind of reticence is expressed in a report to the Council of the Royal Astronomical Society, "Extragalactic Nebulae," by J.H. Reynolds (1938).

SUMMARIES OF THE PERIOD

Summaries of the understanding obtained in this period are found in Tolman 1934a; Hubble 1936c; Heckmann 1942; and Einstein 1945. Perusal of these references will show that the concentration of the period was on the geometric and dynamic aspects of the expanding universe models, with observational predictions and tests being gradually developed; however the idea of the hot big bang and associated physics was missed (Tolman came closest to developing the required theory, with Lemaître also coming close to the general ideas to be developed later but misled by the concept of cosmic rays as relics of the initial state of the universe).

# 3. Foundations for a New Era (1945–1960)

Main Themes of the Period:

1. Observationally: new instruments and observations at new wavelengths.
2. Theoretically: hot big bang as possibility, explaining synthesis of light elements.
3. Steady state as alternative, but radio source evidence against it.

After the second world war, a series of observational and theoretical advances laid the basis for the major expansion of cosmology that was to follow (starting in 1965).

3.1 OBSERVATIONAL EVIDENCE

On the observational side, optical observations increased in quality and quantity, leading to further consolidation of information about the Hubble diagram (Tamman 1984). In particular the 200-inch Hale reflector came into operation in 1949 and enabled Humason to obtain many new redshifts, summarized in the major paper by Humason, Mayall, and Sandage (1956). This was then followed by Sandage's major observational program of the following decades. Hoyle (1959) pointed out that minimum apparent angular diameters would occur in FLRW models with a vanishing cosmological constant, and suggested that this could be used as an observational test for $q_0$.

*3.1.1 The Age Problem and Hubble Constant.* There continued to be a major age problem for simple ($\Lambda = 0$) cosmologies until a revision of the distance scale resulted from the realization that there are two different classes of Cepheid variables with different period-luminosity relations (Baade 1952; Thackeray and Wesselink 1953). This led to an increase in the Hubble distance by a factor of about 2.6, and a revised value of 200 km/sec/Mpc for the Hubble constant, greatly alleviating the age problem. Then Sandage (1958) found that what Hubble had identified as bright stars in distant galaxies were in fact HII regions, leading to a further increase of the distance and time-scale estimates by factor of about 2.2, effectively removing the time-scale problem.

*3.1.2 Deceleration Parameter.* The first direct observationally based value for the deceleration parameter (obtained from the Hubble diagram) was given by Baum (1953): $q_0 = 1 \pm 0.5$. This was followed by a higher value ($3.7 \pm 0.8$) obtained by Humason, Mayall, and Sandage (1956). Later work by Sandage (1968) and Gunn and Oke (1975) reverted to the lower figures, but eventually, following pioneering work by Tinsley that was initially rejected by most workers in the field, realization of the importance of the problem of source evolution has led to pessimism about the direct approach to determining $q_0$.

*3.1.3 Density Estimates.* Oort (1959) estimated the density of luminous matter as about $3 \times 10^{-31}$ gm/cc (a low-density universe). Combined with density estimates based on Baum's values for the deceleration parameter in a universe with vanishing cosmological constant, this led to the problem of "missing mass." Ninety-eight per cent of the matter in the universe is then not observed. What form could this matter take? The problem remains a major problem today, when the "inflationary universe" proposal is usually taken as giving a theoretical reason for expecting that the matter density should take the critical value of about $10^{-29}$ gm/cc. Evidence from galactic rotation curves and from the cluster virial theorem points to a density about one tenth of this value, but there are forms of "dark matter" that are very difficult to detect and that could dominate the mass density of the universe without having been observed.

*3.1.4 Statistics.* Hubble had already studied the distribution of galaxies in depth. Neymann and Scott (1952) and Neymann, Scott, and Shane (1953) undertook the first major statistically based studies of the distributions of galaxies, which later would flower in Peebles's series of studies of the galaxy covariance function (see, e.g., Davis and Peebles 1967). Selection effects and measurement problems related to the surface-brightness distribution across a galaxy were studied by Stock and Schücking (1957), laying the foundation for estimating their effects on the statistics of the magnitude-redshift diagram and on catalogs of sources.

## 3.2 OTHER WAVELENGTHS

Most important was the initiation of new instruments allowing observations of the universe at wavelengths other than the optical and ultraviolet.

*3.2.1 Radio.* Radio observations of extraterrestrial objects had started when Jansky detected radio emission from the galactic center (Jansky 1933a, 1933b). The first extragalactic radio source, Cygnus A, was discovered by Reber in 1934 and confirmed by Hey and coworkers in 1946 (Hey, Parsons, and Phillips 1946). M31 was detected as an extragalactic radio source in 1950 (Hanbury-Brown and Hazard 1950). However, initially the nature of the sources was disputed; for example, Martin Ryle, one of the major pioneers in the area, asserted in 1951 that the detected radio sources are not extragalactic (Ryle 1951; see also Lovell 1987) whereas Gold was convinced that they are (Gold 1951). However in 1954 Baade and Minkowski identified Cygnus A with the brightest member of a faint cluster of galaxies, showing that it is definitely extragalactic (Baade and Minkowski 1954). Detailed observation soon followed, and by 1955 Ryle and his coworkers were deploying radio source counts as evidence against the steady-state theory of Bondi, Gold, and Hoyle (Ryle et al. 1955). Detailed study of radio sources then became a major part of observational cosmology, resulting, for example, in the 3C catalog of sources (Bennett 1962); radio source number counts provided evidence of evolution of the universe, and so gave the first observational evidence against a steady state of the universe. Radio observations led to the discovery in 1963 of the first known quasi-stellar object, 3C 273.

Eventually microwave and radio observations were to detect the cosmic black-body relic radiation from the hot big bang; however this required considerable advances in technology because of the tremendous sensitivity needed.

*3.2.2 X-Ray and Infrared.* The rocket and detector technology that was to pave the way to x-ray extragalactic observations was being developed in the period from 1957 to 1962. The resulting flowering of observations would be a major component of the renaissance of observational cosmology to follow from 1965 on, with the x-ray background being discovered in 1962 and the first x-ray galaxy being detected in 1966. Infrared observations of other galaxies began in 1966–1968, but both technologies would not come to fruition for some decades.

3.3 CREATION OF THE UNIVERSE

Lemaître (1945, 1950) was still developing his idea of the "primeval atom," but regarding it as a big bang expansion starting from a (very small) Einstein static initial state, at late times driven to an exponential expansion by the cosmological constant. However, the decisive steps in understanding the nature of the big bang came in relation to element creation.

*3.3.1 Element Formation and CMWBR.* Developing early ideas of his on the nonequilibrium nature of the abundances of elements, George Gamow (1946b, 1948a, 1948b) led the way in applying the newly understood ideas of nuclear physics to the early universe and in particular relating the expansion of the universe to the formation of the light elements. He proposed creation of heavier elements by nonequilibrium neutron capture from an initial pure neutron state, pointing out (Gamow 1946b) that the required temperature of $10^{9\circ}$ K for this process to happen could be found in the hot early universe. It is ironic that Eddington had a long time earlier realized the need for these high temperatures. In his justly famous book *The Internal Constitution of the Stars* (Eddington 1926), he states:

> The helium ... must have been put together at some time and some place. We will not argue with the critic who urges that stars are not hot enough for this process; we tell him to go and find a *hotter place*.

Eddington did not realize that that hotter place existed in the very early universe when radiation dominated the evolution of FLRW models, as studied by Tolman (1931a). Gamow (1948b) developed the density-temperature-time relation for the radiation-dominated early universe, pointing out that the high expansion rate implied that there was only a short time available for nuclear reactions.

This work was then developed further by Alpher, Herman, and coworkers (1948–1953). Alpher and Herman (1948a) corrected Gamow's pioneering paper, and (Alpher and Herman 1948b) realized the problem of getting around the mass gap at 5 and 8. They predicted (Alpher and Herman 1949, 1950) the cosmic microwave background radiation at 5° K, very close to the 3° K observed in 1965. Their work, apparently repeating calculations by Fermi and Turkevich, seems to have been the first time an electronic computer was used in cosmological calculations. Hayashi (1950) then pointed out the importance of pair production in the early universe and that the neutron abundance would be controlled by the proton-neutron equilibrium ratio. Alpher and Herman (1951) and Alpher, Follin, and Herman (1953) then carried out detailed nucleosynthesis calculations, including the $p/n$ ratio, pair annihilation, and a detailed temperature-time table, resulting in a rather complete theory of the evolution of the elements (see, e.g., Gamow 1956a) and even the prediction of the existence of remnant black-body radiation at a few-degrees Kelvin (Alpher and Herman 1949; Gamow 1956b). Apparently Gamow was told that it would be impossible to detect such radiation. The conclusion was that one could synthesize the light elements in the early universe but not heavy elements.

It was not realized at the time that identification by McKellar (1940) of an interstellar absorption line, measured by Dunham and Adams (Adams 1941) as due to cyanogen excited to 2.7° K, was evidence for cosmic black-body relic radiation, even though a measurement by Adams (1941) confirmed a prediction made by identifying the lines as due to thermal excitation. Dicke and coworkers put upper limits of 20° K on possible background radiation in 1946 (Dicke, Beringer, and Vane 1946) but were unaware of the possible relation to Gamow's work, published in the same volume of the *Physical Review*. The failure to identify positively the relic radiation is one of the celebrated lost opportunities of cosmology, but this judgment, made on looking back with hindsight, must be tempered by a realization of limits of the technology at the time (Harwit 1981, section 2.40).

A while later, as a result of considering the problem of the origin of the elements in the steady-state theory, Burbidge, Burbidge, Fowler, and Hoyle developed in their celebrated paper (1957) the theory of the stellar formation of elements. It became clear that heavy elements could be created but that deuterium could not be formed in sufficient quantities in stars. Peebles picked up this challenge, apparently unaware of previous work on the topic, developing the theory of primordial nucleosynthesis again (Peebles 1966) and predicting the existence of background radiation. The radiation was eventually discovered serendipitously by Penzias and Wilson in 1965 (Penzias and Wilson 1965), just as Peebles and Dicke were setting out to find it (Dicke et al. 1965). It was this discovery above all else that finally convinced physicists that the universe really did have its origin in a hot big bang (and provided evidence against the alternative polyneutron cold big bang theory for element creation in an expanding universe (Peierls, Singwe, and Wroe 1952)).

*3.3.2 Galaxy Formation.* Lemaître's early work on galaxy formation (1933c, 1934) was followed by papers by Gamow and Teller (1939a) on the Jeans mass in a matter-dominated expanding universe and hence the creation of galaxies through gravitational instability (Gamow and Teller 1939b). Gamow (1949a) considered the effect of radiation on condensations, thus already foreshadowing the major themes of much later discussion (see Peebles's review [1984]). Bonnor (1957) showed that a Newtonian approach could be used to develop Jeans's formula for gravitational instability in an expanding universe.

An alternative physical approach was suggested by von Weizsäcker (1951) and Gamow (1952a), who considered the role of turbulence in galaxy formation, again a theme to recur later.

A more mathematical approach was shown in the general relativistic perturbation calculation by Lifschitz (1946) obtaining the growth law for density perturbations in an expanding universe. This outstanding paper introduced methods used by virtually all subsequent investigators; the next major developments were the relation of perturbations to the microwave background radiation (Sachs and Wolfe 1967), on the one hand, and the development of a gauge-free approach to the problem, in particular by Bardeen (1980), on the other hand.

### 3.4 Theoretical Basis for New Developments

Further theoretical advances were to result from the application of more advanced mathematical techniques to the study of cosmology and the study of more general geometries.

*3.4.1 Anisotropic Models.* The fluid in the FLRW models moves in an extremely special way: without rotation, shear, or acceleration. Gamow (1946a) raised the question of whether the universe might be rotating. Investigation of such models called for examination of more general geometries.

The simplest anisotropic universe is the Bianchi I spatially homogeneous model, whose metric (in suitable coordinates) depends on only one variable. The vacuum metric with this symmetry had been investigated by Kasner (1925) and the fluid version by Lemaître (1933a). However this is the most special of the possible symmetries of a spatially homogeneous anisotropic universe, and cannot rotate. The general theory of symmetry groups had been developed by Sophus Lie (1893) and Luigi Bianchi (1897, 1918), who had classified the possible low-dimensional symmetry groups many years before (Bianchi 1918). They were introduced into cosmology by the outstanding logician Kurt Gödel, who discussed relativity with Einstein at the Institute for Advanced Study in Princeton, and gave the first exact solution of the field equations that exhibits cosmological vorticity (Gödel 1949). This was a space-time homogeneous rotating universe model that demonstrated that the existence of causality violations (the existence of closed time-like lines) is compatible with Einstein's field equations, and that there might exist no cosmic time whatever in a rotating universe model. This solution was stationary and predicted no redshifts, but a later paper (Gödel 1952) examined the properties of a family of expanding and rotating spatially homogeneous universe models that could be reasonable models of the observed universe. These were Bianchi

IX universe models; Gödel proved some intriguing theorems about their properties. In a masterly paper at about the same time, Taub (1951) presented the mathematics of all spatially homogeneous vacuum Bianchi models, and exhibited some such specific solutions of the field equations. Together these papers paved the way for many later studies of spatially homogeneous but anisotropic universe models, initially by Heckmann and Schücking (1959a, 1962), who also (1955) determined the Newtonian anisotropic cosmological models analogous to the simple relativistic Bianchi models. Particularly through the work of Thorne (1967), Misner (1968, 1969a, 1969b), Hawking (1968), and Collins and Hawking (1973a, 1973b), the Bianchi universes have provided many new insights into the possible behavior of the universe.

*3.4.2 Inhomogeneous Models.* The real universe is inhomogeneous. Bondi (1948a) and Omer (1949) examined spherically symmetric solutions of the field equations (which had been developed earlier by Lemaître (1933c) and Tolman (1934b)); these have sometimes since been used to examine features of galaxy formation.

A different approach to the study of inhomogeneities in a FLRW model was initiated by Einstein and Strauss (1945), using a "Swiss cheese" model: a number of static Schwarschild solutions imbedded in a FLRW universe, with appropriate junction conditions at the boundaries. They did so to examine an issue raised many years before by de Sitter (1930f): does the expansion of the universe affect the solar system? According to their model, the expansion had no effect whatever. Schücking (1954) developed the theme further; later these models were used by Kantowski and others as useful examples in which to examine gravitational lensing.

Walker (1944a) meanwhile clarified the relation between isotropy and homogeneity by proving that, if a space-time is spherically symmetric everywhere, then it is spatially homogeneous. This was later developed further by Ehlers, Geren, and Sachs (1968) and became one of the foundations of the standard argument for the homogeneity of the universe (see, e.g., Hawking and Ellis 1973).

Finally the study of the general behavior of fluids in general relativity, initiated by Eisenhart (1924) and Synge (1937), was carried further by Lichnerowicz (1955) and Ehlers (1961), examining energy-momentum conservation, vorticity propagation, and the field equations for fluids.

*3.4.3 Singularity Theorems.* One of the recurring themes in cosmology has been whether the FLRW initial singularities (whose existence was implied by the Tolman and Ward 1932 paper) are a result only of the high symmetry of these models. Lemaître (1933a) had extended the result to Bianchi I universes. Now the Indian mathematician Amalkamur Raychaudhuri (1955b) obtained the first general singularity theorem, demonstrating that singularities would occur at the origin of any universe moving without acceleration or rotation, and so would occur in a large class of inhomogeneous and anisotropic universe models. The "Raychaudhuri equation" central to his proof later played a fundamental role in the proof by Hawking and Penrose (Hawking 1967; Hawking and Penrose 1969) of the existence of singularities in general cosmological models (see Tipler, Clarke, and Ellis 1980 for a summary).

Given the existence of singularities in a large class of models, their interpretation was still open to question. Einstein's last words on the singular state were:

> One may ... not assume the validity of the [gravitational field] equations for very high density of field and matter, and one may not conclude that the "beginning of expansion" must mean a singularity in the mathematical sense. (Einstein 1956, p. 129)

He apparently hoped that the presumed unified field equations that would hold under such conditions might not show singularities (see Pais 1982).

*3.4.4 Horizons and Causality.* The issue of the nature of causality and time in cosmology had been raised in an acute form by Gödel (Schilpp 1949) on the basis of his exact rotating solution. Now the confusion surrounding the concept of horizons in cosmology was dispelled by a clear paper by Wolfgang Rindler (1956) defining precisely the concepts of a particle horizon and an event horizon in a FLRW universe. Together with Gödel's papers, this laid the foundations for fundamental studies of the causal structure of space-times, a topic that had been neglected except for work by Robb (1921, 1936) and some remarks by Weyl many years before (Weyl 1924, 1930). Heckmann and Schücking (1959a) and Hoyle (1962) particularly emphasized the restrictions on verifiability in cosmology resulting from our being able to obtain detailed information about distant regions only by null-cone observations. Later Roger Penrose further clarified the nature of horizons in cosmology by using conformal diagrams to examine their causal structure (Penrose 1964); and problems raised by the isotropy of the microwave background radiation, despite the existence of particle horizons, became a central issue leading to the development of the "inflationary universe" idea (Guth 1981).

3.5 THE STEADY-STATE THEORY

The major alternative to the standard theory during this period was the steady state theory (Bondi and Gold 1948; Hoyle 1948). This was proposed because, on the one hand, at that time there was a major time-scale problem (resolved within ten years because of reevaluation of the Hubble constant, see Section 3.1.1); and on the other hand, because it took to its logical conclusion Milne's idea of a cosmological principle (i.e., that the universe must have perfect symmetry), leading to the "perfect cosmological principle" (Bondi 1952): the universe is not only spatially homogeneous and isotropic but is also unchanging in time. If the universe is expanding, this leads uniquely (as was shown by Tolman and Robertson in 1929) to the expanding stationary form of the de Sitter universe. However to have nonzero matter present, one must abandon Einstein's field equations and the usual conservation equations, replacing them with the supposition that matter is continually created as the universe expands, in order to keep the density of matter constant; Hoyle (1948, 1949) proposed an alternative set of field equations incorporating a creation field ("$C$-field") and resulting in modified conservation equations. McCrea (1951) later showed that one could represent the steady-state universe as a general relativity model with the exceptional equation of state $\rho + p = 0$.

The theory was a great stimulus to the study of stellar evolution and the origin of the elements, leading to the paper on stellar nucleosynthesis by Burbidge et al. (1957) and to the realization that primordial nucleosynthesis must be seriously considered as discussed previously (Section 3.3.1). The first significant evidence against the theory was the radio source counts (Ryle et al. 1955). Quasi-stellar counts also were evidence against it, but the final blow that really killed it was the detection of microwave background radiation during the following era (Penzias and Wilson 1965; Dicke et al. 1965), providing a firm consolidation of the expanding universe idea, the microwave radiation observations, and the theory of nucleosynthesis together with measurements of element abundances (Peebles 1966; Wagoner, Fowler, and Hoyle 1968). This was what led finally to widespread acceptance of the standard theory. However the steady-state theory in effect eventually made a new appearance in the guise of Guth's inflationary universe model (Guth 1981) for the very early universe, when the dominance of quantum fields could indeed perhaps lead to the exceptional equation of state and a consequent steady-state exponential expansion.

3.6 OTHER DOUBTS

However, to get a full perspective on this era, we must realize that as well as the full-blown acceptance of the expanding universe model by many at this time (to be shortly confirmed by the discovery of microwave background radiation), there were also those who still doubted this model without proposing a definite alternative such as the steady-state theory. One might perhaps specifically mention Edwin Hubble who, in his George Darwin Lecture "The Law of Redshifts" delivered on May 8, 1953, stated:

> It is important that the law [of redshifts—the correlation between distances of nebulae and displacements of their spectra] be formulated as an empirical relation between observed data out to the limits of the greatest telescope. Then as precision increases the array of possible interpretations permitted by the uncertainties in the observations will correspondingly be reduced. Ultimately when a definite formulation has been achieved free from systematic errors and with reasonably small probable errors, the number of competing interpretations will be reduced to a minimum. (Hubble 1953)

In this lecture Hubble, regarded by many as the person who provided observational vindication for the concept of the expanding universe, in fact did not adopt this model or any other theoretical interpretation of the observational data. Apparently the problem was that the data indicated too high a value for $q_0$, leading him to favor the idea of nonvelocity redshifts.

This kind of doubt continued for some into the later era, even after the discovery of microwave background radiation, because of issues such as the problem of the nature of quasi-stellar objects, and observational evidence of "discordant redshifts," taken to indicate that redshifts are due to some unknown cause rather than to the expansion of the universe (see Field, Arp, and Bahcall 1973). Some alternative theories for nonvelocity redshifts, such as Segal's theory (Segal 1976), have been presented, and it has also been suggested that a hierarchical model is compatible with recent evidence (de Vaucouleurs 1970). Thus, to the present day, the standard theory has not attained universal acceptance, despite the coherence of the evidence supporting this view.

SUMMARIES OF THE PERIOD

Excellent contemporary summaries of the period are in Bondi (1952, 1960), McCrea (1951), McVittie (1956), Heckmann and Schücking (1959b, 1959c), Hoyle (1962), while good popular surveys are in *Scientific American* (September 1956). It is striking how, despite the work of Gamow, Alpher, Herman, and Hayashi, a discussion of the hot big bang and the creation of the elements is not central in any of these accounts except the article by Gamow in *Scientific American* (Gamow 1956a). Rather they concentrated on the issue of the observational tests for $q_0$ and for the steady-state universe, missing the ideas and clues already there (such as the cyanogen excitation measurements) that would be central in the coming major development of theory and observation. McVittie (1956) omitted all mention of the creation of the elements; Bondi (1960) discusses it briefly, but discounts it because it was then realized that heavy elements could be built in stars but not in the early universe. Thus the importance of nucleosynthesis of the light elements in the early universe, and the way this is accessible to observational testing, was not realized. It is also striking how some of the most important work paving the way for later theoretical developments is not mentioned in as excellent a book as Bondi 1960, which, for example, does not refer to Gödel 1949, 1952, Raychaudhuri 1955b, Taub 1951, or Rindler 1956.

## 4. The New Era (From 1960)

Main Themes of the Period:

1. Vindication of the hot big bang idea, with its major observational consequences, particularly the existence of microwave background radiation, confirmed.
2. Observations at many wavelengths led to extensive analysis of matter-radiation interactions in the expanding universe, and of the possible existence of "dark matter" (as yet undetected).

3. Theoretical analyses of anisotropic and inhomogeneous models greatly extended, in particular leading to the very general Hawking-Penrose singularity theorems and their application to cosmology.

After 1965, a revitalization of observational cosmology took place, leading in turn to an exponential growth in the study of physical and astrophysical cosmology. Space limitations prevent more than a cursory mention of these events here; however some key references have been included in the bibliography.

Observational cosmology was revitalized by Sandage's (1961) paper systematically treating the observational consequences of the different FLRW universe models, and leading to extensive observational programs to derive the deceleration parameter from magnitude-redshift observations. However the problem of the unknown nature of source evolution, mentioned by Robertson (1956) and Humason et al. (1956), and then investigated in depth by Tinsley in particular (e.g., Tinsley 1973), led to this program's failing to produce a unique value for $q_0$ and a prediction of the future of the universe.

The most important event during this era was the detection of background radiation by Penzias and Wilson (1965) as a result of measurements undertaken to investigate noise that was interfering with communications satellites, just when Dicke et al. (1965) were setting out to search for this relic black-body radiation from the hot big bang, which had been predicted by Alpher and Herman (1948a, 1949) and was discussed by Doroshkevich and Novikov (1964) and Zeldovich (1965). It was this detection, combined with its relation to the theory of primordial nucleosynthesis (Gamow 1956a, summarizing work of Alpher, Herman, and Gamow; Hoyle and Tayler, 1964; Zeldovich 1965; Peebles 1966, 1971; Wagoner, Fowler, and Hoyle 1968; Weinberg 1972, 1977) that finally convinced many physicists that cosmology is a subject of real physical interest, and led to the application of modern theories of elementary particle physics to the study of the evolution of the early universe (see, e.g., Zee 1984).

Since then there has been a vast explosion in the literature, on the one hand investigating the background radiation at all wavelengths, and its interaction with the matter in the universe; and, on the other hand, applying exotic physical theories of many sorts to the very early universe. Major advances in observational techniques at all wavelengths have led to greatly increased knowledge of galaxies and their clustering. More general (anisotropic and inhomogeneous) geometries have been investigated in depth, as have other theories of gravity; galaxy clustering observations and galaxy formation theories have been the subject of intense activity; gravitational lensing has been observed; and the missing mass and age problems have remained issues central to observational cosmology. Thus the "renaissance of observational cosmology" (Sciama 1971) of the 1960s and 1970s has been followed by a tremendous flourishing of speculative theoretical cosmology in the 1980s, focusing on such issues as grand unification and "inflation" in the early universe, the possible existence of cosmic strings, and the quantum creation of the universe. It still remains to be seen how much of this theory is testable by experiment, and survives that test. This era is still in progress today.

As in the past in the history of cosmology, some of the widely held currently cherished beliefs are based on dogma rather than physics or observational evidence. An example is the widely made claim that an inflationary universe (Guth 1981; Zee 1984) necessarily leads to a present density parameter $\Omega$ that is very close to unity. This is not true: while inflation makes it much more likely that $\Omega$ will be close to unity than if there is no inflation, this does not imply that $\Omega = 1$ today; on the contrary, there are inflationary universe models leading to any desired present-day value for $\Omega$ (Ellis 1988; Madsen and Ellis 1988). The inflationary idea is important because it can solve the horizon problem, but it does not necessarily imply that most of the matter in the universe is "dark matter." Thus a great deal of current astronomical and theoretical activity aimed at searching for such matter (which is quite worthwhile on other grounds) is not necessitated by the inflationary-universe idea, contrary to many claims in the published literature.

Another issue worth noting is the motivation for the inflationary universe proposal: it is said to solve a series of problems now known as the flatness problem, the horizon problem, and the monopole problem (Guth 1981). The interesting thing from a historical point of view is the question: why have these problems apparently come into being in 1981—why did they not exist in the literature before? The monopole problem is different from the other two, because it is a consequence of grand unified theories, and so can have arisen only recently (after these theories were discovered). However the horizon problem should have been recognized soon after the nature of particle horizons was clarified (Rindler 1956), becoming acute once the high degree of isotropy of the microwave background was proven (Partridge and Wilkinson 1967). Indeed Penrose (1968) emphasized aspects of the problem and Misner (1968, 1969a) realized its significance as regards the background radiation almost at once, but most cosmologists seem to have become aware of the issue only much later (once a remedy—namely the inflationary-universe idea—had been proposed). The flatness (or age) problem is more a numerological one, and if it is a problem today it should have been so since estimates of the Hubble constant and the density of matter were available in the 1930s. The issue seems to have been first raised in recent times by Dicke and Peebles (1979) in an interesting, reflective article, but again only seems to have become of general concern after a remedy had been proposed for it (Guth 1981).

Both these examples suggest that working cosmologists should remain very cautious of the bandwagon effect. It may be that the history of cosmology contains cautionary tales useful to us even today.

*Acknowledgments.* I thank Jürgen Ehlers, Malcolm MacCallum, John Stachel, and Don Howard for informative and useful comments that have improved this paper, and Archie Maurellis for assistance with the references.

NOTES

[1] The years 1945 and 1960 have been chosen for definiteness; they could each be varied by a few years, but not much more.

[2] More precisely, $\rho + p = 0$, which for normal fluids implies $\rho = p = 0$.

[3] Based on a time-like Killing vector; the metric form in these coordinates is *not* a FLRW form.

[4] The first textbook was Weyl's *Raum-Zeit-Materie*, 1918.

[5] This is an integrated form of a differential relation given by Ehlers 1961, and Ellis 1971.

[6] Peebles (1984) comments "it is curious that the crucial paragraphs describing how Lemaître estimated $H_0$ and assessed the evidence for linearity were dropped from the 1931 English translation."

[7] This feature is somewhat hidden, perhaps even misrepresented, in Robertson's 1933 survey article.

[8] It is not clear how important they considered this paper, see Chandrasekhar 1983, p. 38.

[9] The early papers were expressed in terms of $R^{\cdot\cdot}$ rather than $q_0$, which was only defined by Robertson in 1955.

SELECTED BIBLIOGRAPHY ON THE HISTORY OF RELATIVISTIC COSMOLOGY

Abbreviations used:

*A. J.*: *Astronomical Journal*
*Adv. Phys.*: *Advances in Physics*
*Am. J. Math.*: *American Journal of Mathematics*
*Am. J. Phys.*: *American Journal of Physics*
*Ann. Math.*: *Annals of Mathematics*
*Ann. d. Physik*: *Annalen der Physik*
*Ann. Soc. Sci. Bruxelles*: *Société Scientifique de Bruxelles. Annales*
*Ap. J.*: *Astrophysical Journal*
*Arkiv. Mat. Ast. Fys.*: *Arkiv foer Matematik, Astronomi, och Fysik*
*Astr. Nach.*: *Astronomische Nachrichten*
*Astrophys. Sp. Sci.*: *Astrophysics and Space Science*
*Bull. Ast. Inst. Neth.*: *Bulletin of the Astronomical Institute of the Netherlands*
*Com. Math. Phys.*: *Communications in Mathematical Physics*
*C. R. Acad. Sci.*: *Académie des Sciences* (Paris). *Comptes Rendus*
*Helv. Phys. Acta*: *Helvetica Physica Acta*
*J. Math. Phys.*: *Journal of Mathematical Physics*
*Mem. RAS*: *Memoirs of the Royal Astronomical Society*
*MNRAS*: *Monthly Notices of the Royal Astronomical Society*
*Naturwiss.*: *Die Naturwissenschaften*
*Nach. Ges. Gött. Wiss.*: *Gesellschaft der Wissenschaften zu Göttingen. Nachrichten*
*Observ.*: *The Observatory*
*PASP*: *Publications of the Astronomical Society of the Pacific*
*Phil. Mag.*: *Philosophical Magazine*
*Phil. Trans.*: *Philosophical Transactions of the Royal Society* (London)
*Phys. Rev.*: *Physical Review*

*Phys. Rev. Lett.*: *Physical Review Letters*
*Phys. Z.*: *Physikalische Zeitschrift*
*PNAS*: *Proceedings of the National Academy of Sciences* (Washington)
*Pop. Ast.*: *Popular Astronomy*
*Preuss. Akad. Wiss. Sitzb.*: *Königlich Preussische Akademie der Wissenschaften* (Berlin). *Sitzungsberichte*
*Proc. Acad. Wet. Amst.*: *Proceedings of the Section of Sciences, Koninklijke Akademie van Wetenschappen te Amsterdam*
*Proc. Edin. Math. Soc.*: *Proceedings of the Edinburgh Mathematical Society*
*Proc. Inst. Radio Eng. Aust.*: *Proceedings of the Institution of Radio and Electronics Engineers of Australia.*
*Proc. Lond. Math. Soc.*: *Proceedings of the London Mathematical Society*
*Proc. Phys. Soc. Lond.*: *Proceedings of the Physical Society London*
*Proc. Roy. Soc.*: *Proceedings of the Royal Society* (London)
*Proc. Roy. Soc. Edin.*: *Proceedings of the Royal Society* (Edinburgh)
*QJRAS*: *Quarterly Journal of the Royal Astronomical Society*
*Qu. J. Math.*: *Quarterly Journal of Mathematics* (Oxford)
*Rev. Mod. Phys.*: *Reviews of Modern Physics*
*Rev. d. Quest. Sci.*: *Revue des Questions Scientifique*
*Sci. Am.*: *Scientific American*
*Trans. IAU*: *Transactions of the International Astronomical Union*
*Trans. Am. Math. Soc.*: *Transactions of the American Mathematical Society*
*Z. Ap.*: *Zeitschrift für Astrophysik*
*Z. Phys.*: *Zeitschrift für Physik*

Adams, W.S. (1941). "Some Results with the Coude Spectrograph of the Mount Wilson Observatory." *Ap. J.* 93: 11–23. [Absorption line from CH, CN molecules which $\Rightarrow 2.3°K$ excitation temperature, see McKellar: CMBR detected but not recognized.]

Alpher R.A., Bethe, H.A., and Gamow, G. (1948). "The Origin of Chemical Elements." *Phys. Rev.* 73: 803–804. [Elements from neutron capture; timescales $\Rightarrow$ abundances: hot big bang.]

Alpher, R.A., Follin, J.W., and Herman, R.C. (1953). "Physical Conditions in the Initial Stages of the Expanding Universe." *Phys. Rev.* 92: 1347–1361. [Detailed element calculations, matter/radiation dynamics and $p/n$ ratio (cf. Hayashi) from computer model. Timetable for HBB (Table IV).]

Alpher, R.A. and Herman, R.C. (1948a). "Evolution of the Universe." *Nature* 162: 774–775. [Corrects Gamow's pioneering paper; initially pure neutrons, successive neutron capture. No beta-decay, but HBB and CMWBR: "The temperature in the universe at the current time is found to be about 5°K".]

——— (1948b). "On the Relative Abundance of the Elements." *Phys. Rev.* 74: 1737–1742. [Elements: included neutron decay. Gap at 5 and 8.]

——— (1949). "Remarks on the Evolution of the Expanding Universe." *Phys. Rev.* 75: 1089–1095. [CMWBR prediction: see equation (12d), "which corresponds to a temperature of 5°K now"; cf. Gamow card to Penzias.]

——— (1950). "Theory of the Origin and Relative Abundance Distribution of the Elements." *Rev. Mod. Phys.* 22: 153–212. [Matter/radiation dynamics. Repeat of Fermi/Turkevich calculations; used Computer. Up to Helium 4, no way around mass gap at 5, 8. Complete references up to 1950.]

——— (1951). "Neutron Capture Theory of Element Formation in an Expanding Universe." *Phys. Rev.* 84: 60–68. [Elements including expansion and beta-decay.]
——— (1972). "Reflections on the Big Bang." In Reines 1972.
Aston, F.W. (1929). "The Mass Spectrum of Uranium Lead and the Atomic Weight of Protoactinium." *Nature* 123: 313. [Age of U235 $\Rightarrow t = 10^{29}$ yrs; see Rutherford.]
Baade, W. (1952). "Extragalactic Nebulae. Report to IAU Commission 28." *International Astronomical Union. Transactions* 8: 397–399. [Two kinds of cepheids $\Rightarrow$ time-scale increase by factor of 2.6.]
——— (1956). "The Period-Luminosity Relation of the Cepheids." *PASP* 68: 5–16. [Time-scale.]
Baade, W., and Minkowski, R. (1954). "Identification of the Radio Sources in Cassiopeia, Cygnus A and Puppis A." *Ap. J.* 119: 206–214. [Brightest member of faint cluster of galaxies $\Rightarrow$ outside milky way.]
Bailian, R., Audouze, J., and Schramm, D.N., eds. (1979). *Physical Cosmology: Les Houches 1979 (Session 32)*. North Holland. [Review articles by Sandage, Tammann, Tinsley, Peebles, Wagoner, Rees, et al.]
Bardeen, J. (1980). "Gauge-Invariant Cosmological Perturbations." *Phys. Rev. D* 22: 1882–1905.
Barrow, J.D., and Tipler, F.J. (1985). *The Anthropic Cosmological Principle*. Oxford and New York: Oxford University Press.
Bass, R.W., and Witten, L. (1957). "Remark on Cosmological Models." *Rev. Mod. Phys.* 29: 452–453. [Topological and global issues.]
Baum, W. (1953). "The Cosmological Distance Scale." *A. J.* 58: 211. [$q_0$ value.]
——— (1957). "Photoelectric Determination of Redshifts Beyond 0.2c." *A. J.* 62: 6–7. [$q_0 = 1 \pm 1/2$.]
Belinskii, V.A., Khalatnikov, I.M., and Lifshitz, E.M. (1970). "Oscillatory Approach to a Singular Point in Relativistic Cosmology." *Adv. Phys.* 19: 523–573. See also *Uspekhi Fizicheskii Nauk* 102: 463. [*Soviet Physics. Uspekhi* 13 (1971): 745.]
Bennett, A.S. (1962). "The Revised 3C Catalogue of Radio Sources." *Mem. RAS* 67: 163–172.
Berendzen, R., Hart, R., and Seeley, D. (1976). *Man Discovers the Galaxies*. New York: Science History Publications.
Berger, A., ed., (1984). *The Big Bang and Georges Lemaître*. Reidel.
Bianchi, L. (1897). "Sugli spazî a tre dimensioni che ammettono un gruppo continuo di movimenti." *Memorie di matematica e di fisica della Società italiana delle scienze* 11: 267–352. [Classifies symmetries of 3-D spaces.]
——— (1918). *Lezioni sulla teoria dei gruppi continui finiti transformazioni*. Pisa: Spoerri.
Bok, B. (1946). "The Timescale of the Universe." *MNRAS* 106: 61–75.
Bolton, J. (1948). "Discrete Sources of Galactic Radio Noise." *Nature* 162: 141–142. [Six discrete radio sources.]
Bondi, H. (1948a). "Spherically Symmetric Models in General Relativity." *MNRAS* 107: 410–425. [Inhomogeneous (Lemaître/Tolman) exact solution.]
——— (1948b). "Review of Cosmology (Council Note on Cosmology)." *MNRAS* 108: 104–120. [Survey.]
——— (1952). *Cosmology*. Cambridge: Cambridge University Press. [Classic text, emphasizes Olbers, role of cosmological principles.]
——— (1955). "Theories of Cosmology." *Advancement of Science* 12: 33.
——— (1960). *Cosmology*. Rev. ed. Cambridge: Cambridge University Press.

Bondi, H., and Gold, T. (1948). "The Steady State Theory of the Expanding Universe." *MNRAS* 108: 252–270. [Steady-state proposal; age discrepancy in standard model.]
——— (1954). "The Steady State Theory of the Homogeneous Expanding Universe." *Observ.* 73: 36–37.
Bondi, H., and McVittie, G.C. (1948). "Observation and Theory in Cosmology." *Observ.* 68: 111–113. [Letter exchange.]
Bonnor, W.B. (1954). "The Stability of Cosmological Models." *Z. Ap.* 35: 10–20. [Tolman-FLRW models joined by O'Brien-Synge boundary conditions.]
——— (1955). "The Instability of the Einstein Universe." *MNRAS* 115: 310–322.
——— (1957). "Jeans Formula for Gravitational Instability." *MNRAS* 117: 104–117. [Newtonian galaxy formation in expanding universe.]
Brans, C. and Dicke, R.H. (1961). "Mach's Principle and a Relativistic Theory of Gravitation." *Phys. Rev.* 124: 925–935. [Variable $G$ theory and cosmology.]
Buc, H.E. (1932). "The Red Shift." *Journal of the Franklin Institute* 21: 197–198. [Tired light.]
Burbridge, E.M., Burbridge, G., Fowler, W.A., and Hoyle, F. (1957). "Synthesis of the Elements in Stars." *Rev. Mod. Phys.* 29: 547–650. [Classic paper on formation of elements in stars.]
Byram, E.T., Chubb, T.A., and Friedman, H. (1966). "Cosmic X-Ray Sources, Galactic and Extra-Galactic." *Science* 152: 66–71. [First x-ray galaxy (M87).]
Callan, C., Dicke, R.H., and Peebles, P.J.E. (1965). "Cosmology and Newtonian Mechanics." *Am. J. Phys.* 33: 105–108.
Chalmers, J.A., and Chalmers, B. (1935). "The Expanding Universe—An Alternative View." *Phil. Mag.* 19: 436–446. [Due to variation of $h$ with time.]
Chandrasekhar, S. (1983). *Eddington: The Most Distinguished Astrophysicist of His Time*. Cambridge and New York: Cambridge University Press.
Chandrasekhar, S., and Heinrich, L.R. (1942). "An Attempt to Interpret the Relative Abundances of the Elements and their Isotopes." *Ap. J.* 95: 288–298. [Origin of elements.]
Charlier, C.W.L. (1908). "Wie eine unendliche Welt aufgebaut sein kann." *Arkiv. Mat. Ast. Fys.* 4 (no. 24): 1–15. [Hierarchical model ⇒ solve Olbers's paradox.]
——— (1922). "How an Infinite World May Be Built Up." *Arkiv. Mat. Ast. Fys.* 16 (no. 22): 1–34. [Hierarchical: see North 1965, pp. 20–22.]
——— (1925). "On the Structure of the Universe." *PASP* 37: 177–191.
Christoffel, E.B. (1869). "On the Transformation of Homogeneous Differential Forms of the Second Order." *Journal für die reine und angewandte Mathematik* 70: 46–70.
Collins, C.B., and Hawking, S.W. (1973a). "The Rotation and Distortion of the Universe." *MNRAS* 162: 307–320.
——— (1973b). "Why is the Universe Isotropic?" *Ap. J.* 180: 317–334.
Combridge, J.T. (1965). *Bibliography of Relativity and Gravitation Theory: 1921 to 1937*. London: King's College, University of London.
Conklin, E.K. (1969). "Velocity of the Earth with Respect to the Cosmic Background Radiation." *Nature* 22: 971–972.
Conklin, E.K., and Bracewell, R.N. (1967). "Isotropy of Cosmic Background Radiation at 10 690 MHz." *Phys. Rev. Lett.* 18: 614. [No observed anisotropy of CMWBR at small scales.]
Curtis, H.D. (1917). "Novae in Spiral Nebulae and the Island Universe Theory." *PASP* 29: 180–182; 206–207.

——— (1933). "The Nebulae." In *Handbuch der Astrophysik*. Vol 2, part 1, secs. 69–79. H.D. Curtis, B. Lindblad, K. Lundmark, and H. Shapley, eds. Berlin: Julius Springer.

Davis, M., and Peebles, P.J.E. (1967). "On the Integration of the BBGKY Equations for the Development of Strongly Non-Linear Clustering in an Expanding Universe." *Ap. J. Suppl.* 34: 425–450. [Theory of origin of covariance function.]

de Cheseaux, J.-P.-L. (1744). *Traité de la comète qui a paru en décembre 1743 et en janvier, février et mars 1744*. Lausanne and Geneva: Bousquet. [Olbers's paradox; see Jaki, Harrison.]

de Sitter, W. (1916a). "Space, Time, and Gravitation." *Observ.* 39: 412–419. [Bending of light.]

——— (1916b). "On Einstein's Theory of Gravitation and Its Astronomical Consequences. I and II." *MNRAS* 76: 699–738; 77: 155–184. [Schwarzschild solution.]

——— (1917a). "On the Curvature of Space." *Proc. Akad. Wet. Amst.* 20: 229–243; 1309–1312.

——— (1917b). "On the Relativity of Inertia: Remarks Concerning Einstein's Latest Hypothesis." *Proc. Akad. Wet. Amst.* 19: 1217–1225. [Original treatment of de Sitter universe, second general relativistic cosmological model static, empty.]

——— (1917c). "On Einstein's Theory of Gravitation and Its Astronomical Consequences III." *MNRAS* 78: 3–28. [Presents Einstein and de Sitter universes, calculates (gravitational) redshift, gives area distance formula, estimates curvature from density.]

——— (1922). "On the Possibility of Statistical Equilibrium of the Universe." *Proc. Akad. Wet. Amst.* 23: 866–888. [Twenty-five redshifts.]

——— (1930a). "Remarks at RAS meeting." *Observ.* 53: 73–39. [Problems with solutions $A$ and $B$; Eddington responds.]

——— (1930b). "On the Distances and Radial Velocities of Extragalactic Nebulae and the Explanation of the Latter by the Relativity Theory of Inertia." *PNAS* 16: 474–488. [Supports Lemaître's dynamic solution: too much matter observed for $B$ to apply, so reject both $A$ and $B$; true solution must be dynamic.]

——— (1930c). "On the Magnitudes, Diameters and Distances of the Extragalactic Nebulae and their Apparent Radial Velocities." *Bull. Ast. Inst. Neth.* 5: 157–171. [Confirms Hubble relation, estimates mass density $10^{-30}$ gm/cc.]

——— (1930d). "The Expanding Universe: Discussion of Lemaître's Solution of the Equations of the Inertial Field." *Bull. Ast. Inst. Neth.* 5: 211–218.

——— (1930e). "Further Remarks on Astronomical Consequences of the Theory of the Expanding Universe." *Bull. Ast. Inst. Neth.* 5: 274–276.

——— (1930f). "Do the Galaxies Expand with the Universe?" *Bull. Ast. Inst. Neth.* 6: 146. [Effect of expansion on solar system.]

——— (1931a). "The Expanding Universe." *Scientia* 49: 1–10.

——— (1931b). "Das sich ausdehnenden Universum." *Naturwiss.* 19: 365–369.

——— (1931c). "Contributions to a British Association Discussion on the Evolution of the Universe." *Nature* 128: 706–709. [Lemaître self-evident.]

——— (1931d). "Some Further Computations Regarding Non-Static Universes." *Bull. Ast. Inst. Neth.* 6: 141–145.

——— (1932a). "On the Expanding Universe." *Proc. Akad. Wet. Amst.* 35: 596–607. [Analysis and classification of all Friedman's worlds with $p = 0$.]

——— (1932b). *Kosmos*. Cambridge, Massachusetts: Harvard University Press.

——— (1933). "On the Expanding Universe and the Time Scale." *MNRAS* 93: 628–634. [Stars could exist for $10^{13}$ yrs.]

——— (1934). "On Distance, Magnitude and Related Quantities in an Expanding Universe." *Bull. Ast. Inst. Neth.* 7: 205–216.
de Vaucouleurs, G. (1970). "The Case for a Hierarchical Cosmology." *Science* 167: 1203–1213.
de Witt, B.S. (1967). "Quantum Theory of Gravity I. The Canonical Theory." *Phys. Rev.* 160: 1113–1148; "Quantum Theory of Gravity II. The Manifestly Covariant Theory." *Phys. Rev.* 162: 1195–1239; "Quantum Theory of Gravity III. Applications of the Covariant Theory." *Phys. Rev.* 162: 1239–1256.
Dicke, R.H., Beringer, R., and Vane, A.B. (1946). "Atmospheric Absorption Measurements with a Microwave Radiometer." *Phys. Rev.* 70: 340–348. [Residual temperature limits at 20°K: CMWBR not detected.]
Dicke, R.H., and Peebles, P.J.E. (1979). "The Big-Bang Cosmology—Enigmas and Nostrums." In *General Relativity: An Einstein Centenary Survey*. S.W. Hawking, and W. Israel, eds. Cambridge: Cambridge University Press, pp. 504–517, 870–871.
Dicke, R.H., Peebles, P.J.E., Roll, P.G., and Wilkinson, D.T. (1965). "Cosmic Black Body Radiation." *Ap. J.* 142: 414–419. [Discovery: see Penzias and Wilson 1965.]
Dingle, H. (1934). "On Isotropic Models of the Universe, with Special Reference to the Stability of the Homogeneous and Static States." *MNRAS* 94: 134–158. [General "nearly homogeneous" universe and its stability.]
——— (1953). "On Science and Modern Cosmology." *MNRAS* 113: 393–407. [Attack on steady-state theory.]
Dirac, P.A.M. (1937). "The Cosmological Constants." *Nature* 139: 323. [Time-dependence of atomic units expressed in constants of nature.]
——— (1938). "A New Basis for Cosmology." *Proc. Roy. Soc. A* 165: 199–208. [Constants.]
Doroshkevich and Novikov (1964). *Doklady Akademii Nauk SSSR* 154: 809. [Background radiation investigations prior to Princeton observations in 1965; cf. Zeldovich 1965.]
Droste, J. (1915). "On the Field of a Single Centre in Einstein's Theory of Gravitation." *Proc. Akad. Wet. Amst.* 17: 998–1011. [Schwarzschild solution.]
——— (1917). "The Field of a Single Centre in Einstein's Theory of Gravitation, and the Motion of a Particle in that Field." *Proc. Akad. Wet. Amst.* 19: 197.
Dyson, F.W., Eddington, A.S., and Davidson, C. (1920). "A Determination of the Deflection of Light by the Sun's Gravitational Field from Observations Made at the Total Eclipse of May 29, 1919." *Phil. Trans.* 220: 291–333. [Successful test of general relativity; implies gravitational lensing possibility.]
Eardley, D., Liang, E.P.T., and Sachs, R.K. (1972). "Velocity Dominated Singularities in Irrotational Dust Cosmologies." *J. Math. Phys.* 13: 99–107.
Eddington, A.S. (1920). *Space, Time and Gravitation*. Cambridge: Cambridge University Press. [Bending of light and lensing; existence of global time in cosmology, p. 163.]
——— (1923) *The Mathematical Theory of Relativity*. Cambridge: Cambridge University Press.
——— (1924a). "On the Relation between the Masses and the Luminosities of the Stars." *MNRAS* 84: 308–332. [Star ages $\Rightarrow t = 2 \times 10^{13}$ yrs.]
——— (1924b). *The Mathematical Theory of Relativity*, 2nd ed. Cambridge: Cambridge University Press. [41 redshifts, 36 positive: favors de Sitter solution *B*, except Einstein solution *A* affords hope of accounting for constants of nature. De Sitter mass horizon is an illusion of the observer at the origin.]

——— (1926). *The Internal Constitution of the Stars.* Cambridge: Cambridge University Press.
——— (1928). *The Nature of the Physical World.* Cambridge: Cambridge University Press. [Heat death of universe.]
——— (1930a). "[Remarks at the Meeting of the Royal Astronomical Society.]" *Observ.* 53: 39–40. [Possibility of nonstatic models?]
——— (1930b). "Space and its Properties." *Nature* 125: 849–850. [Positive comments regarding Lemaître's work, in review of Silberstein's book.]
——— (1930c). "On the Instability of Einstein's Spherical World." *MNRAS* 90: 668–678. [Proof of instability and discussion of Lemaître's expanding universe; see also *Observ.* 53 (1930): 162–164.]
——— (1931a). "Council Note on Expansion of Universe." *MNRAS* 91: 412–416.
——— (1931b). "On the Value of the Cosmical Constant." *Proc. Roy. Soc. A* 133: 605–615.
——— (1931c). "The End of the World from the Standpoint of Mathematical Physics." *Nature* 127: 447–453. [Heat death.]
——— (1932a). "The Expanding Universe." *Proc. Phys. Soc. Lond.* 44: 1–16.
——— (1932b). "The Expanding Universe." *Nature* 129: 421–423.
——— (1933). *The Expanding Universe.* Cambridge: Cambridge University Press. [Age issue, "no abrupt beginning."]
——— (1935a). *Observ.* 58: 37–39.
——— (1935b). "The Speed of Recession of the Galaxies." *MNRAS* 95: 636–638. [Relation between cosmology and wave mechanics. $H = 865$ not 528; number of particles in universe.]
——— (1937a). "The Cosmical Constant and the Recession of the Nebulae." *Am J. Math.* 59: 1–8.
——— (1937b). "The Effect of Redshift on the Magnitudes of Nebulae." *MNRAS* 97: 156–163. [Observational relations.]
——— (1939). *Science Progress* 34: 225. [Abandon Homogeneity?]
——— (1949). *Fundamental Theory.* Cambridge: Cambridge University Press.
Ehlers, J. (1961). "Beiträge zur Mechanik kontinuerlicher Medien." *Akademie der Wissenschaften und der Literatur in Mainz. Mathematisch-naturwissenschaftliche Klasse. Abhandlungen* 11: 1. [General fluid behavior in curved space-time.]
——— (1987). "Hermann Weyl's Contribution to the General Theory of Relativity." Preprint. Munich.
Ehlers, J., Geren, P., and Sachs R.K. (1968). "Isotropic Solutions of the Einstein-Liouville Equations." *J. Math. Phys.* 9: 1344–1349.
Eiesland, J. (1925). "The Group of Motions of an Einstein Space." *Trans. Am. Math. Soc.* 27: 213–245. [Symmetry groups applied to spherical symmetry and Schwarzschild.]
Einstein, A. (1905). "Zur Elektrodynamik bewegter Körper." *Ann. d. Physik* 17: 891–921; English translation in Lorentz et al. 1923, pp. 35–65.
——— (1907). "Über das Relativitätsprinzip und die aus demselben gezogenen Folgerungen." *Jahrbuch der Radioaktivität und Elektronik* 4: 411–462. [First gravitational redshift, bending of light.]
——— (1911). "Über den Einfluss der Schwerkraft auf die Ausbreitung des Lichtes." *Ann. d. Physik* 35: 898–908. English translation in Lorentz et al. 1923, pp. 97–108.
——— (1915a). "Erklärung der Perihelbewegung des Merkur aus der allgemeinen

Relativitätstheorie." *Preuss. Akad. Wiss. Sitzb.* 831–839. [Perihelion of Mercury, correct light bending.]

——— (1915b). "Die Feldgleichungen der Gravitation." *Preuss. Akad. Wiss. Sitzb.* 844–847. [Finally the general theory of relativity is closed as a logical structure.]

——— (1916). "Die Grundlage der allgemeinen Relativitätstheorie." *Ann. d. Physik* 49: 769–822; English translation in Lorentz et al. 1923, pp. 109–164.

——— (1917). "Kosmologische Betrachtungen zur allgemeinen Relativitätstheorie." *Preuss. Akad. Wiss. Sitzb.* 142–152; English translation in Lorentz et al. 1923, pp. 175–188. [First general relativistic cosmological model. Static. Introduces cosmological constant, concept of closed space-sections.]

——— (1918a). "Prinzipielles zur allgemeinen Relativitätstheorie." *Ann. d. Physik* 55: 241–244.

——— (1918b). "Kritisches zu einer von Herrn de Sitter gegebenen Lösung der Gravitationsgleichungen." *Preuss. Akad. Wiss. Sitzb.* 270–272. [Criticizes de Sitter solution as anti-Machian. Queries whether cosmological constant is nonzero.]

——— (1922). "Bemerkung zu der Arbeit von A. Friedmann 'Über die Krümmung des Raumes'." *Z. Phys.* 11: 326. [Claims error in Friedman's expanding solution.]

——— (1923). "Notiz zu der Arbeit von A. Friedmann 'Über die Krümmung des Raumes'." *Z. Phys.* 16: 228. [Withdraws claim of error in Friedman's solution.]

——— (1931). "Zum kosmologischen Problem der allgemeinen Relativitätstheorie." *Preuss. Akad. Wiss. Sitzb.* 235–237. [Preference for zero cosmological constant.]

——— (1936). "Lens-like Action of a Star by the Deviation of Light in the Gravitational Field." *Science* 84: 506–507. [Gravitational lensing by stars?]

——— (1945). *The Meaning of Relativity*, 2nd ed. Princeton, New Jersey: Princeton University Press. [Appendix on cosmology; Olbers, age problem, possibility of dark matter.]

——— (1956). *The Meaning of Relativity*, 5th ed. Princeton, New Jersey: Princeton University Press.

Einstein, A., and de Sitter, W. (1932). "On the Relation between the Expansion and the Mean Density of the Universe." *PNAS* 18: 213–214. [Simplest dynamic case: $k = 0$, $p = 0$, zero cosmological constant, gives relation of expansion and density.]

Einstein, A., and Strauss, E.G. (1945). "The Influence of the Expansion of Space on the Gravitational Fields Surrounding Individual Stars." *Rev. Mod. Phys.* 17: 120–124; 18 (1946): 148–149. [Inhomogeneous "Swiss-Cheese" model universe: FLRW model with Schwarzschild vacuoles.]

Eisenhart, L.P. (1924). "Space-Time Continua of Perfect Fluids in General Relativity." *Trans. Am. Math. Soc.* 26: 205–220. [General fluid motions in general relativity.]

——— (1926). *Riemannian Geometry*. Princeton, New Jersey: Princeton University Press.

Ellis, G.F.R. (1966). "The Dynamics of Pressure-Free Matter in General Relativity." *J. Math. Phys.* 8: 1171–1194. [LRS solutions; shear-free theorem.]

——— (1971). "Relativistic Cosmology." In Sachs 1971, pp. 104–182.

——— (1987). "The Standard Model" *Cosmology and Gravitation*. Proceedings of the 5th Brazilian School on Cosmology and Gravitation. M. Novello, ed. Singapore: World Scientific.

——— (1988). "Does Inflation Imply Omega = 1?" *Classical and Quantum Gravity* 5: 891–901.

Ellis, G.F.R., and King, A.R. (1974). "Was the Big Bang a Whimper?" *Com. Math. Phys.* 31: 209–242). [Cosmological singularities where all physical quantities are finite.]

Ellis, G.F.R., Maartens, R., and Nel, S.D. (1978). "The Expansion of the Universe." *MNRAS* 184: 439–465. [Possibility of cosmological gravitational redshifts.]

Ellis, G.F.R., Nel, S.D., Maartens, R., Stoeger, W., and Whitman, A.P. (1985). "Ideal Observational Cosmology." *Physics Reports* 124: 315–417. [Nonlocal extension of Kristian and Sachs approach to observations.]

Ellis, G.F.R., and Sciama, D.W. (1972). "Global and Non-Global Problems in Cosmology." In *General Relativity: Papers in Honour of J.L. Synge*. L. O'Raifeartaigh, ed. Oxford: Oxford University Press. [Boundary conditions for local physics: Mach, arrow of time.]

Etherington, I.M.H. (1933). "On the Definition of Distance in General Relativity." *Phil. Mag.* 15: 761–773. [Observations in curved space-time: reciprocity theorem.]

Field, G.B. (1969). "Cosmic Background Radiation and its Interaction with Cosmic Matter." *Rivista del Nuovo Cimento* 1: 87–109.

Field, G.B., Arp, H., and Bahcall, J.N. (1973). *The Redshift Controversy*. New York: Benjamin. [Reprint volume with comment.]

Fowler, W.A. (1987). "The Age of the Observable Universe." *QJRAS* 28: 87–108.

Fowler, W.A., and Stephen, W.E. (1968). "Resource Letter OE1 on the Origin of the Elements." *Am. J. Phys.* 36: 289–302.

Freedrichsz, V., and Schechter, A. (1928). "Notiz zur Frage nach der Berechnung der Aberration und der Parallaxe in Einsteins, de Sitters und Friedmanns Welten." *Z. Phys.* 51: 584–592. [Observational relations; N.B. Friedmann aberration and parallax.]

Friedman, A. (1922). "Über die Krümmung des Raumes." *Z. Phys.* 10: 377–386. [First expanding universe model ($k = +1$) with $p \geq 0$; implications not developed.]

——— (1924). "Über die Möglichkeit einer Welt mit konstant negativer Krümmung des Raumes." *Z. Phys.* 21: 326–332. [First expanding model with $k = -1$.]

Fubini, G. (1904). "Sugli spazzi a quattro dimensioni che ammettono un gruppo continuo di movimenti." *Annali di matematica pura ed applicata* 9: 33–90. [Groups of motions, spaces of constant curvature.]

Gamow, G. (1946a). "Rotating Universes?" *Nature* 158: 549. [Rotation?]

——— (1946b). "Expanding Universe and the Origin of the Elements." *Phys. Rev.* 70: 572–573. [Pioneering note on nucleosynthesis in expanding universe. Element abundance curve not equilibrium curve; Lemaître universe temperatures high enough for nuclear reactions; high expansion rate $\Rightarrow$ short time available for reactions?]

——— (1948a). "The Evolution of the Universe." *Nature* 162: 680–682. [Synthesis of deuterium from all-nucleon "ylem" at $10^9\,°K$. See Alpher et al.]

——— (1948b). "The Origin of Elements and Separation of Galaxies." *Phys. Rev.* 74: 505–506. [Temperature and density against time relations. Radiation dominated universe. Jeans length argument.]

——— (1949a). "On Relativistic Cosmogony." *Rev. Mod. Phys.* 21: 367–373. [Effect of radiation on condensations; quotes Fermi and Turkevich on mass gaps at 5 and 8.]

——— (1949b). "Any Physics Tomorrow?" *Physics Today* 2: 16. [Fundamental constants.]

——— (1951). "The Origin and Evolution of the Universe." *American Scientist* 39: 393–406.

——— (1952a). "The Role of Turbulence in the Evolution of the Universe." *Phys. Rev.* 86: 251. [Turbulence $\Rightarrow$ galaxies?]

——— (1952b). *The Creation of the Universe.* New York: Viking Press. [CMWBR at 5°K.]
——— (1954). "Modern Cosmology." *Sci. Am.* 190 (March): 55–63.
——— (1956a). "The Evolutionary Universe." *Sci. Am.* 195 (September): 136–154. [Elements as function of time.]
——— (1956b). "The Physics of the Expanding Universe." *Vistas in Astronomy* 2: 1726–1732.
——— (1970). *My World Line: An Informal Autobiography.* New York: Viking Press.
Gamow, G., and Teller, E. (1939a) "The Expanding Universe and the Origin of the Great Nebulae." *Nature* 143: 116. [Galaxy formation: Jeans mass in matter-dominated expanding model.]
——— (1939b). "On the Origin of Great Nebulae." *Phys. Rev.* 55: 654–657. [Creation of spiral galaxies through gravitational instability.]
Geroch, R.P. (1966). "Singularities in Closed Universes." *Phys. Rev. Lett.* 17: 445–447.
Giacconi, R., Gursky, H., Paolini, F.R., and Rossi, B. (1962). "Evidence for x-Rays from Sources outside the Solar System." *Phys. Rev. Lett.* 9: 439–443. [Discovery of x-ray background.]
Gibbons, G.W., Hawking, S.W., and Siklos, S.T.C., eds. (1983). *The Very Early Universe.* Proceedings of the Nuffield Workshop, Cambridge, June 21, to July 9, 1982. Cambridge and New York: Cambridge University Press. [The inflationary universe.]
Gilbert, C. (1956). "The Gravitational Field of a Star in an Expanding Universe." *MNRAS* 116: 678–683. [Swiss cheese boundary conditions, cf. Einstein and Strauss.]
Godart, O., and Heller, M. (1985). *Cosmology and Lemaître.* History of Astronomy Series, vol 3. Tucson: Pachart Publishing House.
Gödel, K. (1949). "An Example of a New Type of Cosmological Solutions of Einstein's Field Equations of Gravitation." *Rev. Mod. Phys.* 21: 447–450. [First exact rotating cosmological solution. Anomalous causal properties.]
——— (1952). "Rotating Universes in General Relativity Theory." In *Proceedings of the International Congress of Mathematics. Cambridge, Massachusetts.* Vol 1. L.M. Graves et al., eds. [First expanding and rotating Bianchi IX universe models. Shear-free theorem.]
Gold, T. (1951). "The Origin of Cosmic Radio Noise." *Proceedings of the Conference on Dynamics of Ionized Media.* London: Department of Physics, University College, London. [See Bondi 1960. Claims discrete radio sources extragalactic; opposed by Ryle.]
——— (1955). "The 'Horizon' of the Steady State Universe." *Nature* 175: 382.
——— ed. (1967). *The Nature of Time.* Ithaca: Cornell University Press.
Gott, J., Gunn, J.E., Schramm, D.N., and Tinsley, B. (1974). "An Unbound Universe?" *Ap. J.* 194: 543–553. [FLRW models compared with data ⇒ low-density model.]
——— (1976). "Will the Universe Expand Forever?" *Sci. Am.* 234 (March): 62–79.
Gowdy, R.H. (1971). "Gravitational Waves in Closed Universes." *Phys. Rev. Lett.* 27: 826–839; 1102.
Greenstein, J.L., and Schmidt, M. (1964). "The Quasi-Stellar Radio Sources 3C48 and 3C 273." *Ap. J.* 140: 1–34. [Argues against gravitational redshifts.]
Groth, E., and Peebles, P.J.E. (1977). "Statistical Analysis of Catalogs of Extragalactic Objects. VII. Two- and Three-Point Correlation Functions for the High-Resolution Shane-Wirtenan Catalogue of Galaxies." *Ap. J.* 217: 385–405.

Gunn, J.E., and Oke, J.B. (1975). "Spectrophotometry of Faint Cluster Galaxies and the Hubble Diagram: An Approach to Cosmology." *Ap. J.* 195: 255–268.

Gunn, J.E., and Peterson, B.A. (1965). "On the Density of Neutral Hydrogen in Intergalactic Space." *Ap. J.* 142: 1633–1636. [Lyman absorption lines and limits on matter density.]

Guth, A. H. (1981). "Inflationary Universe: A Possible Solution to the Horizon and Flatness Problems." *Phys. Rev. D* 23: 347–356. [Inflation.]

Halliwell, J.J., and Hawking, S.W. (1985). "Origin of Structure in the Universe." *Phys. Rev. D* 31: 1777–1791.

Halley, E. (1720). "Of the Infinity of the Sphere of Fixed Stars, and of the Number, Order, and Light of the Fixed Stars." *Phil. Trans.* 31: 22–24; 24–26. [Olbers's paradox: see Jaki and Harrison.]

Hanbury-Brown, R., and Hazard, C. (1950). "Radio-Frequency Radiation from the Great Nebula in Andromeda (M31)." *Nature* 166: 901–902. [First good evidence of M31 as radio source.]

Harrison, E.R. (1977). "Radiation in Homogeneous and Isotropic Models of the Universe." *Vistas in Astronomy* 20: 341–409.

——— (1981). *Cosmology: The Science of the Universe*. Cambridge: Cambridge University Press.

——— (1984). "The Dark Night Sky Riddle: A 'Paradox' that Resisted Solution." *Science* 226: 941–945. [Olbers's paradox history.]

——— (1985). *Masks of the Universe*. New York: Collier.

——— (1986). "Kelvin and an Old, Celebrated Hypothesis." *Nature* 322: 417–418.

Hartle, J.B., and Hawking, S.W. (1983). "Wave Function of the Universe." *Phys. Rev. D* 28: 2960–2975.

Harwit, M. (1981). *Cosmic Discovery*. Harvester Press. [Historical survey of major discoveries.]

Harzer, P. (1926). "Über die astronomischen Ergebnisse der allgemeinen Relativitätstheorie." *Astr. Nach.* 227: 81–96.

Hawking, S.W. (1966). "Perturbations of an Expanding Universe." *Ap. J.* 145: 544–554.

——— (1967). "The Occurrence of Singularities in Cosmology." *Proc. Roy. Soc. A* 300: 187–201. [Singularity theorems.]

——— (1968). "On the Rotation of the Universe." *MNRAS* 142: 129–141.

Hawking, S.W., and Ellis, G.F.R. (1968). "The Cosmic Black Body Radiation and the Existence of Singularities in our Universe." *Ap. J.* 152: 25–36.

——— (1973). *The Large-Scale Structure of Space-Time*. Cambridge: Cambridge University Press.

Hawking, S.W., and Penrose, R. (1969). "The Singularities of Gravitational Collapse and Cosmology." *Proc. Roy. Soc. A* 314: 529–548.

Hawking, S.W, and Tayler, R.J. (1966). "Helium Production in an Anisotropic Big-Bang Cosmology." *Nature* 209: 1278–1279. [Nucleosynthesis in anisotropic universe.]

Hayashi, C. (1950). "Proton-Neutron Concentration Ratio in the Expanding Universe at the Stage Preceding the Formation of Elements." *Progress of Theoretical Physics* 5: 224–235. [Neutron abundance in early universe not free initial condition but determined by beta-decay; pair production in early universe. See Alpher et al.]

Hazard, C., Mackey, M.B., and Shimmins, A.J. (1963). "Investigations of the Radio Source 3C 273 by the Method of Lunar Occultations." *Nature* 197: 1037–1039. [Twin-source nature of first discovered QSO; see Schmidt 1963.]

Heckmann, O. (1931). "Über die Metrik des sich ausdehnenden Universums." *Nach. Ges. Gött. Wiss.* 127–130. [$k = 0$ and $k = -1$ possibilities?]

——— (1932). "Die Ausdehnung der Welt und ihre Abhängigkeit von der Zeit." *Nach. Ges. Gött. Wiss.* 97–102. [Plots all universes with noninteracting matter and radiation.]

——— (1942). *Theorien der Kosmologie.* Berlin: Julius Springer. [$m/z$ power series in terms of $R$ and derivatives; extensive bibliography, 1933–1940.]

Heckmann, O., and Schücking, E. (1955). "Remarks on Newtonian Cosmology." *Z. Ap.* 38: 95–109. [Anisotropic Newtonian models; boundary conditions, homogeneity.]

——— (1956). "Ein Weltmodell der Newtonschen Kosmologie mit Expansion und Rotation." In Mercier and Kervaire 1956, pp. 114–115.

——— (1959a). "World Models." In *Onzième Conseil de Physique Solvay: La Structure et l'Évolution de l'Univers.* Brussels: Edition Stoops. [Bianchi I cosmological solution.]

——— (1959b). "Newtonsche und Einsteinsche Kosmologie." In *Handbuch der Physik.* Vol. 53. S. Flügge, ed. Berlin: Springer-Verlag, pp. 489–513.

——— (1959c). "Andere kosmologische Theorien." In *Handbuch der Physik.* Vol. 53. S. Flügge, ed. Berlin: Springer-Verlag, pp. 520–537.

——— (1962). "Relativistic Cosmology." In *Gravitation: An Introduction to Current Research.* L. Witten, ed. New York: Wiley, pp. 438–469. [Detailed mathematics of Bianchi universes.]

Hey, J.S. (1946). "Solar Radiation in the 4–6 Metre Radio Wavelength Band." *Nature* 157: 47–48. [Radio emission detected from sunspots.]

——— (1949). "Reports on the Progress of Astronomy: Radio Astronomy." *MNRAS* 109: 179–214. [Extensive survey; extragalactic nature of discrete sources, e.g. Cygnus A, not yet recognized.]

Hey, J.S., Parsons, S.J., and Phillips, J.W. (1946). "Fluctuations in Cosmic Radiation at Radio Frequencies." *Nature* 158: 234. [First extragalactic radio source (Cygnus A) detected, but nature not understood.]

Hilbert, D. (1912). "Begründing der elementaren Strahlungstheorie." *Phys. Z.* 13: 1056–1064. [Reciprocity theorem.]

——— (1913). "Bemerkungen zur Begründing der elementaren Strahlungstheorie." *Phys. Z.* 14: 592–595.

——— (1915). "Die Grundlagen der Physik." *Nach. Ges. Gött. Wiss.* 395. [Field equations of gravity from variational principle.]

Holmes, A., and Lawson, R.W. (1927). "Factors Involved in the Calculation of the Ages of Radioactive Matter." *American Journal of Science* 13: 327–344. [Uranium/lead $\Rightarrow$ ages: $1.3 \times 10^9$ yr.]

Hoyle, F. (1946). "The Synthesis of the Elements from Hydrogen." *MNRAS* 106: 343–383. [Origin of elements.]

——— (1948). "A New Model for the Expanding Universe." *MNRAS* 108: 372–382. [Steady-state theory; see Bondi, Gold 1948, 1954.]

——— (1949). "On the Cosmological Problem." *MNRAS* 109: 365–371. [Steady state.]

——— (1959). "The Relation of Radio Astronomy to Cosmology." In *Paris Symposium on Radio Astronomy.* R.N. Bracewell, ed. Stanford: Stanford University Press, pp. 529–532. [Proposed angular diameter redshift test; minimum angular diameters.]

—— (1962). "Cosmological Tests of Gravitational Theories." In *Evidence for Gravitational Theories*. Proceedings of the International School of Physics "Enrico Fermi" Course 20. C. Møller, ed. New York and London: Academic Press, pp. 141–173. [Minimum angular diameters.]

Hoyle, F., and Sandage, A. (1956). "The Second-Order Term in the Redshift Magnitude Relation." *PASP* 68: 301–307. [Deceleration parameter.]

Hoyle, F., and Tayler, R.J. (1964). "The Mystery of the Cosmic Helium Abundance." *Nature* 203: 1108–1110.

Hubble, E.P. (1924). *Annual Report of Mt. Wilson Observatory*. [Found Cepheids in M31—first object proved outside our galaxy.]

—— (1925a). "NGC 6822, a Remote Stellar System." *Ap. J.* 62: 409–433. [Distance ⇒ extragalactic.]

—— (1925b). "Cepheids in Spiral Nebulae." *Observ.* 48: 139–142. [Distance to M31, M33.]

—— (1925c). "Cepheids in Spiral Nebulae." *Pop. Ast.* 33: 252–255.

—— (1926a). "A Spiral Nebula as a Stellar System, M33." *Ap. J.* 63: 236–274. [Extragalactic nature.]

—— (1926b). "Extra-Galactic Nebulae." *Ap. J.* 64: 321–369. [Classification of galaxies, observational proof of uniformity, and density estimate leading to estimate of radius of Einstein static universe $A$.]

—— (1929a). "The Exploration of Space." *Harper's Magazine* 58: 732–738. [$A$ or $B$?]

—— (1929b). "A Relation between Distance and Radial Velocity among Extragalactic Nebulae." *PNAS* 15: 169–173. [First accepted proof of linear redshift-distance relation, but not related to expansion of universe.]

—— (1929c). "A Spiral Nebula as a Stellar System, M31." *Ap. J.* 69: 103–158. [Distance to Andromeda ⇒ extragalactic.]

—— (1932). "The Surface Brightness of Threshold Images." *Ap. J.* 76: 106–116. [Selection effects.]

—— (1934). "The Distribution of Extragalactic Nebulae." *Ap. J.* 79: 8–76. [Test of uniformity; proof of existence of galaxy clusters; estimate of density.]

—— (1936a). "The Luminosity Function of Nebulae." *Ap. J.* 84: 158–179, 270–296. [Hubble constant 526 km/sec/Mpc from fifth brightest cluster member.]

—— (1936b). "Effects of Redshift on the Distribution of Nebulae." *Ap. J.* 84: 517–554. [Number density/power series ($q_0$ too large ⇒ velocity shifts?).]

—— (1936c). *The Realm of the Nebulae*. New Haven, Connecticut: Yale University Press. [Redshift-distance relation.]

—— (1937a). "Redshifts and the Distribution of Nebulae." *MNRAS* 97: 506–513. [Apparent luminosity.]

—— (1937b). *The Observational Approach to Cosmology*. Oxford: Clarendon Press.

—— (1953). "The Law of Redshifts." *MNRAS* 113: 658–666. [George Darwin lecture.]

Hubble, E.P., and Humason, M.L. (1931). "The Velocity-Distance Relation among Extra-Galactic Nebulae." *Ap. J.* 74: 43–80. [Linear relation with Hubble constant of 558 km/sec/Mpc.]

Hubble, E.P., and Tolman, R.C. (1935). "Two Methods of Investigating the Nature of the Nebular Redshift." *Ap. J.* 82: 302–337. [Galactic apparent magnitudes and brightness profiles in FLRW and in tired-light models; intensity relation.]

Humason, M.L. (1929). "The Large Radial Velocity of NGC 7619." *PNAS* 15: 167–168.

———— (1931). "Apparent Velocity Shifts in the Spectra of Faint Nebulae." *Ap. J.* 74: 35–42.

———— (1936). "The Apparent Radial Velocities of 100 Extragalactic Nebulae." *Ap. J.* 83: 10–22.

Humason, M.L., Mayall, N.U., and Sandage, A.R. (1956). "Redshifts and Magnitudes of Extragalactic Nebulae." *A.J.* 61: 97–162. [$q_0 = 3.7 \pm 0.8$].

Jaki, S.L. (1969). *The Paradox of Olber's Paradox: A Case History of Scientific Thought.* New York: Herder and Herder.

Jansky, K.G. (1933a). "Electrical Disturbances Apparently of Extraterrestrial Origin." *Proc. Inst. Radio Eng. Aust.* 21: 1387–1398. [Discovery of 14.6-m radio waves from the galactic center.]

———— (1933b). "Electrical Phenomena that Apparently are of Interstellar Origin." *Pop. Ast.* 41: 548–558. [Radio signal detection.]

———— (1935). "A Note on the Source of Interstellar Interference." *Proc. Inst. Radio Eng. Aust.* 23: 1158–1163.

Jeans, J. (1902). "The Stability of a Spherical Nebula." *Phil. Trans.* 199: 1–53. [Instability to gravitational fluctuations of static mass of uniform density; Jeans mass.]

———— (1923). Review of A.S. Eddington, *The Mathematical Theory of Relativity.* *Observ.* 46: 191–193. [Redshifts *A* or *B*?]

———— (1928a). *Astronomy and Cosmogony.* Cambridge: Cambridge University Press. [Galaxies, star ages.]

———— (1928b). "The Physics of the Universe." *Nature* 122: 689–700. [Heat death; age of stars = $10^{13}$ yrs.]

———— (1929). *The Universe Around Us.* Cambridge: Cambridge University Press. [Heat death.]

———— (1930). *The Mysterious Universe.* Cambridge: Cambridge University Press. [Olbers's paradox.]

———— (1931). "Contributions to a British Association Discussion on the Evolution of the Universe." *Nature* 128 (Supplement): 701–704. [Atomic size decreasing with time $\Rightarrow$ appearance of expanding universe.]

———— (1939). "The Expanding Universe and the Origin of the Great Nebulae." *Nature* 143: 158–159. [Response to Gamow and Teller.]

Jeffreys, H. (1928). *The Earth: Its Origin, History and Physical Constitution.* Ch. 5. Cambridge: Cambridge University Press. [Ages from lead/uranium $\Rightarrow 2.6 \times 10^9$ yrs.]

Johnson, H.L. (1966). "Infrared Photometry of Galaxies." *Ap. J.* 143: 187–191. [First infrared observations of other galaxies.]

Jordan, P. (1937). "Die physikalischen Weltkonstanten." *Naturwiss.* 25: 513–517.

———— (1938). "Zur empirischen Kosmologie." *Naturwiss.* 26: 417–421. [Time-dependent constants of nature; cf. Dirac.]

———— (1949). "Formation of the Stars and Development of the Universe." *Nature* 164: 637–640.

———— (1952). *Gravitation and the Universe.* Braunschweig: Friedrich Vieweg und Sohn.

Jüttner, F. (1911). "Das Maxwellsche Gesetz der Geschwindigkeitsverteilung in der Relativtheorie." *Ann. d. Physik* 34: 856–882. [Relativistic generalization of Maxwell equilibrium distribution function for particles.]

———— (1928). "Die relativistische Quantentheorie des Idealen Gases." *Z. Phys.* 47:

542–566. [Relativistic equilibrium distribution function for systems of bosons and fermions.]

Kahn, C., and Kahn, F. (1975). "Letters from Einstein to de Sitter on the Nature of the Universe." *Nature* 257: 451–454.

Kantowski, R., and Sachs, R.K. (1967). "Some Spatially Homogeneous Anisotropic Relativistic Cosmological Models." *J. Math. Phys.* 7: 443–446.

Kasner, E. (1921). "Geometrical Theorems on Einstein's Cosmological Equations." *Am. J. Math.* 43: 217–221.

——— (1925). "Solutions of Einstein's Equations Involving Functions of Only One Variable." *Trans. Am. Math. Soc.* 27: 155–162. [Anisotropic (Bianchi I) solution.]

Kennedy, R.J., and Barkas, W. (1936). "The Nebular Redshift." *Phys. Rev.* 49: 449–452. [Kinematic relativity.]

Kermack, W.O., and McCrea, W.H. (1933). "On Milne's Theory of World-Structure." *MNRAS* 93: 519–529. [Kinematic relativity.]

Kermack, W.O., McCrea, W.H., and Whittaker, E.T. (1933). "On Properties of Null Geodesics and their Application to the Theory of Radiation." *Proc. Roy. Soc. Edin.* 53: 31–47. [Observations.]

Killing, W. (1892). "On the Foundations of Geometry." *Journal für die reine und angewandte Mathematik* 109: 121–186. [Symmetry properties of curved spaces.]

Kohler, M. (1933). "Beiträge zum kosmologischen Problem und zur Lichtausbreitung in Schwerefeldern." *Ann. d. Physik* 16: 129–161. [Energy and entropy of radiation; $z$-$m$ relation; horizons.]

Kristian, J., and Sachs, R.K. (1966). "Observations in Cosmology." *Ap. J.* 143: 379–399. [First detailed paper of observations in general universe models.]

Kristian, J., Sandage, A., and Westphal, J.A. (1976). "The Extension of the Hubble Diagram. II. New Redshifts and Photometry of Very Distant Galaxy Clusters: First Indication of a Deviation of the Hubble Diagram from a Straight Line." *Ap. J.* 221: 383–394.

Kundt, W. (1956). "Trägheitsbahnen in einem von Gödel angegeben kosmologischen Model." *Z. Phys.* 145: 611–620. [Geodesics in Gödel's universe.]

Lanczos, C. (1922). "Bemerkungen zur de Sitterschen Welt." *Phys. Z.* 23: 539–543. [First nonstationary form ($k = +1$) for de Sitter.]

——— (1923). "Über die Rotverschiebung in der de Sitterschen Welt." *Z. Phys.* 17: 168–189. [Determines redshift in expanding ($k = +1$) form; elimination of mass horizon.]

——— (1924). "Über eine stationäre Kosmologie im Sinne der de Sitterschen Welt." *Z. Phys.* 21: 73–110.

Lang, K.R., and Gingerich, O. (1979). *A Source Book in Astronomy and Astrophysics, 1900–1975*. Cambridge, Massachusetts: Harvard University Press. [Reprints of classic articles.]

Layzer, D. (1954). "On the Significance of Newtonian Cosmology." *A. J.* 59: 268–270.

Lemaître, G. (1925a). "Note on de Sitter's Universe." *Phys. Rev.* 25: 903. [Second ($k = 0$, steady-state) expanding form for de Sitter.]

——— (1925b). "Note on de Sitter's Universe." *Journal of Mathematics and Physics* (MIT) 4: 189–192. [Linear $d$-$z$ relation.]

——— (1927). "A Homogeneous Universe of Constant Mass and Increasing Radius Accounting for the Radial Velocity of Extragalactic Nebulae." *Ann. Soc. Sci. Bruxelles A* 47: 49–59. In French, translated in Lemaître 1931a. [Expanding universe model, $k = +1$, cosmological constant $\geq 0$. Conservation laws and field equations with $p \geq 0$, expansion from Einstein static state. First redshift derivation in these universes, related to Hubble's constant. (Olbers).]

——— (1929). "La Grandeur de l'espace." *Rev. d. Quest. Sci.* March 16: 189–216. [Advantages of closed model.]

——— (1930). "On the Random Motion of Material Particles in an Expanding Universe." *Bull. Ast. Inst. Neth.* 5: 273–274. [$k = +1$: kinetic theory?]

——— (1931a). "A Homogeneous Universe of Constant Mass and Increasing Radius Accounting for the Radial Velocity of Extragalactic Nebulae." *MNRAS* 91: 483–490. [Translation of Lemaître 1927 paper.]

——— (1931b). "The Expanding Universe." *MNRAS* 91: 490–501, 703. [Start by condensations?]

——— (1931c). "The Beginning of the World from the Point of View of Quantum Theory." *Nature* 127: 706. [Quantum origin of universe.]

——— (1931d). "The Evolution of The Universe: Discussion." *Nature* 128: 704–706. (British Association discussion). [Expansion: cosmic rays as relics?]

——— (1932). "La Expansion de l'espace." *Rev. d. Quest. Sci.* 20: 391–410.

——— (1933a). "L'Univers en expansion." *Ann. Soc. Sci. Bruxelles A* 53: 51–85. [Review; Bianchi I universe does not avoid singularity.]

——— (1933b). "Condensations sphériques dans l'univers en expansion." *C. R. Acad. Sci.* 196: 903–904.

——— (1933c). "La formation des nebuleuses dans l'univers en expansion." *C. R. Acad. Sci.* 196: 1085–1087. [Inhomogeneous (Tolman/Bondi) solution.]

——— (1934). "Evolution of the Expanding Universe." *PNAS* 20: 12–17. [Galaxy formation?]

——— (1945). "L'hypothèse de l'atome primitif." *Actes de la Société Helvétique des Science Naturelles*: 77–96.

——— (1949a). "Cosmological Applications of Relativity." *Rev. Mod. Phys.* 21: 357–366.

——— (1949b). "The Cosmological Constant." In Schilpp 1949, pp. 437–456.

——— (1950). *The Primeval Atom: An Essay on Cosmogony.* New York: Van Nostrand.

Levi-Civita, T. (1917). "Realta fisica di alcuni spazi normali del bianchi." *Reale Accademia dei Lincei. Classe di scienze fisiche, matematiche e naturali. Rendiconti* 26: 519–531. [Einstein static universe with pressure, p. 521.]

Lichnerowicz, A. (1955). *Théories relativiste de la gravitation et de l'électomagnétisme. Relativité générale et théories unitaires.* Paris: Masson.

——— (1959). "On the Quantization of the Gravitational Field for a Space-Time of Constant Curvature." *C. R. Acad. Sci.* 249: 2287–2289.

Lichnerowicz, A., and Marrot, R. (1940). "Sur l'équation intégrodifférentielle de Boltzmann." *C. R. Acad. Sci.* 210: 531–533. [First relativistic generalization of complete Boltzmann equations; cf. Walker, Marrot.]

Lie, S. (1893). *Vorlesungen über kontinuierlichen Gruppen.* Leipzig: B.G. Teubner.

Lifschitz, E.M. (1946). "On the Gravitational Stability of the Expanding Universe." *Journal of Physics. USSR* 10: 116; (*JETP* 16: 587). [First relativistic treatment of galaxies in expanding universe.]

Lifschitz, E.M., and Khalatnikov, I.M. (1963). "Investigations in Relativistic Cosmology." *Adv. Phys.* 12: 185–249. [Detailed dynamical investigation of early universe.]

Limber, D.N. (1953). "The Analysis of Counts of the Extragalactic Nebulae in Terms of a Fluctuating Density Field." *Ap. J.* 117: 134–144. [Galaxy counts: mean density and fluctuations.]

——— (1954). "Analysis of Counts of the Extragalactic Nebulae in Terms of a Fluctuating Density Field. II." *Ap. J.* 119: 655–681.
Lock, C.N.H. (1930). *Proc. Phys. Soc. Lond.* 42: 264. [Equations of motion of a viscous fluid in tensor notation.]
Lodge, O. (1919). "Connection between Light and Gravitation." *Phil. Mag.* 38: 737.
——— (1921). "On the Supposed Weight and Ultimate Fate of Radiation." *Phil. Mag.* 41: 549–557.
Lorentz, H.A., Einstein, A., Minkowski, H., and Weyl, H. (1923). *The Principle of Relativity: A Collection of Original Memoirs on the Special and General Theory of Relativity*. W. Perrett and G.B. Jeffery, trans. London: Methuen; reprint New York: Dover, 1952. [Reprints of papers, including Einstein 1911, 1916, and 1917.]
Lovell, B. (1987). "The Emergence of Radio Astronomy in the UK after World War II." *QJRAS* 28: 1–9.
Lundmark, K.E. (1920). *Stockholm Academy. Handl.* 50: 81.
——— (1924). "The Determination of the Curvature of Space-Time in de Sitter's World." *MNRAS* 84: 747–770. [No systematic $z$-$d$ relation observed (cf. Silberstein).]
——— (1925). "The Motions and Distances of Spiral Nebulae." *MNRAS* 85: 865–894. [Redshift-distance relation with quadratic term (cf. Wirtz).]
MacCallum, M.A.H. (1979). "Anisotropic and Inhomogeneous Relativistic Cosmologies." In *General Relativity: An Einstein Centenary Survey*. S.W. Hawking, and W. Israel, eds. Cambridge: Cambridge University Press, pp. 533–580, 874–883.
MacCallum, M.A.H., and Taub, A.H. (1972). "Variational Principles and Spatially Homogeneous Universes, Including Rotation." *Com. Math. Phys.* 25: 173–189.
Mach, E. (1883). *Die Mechanik in ihrer Entwicklung, historisch-kritisch dargestellt*. Leipzig: Brockhaus. (Sec. 6, ch. 2, 8th Ed., 1921.) [Gravitation and the origin of inertia.]
——— (1902). *The Science of Mechanics: A Critical and Historical Account of its Development*, 2nd ed. T.J. McCormack, trans. Chicago: Open Court.
MacMillan, W.D. (1918). "On Stellar Evolution." *Ap. J.* 48: 35–49. [A steady-state theory; see Harrison 1977.]
——— (1923). "Some Mathematical Aspects of Cosmology." *Science* 62: 63, 96: 121–127.
——— (1932). "Velocities of the Spiral Nebulae." *Nature* 129: 93. [Tired light proposal, cf. Zwicky 1929a, 1929b.]
Madsen, M.S., and Ellis, G.F.R. (1988). "The Evolution of Omega in Inflationary Universes." *MNRAS* 234: 67–77.
Marrot, R. (1946). "Sur l'équation intégrodifférentielle de Boltzmann." *Journal des Mathématiques Pures et Appliquées* 25: 93–159. [General relativity Boltzmann equation; macroscopic laws of mass and energy-momentum and H-theorem from microscopic laws, cf. Lichnerowicz and Marrot, Walker.]
Mason, W.R. (1932). "A Newtonian Gravitational System and the Expanding Universe." *Phil. Mag.* 14: 386–400. [Relation of Hubble constant, density of matter.]
Mattig, W. (1958). "Über den Zusammenhang zwischen Rotverschiebung und scheinbarer Helligkeit." *Astr. Nach.* 284: 109–111. [First analytical area distance formula.]
——— (1959). "Über den Zusammenhang zwischen der Anzahl der extragalaktischen Objekte und der scheinbaren Helligkeit." *Astr. Nach.* 285: 1–2. [Number counts; independent of $q_0$ to first order.]

Mayer, M.G., and Teller, E. (1949). "On the Origin of the Elements." *Phys. Rev.* 76: 1226–1231. [Static (poly-neutron) cold element synthesis.]

McCrea, W.H. (1931). "A 'Cubical' Universe." *Proc. Edin. Math. Soc.* 2: 158.

——— (1935). "Observable Relations in Relativistic Cosmology." *Z. Ap.* 9: 290–314. [Number, magnitude relation.]

——— (1939). "Observable Relations in General Relativistic Cosmology II." *Z. Ap.* 18: 98–115. [Observable relations in general models (inhomogeneous).]

——— (1951). "Relativity Theory and the Creation of Matter." *Proc. Roy. Soc. A* 206: 562–575. [FLRW Raychaudhuri equation ⇒ Friedmann; (density + 3 × pressure) is active gravitational mass.]

——— (1953). "Cosmology." *Reports on Progress in Physics* 16: 321–363. [Review.]

——— (1955a). "Newtonian Cosmology." *Nature* 175: 446.

——— (1955b). "On the Significance of Newtonian Cosmology." *A. J.* 60: 271–274. [Newtonian theory on sound basis.]

McCrea, W.H., and McVittie, G.C. (1931a). "On the Contraction of the Universe." *MNRAS* 91: 128–133. [Inhomogeneity ⇒ contraction from static state.]

——— (1931b). "The Expanding Universe." *MNRAS* 92: 7–12.

McCrea, W.H., and Milne, E.A. (1934). "Newtonian Universes and the Curvature of Space." *Qu. J. Math.* 5: 73–80. [First Newtonian universe models.]

McKellar, A. (1940). "Evidence for the Molecular Origin of Some Hitherto Unidentified Interstellar Lines." *PASP* 52: 187–192. [CN absorption lines represent 2.5°K excitation: unrecognized observation of CMWBR. See Adams.]

——— (1941). *Publ. Dansk Ast. Obs.* 7: 251.

McVittie, G.C. (1931). "The Problem of $n$ Bodies and the Expansion of the Universe." *MNRAS* 91: 274–283. [Approximate metric for particles in FLRW.]

——— (1932). "Condensations in an Expanding Universe." *MNRAS* 92: 500–518. [Revision of Jeans's work on condensation of matter into nebulae.]

——— (1933). "The Mass-Particle in an Expanding Universe." *MNRAS* 93: 325–339. [Particles in an expanding universe.]

——— (1934). "Remarks on the Geodesics of Expanding Space-Time." *MNRAS* 94: 476–483.

——— (1937). *Cosmological Theory*. London: Methuen. ["The status of the timescale remains an outstanding problem."]

——— (1938a). "The Distances of the Extragalactic Nebulae." *MNRAS* 98: 384–397.

——— (1938b). "Corrections to the Apparent Photographic Magnitudes of Extragalactic Nebulae." *Observ.* 61: 209–214. [Observational power series; magnitude, flux relation.]

——— (1939). "Observations and Theory in Cosmology." *Proc. Phys. Soc. Lond.* 51: 529–538.

——— (1940). "Kinematical Relativity." *Observ.* 63: 273–281. [Newtonian.]

——— (1954). "Relativistic and Newtonian Cosmology." *A. J.* 59: 173–180.

——— (1956). *General Relativity and Cosmology*. London: Chapman and Hall.

——— (1967). "Georges Lemaître." *QJRAS* 8: 294–297.

McVittie, G.C., and Wyatt, S.P. (1959). "The Background Radiation in a Milne Universe." *Ap. J.* 130: 1–11. [Integrated radiation from galaxies as cosmological test?]

Mercier, A., and Kervaire, M., eds. (1956). *Fünfzig Jahre Relativitätstheorie. Bern, 11.–16. Juli 1955. Verhandlungen.* (*Helv. Phys. Acta* Suppl. 4.) Basel: Birkhäuser.

Milne, E.A. (1933a). "World Structure and the Expansion of the Universe." *Z. Ap.* 6: 1–95, 244; 7: 185. [First presentation of Milne's kinematic cosmology. Background radiation limits (Olbers); evolution of sources; Cosmological Principle.]
——— (1933b). "Remarks on World Structure." *MNRAS* 93: 668–680.
——— (1934a). "A Newtonian Expanding Universe." *Qu. J. Math.* 5: 64–72.
——— (1934b). "World Models and the World Picture." *Observ.* 57: 24–27.
——— (1935). *Relativity, Gravitation and World-Structure.* Oxford: Clarendon Press. [FLRW symmetry ⇒ Hubble law. Olbers's paradox.]
——— (1941). "Kinematical Relativity—A Discussion." *Observ.* 64: 11–16.
——— (1948). *Kinematic Relativity.* Oxford: Clarendon Press.
Minkowski, H. (1909). "Raum und Zeit." *Phys. Z.* 10: 104–111. [Space-time concept; English translation in Lorentz et al. 1923.]
Misner, C.W. (1968). "The Isotropy of the Universe." *Ap. J.* 151: 431–457. [Neutrino viscosity.]
——— (1969a). "Mixmaster Universe." *Phys. Rev. Lett.* 22: 1071–1074.
——— (1969b). "Quantum Cosmology I." *Phys. Rev.* 186: 1319–1327.
——— (1972). "Minisuperspace." In *Magic Without Magic: John Archibald Wheeler. A Collection of Essays in Honor of His Sixtieth Birthday.* J.R. Klauder, ed. San Francisco: Freeman, pp. 441–473.
Misner, C.W., and Taub, A.H. (1969). "A Singularity-Free Empty Universe." *Soviet Physics. JETP* 28: 122–133.
Misner, C.W., Thorne, K.S., and Wheeler, J.A. (1973). *Gravitation.* San Francisco: Freeman. [Textbook, extensive bibliography.]
Moessard, M. (1892). "Sur la méthode Doppler-Fizeau." *C. R. Acad. Sci.* 114: 1471–1473. [Doppler shift interpretation, cf. Curtis 1933.]
Nakagami, M., and Miga, K. (1939). *Electrotechnical Journal* 3: 216. [Radio emission from sun detected but source not realized; see Lovell 1987.]
Neumann, C. (1874). "Über das Webersche Gesetz." *Königlich Sächsische Gesellschaft der Wissenschaften zu Leipzig. Mathematisch-physikalische Klasse. Abhandlungen* 21: 77–200. [Difficulties with Newtonian cosmology; see p. 97.]
——— (1896). *Allgemeine Untersuchungen über das Newton'sche Princip der Fernwirkungen, mit besonderer Rücksicht auf die elektrischen Wirkungen.* Leipzig: B.G. Teubner.
Neymann, J., and Scott, E.L. (1952). "A Theory of the Spatial Distribution of Galaxies." *Ap. J.* 116: 144–163. [Statistical analysis of galaxy distribution.]
Neymann, J., Scott, E.L., and Shane, C.D. (1953). "On the Spatial Distribution of Galaxies: A Specific Model." *Ap. J.* 117: 92–133.
North, J. D. (1965). *The Measure of the Universe: A History of Modern Cosmology.* Oxford: Oxford University Press.
Okaya, T. (1928). "Sur les équations cosmique." *Académie Royale de Belgique. Classe des Sciences. Mémoires* 10, no. 8.
Olbers, H.W.M. (1823). "Ueber die Durchsichtigkeit des Weltraums." *Astronomisches Jahrbuch.* J.E. Bode, ed. Berlin: Spathen, 1826. [Dark night sky paradox: see Jaki 1969 and Harrison 1986.]
Olive, K.A., Schramm, D.N., Steigman, G., Turner, M.S., and Yang, J. (1981). "Big Bang Nucleosynthesis as a Probe of Cosmology and Particle Physics." *Ap. J.* 246: 557–568.
Omer, G.C. (1949). "A Non-Homogeneous Cosmological Model." *Ap. J.* 109: 164–176. [Inhomogeneous cosmology? (abandon cosmological principle).]

Oort, J. (1927). "Investigations Concerning the Rotational Motion of the Galactic System." *Bull. Ast. Inst. Neth.* 4: 79–89. [Estimates mass of galaxy.]

—— (1931). "Some Problems Concerning the Distribution of Luminosities and Peculiar Velocities of Extragalactic Nebulae." *Bull. Ast. Inst. Neth.* 6: 155–160. [Revises Hubble-Humason value of Hubble constant by 1/2.]

—— (1959). "Distribution of Galaxies and the Density of the Universe." In *Onzième Conseil de Physique Solvay: La Structure et l'Évolution de l'Univers*. Brussels: Edition Stoops. [Density = $2 \times 10^{-31}$ ⇒ "low density" universe: missing matter?]

Paddock, G.F. (1916). "The Relation of the System of Stars to the Spiral Nebulae." *PASP* 28: 109–115. [$K$-term in redshift: expansion.]

Pais, A. (1982). *"Subtle is the Lord ...": The Science and the Life of Albert Einstein*. New York: Oxford University Press.

Parker, L. (1968). "Particle Creation in Expanding Universes." *Phys. Rev. Lett.* 21: 562–564.

—— (1969). "Quantized Fields and Particle Creation in Expanding Universes I." *Phys. Rev.* 183: 1057–1068.

—— (1971). "Quantized Fields and Particle Creation in Expanding Universes II." *Phys. Rev. D* 3: 346–356, 2546.

—— (1972a). "Backscattering Caused by Expansion of the Universe." *Phys. Rev. D* 5: 2905–2908.

—— (1972b). "Particle Creation in Isotropic Cosmologies." *Phys. Rev. Lett.* 28: 705–708.

Parker, L., and Fulling, S. (1973). "Quantized Matter and the Avoidance of Singularities in General Relativity." *Phys. Rev. D* 7: 2357–2374.

Partridge, R.B., and Wilkinson, D.T. (1967). "Isotropy and Homogeneity of the Universe from Measurements of the Cosmic Microwave Background." *Phys. Rev. Lett.* 18: 557–559. [Isotropy of CMWBR.]

Peebles, P.J.E. (1965). "The Blackbody Radiation Content of the Universe and the Formation of Galaxies." *Ap. J.* 142: 1317–1326.

—— (1966). "Primordial Helium Abundance and the Primordial Fireball II." *Ap. J.* 146: 542–552. [Nucleosynthesis in the HBB.]

—— (1971). *Physical Cosmology*. Princeton, New Jersey: Princeton University Press. [Excellent textbook.]

—— (1976). "A Cosmic Virial Theorem." *Astrophys. Sp. Sci.* 45: 3–19.

—— (1984). "Impact of Lemaître's Ideas on Modern Cosmology." In Berger 1984.

Peierls, R.E., Singwi, K.S., and Wroe, D. (1952). "The Polyneutron Theory of the Origin of the Elements." *Phys. Rev.* 87: 46–50. [Cold big bang (polyneutron) theory of element formation: universe of cold matter expanding from initial singularity.]

Penrose, R. (1964). "Conformal Treatment of Infinity." In *Relativity, Groups and Topology*. C. de Witt and B. de Witt, eds. New York and London: Gordon and Breach, pp. 563–584. [Conformal view of horizons in cosmology.]

—— (1968). "Structure of Space-Time." In *Battelle Rencontres*. C.M. de Witt and J.A. Wheeler, eds. New York: Benjamin, pp. 121–235.

—— (1979). "Singularities and Time-Asymmetry." In *General Relativity: An Einstein Centenary Survey*. S.W. Hawking, and W. Israel, eds. Cambridge: Cambridge University Press, pp. 581–638, 883–886.

Penzias, A.A., and Wilson, R.W. (1965). "A Measurement of Excess Antenna Temperature at 4080 Mc/s." *Ap. J.* 142: 419–421. [First detection of CMWBR; see Dicke et al. 1965.]

Perko, T.E., Matzner, R.A., and Shepley, L.C. (1972). "Galaxy Formation in Anisotropic Cosmologies." *Phys. Rev. D* 6: 969–983.

Petrov, A.Z. (1955). "On Spaces of Maximal Mobility which Define a Gravitational Field." *Doklady Akademii Nauk SSSR* 105: 905. [Gravitational fields classified by symmetries.]

Pierpont, J. (1927). "Optics in a Space of Constant Non-Vanishing Curvature." *Am. J. Math.* 49: 343–354.

Pirani, F.A.E. (1954a). "On the Influence of the Expansion of Space on the Gravitational Field Surrounding an Isolated Body." *Cambridge Philosophical Society. Proceedings* 50: 637–638.

—— (1954b). "The Steady State Theory of the Homogeneous Expanding Universe." *Observ.* 74: 172–173.

—— (1955). "On the Energy Momentum Tensor and The Creation of Matter in General Relativity." *Proc. Roy. Soc. A* 228: 455–462.

Polubarinova-Kochina, P.Ya. (1964). "A.A. Friedmann." *Soviet Physics. Uspekhi* 20: 467–472.

RAS (Royal Astronomical Society) (1919). "Report of RAS and RS Meeting on Solar Eclipse Expedition." *Observ.* 42: 389–398.

—— (1930). "Report of the RAS Meeting in February 1930." *Observ.* 53: 33–44.

Raychaudhuri, A.K. (1952). "Condensations in Expanding Cosmological Models." *Phys. Rev.* 86: 90–92.

—— (1955a). "Perturbed Cosmological Models." *Z. Ap.* 37: 103–107.

—— (1955b). "Relativistic Cosmology." *Phys. Rev.* 98: 1123–1126. [General expansion equation and singularity theorem (shearing and expanding motion).]

—— (1957). "Relativistic and Newtonian Cosmology." *Z. Ap.* 43: 161–164.

—— (1958). "An Anisotropic Cosmological Solution in General Relativity." *Proc. Phys. Soc. Lond.* 72: 263–264. [Bianchi I model.]

Reber, G. (1940). "Cosmic Static." *Ap. J.* 91: 621–624. [Radio emission from Cygnus A detected in 1938: see Hey et al. and Lovell.]

—— (1944). "Cosmic Static." *Ap. J.* 100: 279–287.

Rees, M.J., and Sciama, D.W. (1968). "Large Scale Density Inhomogeneities in the Universe." *Nature* 217: 511–516. [Effect on CMWBR.]

Reeves, H., Audouze, J., Fowler, W.A., and Schramm, D.N. (1973). "On the Origin of Light Elements." *Ap. J.* 179: 909–930.

Reines, F. (1972). *Cosmology, Fusion and Other Matters: George Gamow Memorial Volume.* Bristol: Adam Hilger.

Reynolds, J.H. (1932). "Physical and Observational Evidence for an Expanding Universe." *Nature* 130: 458–462.

—— (1938). "Extragalactic Nebulae." *MNRAS* 98: 334–345.

Ricci, G., and Levi-Civita, T. (1901). "Méthodes de Calcul différential absolu et leurs applications." *Mathematische Annalen* 54: 125–201, 608. [Survey of differential geometry.]

Riemann, B. (1854). "Über die Hypothesen, welche der Geometrie zu Grunde liegen." *Königliche Gesellschaft der Wissenschaften zu Göttingen. Abhandlungen* 13 (1867): 133–152. English translation in *A Source Book in Mathematics.* Vol. 2. D.E. Smith, ed. New York and London: McGraw-Hill, 1929, pp. 411–425. [Groundbreaking paper suggesting idea of curved spaces.]

Rindler, W. (1955). "On the Coordination of the Riemannian and Kinematic Techniques in Theoretical Cosmology, with Particular Reference to the Shift-Distance Law." *MNRAS* 116: 335–350.

——— (1956). "Visual Horizons in World Models." *MNRAS* 116: 662–677. [First clear exposition of nature of particle horizons in FLRW models.]
Ritchey, G.W. (1917a). "Novae in Spiral Nebulae." *PASP* 29: 210–212. [First extragalactic nova ⇒ distance estimates.]
——— (1917b). "Another Faint Nova in the Andromeda Nebulae." *PASP* 29: 257.
Robb, A.A. (1921). *The Absolute Relations of Space and Time*. Cambridge: Cambridge University Press.
——— (1936). *Geometry of Space and Time*. Cambridge: Cambridge University Press.
Robertson, H.P. (1927). "Dynamical Space-Times which Contain a Conformal Euclidean 3-Space." *Trans. Am. Math. Soc.* 29: 481–490. [Foundations of FLWR models.]
——— (1928). "On Relativistic Cosmology." *Phil. Mag.* 5: 835–848. [Finds expanding ($k = 0$) coordinates for de Sitter universe; states data fit linear relationship.]
——— (1929). "On the Foundations of Relativistic Cosmology." *PNAS* 15: 822–829. [Finds general FLRW metric (all $k$) but does not realize its significance. Shows Einstein and de Sitter are only static solutions: cf. Tolman.]
——— (1932). "The Expanding Universe." *Science* 76: 221–226. [Expansion but no big bang.]
——— (1933). "Relativistic Cosmology." *Rev. Mod. Phys.* 5: 62–90. [Excellent review; bibliography to 1933.]
——— (1935). "Kinematics and World Structure I." *Ap. J.* 82: 284–301. [Geometry of kinematic relativity, homogeneity and isotropy ⇒ FLRW models: group theory applied.]
——— (1936a). Review of A.E. Milne, *Relativity, Gravitation and World Structure*. *Ap. J.* 83: 61–66.
——— (1936b). "Kinematics and World Structure II and III." *Ap. J.* 83: 187–201, 257–271. [Kinematics.]
——— (1938). "The Apparent Luminosity of a Receding Nebula." *Z. Ap.* 15: 69–81. [Apparent and absolute magnitudes of light source in expanding universe from energy-momentum conservation for photons. Luminosity formulas corrected, $m$-$z$ relation.]
——— (1949). "Geometry as a Branch of Physics." In Schilpp 1949, pp. 315–332.
——— (1956). "Cosmological Theory." In Mercier and Kervaire 1956, pp. 128–146.
Robertson, H.P., and Walker, A.G. (1955). "Theoretical Aspects of the Nebular Redshift." *PASP* 67: 82–98. [Evolution problem; definition of $q_0$.]
Roll, P.G., and Wilkinson, D.T. (1966). "Cosmic Background Radiation at 3.2 cm—Support for Cosmic Blackbody Radiation." *Phys. Rev. Lett.* 16: 405–407. [Isotropy to 10%.]
Rubin, V.C. (1954). "Fluctuations in the Space Distribution of Galaxies." *PNAS* 40: 541. [Galaxy statistics, following Limber 1953.]
Rutherford, E. (1929). "Origin of Actinium and the Age of the Earth." *Nature* 123: 313–314. [Uranium ages ⇒ earth formed after sun. $t = 3.4 \times 10^{11}$ yrs; established use of radioactive nuclei in geo-chronometry and cosmochronology (Fowler).]
Ruse, H.S. (1933). "On the Measurement of Spatial Distance in Curved Spaces."*Proc. Roy. Soc. Edin.* 53: 79–88.
Ryan, M.P., and Shepley, L.C. (1975). *Homogeneous Relativistic Cosmologies*. Princeton, New Jersey: Princeton University Press.
——— (1976). "Resource Letter RC-1: Cosmology." *Am. J. Phys.* 44: 223–230.
Ryle, M. (1951). *Proceedings of the Conference on Dynamics of Ionized Media*. London:

Department of Physics, University College, London. [See Bondi 1960. Claims discrete radio sources not extragalactic, opposed by Gold.]
——— (1958). "The Nature of the Cosmic Radio Sources. (Bakerian Lecture.)" *Proc. Roy. Soc. A* 248: 289–308.
Ryle, M., and Scheuer, P.A.G. (1955). "The Spatial Distribution and the Nature of Radio Stars." *Proc. Roy. Soc. A* 230: 448–462. [First use of radio source counts for cosmology.]
Ryle, M., Shakeshaft, J.R., Baldwin, J.E., Elsmore, B., and Thomson, J.H. (1955). "A Survey of Radio Sources between Declination $-38°$ and $+38°$." *Mem. RAS* 67: 106–154. [Radio sources as evidence against steady-state universe.]
Sachs, R.K., ed. (1971). *General Relativity and Cosmology*. Proceedings of the International School of Physics "Enrico Fermi," Course 47. New York and London: Academic Press. [Articles on cosmology by Ehlers, Ellis, Sciama, Rees, et al.]
Sachs, R.K., and Wolfe, A.M. (1967). "Perturbations of a Cosmological Model and Angular Variations of the Microwave Background." *Ap. J.* 147: 73–90. [CMWBR anisotropy theory.]
Sakharov, A. (1967). "Violation of CP Invariance, C Assymmetry, and Baryon Asymmetry of the Universe." *JETP Letters* 5: 24–27. [Possibility of baryosynthesis.]
Sandage, A. (1956). "The Redshift." *Sci. Am.* 195 (September): 170–182.
——— (1958). "Current Problems in the Extra-Galactic Distance Scale." *Ap. J.* 127: 513–526. [HII regions $\Rightarrow$ revision of distance scale by a factor of 2.]
——— (161). "The Ability of the 200-Inch Telescope to Discriminate between Selected World-Models." *Ap. J.* 133: 355–392. [Possible optical measurements.]
——— (1968). "Observational Cosmology." *Observ.* 88: 91–106.
Sato, K. (1981). "First Order Phase Transition of a Vacuum and the Expansion of the Universe." *MNRAS* 195: 467–479. [Inflation.]
Schatzmann, E., ed. (1973). *Cargese Lectures in Physics*. Vol 6. London: Gordon and Breach. [Articles on cosmology by Ellis, MacCallum, Harrison, Longair, Rees, Silk, et al.]
Schilpp, P.A., ed. (1949). *Albert Einstein: Philosopher-Scientist*. Evanston, Illinois: The Library of Living Philosophers. [Includes articles by Gödel, Lemaître, Milne, Robertson.]
Schmidt, M. (1963). "3C 273: A Star-Like Object with a Large Redshift." *Nature* 197: 1040. [First QSO redshift: see Hazard et al.]
Schouten, J.A. (1923). *Der Ricci-Kalkül. Eine Einführung in die neueren Methoden und Probleme der mehrdimensionalen Differentialgeometrie*. Berlin: Julius Springer.
——— (1954). *Ricci-Calculus. An Introduction to Tensor Analysis and its Geometrical Applications*, 2nd ed. Berlin, Göttingen, and Heidelberg: Springer-Verlag.
Schrödinger, E. (1939). "The Proper Vibrations of the Expanding Universe." *Physics* 6: 899–912.
——— (1950). *Expanding Universes*. Cambridge: Cambridge University Press.
Schücking, E. (1954). "Das Schwarzschildsche Linienelement und die Expansion des Weltalls." *Z. Phys.* 137: 595–603. [Swiss Cheese model, cf. Einstein and Strauss.]
——— (1957). "Homogeneous Shear-Free World Models in Relativistic Cosmology." *Naturwiss.* 44: 507.
Schwarzschild, K. (1900). "Über das zulässige Krümmungsmaass des Raumes." *Vierteljahrsschrift der Astronomischen Gesellschaft* 35: 337–347. [Elliptical universe, parallax, size estimates.]
——— (1907). "Über Lamberts kosmologische Briefe." *Nach. Ges. Gött. Wiss.*

——— (1916a). "Über das Gravitationsfeld eines Massenpunktes nach der Einsteinschen Theorie." *Preuss. Akad. Wiss. Sitzb.* 189–196. [Static spherically symmetric vacuum solution; cf. Droste.]

——— (1916b). "Über das Gravitationsfeld einer Kugel aus inkompressibler Flüussigkeit nach der Einsteinschen Theorie." *Preuss. Akad. Wiss. Sitzb.* 424–434. [Static spherically symmetric fluid solution.]

Schwarzschild, M. (1954). "Mass Distribution and Mass-Luminosity Values in Galaxies." *A. J.* 59: 273–284. [Virial theorem $\Rightarrow M/L$ too large.]

Sciama, D.W. (1971). "Astrophysical Cosmology." In Sachs 1971, pp. 183–236.

*Scientific American* (1956). 195 (September). [Special issue on cosmology].

Scott, E.L. (1956). "The Brightest Galaxy in a Cluster as a Distance Indicator." *Ap. J.* 62: 248–265. [Selection effects.]

Segal, I.E. (1976). *Mathematical Cosmology and Extragalactic Astronomy*. New York: Academic Press.

Sexl, R.U., and Urbantke, H.K. (1969). "Production of Particles by Gravitational Fields." *Phys. Rev.* 179: 1247–1250. [Cf. Parker.]

Shapley, H. (1917a). "Note on the Magnitudes of Novae in Spiral Galaxies." *PASP* 29: 213–217.

——— (1917b). "Studies Based on the Colours and Magnitudes in Stellar Clusters II." *Ap. J.* 45: 123–140. [See p. 139: Olbers $\Rightarrow$ finite stellar system (Berendzen et al., p. 183).]

——— (1918a). "VI: On the Determination of the Distances of Globular Clusters." *Ap. J.* 48: 89–124. [Galaxy distance scale (Cepheids).]

——— (1918b). "The Distribution, Distances in Space and Dimensions of 69 Globular Clusters." *Ap. J.* 48: 154–181.

——— (1923). "Note on the Distance of NGC 6822." *Harvard College Observatory. Bulletin* 796. [Distance to NGC 6822.]

——— (1929). "Note on the Velocities and Magnitudes of External Galaxies." *PNAS* 15: 565–570. [Criticizes Hubble relation.]

——— (1933). "Luminosity Distributions and Average Density of Matter in Twenty-Five Groups of Galaxies." *PNAS* 19: 591–596. [Density of matter $10^{-30}$ gm/cc.]

Shapley, H., and Ames, A. (1932). "A Survey of the External Galaxies Brighter than the Thirteenth Magnitude." *Harvard College Observatory. Annals* 88: 43–75. [Catalogue; uniformity.]

Shapley, H., and Curtis, H.D. (1921). "The Scale of the Universe." *National Research Council. Bulletin* 2 (Pt. 3) no. 11: 171. ["Great Debate"; physical laws same everywhere, Berendzen et al., p. 41.]

Shapley, H., and Shapley, M. (1919). "Studies Based on Colors and Magnitudes in Stellar Clusters. Fourteenth paper: Further Remarks on the Structure of the Galactic System." *Ap. J.* 50: 107–140. [$m$-$z$ relation in de Sitter.]

Shepley, L.C., and Strassenberg, A.A. (1979). *Cosmology: Selected Reprints*. Stony Brook, New York: American Association of Physics Teachers, SUNY.

Silberstein, L. (1924). "The Curvature of de Sitter's Space-Time Derived from Globular Clusters." *MNRAS* 84: 363–366. [One of controversial series of papers.]

Silk, J. (1967). "Fluctuations in the Primordial Fireball." *Nature* 215: 1155–1156.

Slipher, V.M. (1914). "The Radial Velocity of the Andromeda Nebula." *Lowell (Flagstaff) Observatory Bulletin* 58: 56–57. [First extragalactic redshift.]

——— (1915). "Spectrographic Observations of Nebulae." *Pop. Ast.* 23: 21–24. [Eleven of fifteen velocities $> 0$.]

——— (1922). Quoted in Eddington 1924b, p. 162. [40 radial velocities.]
Smith, R.W. (1979). "The Origins of the Velocity-Distance Relation." *Journal for the History of Astronomy* 10: 133–164.
——— (1982). *The Expanding Universe*. Cambridge: Cambridge University Press.
Smith, S. (1936). "The Mass of the Virgo Cluster." *Ap. J.* 83: 23–30. [Stability of clusters of galaxies, virial theorem and mean density (cf. Zwicky).]
Smoot, G., Gorenstein, M.V., and Muller, R.A. (1977). "Detection of Anisotropy in the Cosmic Blackbody Radiation." *Phys. Rev. Lett.* 39: 898–901. [First CMWBR (dipole) anisotropy detection.]
Stachel, John (1986). "Eddington and Einstein." In *The Prism of Science*. E. Ullmann-Margalit, ed. Dordrecht and Boston: D. Reidel, pp. 225–250.
Stebbins, J., and Whitford, A.E. (1937). "Photoelectric Magnitudes and Colors of Extragalactic Nebulae." *Ap. J.* 86: 247–273.
——— (1948). "Six-Colour Photometry of Stars. IV: The Colours of Extra-Galactic Nebulae." *Ap. J.* 108: 413–428.
Sterne, T.E. (1933). "The Equilibrium Theory of the Abundance of the Elements: A Statistical Investigation of Assemblies in Equilibrium in which Transmutations Occur." *MNRAS* 93: 736–767. [Equilibrium element synthesis ($e$-process) from neutrons, protons, and electons at constant temperatures.]
Stewart, J.Q. (1931). "Nebular Redshift and Universal Constants." *Phys. Rev.* 38: 2071. [Hubble constant determined by fundamental constants?]
Stock, J., and Schücking, E. (1957). "Remarks on the Magnitude of Extragalactic Nebulae." *A. J.* 62: 98–104. [Surface brightness effects in galaxy observations.]
Straubel, R. (1903). "Über einen allgemeinen Satz der geometrischen Optik und einige Anwendungen." *Phys. Z.* 4: 114–117. [Reciprocity theorem, cf. Hilbert.]
Stromberg, G. (1925). "Analysis of Radial Velocities of Globular Clusters and Non-Galactic Nebulae." *Ap. J.* 61: 353–362. [No clear velocity-distance relation.]
Sunyaev, R., and Zeldovich, Ya. B. (1970). "Small-Scale Fluctuations of Relic Radiation." *Astrophys. Sp. Sci.* 7: 3–19.
——— (1972). "The Observation of Relic Radiation as a Test of the Nature of X-Ray Radiation from the Clusters of Galaxies." *Comments Astrophys. Sp. Sci.* 4: 173–178.
Synge, J.L. (1934a). "On the Expansion or Contraction of a Symmetrical Cloud under the Influence of Gravity." *PNAS* 20: 635–640. [Raychaudhuri equation at center.]
——— (1934b). "On the Deviation of Geodesics and Null Geodesics, Particlarly in Relation to the Properties of Spaces of Constant Curvature and Indefinite Line-Element." *Ann. Math.* 35: 705–713. [Geodesic deviation equation.]
——— (1936). "Limitations on the Behaviour of an Expanding Universe." *Royal Society of Canada. Transactions* 30 165–178. [Inequalities of state; phase planes.]
——— (1937). "Relativistic Hydrodynamics." *Proc. Lond. Math. Soc.* 43: 36–416. [General fluid behavior in curved space-times.]
——— (1957). *The Relativistic Gas*. Amsterdam: North Holland. [Special relativistic kinetic theory.]
——— (1960). *Relativity: The General Theory*. Amsterdam: North Holland. [Extensive references.]
Synge, J.L., and Schild, A. (1949). *Tensor Calculus*. Toronto: University of Toronto Press.
Takahashi, Y., and Umezawa, H. (1957). "A General Treatment of Expanding Systems." *Il Nuovo Cimento* 6: 1324–1334; 1382–1391. [Particle creation by expansion, cf. Parker.]

Tammann, G.A. (1984). "The Hubble Diagram." In *Clusters and Groups of Galaxies*. F. Mardirossian et al., eds. Dordrecht: D. Reidel.

Taub, A.H. (1937). "Quantum Equations in Cosmological Spaces." *Phys. Rev.* 51: 512–525.

——— (1948). "Relativistic Rankine-Hugoniot Equations." *Phys. Rev.* 74: 328–334. [Macroscopic conservation laws from kinetic theory.]

——— (1951). "Empty Space-Times Admitting a 3-Parameter Group of Motions." *Ann. Math.* 53: 472–490. [Symmetries and equations of empty Bianchi universes.]

——— (1959). "On Circulation in Relativistic Hydrodynamics." *Archive for Rational Mechanics and Analysis* 3: 312–324.

Tauber, G.E., and Weinberg, J.W. (1961). "Internal State of a Self-Gravitating Gas." *Phys. Rev.* 122: 1342–1365. [General relativistic $H$-theorem and equilibrium conditions.]

Tayler, R.J. (1983). "The Neutron in Cosmology." *QJRAS* 24: 1–9.

Teller, E. (1948). "On the Change of Physical Constants." *Phys. Rev.* 73: 801–802. [Greater stellar evolution rate in past?]

Temple, G. (1938). "New Systems of Normal Coordinates for Relativistic Optics." *Proc. Roy. Soc. A* 168: 122–148. [Observations in general space-time.]

——— (1939). "Relativistic Cosmology." *Proc. Phys. Soc. Lond.* 51: 465–478. [Survey.]

Ten Bruggencate, P. von (1937). "Dehnt sich das Weltall aus?" *Naturwiss.* 35: 561–566.

ter Haar, D. (1950). "Cosmogonical Problems and Stellar Energy." *Rev. Mod. Phys.* 22: 119–152. [Review of element creation.]

Thackeray, A.D., and Wesselink, A.J. (1953). "Distances of the Magellanic Clouds." *Nature* 171: 693. [Hubble constant revision, see Baade.]

Thirring, H. (1918). "Über die Wirkung rotierender ferner Massen in der Einsteinschen Gravitationstheorie." *Phys. Z.* 19: 33–39. [Machian effects of rotation.]

Thirring, H., and Lense, J. (1918). "Über den Einfluss der Eigenrotation der Zentralkörper auf die Bewegung der Planeten und Monde nach der Einsteinschen Gravitationstheorie." *Phys. Z.* 19: 156–163.

Thomson, W. (Lord Kelvin) (1901). "Note on the Possible Density of the Luminiferous Medium and on the Mechanical Value of a Cubic Mile of Starlight." *Phil. Mag.* 2: 161–177. [Olbers's paradox; see Harrison 1986.]

Thorne, K.S. (1967). "Primordial Element Formation, Primordial Magnetic Fields, and the Isotropy of the Universe." *Ap. J.* 148: 51–68. [Bianchi I nucleosynthesis and CMWBR anisotropy.]

Tinsley, B.M. (1973). "Analytical Approximations to the Evolution of Galaxies." *Ap. J.* 186: 35–49.

Tipler, C., Clarke, C.J.S., and Ellis, G.F.R. (1980). "Singularities and Horizons: A Review Article." In *General Relativity and Gravitation*. Vol. 2. A. Held, ed. New York: Plenum Press. [Extensive bibliography.]

Tolman, R.C. (1922). "Thermodynamic Treatment of the Possible Formation of Helium from Hydrogen." *Journal of the American Chemical Society* 44: 1902–1908. [Creation of elements?; very early attempt.]

——— (1928a). "On the Extension of Thermodynamics to General Relativity." *PNAS* 14: 268–272. [Thermodynamics in curved space-times.]

——— (1928b). "On the Energy and Entropy of Einstein's Closed Universe." *PNAS* 14: 348–353.

——— (1928c). "On the Equilibrium between Matter and Radiation in Einstein's Closed Universe." *PNAS* 14: 353–356.

——— (1928d). "Further Remarks on the 2nd Law of Thermodynamics in General Relativity." *PNAS* 14: 701–706. [Thermodynamics in curved space-times.]

——— (1929a). "On the Possible Line Elements for the Universe." *PNAS* 15: 297–304. [Shows Einstein $A$ and de Sitter $B$ are only static universes.]

——— (1929b). "On the Astronomical Implications of de Sitter Line Element for the Universe." *Ap. J.* 69: 245–274. [Detailed comparison of de Sitter universe $B$ with redshift observations. Does not satisfactorily explain observed Doppler shifts.]

——— (1930a). "On the Effect of Annihilation of Matter on the Wavelength of Light from the Nebulae." *PNAS* 16: 320–337. [Variable radius evolving universe.]

——— (1930b). "More Complete Discussion of the Time Dependence of the Non-Static Line Element for the Universe." *PNAS* 16: 409–420. [Power series.]

——— (1930c). "On the Estimation of Distances in a Curved Universe with a Non-Static Line Element." *PNAS* 16: 511–520. [First FLRW area distance definition and relation to luminosity of distant galaxies (background radiation?).]

——— (1930d). "Discussion of Various Treatments which Have Been Given to the Non-Static Line Element for the Universe." *PNAS* 16: 582–594. [Classification of nonstatic line-elements.]

——— (1931a). "On the Problem of Entropy of the Universe as a Whole." *Phys. Rev.* 37: 1639–1660. [Second law of thermodynamics; universe filled with black-body radiation.]

——— (1931b). "Non-Static Model of the Universe with Reversible Annihilation of Matter." *Phys. Rev.* 38: 797–814. [Relativistic thermodynamics applied to monatomic gas in equilibrium with black-body radiation.]

——— (1931c). "On the Theoretical Requirements for a Periodic Behavior of the Universe." *Phys. Rev.* 38: 1758–1771. [Nonexistence of strictly periodic reversible universes; singularity due to high symmetry.]

——— (1932a). "Models of the Physical Universe." *Science* 75: 367–373. [Drops cosmological constant; accepts singularity.]

——— (1932b). "Possibilities in Relativistic Thermodynamics for Irreversible Processes Without Exhaustion of Free Energy." *Phys. Rev.* 39: 320–336.

——— (1934a). *Relativity, Thermodynamics, and Cosmology.* Oxford: Clarendon Press. [Radiation and entropy in cosmology; Olbers in chapter 10, part 2.]

——— (1934b). "Effect of Inhomogeneity on Cosmological Models." *PNAS* 20: 169–176. [Inhomogeneous Lemaître-Bondi pressure-free solution; Raychaudhuri equation at center.]

——— (1949). "The Age of the Universe." *Rev. Mod. Phys.* 21: 374–378. [Abandon cosmological principle?]

Tolman, R.C., and Ward, M. (1932). "On the Behavior of Non-Static Models of the Universe when the Cosmological Constant is Omitted." *Phys. Rev.* 39: 835–843. [First singularity theorem: proves FLRW singularity exists if $p \geq 0$, $k = +1$.]

Trumpler, R. (1930). "Preliminary Results on the Distances, Dimensions and Space Distribution of Open Star Clusters." *Lick Observatory Bulletin* 14: 154–188. [Demonstrates galactic absorption.]

Vaidya, P.C., and Shah, K.B. (1957). "A Radiating Mass Particle in an Expanding Universe." *Proc. Nat. Inst. Sci. India A* 23: 534–539.

van Maanen, A. (1922). "Investigations on Proper Motions. VIII Paper: Internal Motions in the Spiral Nebula M94 = NGC 4736." *Ap. J.* 56: 208–216. [$m$-$z$ for de Sitter?]

von Laue, M. (1931). "Die Lichtfortpflanzung in Räumen mit zeitlich veränderlicher Krümmung nach der allgemeinen Relativitätstheorie." *Preuss. Akad. Wiss. Sitzb.* 7: 123–131. [Waves as opposed to geometric optics.]

——— (1938). "The Apparent Luminosity of Receding Nebula." *Z. Ap.* 15: 160–161.

von Seeliger, H. (1895). "On Newton's Law of Gravitation." *Astr. Nach.* 137: 129–136. [Problems with Newtonian cosmology ⇒ exponential cutoff? Berendzen et al., North.]

——— (1896). *Bayerische Akademie der Wissenschaften* (Munich). *Mathematisch-naturwissenschaftliche Klasse. Sitzungsberichte.*

von Weizsäcker, C.F. (1937). "Über Elementenwandlungen im Innern der Sterne I." *Phys. Z.* 36: 176–191.

——— (1938). "Über Elementenwandlungen im Innern der Sterne II." *Phys. Z.* 39: 633–646. [Element formation: equilibrium theory.]

——— (1947). "Zur Kosmogonie." *Z. Ap.* 24: 181–206.

——— (1951). "The Evolution of Galaxies and Stars." *Ap. J.* 114: 165–186. [Turbulence.]

Wagoner, R.V., Fowler, W.A., and Hoyle, F. (1968). "On the Synthesis of Elements at Very High Temperature." *Ap. J.* 148: 3–49.

Walker, A.G. (1933). "Spatial Distance in General Relativity." *Qu. J. Math.* 4: 71–80.

——— (1934). "Distance in an Expanding Universe." *MNRAS* 94: 159–167.

——— (1935). "On Riemannian Spaces with Spherical Symmetry about a Line and the Conditions for Isotropy in General Relativity." *Qu. J. Math.* 6: 81–93. [The only universes in which there is local spherical symmetry about each fundamental observer are the FLRW models.]

——— (1936a). "The Boltzmann Equation in General Relativity." *Proc. Edin. Math. Soc.* 4: 238–253. [First Boltzmann equation in general relativity(?), cf. Marrot.]

——— (1936b). "On Milne's Theory of World Structure." *Proc. Lond. Math. Soc.* 42: 90–127. [Group theory applied to cosmology. Kinematical relativity ⇒ FLRW geometry.]

——— (1940). "Relativistic Mechanics." *Proc. Lond. Math. Soc.* 46: 113–154. [Newtonian.]

——— (1941). "[Kinematical Relativity—A Discussion.]" *Observ.* 64: 17–23.

——— (1944a). "Completely Symmetric Spaces." *Journal of the London Mathematical Society* 19: 219–226, 227(?). [Isotropy everywhere ⇒ homogeneity.]

——— (1944b). "Foundations of Kinematical Relativity." *Observ.* 65: 242–243.

Walsh, D., Carswell, R.F., and Weymann, R.J. (1979). "0956+561 A, B: Two Quasi-Stellar Objects or Gravitational Lens?" *Nature* 279: 381. [Gravitational lens observation.]

Weinberg, S. (1972). *Gravitation and Cosmology: Principles and Applications of the General Theory of Relativity.* New York: Wiley. [Excellent textbook.]

——— (1977). *The First Three Minutes.* New York: Basic Books.

——— (1979). "Cosmological Production of Baryons." *Phys. Rev. Lett.* 42: 850–853.

Weyl, H. (1917). "Zur Gravitationstheorie." *Ann. d. Physik* 54: 117–145. [General relativistic conservation equations, see Pais.]

——— (1918). *Raum-Zeit-Materie.* Berlin: Julius Springer; 2nd ed., 1919; 3rd ed., 1920; 4th ed., 1921; 5th ed., 1923. [Latter not translated into English, contains first statisfactory treatment of redshift coupled with estimates of the expansion timescale; see Ehlers 1987.]

——— (1919). "Eine neue Erweiterung der Relativitätstheorie." *Ann. d. Physik* 59: 101–133. [Large number coincidences, p. 129.]

——— (1922). *Space-Time-Matter*. Translation of Weyl 1918 (based on 4th German ed., 1921). London: Methuen.

——— (1923). "Zur allgemeinen Relativitätstheorie." *Phys. Z.* 24: 230–232. [Redshift as ratio of proper times; linear velocity-distance ratio in expanding de Sitter universe ($k = +1$?) with coherency postulate.]

——— (1924). "Massenträgheit und Kosmos. Ein Dialog." *Naturwiss.* 12: 197–204. [Olbers/causality.]

——— (1930). "Redshift and Relativistic Cosmology." *Phil. Mag.* 9: 936–943. [Redshift and causality in expanding ($k = 0$) de Sitter.]

——— (1934). "Universum und Atom." *Naturwiss.* 22: 145–149. [Large numbers]

Wheeler, J.A. (1968a). "Superspace and the Nature of Quantum Geometrodynamics." In *Batelle Rencontres*. C.M. de Witt and J.A. Wheeler, ed. New York: Benjamin, pp. 242–307.

——— (1968b). *Einsteins Vision. Wie steht es heute mit Einsteins Vision, alles als Geometrie aufzufassen?* Berlin and New York: Springer-Verlag. [Closed universe concept and boundary conditions.]

Wheeler, J.A., and Feynman, R. (1945). "Interaction with the Absorber as the Mechanism for Radiation." *Rev. Mod. Phys.* 17: 157–181. [Cosmology and time asymmetry of electrodynamics.]

——— (1949). "Classical Electrodynamics in Terms of Direct Particle Interaction." *Rev. Mod. Phys.* 21: 425–433.

Whittaker, E.T. (1931). "On the Definition of Distance in Curved Spaces and the Displacement of the Spectral Lines of Distant Sources." *Proc. Roy. Soc. A* 133: 93–105. [Definition of area distance; applied to de Sitter universe.]

——— (1955). "On Gauss' Theorem and the Concept of Mass in General Relativity." *Proc. Roy. Soc. A* 149: 384–395. [Active gravitational mass is (density + 3 × pressure).]

Whitrow, G.J. (1935). "On Equivalent Observers." *Qu. J. Math.* 6: 249–260.

——— (1936a). "Photons, Energy and Red-Shifts in the Spectra of Nebulae." *Qu. J. Math.* 7: 271–276.

——— (1936b). "World Structure and the Sample Principle." *Z. Ast.* 12: 47–55.

——— (1937). "World Structure and the Sample Principle II." *Z. Ast.* 13: 113–125.

——— (1955). "Why Physical Space has Three Dimensions." *British Journal for the Philosophy of Science* 6: 13–31.

Wirtz, C. (1918). "Über die Bewegung der Nebelflecke." *Astr. Nachr.* 206: 109–116. [$K$-Term to represent systematic $z$-effect.]

——— (1921). "Einiges zur Statistik der Radialbewegung von Spiralnebeln und Kugelsternhaufen." *Astr. Nachr.* 215: 349–354. [De Sitter $m$-$z$?]

——— (1922). "Notiz zur Radialbewegung der Spiralnebel." *Astr. Nachr.* 216: 451. [$K$-term represents systematic recession?]

——— (1924). "De Sitter's Kosmologie und die Radialbewegungen der Spiralnebel." *Astr. Nachr.* 222: 21–26. [Recession depends on distance?]

Yoshimura, M. (1978). "Unified Gauge Theories and the Baryon Number of the Universe." *Phys. Rev. Lett.* 41: 281–284. [Baryosynthesis.]

Zee, A., ed. (1984). *Unity of Forces in the Universe*. Vol. 2. Singapore: World Scientific. [Reprint volume.]

Zeldovich, Ya. B. (1965). "Survey of Modern Cosmology." *Advances in Astronomy and Astrophysics* 3: 241–379. [Written in 1964. Discusses hot version of the initial stage of the universe as developed by Gamow et al.; states "electromagnetic radiation must have remained with a temperature ranging from several degrees to 20–30°K," but misinterprets radio astronomy data to have shown that "the temperature of thermal radio waves does not exceed 1°K (Ohm 1961)" and incorrectly states that while theory predicts primordial production of 35% of helium by weight, "the investigations of steller matter testify to the fact that the initial helium content was below 10–20%."]

—— (1970). "Particle Production in Cosmology." *JETP Letters* 12: 307. [Particle creation by anisotropic expansion.]

Zwicky, F. (1929a). "On the Redshift of Spectral Lines through Interstellar Space." *PNAS* 15: 773–779. [Tired light proposal.]

—— (1929b). "On the Redshift of Spectral Lines through Interstellar Space." *Phys. Rev.* 33: 1077. [Tired light.]

—— (1933). "Die Rotverschiebung von extragalaktischen Nebeln." *Helv. Phys. Acta* 6: 110–127. [Virial theorem.]

—— (1935). "Remarks on Redshift from Nebulae." *Phys. Rev.* 48: 802–806. [Tired light.]

—— (1938). "On the Clustering of Nebulae." *PASP* 50: 218–220. [Galaxy clustering.]

—— (1937a). "Nebulae as Gravitational Lenses." *Phys. Rev. Lett.* 51: 290. [First suggestion of extragalactic lensing.]

—— (1937b). "On the Probability of Detecting Nebulae which Act as Gravitational Lenses." *Phys. Rev.* 51: 679. [Lensing.]

—— (1937c). "On the Masses of Nebulae and of Clusters of Nebulae." *Ap. J.* 86: 217–246. [Estimates of masses of nebulas from the dynamics of clusters of nebulas.]

—— (1939). "On the Theory and Observation of Highly Collapsed Stars." *Phys. Rev.* 55: 726–743. [Large numbers and physics.]

# Contributors

*Peter G. Bergmann*, Department of Physics, New York University, 4 Washington Place, New York, New York 10003, USA.

*Michel Biezunski*, 1, Boulevard du Temple, Paris 75003, France.

*Carlo Cattani*, Dipartimento di Matematica, Università degli Studi di Roma "La Sapienza," Piazzale Aldo Moro, 2, 00185 Roma, Italy.

*Michelangelo De Maria*, Dipartimento di Fisica, Università degli Studi di Roma "La Sapienza," Piazzale Aldo Moro, 2, 00185 Roma, Italy.

*Jean Eisenstaedt*, Laboratoire de Physique Théorique, Institut Henri Poincaré, 11, rue Pierre et Marie Curie, 75231 Paris Cedex 05, France.

*George F.R. Ellis*. SISSA, Strada Costiere 11, Miramare, Trieste 34014, Italy.

*Peter Havas*, Department of Physics, Temple University, Philadelphia, Pennsylvania 19122, USA.

*Don Howard*, Department of Philosophy, University of Kentucky, Lexington, Kentucky 40506, USA.

*Pierre Kerszberg*, Department of General Philosophy, University of Sydney, Sydney, New South Wales 2006, Australia.

*A.J. Kox*, Instituut voor Theoretische Fysica, Universiteit van Amsterdam, Valckenierstraat 65, 1018 XE Amsterdam, The Netherlands.

*John Norton*, Department of History and Philosophy of Science, 1017 Cathedral of Learning, University of Pittsburgh, Pittsburgh, Pennsylvania 15260, USA.

*John Stachel*, Department of Physics, Boston University, 590 Commonwealth Avenue, Boston, Massachusetts 02215, USA.

*Vladimir Vizgin*, Institute of History of Science and Technology, Academy of Sciences. Staropansky per., 1/5, 103012 Moscow, USSR.

# Index

Abraham, Max, xi, 3, 45, 59, 61, 160–174, 183–185, 196–197, 199–200, 305
Abrams, L.S., 216, 228, 231
Adams, Edwin P., 364
Adams, Walter S., 394, 403, 419
Adelman, George, ix
Alpher, Ralph A., 393–394, 399–400, 403–404, 412
Anderson, Clelia, ix
Anderson, James L., 45, 299
Arnowitt, Richard, 298
Arp, Halten C., 399, 410
Arzeliès, Henri, 60–61, 254, 270
Ashtekar, Abhay, vii–viii, 299
Aston, Francis W., 389, 404
Audouze, Jean, 404, 422

Baade, Walter, 391–392, 404, 427
Bach, Rudolf, 241, 245–246, 270
Bahcall, James N., 399, 410
Bailian, R., 404
Baldwin, Jack E., 424
Bardeen, John, 395, 404
Bargmann, Sonja, 46, 158
Bargmann, Valentin, 257, 259
Barkas, Walter, 416
Barrow, John, 368, 404
Bass, R.W., 404
Battimelli, Giovanni, 172–174, 196, 199
Baum, William A., 391–392, 404
Bażański, Stanisław, 235, 256, 269–270, 275

Becquerel, Jean, 29, 43, 218, 222–223, 228, 246, 270
Beer, Arthur, 292
Belinskii, V.A., 404
Bembo, Tomaso, 274
Bennett, A.S., 392, 404
Berendzen, Richard, 368, 376–377, 404, 425, 429
Berger, André, 368, 404, 421
Bergmann, Peter G., vii–viii, xii, 2, 4, 257, 259, 282, 284, 287, 290–291, 293, 433
Beringer, Robert, 394, 407
Besso, Michele Angelo, 43, 47, 68, 81, 84–86, 88, 95, 100, 121–122, 131–133, 147, 153–154, 159, 171, 173–174, 178, 184, 196, 200, 337, 340–341, 345, 366
Bethe, Hans A., 403
Białobrzeski, Czesław, 251
Bianchi, Luigi, 382, 395, 404
Bičak, Jiri, 61, 98
Biesbroeck, George van, 282
Biezunski, Michel, viii, xii, 4, 315, 322–323, 433
Bishop, Richard L., 93, 97
Bode, Johann Elert, 420
Bohr, Niels, 310
Bok, Bart J., 404
Bolton, J., 404
Boltzmann, Ludwig, 307
Bondi, Hermann, 285–286, 290, 373, 385, 392, 396–397, 399, 404–405, 411, 413, 417, 424, 428

436  Index

Bonnor, William B., 395, 405
Born, Hedwig, 270, 290
Born, Irene, 270, 290
Born, Max, 49, 60, 122, 157, 257, 269–270, 277, 280, 282, 289–290
Bracewell, Ronald N., 405, 413
Brans, Carl H., 405
Brecher, Kenneth, viii
Brillouin, Marcel, 264
Brose, Henry L., 47, 276, 365–366
Brouwer, Dirk, 307
Buc, H.E., 405
Bunge, Mario, 273
Burbridge, E. Margaret, 394, 398, 405
Burbridge, Geoffrey, 394, 398, 405
Byram, E.T., 405


Callan, Curtis G., Jr., 405
Campbell, William Wallace, 291
Carmeli, Moshe, 62
Carswell, R.F., 429
Cartan, Elie, xii, 4, 238, 299, 309, 312, 315–323, 387
Cassidy, David, viii
Cattani, Carlo, viii, xi, 3, 160, 172, 174–175, 195–196, 199, 433
Ceccherini, V., 174, 196
Chalmers, Bruce, 405
Chalmers, J.A., 405
Champion, Leonard, 54
Chandrasekhar, Subrahmanyan, 243, 270, 383, 405
Chandrasekharan, Komaravolu, 314, 402
Charlier, Carl W.L., 385–386, 390, 405
Chazy, Jean, 255, 260, 270
Christian, Joy, viii
Christoffel, Elwin Bruno, 107, 279, 290, 370, 405
Chubb, Talbot A., 405
Clarke, Christopher J.S., 368, 397, 427
Clark, Gordon L., 256, 260, 268, 270–271
Cohen, Robert S., ix, 292
Collins, C. Barry, 396, 405
Colodny, Robert G., 100, 276
Combridge, J.T., 405
Conklin, E.K., 405
Contu, Raffaele, 274

Coolidge, Julian Lowell, 93, 97
Cowles, Lauren, ix
Crelinsten, Jeffrey, 290–291
Crommelin, Andrew, 279, 308
Curtis, Heber D., 376, 386, 390, 405–406, 420, 425


Damour, Thibault, 270
Darwin, George, 398, 414
Davidson, Charles, 407
Davis, Morton J., 392, 406
Debever, Robert, 322–323
de Broglie, Louis, 310, 313
de Cheseaux, Jean-Philippe-Loys, 385, 406
De Donder, Theophile, 215–216, 228, 231, 238, 243–245, 270–271
De Jans, Carlo, 224–225, 230–231, 254, 271
Dekker, E., 363
De Maria, Michelangelo, viii, xi, 3, 160, 172–175, 195–196, 199, 433
Descartes, René, 276
Deser, Stanley, 298
de Sitter, Willem, xii, 4, 155, 208, 219–221, 229–231, 243–244, 255–256, 258, 260, 268, 271, 325–328, 331–334, 336–341, 344, 347–365, 368–381, 383–385, 387–389, 396, 406–407
de Vaucouleurs, Gerard, 399, 407
de Witt, Cécile M., 292, 299, 421, 430
de Witt, Bryce S., 407, 421
Dicke, Robert H., 283, 291, 394, 398, 400–401, 405, 407, 421
Dieke, Sally H., 228, 231
Dingle, Herbert, 385, 407
Dirac, Paul A.M., 155, 157, 252–254, 271, 294–295, 297–299, 310–311, 387, 407, 415
Dirichlet, Gustave P.L., 303
Doroshkevich, A.G., 400, 407
Douglas, Vibert A., 243, 271
Droste, Johannes, 207, 210, 212, 216–219, 222–224, 226–227, 229–231, 243, 254–256, 258, 271, 274, 286, 291, 368, 407, 424–425
Drude, Paul, 305

Drumaux, Paul, 226, 231
Dukas, Helen, 48, 256, 266
Dunham, Theodore, 394
Dyson, Frank W., 279, 281, 311, 407

Eardley, Douglas, 407
Earman, John, vii–viii, 3, 42, 45, 59–61, 63, 72, 89–90, 98, 156–158, 172–174, 195–196, 199, 212, 228, 231, 281, 290–291
Eddington, Arthur Stanley, 5, 45, 54, 61, 211, 220–221, 226–229, 232, 243–246, 252–254, 256, 258, 260, 268–271, 275, 279–282, 285, 288, 291, 300, 308, 312, 333, 342, 358–359, 362, 364, 371, 373, 375, 377, 379–384, 387–389, 393, 405–408, 415, 426
Ehlers, Jürgen, 273, 380, 396, 401–402, 408, 424, 429
Ehrenfest, Paul, 16, 43, 49, 57, 59–62, 70, 81, 86–87, 106, 127, 152, 154, 156, 164, 166, 172, 204–205, 208–209, 248–249, 338, 341, 351–352, 372
Eiesland, John, 382, 408
Einstein, Albert, v-xii, 1–4
  and Abraham, 160–174
  and the canonical formulation of general-relativistic theories, 293, 298–299
  and de Sitter, 325–366
  and the development of general relativity, 63–74, 77–90, 92–100, 101–159
  and distant parallelism, 315–324
  and the equivalence principle, 5–35, 37–47
  and Levi-Civita, 175–200
  and Lorentz, 201–212
  and the mid-century decline of interest in general relativity, 277–292
  and the origins of field theory, 300–301, 303–313
  and the problem of motion, 234–276
  and relativistic cosmology, 368–373, 375–376, 379–387, 389–390, 393, 395–397, 405–414, 416–418, 421, 424, 426–428, 430

  and the rotating disk, 48–62
  and the Schwarzschild solution, 213–216, 219–223, 225, 228, 230–231, 233
Eisenhart, Luther P., 342, 364, 386, 396, 409
Eisenstaedt, Jean, viii, xii, 1, 4, 213, 229–230, 232, 254, 272, 277, 290–291, 433
Elkana, Yehuda, 158
Ellis, George F.R., viii, xii, 4, 42, 47, 93–95, 97, 99, 367–368, 372, 388, 396–397, 401–402, 409–410, 412, 418, 424, 427, 433
Elsmore, B., 424
Etherington, I.M.H., 388, 410
Euclid, 62
Euler, Leonhard, 297

Fermi, Enrico, 394, 403, 410, 414
Feynman, Richard P., 430
Fichte, Johann Gottlieb, 307
Fichtenholz, I. Gregorius, 256, 272
Fickler, Stuart I., 62
Field, George B., 47, 275, 399, 410
Finlay-Freundlich, Erwin, 54, 88, 99, 138, 142, 281–82
Flamm, Ludwig, 217–218, 221, 226, 232
Flügge, Siegfried, 413
Fock, Vladimir A., 262–263, 265, 267–268, 270, 272, 310, 312
Fokker, Adriaan D., 81–83, 99, 154–155, 158, 209–211, 238, 243, 255, 272, 274–275
Follin, James W., 394, 403
Forman, Paul, 151, 310, 313
Fowler, William A., 394, 398, 400, 405, 410, 422–423, 428
Frank, Philipp, 60, 62
Frederiks, V.K., 312, 376, 387, 410
Freedrichsz, V. see Frederiks, V.K.
Freundlich, Erwin. see Finlay-Freundlich, Erwin
Friedman, Aleksander A. (Friedmann), 308, 312, 326–327, 358–359, 361, 364, 368, 375–376, 378–379, 381, 385–387, 406, 409–410, 419, 422
Friedman, Herbert, 405

Friedman, Michael, 11, 42–43, 47
Friedrich, Carl, 65, 238
Fubini, Guido, 410
Fulling, Stephen A., 421

Galilei, Galileo, 8, 239
Gamow, George, 375, 382, 393–395,
    399–400, 403, 410–411, 415, 431
Gauss, Carl Friedrich, 26, 51–52, 58,
    106, 279, 290, 303, 430
Geiser, Karl Friedrich, 58, 60
Geren, P., 396, 408
Geroch, Robert P., 37, 44, 411
Giacconi, Riccardo, 411
Gibbons, Gary W., 411
Gilbert, C., 411
Gill, Irena, 269
Gillispie, Charles Coulston, 231
Gingerich, Owen, 416
Glaus, Beat, 269
Glick, Thomas F., 323
Glymour, Clark, vii, 3, 42, 45, 59–61,
    63, 72, 89–90, 98, 154, 156–158,
    172–174, 195–196, 199, 228, 231,
    281, 290–291
Godart, Odon, 411
Gödel, Kurt, 375, 395–397, 399, 411,
    416, 424
Goenner, Hubert, 254, 272
Gold, Thomas, 373, 392, 397, 405, 411,
    413, 424
Goldberg, Joshua N., vii–viii, 234–235,
    239, 272–273
Goldberg, Samuel I., viii, 93, 97
Goldberg, Stanley, 208, 211
Goodstein, Judith, 195, 198, 200, 269
Gordon, Walter, 301, 310
Gorenstein, Marc V., 426
Gott, James, 411
Gowdy, Robert H., 411
Graves, Lawrence M., 411
Greenstein, Jesse L., 411
Grommer, Jakob, 235, 240, 246–248,
    250, 253–254, 258, 265, 272, 370
Grøn, Ø., 59, 62
Grossmann, Marcel, 7, 19, 43, 47, 56,
    58–61, 64–70, 72–74, 81, 83–84,
    86, 88–89, 92–93, 95, 99–100,

101–104, 106–116, 119, 124–125,
128–130, 132–135, 141, 153–154,
158, 167–168, 172, 174, 175–179,
183, 185, 195, 199–200, 201–202,
208, 211, 232, 304–305
Groth, Edward J., 411
Grünbaum, Adolf, vii, 59, 62
Gullstrand, Allvar, 222, 280
Gunn, James E., 391, 411–412
Gursky, Herbert, 411
Guth, Eugene, 60, 62
Guth, Alan H., 373, 397–398, 401, 412

Haas, Arthur, 312
Habicht, Conrad, 138
Hacker, Walter, 273
Hadamard, Jacques, 214, 216, 252, 316
Hagihara, Yusuke, 230
Hale, George E., 391
Halley, Edmund, 385, 412
Halliwell, Jonathan J., 412
Hamilton, William Rowan, 211
Hanbury-Brown, R., 392, 412
Harrison, B. Kent, 286, 291, 424
Harrison, Edward R., 368, 385, 406,
    412, 418, 420, 427
Hart, Richard C., 368, 376, 404
Hartle, James B., 412
Harwit, Martin, 368, 394, 412
Harzer, Paul, 412
Havas, Peter, viii, xii, 4, 234–236, 238–
    239, 253, 269–270, 272–273, 433
Hawking, Stephen W., 42, 47, 93–95,
    97, 99, 288, 291, 371, 396–397,
    405–406, 411–412, 418, 421
Hayashi, Chushiro, 394, 399, 403, 412
Hazard, Cyril, 392, 412, 424
Heckmann, Otto H.L., 382, 388, 390,
    396–397, 399, 413
Heinrich, Louis R., 405
Heisenberg, Werner, 250
Held, Alan, 47, 159, 427
Heller, Michael, 411
Helmer, Olaf, 100
Helmholtz, Hermann von, 93, 342
Herglotz, Gustav, 246
Herman, Robert C., 393–394, 399–400,
    403

Hermann, Armin, 58–59, 62, 85, 99, 153, 155–156, 158, 173–174, 362, 365
Herneck, Friedrich, 59, 62, 71, 91–92, 99
Hertz, Paul, 119, 153–154
Hertzsprung, Ejnar, 213, 232
Hesse, Mary B., 234, 273
Hey, J.S., 392, 412–414, 422
Hilbert, David, xii, 2, 61, 98–99, 105, 122, 141, 150–151, 153, 155–158, 160, 172, 199, 206, 211, 218, 220, 231–232, 238, 243, 269, 273, 275, 300–301, 303–307, 311–314, 370, 413, 426
Hirosige, Tetu, 208, 210–211
Hitler, Adolf, 252, 268
Hoffmann, Banesh, 63, 89–90, 99, 152–153, 158, 235, 257, 259, 262, 269–270, 284
Holmes, Arthur, 389, 413
Holton, Gerald, 158
Hopf, Ludwig, 70–71, 154, 171
Horn, Leo van den, 207
Howard, Don, vi, 228, 290, 311, 383, 401, 433
Hoyle, Fred, 373, 391–392, 394, 397–400, 405, 413–414, 429
Hubble, Edwin P., 369, 376–381, 383–384, 386, 388–392, 397–398, 401, 406, 412, 414, 416–417, 421, 425–427
Humason, Milton L., 378, 388–391, 400, 414–415, 421
Humm, Rudolf J., 245, 269, 273
Husserl, Edmund, 307, 311

Illy, József, 60, 62, 210
Infeld, Leopold, 52, 58, 62, 153, 158, 235–236, 253–254, 257–260, 262–270, 272–273, 284, 288–289, 291
Ishiwara, Jun, 307
Israel, Werner, 288, 291, 407, 418, 421
Ivanenko, Dimitriy, 310, 312

Jaeggli, Alvin, 88
Jaki, Stanley L., 385, 406, 412, 415, 420
Janis, Allen, I., 42, 45, 59, 62

Jansky, Karl G., 392, 415
Jeans, James, 379, 385, 394–395, 405, 410–411, 415, 419
Jeffery, George B., 47, 61, 98, 271, 364–365, 418
Jeffreys, Harold, 389, 415
Johnson, Harold L., 415
Jordan, Pascual, 387, 415
Jost, Res, 60, 88
Jüttner, Ferencz, 415

Kahn, Carla, 326, 365, 373, 416
Kahn, Franz, 326, 365, 373, 416
Kaluza, Theodor, 59, 62, 249, 308, 312, 316
Kant, Immanuel, 307
Kantowski, Robert, 396, 416
Kasner, Edward, 229, 232, 381, 386, 395, 416
Kaufmann, Walter, 305
Kelvin. see Thomson, William
Kemmer, Nicholas, 272
Kennedy, Roy J., 416
Kermack, William O., 388, 416
Kerr, Roy P., 265, 273, 275
Kerszberg, Pierre, viii, xii, 4, 229–230, 232, 325, 359, 365, 433
Kervaire, Michel, 413, 419, 423
Khalatnikov, I.M., 404, 417
Killing, Wilhelm, 416
King, Andrew R., 372, 409
Kingsepp, James, ix
Kirchhoff, Gustav Robert, 166
Klauder, John R., 420
Klein, Felix, 160, 303, 305, 312, 322–323, 343, 345, 358, 360, 365
Klein, Martin J., 59, 62, 207
Klein, Oskar, 249, 301, 310, 316
Klug, Ernst, 44
Kneser, Adolf, 161
Kohler, Max, 387, 390, 416
Komar, Arthur, 294, 298
Kopff, August, 246, 273
Kottler, Friedrich, 7, 8, 13, 19, 26, 32
Kox, A.J., viii, xii, 3, 201, 228, 433
Kretschmann, Erich, 94, 99, 363
Kristian, Jerome, 388, 409, 416
Krutkov, G. (Krutkoff), 375

Kuhn, Thomas S., 300
Kundt, Wolfgang, 416
Kusaka, Shuichi, 62

Laan, H. van der, 363
Lakatos, Imre, 300
Lanczos, Alice, 269
Lanczos, Cornelius (Lancius), 60, 62–63, 89, 99, 152, 158, 218–219, 232, 248–250, 254, 258, 261–262, 265, 269, 274, 358, 365, 373, 416
Lang, Kenneth R., 416
Langevin, Paul, 244, 280
Laue, Max von, 14, 31–32, 39–40, 43–45, 89, 218, 226–227, 233, 246, 274, 279–280, 292, 387, 429
Lawson, Robert W., 46, 61, 158, 364, 389, 413
Layzer, David, 416
Leão, João, viii
Leavitt, Henrietta, 376
Lecher, Ernst, 209, 211
Lee, David L., 42, 47
Leich, Walter, 88
Lemaître, Georges, 218, 226–227, 232, 288, 326, 358–359, 365, 368, 373, 378–387, 389–390, 393–396, 402, 404, 406, 408, 410–411, 416–417, 419, 424, 428
Lenard, Philipp, 280
Lense, Josef, 427
Leroy, Jules, 323
Leverrier, Urbain J.J., 289, 291
Levi-Civita, Tullio, xi, 3, 42, 58, 60, 65, 88, 107, 138, 154, 160–161, 165, 171–177, 181, 183–200, 241, 245, 256, 260–261, 268, 270, 274–275, 279, 284–285, 290–291, 306, 370, 372, 417, 422
Levinson, Horace, 263–264, 274
Levinson, Norman, 263
Levy, Hyman, 56–57, 62, 273
Liang, Edison P.-T., 407
Lichnerowicz, André, 396, 417–418
Lie, Sophus, 395, 417
Lieber, Hugh G., 42
Lieber, Lillian R., 42
Lifschitz, Evgenii M., 395, 404, 417

Lightman, Alan P., viii, 42, 47
Limber, D. Nelson, 417–418
Lindblad, Bertil, 406
Lock, C.N.H., 418
Lodge, Oliver, 418
Loedel-Palumbo, Enrique, 229, 232
London, Fritz, 312
Longair, Malcolm S., 424
Lorentz, Hendrik Antoon, xii, 3, 45–47, 54, 61, 98, 101, 126–127, 138–140, 151–154, 156–158, 160, 164, 166, 195, 197, 201–212, 216, 234, 238, 243–244, 255–256, 258, 271, 274, 307, 312, 362, 364–365, 408–409, 418, 420
Lovell, Bernard, 392, 418, 420, 422
Lubański, Joseph Kazimir, 267
Lummer, Otto R., 100
Lundmark, Knut E., 377, 406, 418

Maanen, Adriaan van, 376–377, 428
Maartens, Roy, 372, 374, 410
MacCallum, Malcolm A.H., 401, 418, 424
Mach, Ernst, 41, 45–46, 59, 62, 71, 99, 166, 169, 219, 307, 328–330, 333, 354, 357, 365, 410, 418
Mackey, M.B., 412
MacMillan, William D., 386, 418
Madsen, Mark S., 401, 418
Malament, David, 42, 44
Manin, Yuri I., 311, 312
Mardirossian, Fabio, 427
Marrot, R., 417–418, 429
Mason, W.R., 418
Mathisson, Myron, 250–254, 257–258, 260, 264, 267–269, 274–275
Mattig, Wolfgang, 388, 418
Matzner, Richard A., 422
Maurellis, Archie, 401
Mayall, Nicholas U., 391, 415
Mayer, Maria G., 419
Mayer, Walter, 250, 253, 320–323
McCormack, Thomas J., 365, 418
McCormmach, Russell, 60, 62, 208, 210, 212, 307, 312
McCrea, William H., 383, 387–388, 398–399, 416, 419

McKellar, A., 394, 403, 419
McVittie, George C., 226, 232, 288, 291, 380, 383, 385, 388–389, 399, 405, 419
Medicus, Heinrich A., 150, 158, 228, 232
Mehra, Jagdish, 63, 89–90, 99, 150, 152, 155, 158, 238, 275
Mercier, André, 413, 419, 423
Merleau-Ponty, Jacques, 325–326, 365
Meyerson, Émile, 159
Mie, Gustav, 150, 167, 169, 216, 228, 232, 246, 275, 301, 304–306, 311
Miga, K., 420
Miller, Arthur I., 60, 62, 173–174
Mills, Robert L., 310
Milne, Edward A., 288, 343, 365, 381, 385–387, 397, 416, 419, 423–424, 429
Minkowski, Hermann, 87, 99, 106, 303, 313, 349, 370, 418, 420
Minkowski, Richard, 392, 404
Misner, Charles W., 83, 99, 231, 233, 298, 396, 401, 420
Moessard, M., 386, 390, 420
Moffat, John, 265, 275
Møller, Christian, 293, 414
Morduch, G.E., 286, 292
Mössbauer, Rudolf L., 283
Moszkowski, Alexander, 349, 365
Motte, Andrew, 44
Moulton, Forrest Ray, 263
Muller, Richard A., 426

Nakagami, M., 420
Nathan, Otto, 48, 88, 99
Naumann, Friedrich, 139
Nel, Stanley D., 372, 374, 410
Nelkowski, Horst, 100, 159, 233
Nersessian, Nancy J., 210
Neumann, Carl, 369, 420
Newman, Maxwell H.A., 306, 313
Newton, Isaac, 11, 47, 75, 77, 119, 206, 210, 219, 284–285, 290
Neymann, Jerzy, 392, 420
Noether, Emmy, 294, 303
Norden, Heinz, 88, 99
Nordmann, Charles, 216, 233

Nordström, Gunnar, 27, 81–82, 99, 154, 167, 307
North, John D., 229, 233, 326, 365, 369, 405, 420, 429
Norton, John, viii, xi, 3, 5, 19, 29, 47, 101, 173–174, 175–176, 194–197, 208, 212, 228, 233, 316, 323, 433
Novello, Mario, 409
Novikov, Igor D., 400, 407

O'Brien, S., 405
Ohanian, Hans C., 6, 44
Okaya, T., 420
Oke, John B., 391, 412
Olbers, Heinrich W.M., 373, 379, 385–386, 404–406, 412, 415, 417, 420, 427, 430
Olive, Keith A., 420
Omer, Guy C., 396, 420
Oort, Jan, 388, 391, 421
Oppenheimer, J. Robert, 227, 283, 291
O'Raifeartaigh, Lochlainn S., 62, 99, 158, 275, 410

Paddock, G.F., 377, 421
Page, Leigh, 226, 233
Painlevé, Paul, 44, 222–223, 233
Pais, Abraham, 18, 47, 150, 156, 159, 209, 212, 228, 233, 269, 275, 365, 368, 370, 383, 397, 421, 429
Paolini, Frank R., 411
Papapetrou, Achille, 262, 275
Parker, Leonard, 421, 425–26
Parsons, S.J., 392, 413
Partridge, Robert B., 401, 421
Pauli, Wolfgang, 5–8, 33–34, 36, 39, 42, 47, 245–246, 268–269, 275, 280, 302–303, 306, 308, 313
Peebles, Phillip J.E., 325, 363, 365, 368, 384–385, 392, 394–395, 398, 400–402, 404–405, 407, 411, 421, 431
Peierls, Rudolf E., 394, 421
Penrose, Roger, 373, 397, 401, 412, 421
Penzias, Arno A., 394, 398, 400, 403, 407, 421
Perko, T.E., 422

Perrett, W., 47, 61, 98, 271, 364, 365–418
Peterson, Bruce A., 412
Petrov, Aleksei Z., 313, 422
Petrova, N.M., 262, 275
Petzoldt, Joseph, 52–53, 62
Phillips, J.W., 392, 413
Pick, Georg, 60
Pierpont, James, 422
Pirani, Felix A.E., 285, 291, 297, 422
Planck, Max, 307
Plebański, Jerzy, 235, 256, 259, 265, 267, 269, 273, 275
Poincaré, Henri, 160–161, 239, 293, 307, 342, 366
Polak, L.S., 313
Polubarinova-Kochina, Pelageia Yakovlevna, 422
Popper, Karl, 279
Pound, Robert V., 283
Pyenson, Lewis, 216, 233, 273, 303–304, 313

Rabe, Eugene K., 225, 233
Raman, Varadaraja V., 310, 313
Raychaudhuri, Amalkamur K., 380, 396, 399, 422, 428
Reber, G., 422
Rebka, Glen A., 283
Rees, Martin J., 365, 404, 422, 424
Reeves, Hubert, 422
Rehtz, Alfred, 42–43
Reich, Karin, 88, 93, 99
Reichenbächer, Ernst, 307
Reid, Constance, 306, 313
Reines, Frederick, 368, 404, 422
Renn, Jürgen, viii
Reynolds, John H., 390, 422
Reyntjens, J., 44
Ricci-Curbastro, Gregorio, 58, 60, 65, 88, 107, 154, 256, 275, 279, 290, 370, 422
Rice, James, 216, 228, 233
Riemann, Bernhard, 58, 60, 65, 76, 88, 99, 279, 290, 303, 338–339, 342, 365–366, 370, 422

Rindler, Wolfgang, 328, 365, 397, 399, 401, 423
Ritchey, George W., 376, 423
Ritter, James, 321, 323
Robb, Alfred A., 397, 423
Robertson, Howard P., 230, 253–254, 256–258, 260–262, 269, 275, 288–289, 326, 358, 365, 367–368, 372–373, 375, 377, 379–382, 386, 388, 397, 400, 402, 423–424
Robinson, Ivor, 275
Robinson, Joanna R., 275
Roll, Peter G., 407, 423
Rosen, George, 62
Rosen, Nathan, 45, 47, 265, 272, 275, 287, 291
Rosenfeld, Leon, 293–294
Roseveare, N.T., 290–291
Rossi, Bruno B., 411
Rubin, Vera C., 423
Ruffini, Remo, 100
Ruse, Harold S., 423
Russell, Bertrand, 76, 78, 100
Rutherford, Ernest, 389, 404, 423
Ryan, Michael P., 368, 423
Ryle, Martin, 392, 398, 411, 424

Sachs, Rainer K., 93, 97, 100, 388, 395–396, 407–408, 410, 416, 424–425
Sagnac, Georges, 209, 212
Sakharov, Andrei, 424
Sakurai, Jun J., 310
Sandage, Allan R., 391, 400, 404, 414–416, 424
Sato, Katsuniko, 424
Schäfer, G., 270
Schaffner, Kenneth F., 208, 212, 292
Schatz, Henriette, 207
Schatzmann, Evry, 424
Schechter, A., 376, 387, 410
Scheinerm, Julius, 377
Scheuer, Peter A.G., 424
Schild, Alfred, 235, 253–254, 273, 283, 292, 297, 426
Schilpp, Paul Arthur, 46, 61, 98, 272, 397, 417, 423–424

Schlick, Moritz, 34–39, 44, 47, 54, 85–86
Schmidt, Maarten, 411–412, 424
Schott, George A., 252, 275
Schouten, Jan Arnoldus, 93, 100, 306, 313, 424
Schramm, David N., 404, 411, 420, 422
Schrödinger, Erwin, 93, 100, 216, 228, 233, 242, 250, 310, 313, 372, 424
Schücking, Engelbert, 392, 396–397, 399, 413, 424, 426
Schur, Friedrich H., 343, 365
Schwarzschild, Karl, xii, 213–219, 223, 227–230, 232–233, 242, 254, 275–276, 344, 368, 372, 424–425
Schwarzschild, Martin, 425
Sciama, Dennis W., 400, 410, 422, 424–425
Scott, Elizabeth L., 392, 420, 425
Seeley, Daniel, 368, 376, 404
Seelig, Carl, 46, 99, 158, 173–174, 250, 262, 265, 269, 276, 305–306, 313
Seeliger, Hugo von, 369, 428
Segal, Irving E., 399, 425
Sexl, Roman U., 425
Shah, K.B., 428
Shakeshaft, John R., 424
Shane, Charles D., 392, 420
Shapley, Harlow, 370, 375–376, 378, 388–389, 406, 425
Shapley, Martha, 425
Shearman, Margo, ix
Shepley, Lawrence C., 368, 422–423, 425
Shimmins, A.J., 412
Siklos, Stephen T.C., 411
Silberstein, Ludwik, 33, 47, 273, 377, 408, 418, 425
Silk, Joseph, 424–425
Silliman, Benjamin, 390
Singwi, Kundan Singh, 394, 421
Slipher, Vesta M., 372, 377, 425–426
Smith, David E., 100, 365, 422
Smith, Robert W., 349, 366, 368, 371, 373, 375–379, 383, 426
Smith, Sinclair, 426
Smoluchowski, Marian, 60, 62
Smoot, George F., 426

Smorodinskiĭ, Ya. A., 63, 66, 89, 100, 152
Sommerfeld, Arnold, 47–49, 58–59, 61–62, 85, 99, 119, 138, 142, 145, 152–153, 155–156, 164, 173–174, 275, 310, 313, 362, 365, 418
Speziali, Pierre, 42, 47, 68, 84, 86, 100, 153, 159, 173–174, 196, 200, 337, 345, 347, 366
Średniawa, Bronislaw, 252, 267, 276
Stachel, Deborah, ix
Stachel, John, v–viii, xi, 1, 3, 42, 47–48, 59, 63–74, 76–78, 82, 85–86, 88–89, 100, 102, 107, 111, 131, 151–152, 159, 173–174, 194–195, 200, 207, 219, 228, 233, 269–270, 273, 276, 287, 290, 292, 311, 316, 323, 363, 376, 381, 383, 401, 426, 433
Stalin, Josef, 252, 268
Stebbins, Joel, 426
Steigmann, Gary, 420
Stein, Howard, 210, 212
Stephen, William E., 410
Sterne, Theodore E., 384, 426
Stewart, John Q., 426
Stock, J., 392, 426
Stoeger, William, 410
Straneo, Paolo, 216, 228, 233
Strassenberg, A.A., 425
Straubel, Rudolf, 426
Strauss, Ernst G., 150, 396, 409, 411, 424
Stromberg, Gustaf B., 377, 426
Sunyaev, R.Z., 426
Suttorp, Leendert, 207
Synge, John Lighton, 5, 7, 39–40, 45, 47, 62, 99, 158, 275, 284–286, 289, 292, 368, 386, 396, 405, 410, 426
Szilard, Leo, 250


Takahashi, Yasushi, 426
Tammann, Gustav A., 388–389, 391, 404, 427
Taub, Abraham H., 396, 399, 418, 420, 427

Tauber, Gerald E., 427
Tayler, Roger J., 400, 412, 414, 427
Teller, Edward, 394, 411, 415, 419, 427
Temple, George F.J., 388, 427
von Ten Bruggencate, P., 427
ter Haar, D., 427
Teske, Armin, 62
Thackeray, A. David, 391, 427
Thiele, Joachim, 53, 54, 62
Thirring, Hans, 427
Thomson, J.H., 424
Thomson, William (Lord Kelvin), 385, 412, 427
Thorne, Kip S., 42, 47, 83, 99, 231, 233, 291, 396, 420, 427
Timofeyeva, I.S., 313
Tinsley, Beatrice, 391, 400, 404, 411, 427
Tipler, Frank J., 368, 397, 404, 427
Titze, Hubert, 46
Tobies, Renata, 311
Tolle, Kimberly D., ix
Tolman, Richard C., 282, 288, 292, 373, 379–382, 384–388, 390, 393, 396–397, 404–405, 414, 417, 423, 427–428
Torretti, Roberto, 42, 44, 47, 96, 151, 342, 366
Trautman, Andrzej, 270, 283, 292
Treder, Hans-Jürgen, 310, 313
Trumpler, Robert J., 389, 428
Turkevich, Anthony, 403, 410
Turner, Michael S., 420

Ullmann-Margalit, Edna, 426
Umezawa, Hiroomi, 426
Urbantke, Helmuth K., 425

Vaidya, P.C., 428
Vane, Arthur B., 394, 407
Varičak, V., 49, 62
Vizgin, Vladimir P., viii, xii, 2, 4, 63, 66, 89, 100, 152, 159, 300–301, 303, 305, 309–310, 313, 433
Voigt, Woldemar, 307

Wagoner, Robert V., 398, 400, 404, 429
Wakano, Masami, 291
Walker, Arthur Geoffrey, 326, 368, 386, 396, 417–418, 423, 429
Wallace, Philip R., 254, 273
Walsh, David, 429
Ward, Morgan, 384, 386, 396, 428
Weber, Wilhelm, 303
Weinberg, Joseph W., 93, 100, 400, 427
Weinberg, Steven, 368, 400, 429
Weitzenböck, Roland, 316–317, 323–324
Weizsäcker, Carl Friedrich von, 384, 395, 429
Wertheimer, Max, 60, 62
Wesselink, Adriaan J., 391, 427
Westphal, James A., 416
Weyl, Hermann, xii, 2, 76, 99, 100, 211, 217, 221–222, 229, 231, 233, 238, 241, 245–249, 251–254, 258, 262, 266, 269–270, 276, 280, 284, 292, 300–303, 305–308, 310–314, 339, 358–359, 365–366, 371, 373, 375, 377, 379, 385, 387, 389, 397, 402, 408, 418, 429–430
Weyl-Baer, Ellen, 269
Weymann, Ray J., 429
Weyssenhoff, Jan, 252
Wheeler, John Archibald, 83, 99, 214, 231, 233, 285, 291–292, 323–324, 371, 420–421, 430
White, Henry S., 99, 365
Whitehead, Alfred North, 232, 279
Whitford, Albert E., 426
Whitman, A.P., 410
Whitrow, Gerald J., 286, 292, 325, 366, 430
Whittaker, Sir Edmund T., 54, 62, 226–227, 233–234, 276, 286, 292, 387–388, 416, 430
Wiechert, Emil, 305, 307
Wigner, Eugene P., 313
Wilkinson, David T., 407, 421, 423
Wilson, Robert W., 394, 398, 400, 407, 421
Wirtz, Carl, 377, 418, 430
Witten, Louis, 62, 404, 413
Wolfe, Arthur M., 395, 423
Woolf, Harry, 324, 365

Wroe, D., 394, 421
Wu, Hung-Hsi, 93, 97, 100
Wünsch, V., 311
Wyatt, S.P., 419

Yang, Chen Ning, 310
Yang, Jongmann, 420
Yoshimura, Motohiko, 430

Zahar, Elie, 152, 159
Zangger, Heinrich, 173
Zee, Anthony, 400–401, 430
Zeeman, Pieter, 274
Zeisler, Ernest B., 264, 274
Zeldovich, Yakov B., 400, 407, 426, 431
Zwicky, Fritz, 386, 389, 418, 426, 431